Applications of Unitary Symmetry and Combinatorics

Applications of Unitary Symmetry and Combinatorics

James D Louck
Los Alamos National Laboratory Fellow
Santa Fe, New Mexico, USA

World Scientific

NEW JERSEY · LONDON · SINGAPORE · BEIJING · SHANGHAI · HONG KONG · TAIPEI · CHENNAI

Published by

World Scientific Publishing Co. Pte. Ltd.
5 Toh Tuck Link, Singapore 596224
USA office: 27 Warren Street, Suite 401-402, Hackensack, NJ 07601
UK office: 57 Shelton Street, Covent Garden, London WC2H 9HE

British Library Cataloguing-in-Publication Data
A catalogue record for this book is available from the British Library.

APPLICATIONS OF UNITARY SYMMETRY AND COMBINATORICS
Copyright © 2011 by World Scientific Publishing Co. Pte. Ltd.

All rights reserved. This book, or parts thereof, may not be reproduced in any form or by any means, electronic or mechanical, including photocopying, recording or any information storage and retrieval system now known or to be invented, without written permission from the Publisher.

For photocopying of material in this volume, please pay a copying fee through the Copyright Clearance Center, Inc., 222 Rosewood Drive, Danvers, MA 01923, USA. In this case permission to photocopy is not required from the publisher.

ISBN-13 978-981-4350-71-6
ISBN-10 981-4350-71-0

Printed in Singapore.

*In recognition of contributions to the generation
and spread of knowledge*

William Y. C. Chen

Tadeusz and Barbara Lulek

And to the memory of

Lawrence C. Biedenharn

Gian-Carlo Rota

Preface and Prelude

We have titled this monograph "Applications of Unitary Symmetry and Combinatorics" because it uses methods developed in the earlier volume "Unitary Symmetry and Combinatorics," World Scientific, 2008 (hereafter Ref. [46] is referred to as [L]). These applications are highly topical, and come in three classes: (i) Those still fully mathematical in content that synthesize the common structure of doubly stochastic, magic square, and alternating sign matrices by their common expansions as linear combinations of permutation matrices; (ii) those with an associated physical significance such as the role of doubly stochastic matrices and complete sets of commuting Hermitian operators in the probabilistic interpretation of nonrelativistic quantum mechanics, the role of magic squares in a generalization of the Regge magic square realization of the domains of definition of the quantum numbers of angular momenta and their counting formulas (Chapter 6), and the relation between alternating sign matrices and a class of Gelfand-Tsetlin patterns familiar from the representation of irreducible representations of the unitary groups (Chapter 7) and their counting formulas; and (iii) a physical application to the diagonalization of the Heisenberg magnetic ring Hamiltonian, viewed as a composite system in which the total angular momentum is conserved.

A uniform viewpoint of rotations is adopted at the outset from [L], based on the method of Cartan [15] (see also [6]), where it is fully defined and discussed. It is not often made explicit in the present volume:

A unitary rotation of a composite system is its redescription under an $SU(2)$ unitary group frame rotation of a right-handed triad of perpendicular unit vectors $(\mathbf{e}_1, \mathbf{e}_2, \mathbf{e}_3)$ that serves as common reference system for the description of all the constituent parts of the system.

This Preface serves three purposes: A prelude and synthesis of things to come based on results obtained in [L], now focused strongly on the basic structural elements and their role in bringing unity to the understanding of the angular momentum properties of complex systems viewed as composite wholes; a summary of the contents by topics; and the usual elements of style, readership, acknowledgments, etc.

OVERVIEW AND SYNTHESIS OF BINARY COUPLING THEORY

The theory of the binary coupling of n angular momenta is about the pairwise addition of n angular momenta associated with n constituent parts of a composite physical system and the construction of the associated state vectors of the composite system from the $SU(2)$ irreducible angular momentum multiplets of the parts. Each such possible way of effecting the addition is called a *binary coupling scheme*.

We set forth in the following paragraphs the underlying conceptual basis of such binary coupling scheme. Each binary coupling scheme of order n may be described in terms of a sequence having two types of parts: n points ∘ ∘ ⋯ ∘ and $n-1$ parenthesis pairs ( ), ( ), ..., ( ). A parenthesis pair ( ) constitutes a single part. Thus, the number of parts in the full sequence is $2n-1$. By definition, the binary bracketing of order 1 is ∘ itself, the binary bracketing of order 2 is (∘ ∘), the two binary bracketings of order 3 are $\bigl((\circ\circ)\bigr)$ and $\bigl(\circ\,(\circ\circ)\bigr)$, ... In general, we have the definition:

A binary bracketing $\mathbb{B}_n$ of order $n \geq 2$ is any sequence in the n points ∘ and the $n-1$ parenthesis pairs ( ) that satisfies the two conditions: (i) It contains a binary bracketing of order 2, and (ii) the mapping (∘ ∘) ↦ ∘ gives a binary bracketing of order $n-1$.

Then, since the mapping (∘ ∘) ↦ ∘ again gives a binary bracketing for $n \geq 3$, the new binary bracketing of order $n-1$ again contains a binary bracketing of order 2. This implies that this mapping property can be used repeatedly to reduce every binary bracketing of arbitrary order to ∘, the binary bracketing of order 1.

The appropriate mathematical concept for diagramming all such binary bracketings of order n is that of a binary tree of order n. We have described in [L] a "bifurcation of points" build-up principle for constructing the set $\mathbb{T}_n$ of all binary trees of order n in terms of *levels* (see [L, Sect. 2.2]). This is a standard procedure found in many books on combinatorics. It can also be described in terms of an assembly of four basic objects called *forks* that come in four types, as enumerated by

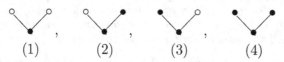

(1)   (2)   (3)   (4)

The • point at the bottom of these diagrams is called the *root* of the fork, and the other two point are called the *endpoints* of the fork. The assembly rule for forks into a binary tree of order n can be formulated in term of the "pasting" together of forks.

PREFACE AND PRELUDE

Our interest in viewing a binary tree as being composed of a collection of pasted forks of four basic types is because the configuration of forks that appears in the binary tree encode exactly how the pairwise addition of angular momenta of the constituents of a composite system is to be effected. Each such labeled fork has associated with it the elementary rule of addition of the two angular momenta, as well as the Wigner-Clebsch-Gordan (WCG) coefficients $C^{j_1 \; j_2 \; k}_{m_1 m_2 \mu}$ that effect the coupling of the state vectors of two subsystems of a composite system having angular momentum $\mathbf{J}(1)$ and $\mathbf{J}(2)$, respectively, to an intermediate angular momentum $\mathbf{J}(1) + \mathbf{J}(2) = \mathbf{K}$, as depicted by the labeled fork:

$$\underset{k}{\overset{j_1 \quad j_2}{\vee}} \quad : \quad k = j_1 + j_2, j_1 + j_2 - 1, \ldots, |j_1 - j_2|.$$

Similarly, the labeled basic fork 2 given by

$$\underset{k'}{\overset{j_i \quad k}{\vee}} \quad : \quad k' = j_i + k, j_i + k - 1, \ldots, |j_i - k|$$

encodes the addition $\mathbf{J}(i) + \mathbf{K} = \mathbf{K}'$ of an angular momentum $\mathbf{J}(i)$ of the constituent system and an intermediate angular momentum $\mathbf{K}$ to a "total" intermediate angular momentum $\mathbf{K}'$, as well as the attended WCG coefficients $C^{j_i \; k \; k'}_{m_i \mu \, \mu'}$ that effect the coupling. Labeled forks 3 and 4 have a similar interpretation. These labeled forks of a standard labeled binary tree of order n encode the constituent angular momenta that enter into the description of the basic $SU(2)$ irreducible state vectors of a composite system.

A build-up rule for the pasting of forks can be described as follows:

1. Select a fork from the set of four forks above and place the • root point over any • endpoint of the four forks, merging the two • points to a single • point. Repeat this *pasting process* for each basic fork. This step gives a set of seventeen distinct graphs, where we include the basic fork containing the two o points in the collection:

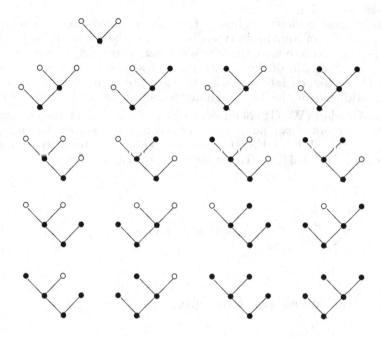

2. Select a single graph from the set of seventeen graphs generated at Step 1 and repeat the pasting process with each of the four basic forks. This gives back the three graphs from the collection above having only endpoints of type ○, which includes all binary tree graphs of order 2 and order 3, and, in addition, ninety-six more graphs as follows: (1)(4)(6) from the six graphs above having one • endpoints; (2)(4)(6) from the six graphs having two • endpoints, and (3)(4)(2) from the three graphs having three • endpoints. From this large collection, put aside those having only ○ endpoints, including no repetitions. This gives the set of all binary trees or order 2, 3, 4:

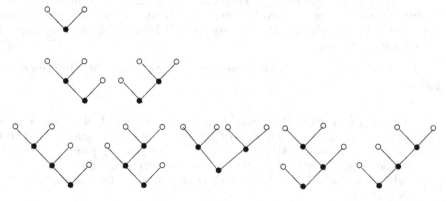

PREFACE AND PRELUDE

3. Select a single graph from the full set generated at Step 2 and repeat the pasting process. putting aside all those having only ∘ endpoints, including no repetitions. Repeat this process for the next set of graphs, etc. At Step h of this pasting process of basic forks to the full collection generated at Step $h-1$, there is obtained a huge multiset of graphs that includes all binary trees of order $2, 3, \ldots, h+2$ — all those with ∘ endpoints. A very large number of graphs with 1• endpoint, 2• endpoints, ... is also obtained, these being needed for the next step of the pasting process. All binary trees of order n are included in the set of graphs generated by the pasting process at step $h = n - 2$, these being the ones put aside.

   The pasting process is a very inefficient method for obtaining the set of binary trees of order n. A more efficient method is to generate this subset recursively by the following procedure: Suppose the set of binary trees of order n has already been obtained, and that it contains a_n members. Then, consider the new set of na_n graphs obtained by replacing a single ∘ endpoint by a • point in each of the a_n binary tree graphs. In the next step, paste the • root of basic fork 1 over each of these • endpoints, thus obtaining na_n binary trees of order $n + 1$. This **multiset** of cubic graphs of order $n + 1$ then contains the set of a_{n+1} binary trees of order $n + 1$. Thus, starting with fork 1 from the basic set of four, we generate all binary trees of arbitrary order by this pasting procedure. (This procedure works because every graph having one • endpoint is included in the huge set generate in Items 1, 2, 3 above.)

   Only binary trees having ∘ endpoints enter into the binary coupling theory of angular momenta because each such coupling scheme must correspond to a binary bracketing. The set of binary bracketings of order n is one-to-one with the set of binary trees of order n, and the bijection rule between the two sets can be formulated exactly. The cardinality of the set $\mathbb{T}_n$ is the Catalan number a_n. We repeat, for convenience, many examples of such binary bracketings and corresponding binary trees from [L], and also give in this volume many more, including examples for arbitrary n. As in [L], we call the binary bracketing corresponding to a given binary tree the *shape of the binary tree:*

   The shape Sh_T of a binary tree $T \in \mathbb{T}_n$ is defined to be the binary bracketing of order n corresponding to T.

   The concept of shape transformation is so basic that we discuss it here in the Preface in its simplest realization: There are two binary trees of order three, as given above. We now label the endpoints of these binary trees by a permutation of some arbitrary set of elements x_1, x_2, x_3 :

We initially ignore the labels. The first graph corresponds to the binary bracketing $((\circ\circ)\circ)$, and the second one to $(\circ(\circ\circ))$. The binary bracketing, now called the shape of the binary tree, is defined in the obvious way by reading left-to-right across the binary tree, and inserting a parenthesis pair for each • point:

$$\mathrm{Sh}_T = ((\circ\circ)\circ), \quad \mathrm{Sh}_{T'} = (\circ(\circ\circ)).$$

These shapes are called *unlabeled shapes*. If the ○ endpoints are labeled by a permutation of distinct symbols such as x_1, x_2, x_3, the ○ points of the shape are labeled by the corresponding symbols to obtain the *labeled shapes*

$$\mathrm{Sh}_T(x_1, x_2, x_3) = ((x_1\, x_2)\, x_3), \quad \mathrm{Sh}_{T'}(x_2, x_1, x_3) = (x_2\, (x_1\, x_3)).$$

What is interesting now is that the first labeled shape can be transformed to the second labeled shape by the elementary operations C and A of commutation and association:

$$((x_1\, x_2)\, x_3) \xrightarrow{C} ((x_2\, x_1)\, x_3) \xrightarrow{A} (x_2\, (x_1\, x_3)),$$

where the action of commutation C and that of association A have their usual meaning:

$$C: (x\,y) \mapsto (y\,x), \quad A: (x\,y)\,z \mapsto x\,(y\,z).$$

Thus, we have the shape transformation:

$$\mathrm{Sh}_T(x_1, x_2, x_3) \xrightarrow{AC} \mathrm{Sh}_{T'}(x_2, x_1, x_3),$$

where the convention for the action of commutation and association is that C acts first on the shape $\mathrm{Sh}_T(x_1, x_2, x_3)$ to effect the transformation to the shape $\mathrm{Sh}_T(x_2, x_1, x_3)$ followed by the action of A on $\mathrm{Sh}_T(x_2, x_1, x_3)$ to give the shape $\mathrm{Sh}_{T'}(x_2, x_1, x_3)$.

We state at the outset the generalization of this result:

The set $\mathbb{T}_n$ of binary trees of order n is unambiguously enumerated by its set of shapes Sh_T, $T \in \mathbb{T}_n$, the number of which is given by the Catalan numbers a_n. Let $x_1, x_2, \ldots, x_n$ be a collection of n arbitrary distinct objects and $x_\pi = (x_{\pi_1}, x_{\pi_2}, \ldots, x_{\pi_n})$ an arbitrary permutation of the x_i, where $\pi = (\pi_1, \pi_2, \ldots, \pi_n)$ is an arbitrary permutation in the group S_n of all permutations of the reference set $1, 2, \ldots, n$.

Then, there exists a shape transformation $w(A, C)$ such that the labeled shape $Sh_T(x_\pi)$ is transformed to the labeled shape $Sh_{T'}(x_{\pi'})$, for each $T \in \mathbb{T}_n, \pi \in S_n, T' \in \mathbb{T}_n, \pi' \in S_n$; that is,

$$Sh_T(x_\pi) \xrightarrow{w(A,C)} Sh_{T'}(x_{\pi'}),$$

where the shape transformation $w(A, C)$ is a word in the two letters A and C.

The content of this result will be amplified by many explicit examples. It is a basic unifying result for the binary theory of the coupling of n angular momenta.

The pairwise addition of n angular momenta $\mathbf{J}(i) = J_1(i)\mathbf{e}_1 + J_2(i)\mathbf{e}_2 + J_3(i)\mathbf{e}_3, i = 1, 2, \ldots, n$, with components referred to a common right-handed inertial reference frame $(\mathbf{e}_1, \mathbf{e}_2, \mathbf{e}_3)$, to a total angular momentum $\mathbf{J} = J_1\mathbf{e}_1 + J_2\mathbf{e}_2 + J_3\mathbf{e}_3 = \mathbf{J}(1) + \mathbf{J}(2) + \cdots + \mathbf{J}(n)$ is realized in all possible ways by the standard labeling of each binary tree $T \in \mathbb{T}_n$ as given in terms of its shape by $Sh_T(\mathbf{J}(\pi_1), \mathbf{J}(\pi_2), \ldots, \mathbf{J}(\pi_n))$, where $\pi = (\pi_1, \pi_2, \ldots, \pi_n)$ is any permutation $\pi \in S_n$. The pairwise addition encoded in a given binary tree for any given permutation $\pi \in S_n$ is called a *coupling scheme*. For example, the two coupling schemes encoded by the shapes of the binary trees of order 3 given above are:

$$\left(\big(\mathbf{J}(1) + \mathbf{J}(2)\big) + \mathbf{J}(3)\right), \quad \left(\mathbf{J}(3) + \big(\mathbf{J}(2) + \mathbf{J}(1)\big)\right).$$

For general n, there are $n! a_n$ distinct coupling schemes for the pairwise addition of n angular momenta. The rule whereby we assign the n angular momenta $(\mathbf{J}(1), \mathbf{J}(2), \ldots, \mathbf{J}(n))$ of the n constituents of a composite system to the endpoints of a binary tree $T \in \mathbb{T}_n$ by the left-to-right assignment to the corresponding points in the shape Sh_T of T is called the *standard rule labeling* of the ∘ points (all endpoints) of the binary tree T. In many instances, we need also to apply this standard rule labeling for the assignment of a permutation $(\mathbf{J}(\pi_1), \mathbf{J}(\pi_2), \ldots, \mathbf{J}(\pi_n))$ of the angular momenta of the constituent parts, as illustrated in the example above. Caution must be exercised in the interpretation of all these pairwise additions of angular momenta in terms of the tensor product space in which these various angular momenta act. This is reviewed from [L] in Chapters 1 and 5 in the context of the applications made here. A standard labeling rule is required to define unambiguously objects such as Wigner-Clebsch-Gordan coefficients and triangle coefficients to labeled binary trees, objects that given numerical content to the theory.

To summarize: In the context of the binary coupling theory of angular momenta, we deal with standard labeled binary trees, their shapes, and the transformations between shapes. Out these few simple underlying structural elements there emerges a theory of almost unlimited, but manageable, complexity: Mathematically, it is theory of relations between $3(n-1) - j$ coefficients and Racah coefficients and their cubic graphs; physically, it is a theory of all possible ways to compound, pairwise, the individual angular momenta of the n constituent parts of a complex composite system to the total angular momentum of the system.

The implementation of the binary coupling theory of angular momenta leads directly to the Dirac concept [24] of characterizing the Hilbert space state vectors of each coupling scheme in terms of complete sets of commuting Hermitian operators. This characterization is described in detail in [L], and reviewed in the present volume in Chapters 1 and 5; it is described broadly as follows. The complete set of $2n$ mutually commuting Hermitian operators for each coupling scheme $T \in \mathbb{T}_n$ is given by

$$\mathbf{J}^2(1), \mathbf{J}^2(2), \ldots, \mathbf{J}^2(n), \mathbf{J}^2, J_3;$$
$$\mathbf{K}_T^2(1), \mathbf{K}_T^2(2), \ldots, \mathbf{K}_T^2(n-2).$$

The first line of operators consists of the n total angular momentum squared of each of the constituent systems, together with the total angular momentum squared of the composite system and its 3−component. The second line of operators consists of the squares of the $n-2$ so-called intermediate angular momenta, $\mathbf{K}_T(i) = (K_{T,1}(i), K_{T,2}(i), K_{T,3}(i)), i = 1, 2, \ldots, n-2$. There is a distinct set of such intermediate angular momenta associated with each binary tree $T \in \mathbb{T}_n$, where each $\mathbf{K}_T(i)$ is a $0-1$ linear combination of the constituent angular momenta $\mathbf{J}(i), i = 1, 2, \ldots, n$, in which the 0 and 1 coefficients are uniquely determined by the shape of the binary tree $T \in \mathbb{T}_n$. Thus, *each standard labeled binary coupling scheme has associated with it a unique complete set of $2n$ mutually commuting Hermitian operators*, as given generally above. For example, the complete sets of mutually commuting Hermitian operators associated with the labeled binary trees T and T' of order 3 given above are the following, respectively:

scheme T: $\mathbf{J}^2(1), \mathbf{J}^2(2), \mathbf{J}^2(3), \mathbf{J}^2, J(3); \ \mathbf{K}_T^2(1) = \left(\mathbf{J}(1) + \mathbf{J}(2)\right)^2.$

scheme T': $\mathbf{J}^2(1), \mathbf{J}^2(2), \mathbf{J}^2(3), \mathbf{J}^2, J(3); \ \mathbf{K}_{T'}^2(1) = \left(\mathbf{J}(2) + \mathbf{J}(1)\right)^2.$

There is a set of *simultaneous eigenvectors* associated with each complete set of mutually commuting Hermitian operators defined above for each binary tree $T \in \mathbb{T}_n$ of labeled shape $Sh_T(\mathbf{J}(\pi_1), \mathbf{J}(\pi_2), \ldots, \mathbf{J}(\pi_n))$. It is convenient now to denote this shape by $Sh_T(\mathbf{j}_\pi)$, where we use the angular momentum quantum number j_i in place of the angular momentum operator $\mathbf{J}(i)$ in the labeled shape. These are the quantum numbers j_i associated with the eigenvalue $j_i(j_i+1)$ of the squared "total" angular momentum $\mathbf{J}^2(i)$, each $i = 1, 2, \ldots, n$, of each of the n constituents of the composite physical system in question. The simultaneous eigenvectors in this set are denoted in the Dirac ket-vector notation by

$$|T(\mathbf{j}_\pi \mathbf{k})_{jm}\rangle,$$

where $\mathbf{j}_\pi = (j_{\pi_1}, j_{\pi_2}, \ldots, j_{\pi_n})$, and where j denotes the total angular momentum quantum number of the eigenvalue $j(j+1)$ of the squared total angular momentum $\mathbf{J}^2$ and m the eigenvalue of the 3-component J_3. We generally are interested in the finite set of vectors enumerated, for specified $\mathbf{j} = (j_1, j_2, \ldots, j_n)$, each $j_i \in \{0, 1/2, 1, 3/2, 2, \ldots\}$, by the range of values of the total angular momentum quantum numbers $j\,m$; and by the range of values of all the intermediate quantum numbers $\mathbf{k} = (k_1, k_2, \ldots, k_{n-2})$. The latter are associated with the eigenvalues $k_i(k_i + 1)$ of the squared intermediate angular momenta for given $T \in \mathbb{T}_n$. This gives the set of eigenvectors associated with the labeled shape $Sh_T(\mathbf{j}_\pi)$, and defines an orthonormal basis of a finite-dimensional tensor product Hilbert space denoted by $\mathcal{H}_\mathbf{j} = \mathcal{H}_{\mathbf{j}_1} \otimes \mathcal{H}_{\mathbf{j}_2} \otimes \cdots \otimes \mathcal{H}_{\mathbf{j}_n}$ of dimension equal to $(2j_1 + 1)(2j_2 + 1) \cdots (2j_n + 1)$. The domain of definition of $j\,m$ for each coupled state vector $|T(\mathbf{j}_\pi \mathbf{k})_{jm}\rangle$ corresponding to the binary tree T of shape $Sh_T(\mathbf{j}_\pi)$ is $j \in \{j_{\min}, j_{\min} + 1, \ldots, j_{\max}\}$, where $j_{\min}$ is the least nonnegative integer or positive half-odd integer among the 2^n sums of the form $\pm j_1 \pm j_2 \pm \cdots \pm j_n$, and $j_{\max} = j_1 + j_2 + \cdots + j_n$. The domain of definition of the intermediate quantum number k_i depends on the labeled shape of the binary tree T; it belongs to a uniquely defined set as given by $\mathbf{k} \in \mathbb{K}_T^{(j)}(\mathbf{j}_\pi)$, the details of which are not important here. Thus, an orthonormal basis of the space $\mathcal{H}_\mathbf{j}$ is given, for each $T \in \mathbb{T}_n$ and each labeled shape $Sh_T(\mathbf{j}_\pi)$, by

$$\mathbf{B}_T(\mathbf{j}_\pi) = \left\{ |T(\mathbf{j}_\pi \mathbf{k})_{jm}\rangle \,\Big|\, j \in \{j_{\min}, j_{\min}+1, \ldots, j_{\max}\}; \text{and for each } j, \right.$$

$$\left. \mathbf{k} \in \mathbb{K}_T^{(j)}(\mathbf{j}_\pi);\, m \in \{j, j-1, \ldots, -j\} \right\}.$$

In all, we have $n!a_n$ sets of coupled orthonormal basis vectors, each set $\mathbf{B}_T(\mathbf{j}_\pi)$ giving a basis of the same tensor product space $\mathcal{H}_\mathbf{j}$.

We digress a moment to recall that the basic origin of the tensor product Hilbert space $\mathcal{H}_\mathbf{j}$ is just the vector space formed from the tensor product of the individual Hilbert spaces $\mathcal{H}_{\mathbf{j}_i}$ of dimension $2j_i + 1$ on which the angular momentum $\mathbf{J}(i)$ has the standard action, with the commuting Hermitian operators $\mathbf{J}^2(i), J_3(i)$ being diagonal with eigenvalues $j_i(j_i+1)$ and m_i. The orthonormal basis of the space $\mathcal{H}_\mathbf{j}$ is now the *uncoupled basis* $|\mathbf{j}\,\mathbf{m}\rangle = |j_1\,m_1\rangle \otimes |j_2\,m_2\rangle \otimes \cdots \otimes |j_n\,m_n\rangle$, in which we can have $j_i \in \{0, 1/2, 1, 3/2, 2, \ldots\}$, and, for each selected j_i, the so-called projection quantum number m_i assumes all values $m_i = j_i, j_i - 1, \ldots, -j_i$. Thus, the uncoupled orthonormal basis $\mathbf{B}_\mathbf{j} = \{|\mathbf{j}\,\mathbf{m}\rangle \mid \text{each } m_i = j_i, j_i - 1, \ldots, -j_i\}$ of $\mathcal{H}_\mathbf{j}$ is the space of simultaneous eigenvectors of the complete mutually commuting Hermitian operators $\mathbf{J}^2(i), J_3(i), i = 1, 2, \ldots, n$, where the components of each of the angular momentum operators $\mathbf{J}(i)$ has the standard action on $\mathcal{H}_{\mathbf{j}_i}$. These basic relations are presented in great detail in Ref. [6], in [L], and reviewed in Chapter 1 of this volume. The important point for the present work is:

Each simultaneous eigenvector $|T(\mathbf{j}_\pi \mathbf{k})_{jm}\rangle$ of the set of $2n$ mutually commuting Hermitian operators corresponding to each binary tree $T \in \mathbb{T}_n$ of shape $Sh_T(\mathbf{j}_\pi)$ is a real orthogonal transformation of the eigenvectors $|\mathbf{j}\,\mathbf{m}\rangle \in \mathbf{B}_\mathbf{j}$. The coefficients in each such transformation are generalized Wigner-Clebsch-Gordan (WCG) coefficients, which themselves are a product of known ordinary WCG coefficients, where the product is uniquely determined by the shape of the labeled binary tree $Sh_T(\mathbf{j}_\pi)$.

This is, of course, just the expression of the property that we have constructed $n! a_n$ uniquely defined coupled orthonormal basis sets of the space $\mathcal{H}_\mathbf{j}$ from the uncoupled basis of the same space $\mathcal{H}_\mathbf{j}$.

There is a class of subspaces of $\mathcal{H}_\mathbf{j}$ of particular interest for the present work. This class of subspaces is obtained from the basis vectors $\mathbf{B}_T(\mathbf{j}_\pi)$ of $\mathcal{H}_\mathbf{j}$ given above by selecting from the orthonormal basis vectors $|T(\mathbf{j}_\pi \mathbf{k})_{jm}\rangle$ those that have a prescribed total angular momentum quantum number $j \in \{j_{\min}, j_{\min}+1, \ldots, j_{\max}\}$ and, for each such j, a prescribed projection quantum number $m \in \{j, j-1, \ldots, -j\}$. Thus, the orthonormal basis vectors in this set are given by

$$\mathbf{B}_T(\mathbf{j}_\pi, j, m) = \{|T(\mathbf{j}_\pi \mathbf{k})_{jm}\rangle \mid \text{each } \mathbf{k} \in \mathbb{K}_T^{(j)}(\mathbf{j}_\pi)\}.$$

We further specialize this basis set to the case $\pi = $ identity permutation:

$$\mathbf{B}_T(\mathbf{j}, j, m) = \{|T(\mathbf{j}\,\mathbf{k})_{jm}\rangle \mid \text{each } \mathbf{k} \in \mathbb{K}_T^{(j)}(\mathbf{j})\}.$$

This basis set of orthonormal vectors then defines a subspace $\mathcal{H}(\mathbf{j}, j, m) \subset \mathcal{H}_\mathbf{j}$, which is the same vector space for each $T \in \mathbb{T}_n$ and for every permutation $\mathbf{j}_\pi$ of $\mathbf{j}$; that is, the following direct sum decomposition of the tensor product space $\mathcal{H}_\mathbf{j}$ holds:

$$\mathcal{H}_\mathbf{j} = \sum_{j=j_{\min}}^{j_{\max}} \sum_{m=-j}^{j} \oplus \mathcal{H}(\mathbf{j}, j, m).$$

We repeat: The important structural result for this vector space decomposition is:

Each basis set $\mathbf{B}_T(\mathbf{j}_\pi, j, m), T \in \mathbb{T}_n, \pi \in S_n$, is an orthonormal basis of one and the same space $\mathcal{H}(\mathbf{j}, j, m)$.

The dimension of the tensor product space $\mathcal{H}(\mathbf{j}, j, m) \subset \mathcal{H}_\mathbf{j}$ is $N_j(\mathbf{j})$, the Clebsch-Gordan (CG) number. This important number is the number of times that a given $j \in \{j_{\min}, j_{\min}+1, \ldots, j_{\max}\}$ is repeated for specified $\mathbf{j}$. They can be calculated explicity by repeated application of the elementary rule for the addition of two angular momenta, as described in detail in [L]. The CG number is shape independent; that is, $N_j(\mathbf{j})$ counts the number of orthonormal basis vectors in the basis set $\{|T(\mathbf{j}_\pi\,\mathbf{k})_{jm}\rangle\,|\text{for specified}\,j\,m\}$ of the space $\mathcal{H}(\mathbf{j},j,m)$ for each binary tree $T \in \mathbb{T}_n$ and each $j_\pi,\,\pi \in S_n$.

It may seem highly redundant to introduce such a variety of orthonormal basis sets of vectors that span the same space $\mathcal{H}_\mathbf{j}$, but it is within these vector space structures that resides the entire theory of $3(n-1)-j$ coefficients. This aspect of the theory is realized through shape transformations applied to the binary trees whose standard labels appear in the binary coupled state vectors in the various basis sets $\mathcal{H}_T(\mathbf{j}_\pi, j, m)$, $T \in \mathbb{T}_n$, $\pi \in S_n$. The implementation of such shape transformations into numerical-valued transformations between such state vectors uses the notion of a recoupling matrix.

The matrix with elements that give the real orthogonal transformation matrix between distinct sets $\{|T(\mathbf{j}_\pi\,\mathbf{k})_{jm}\rangle\,|\,\mathbf{k} \in \mathbb{K}_T^{(j)}(\mathbf{j}_\pi)\}$ and $\{|T(\mathbf{j}_{\pi'}\,\mathbf{k}')_{jm}\rangle\,|\,\mathbf{k} \in \mathbb{K}_T^{(j)}(\mathbf{j}_{\pi'})\}$ of simultaneous basis eigenvectors, each of which is an orthonormal basis of the vector space $\mathcal{H}(\mathbf{j}, j, m)$, is called a *recoupling matrix*. Thus, we have that

$$|T'(\mathbf{j}_{\pi'}\,\mathbf{k}')_{jm}\rangle = \sum_{\mathbf{k}\in\mathbb{K}_T^{(j)}(\mathbf{j}_\pi)} \left(R^{S;S'}\right)_{\mathbf{k},j;\mathbf{k}',j} |T(\mathbf{j}_\pi\,\mathbf{k})_{jm}\rangle,$$

where the recoupling matrix is the real orthogonal matrix, denoted $R^{S;S'}$, with elements given by the inner product of state vectors:

$$\left(R^{S;S'}\right)_{\mathbf{k},j;\mathbf{k}',j} = \left\langle T(\mathbf{j}_\pi\,\mathbf{k})_{jm} \middle| T'(\mathbf{j}_{\pi'}\,\mathbf{k}')_{jm}\right\rangle.$$

Here we have written the labeled shapes in the abbreviated forms:

$$S = Sh_T(\mathbf{j}_\pi), \quad S' = Sh_{T'}(\mathbf{j}_{\pi'}).$$

As this notation indicates, these transformations are independent of m; that is, they are invariants under $SU(2)$ frame transformations. The matrix elements are fully determined in each coupling scheme in terms of known generalized WCG coefficients, since each coupled state vector in the inner product is expressed as a linear combination of the orthonormal basis vectors in the set $\mathbf{B}_\mathbf{j}$ of the tensor product space $\mathcal{H}_\mathbf{j}$ with coefficients that are generalized WCG coefficients. Each recoupling matrix is a fully known real orthogonal matrix of order $N_j(\mathbf{j})$ in terms of its elements, the generalized WCG coefficients. Since the inner product is real, it is always the case that the recoupling matrix satisfies the relation

$$R^{S;S'} = \left(R^{S';S}\right)^{tr},$$

where tr denotes the transpose of the matrix. But the most significant property of a recoupling matrix that originates from the completeness of the mutually commuting Hermitian operators that define each binary coupled state is the multiplication property:

$$R^{S_1;S_3} = R^{S_1;S_2} R^{S_2;S_3},$$

which holds for arbitrary binary trees $T_1, T_2, T_3 \in \mathbb{T}_n$ and for all possible labeled shapes $S_1 = Sh_{T_1}(\mathbf{j}_{\pi^{(1)}})$, $S_2 = Sh_{T_2}(\mathbf{j}_{\pi^{(2)}})$, $S_3 = Sh_{T_3}(\mathbf{j}_{\pi^{(3)}})$, where each permutation $\pi^{(i)} \in S_n$. It is this elementary multiplication rule that accounts fully for what are known as the Racah sum-rule between Racah coefficients and the Biedenharn-Elliott identity between Racah coefficients. Indeed, when iterated, this multiplication rule generates infinite classes of relations between $3(n-1)-j$ coefficients, $n \geq 3$.

We can now bring together the notion of general shape transformations as realized in terms of words in the elementary association actions A and commutation actions C to arrive at the fundamental relation underlying the properties of the set of binary coupled angular momenta state vectors:

Let
$$w(A, C) = L_1 L_2 \cdots L_r,$$

where each L_h is either an elementary association operation A or a elementary commutation operation C, give a word $w(A, C)$ that effects the shape transformation given by

$$S_1 \xrightarrow{L_1} S_2 \xrightarrow{L_2} S_3 \xrightarrow{L_3} \cdots \xrightarrow{L_r} S_{r+1}.$$

The abbreviated shapes are defined by $S_h = Sh_{T_h}(\mathbf{j}_{\pi^{(h)}})$, $h = 1, 2, \ldots, r$, with corresponding elementary shape transformations given by

$$S_h \xrightarrow{L_h} S_{h+1}, h = 1, 2, \ldots, r.$$

Thus, the transformation from the initial shape S_1 to the final shape S_{r+1} is effected by a succession of elementary shape transformations via $S_2, S_3, \ldots, S_r$. The matrix elements of the product of recoupling matrices given by

$$R^{S_1;S_{r+1}} = R^{S_1;S_2} R^{S_1;S_2} R^{S_2;S_3} \cdots R^{S_r;S_{r+1}}$$

is equal to the inner product of binary coupled state vectors given by

$$\left(R^{S_1;S_{r+1}}\right)_{\mathbf{k}^{(1)},j;\mathbf{k}^{(r+1)},j} = \left\langle T_1(\mathbf{j}_{\pi^{(1)}}\,\mathbf{k}^{(1)})_{jm} \,\middle|\, T_{r+1}(\mathbf{j}_{\pi^{(r+1)}}\,\mathbf{k}^{(r+1)})_{jm} \right\rangle.$$

Then, the main result is: There exists such a shape transformation by elementary shape operations between every pair of specified initial and final shapes, S_1 and S_{r+1}.

PREFACE AND PRELUDE   xix

We introduce yet another very useful notation and nomenclature for the matrix elements of a general recoupling matrix $R^{S;S'}$, which are given by the inner product of state vectors above. This is the concept of a *triangle coefficient*, which encodes the detailed coupling instructions of its labeled forks discussed above. A triangle coefficient has a left-triangle pattern and a right-triangle pattern. The left-triangle pattern is a $3\times(n-1)$ matrix array whose 1×3 columns, $n-1$ in number, are the quantum numbers that encode the elementary addition of two angular momenta of the $n-1$ labeled forks that constitute the fully labeled binary tree $T(\mathbf{j}_\pi\,\mathbf{k})_j$. For example, the column corresponding to the labeled fork 1 given earlier is $\mathrm{col}(j_1\,j_2\,k)$ with a similar rule for labeled forks of the other three types. The left-triangle pattern of a triangle coefficient is this collection of $n-1$ columns, read off the fully labeled binary tree, and assembled into a $3\times(n-1)$ matrix pattern by a standard rule. The right-triangle pattern is the $3\times(n-1)$ triangular array constructed in the same manner from the fully labeled binary tree $T'(\mathbf{j}_{\pi'}\,\mathbf{k}')_j$. These two triangle patterns denoted, respectively, by $\Delta_T(\mathbf{j}_\pi\,\mathbf{k})_j$ and $\Delta_{T'}(\mathbf{j}_{\pi'}\,\mathbf{k}')_j$, are now used to define the triangle coefficient of order $2(n-1)$:

$$\left\{\Delta_T(\mathbf{j}_\pi\,\mathbf{k})_j\,\middle|\,\Delta_{T'}(\mathbf{j}_{\pi'}\,\mathbf{k}')_j\right\} = \left(R^{S;S'}\right)_{\mathbf{k},j;\mathbf{k}',j} = \left\langle T(\mathbf{j}_\pi\,\mathbf{k})_{jm}\,\middle|\,T'(\mathbf{j}_{\pi'}\,\mathbf{k}')_{jm}\right\rangle.$$

The main result for the present discussion is: The triangle coefficients (matrix elements of recoupling matrices) for the matrix elements of the elementary commutation operation C and association operation A in the product of recoupling matrices is a basic phase factor of the form $(-1)^{a+b-c}$ for a C-transformation, and a definite numerical object of the form $(-1)^{a+b-c}\sqrt{(2k+1)(2k'+1)}\,W$ for an A-transformation, where W is a Racah coefficient. It follows that the matrix elements $\left(R^{S_1;S_{r+1}}\right)_{\mathbf{k}^{(1)},j;\mathbf{k}^{(r+1)},j}$ of the recoupling matrix is always a summation over a number of Racah coefficients equal to the number of associations A that occur in the word $w(A,C)$, where the details of the multiple summations depend strongly on the shapes of the underlying pair of binary trees related by the A-transformations.

Each word $w(A,C) = L_1 L_2 \cdots L_r$ that effects a shape transformation between binary coupling schemes corresponding to a shape S_1 and S_{r+1} has associated with it a *path*, which is defined to be

$$\mathrm{path} = S_1 \xrightarrow{L_1} S_2 \xrightarrow{L_2} \cdots \xrightarrow{L_r} S_{r+1}.$$

But there are many distinct words $w_1(A,C), w_2(A,C), \ldots$ that effect the same transformation

$$S_1 \xrightarrow{w_i(A,C)} S_{r+1},\, i = 1,2,\ldots$$

via different intermediate shapes. Hence, there are correspondingly many paths of the same or different lengths between the same given pair of shapes, where the *length of a path* is defined to be the number of associations A in the path. There is, of course, always a path of minimum length. It is this many-fold structure of paths that gives rise to different expressions of one and the same $3(n-1) - j$ coefficient, as well as to a myriad of relations between such coefficients.

An arbitrary triangle coefficient of order $2(n-1)$ is always expressible as a product of recoupling matrices related by elementary shape transformations that give either simple phase transformations or a triangle coefficient of order four, since irreducible triangle coefficients of order four (those not equal to a phase factor or zero) are always of the form $(-1)^{a+b-c}\sqrt{(2k+1)(2k'+1)}\,W$. Thus, triangle coefficients of order four may be taken as the fundamental objects out of which are built all triangle coefficients. Triangle coefficients provide a universal notation for capturing the structure of all recoupling matrices. They possess some general simplifying structural properties that are inherited from the pair of standard labeled binary trees whose labeled fork structure they encode. For the description of this structure, we introduce the notion of a common fork: *Two standard labeled binary trees are said to have a common fork if each binary tree contains a labeled fork having endpoints with the same pair of labels, disregarding order.* Then, the left and right patterns of the triangle coefficient corresponding to a pair of standard labeled binary tree with a common fork has a column in its left pattern and one in its right pattern for which the entries in the first two rows are either the same or the reversal of one another: If the order of the labels in the two columns is the same, then the triangle coefficient is equal to the reduced triangle coefficient obtained by removal of the column from each pattern and multiplying the reduced pattern by a Kronecker delta factor in the intermediate quantum labels in row three of the common fork column. A similar reduction occurs should the quantum labels of the common fork be reversed, except now a basic phase factor multiplies the reduced triangle coefficient. Of course, if the resulting reduced triangle coefficient contains still a pair of columns corresponding to a common fork, then a further reduction takes place. This continues until a triangle coefficient of order $2(n-1)$ containing s columns corresponding to s common forks is reduced to a product of basic phase factors times an irreducible triangle coefficient of order $2(n-s)$. An *irreducible triangle coefficient* is one for which the corresponding pair of labeled binary trees has no common fork — the triangle coefficient has no common columns.

But the structure of irreducible triangle coefficients does not stop here. Each irreducible triangle coefficient defines a cubic graph. In particular, irreducible triangle coefficients of order $2(n-1)$ enumerate all possible "types" of cubic graphs of order $2(n-1)$ that can occur in the binary theory of the coupling of angular momenta. The cubic graph

$C^*_{2(n-1)}$ of an irreducible triangle coefficient of order $2(n-1)$ is obtained by the very simple rule: Label $2(n-1)$ points by the $2(n-1)$ triplets (triangle) of quantum labels constituting the columns of the triangle coefficient, and draw a line between each pair of points that is labeled by a triangle containing a common symbol. This defines a graph with $2(n-1)$ points, $3(n-1)$ lines, with three lines incident on each point, which is the definition of a cubic graph $C^*_{2(n-1)}$ of "angular momentum type" of order $2(n-1)$. While cubic graphs do not enter directly into such calculations, they are the objects that are used to classify a given collection of $3(n-1)-j$ coefficients into types.

In summary: *We have crafted above a conceptual and graphical framework that gives a uniform procedure for computing all $3(n-1)-j$ coefficients, based in the final step on computing the matrix elements of a recoupling matrix expanded into a product of recoupling matrices corresponding to a path of elementary shape transformations. The reduction process, applied to the matrix elements in this product, then automatically reduces in consequence of common forks to give the desired expression for the matrix elements of an arbitrary recoupling matrix (triangle coefficient), which is equal to the inner product of state vectors fully labeled by the simultaneous eigenvectors of the respective pair of complete sets of $2n$ commuting Hermitian operators. Moreover, properties of recoupling matrices can be used to generate arbitrarily many relations among irreducible triangle matrices of mixed orders, and expressions for one and the same $3(n-1)-j$ in terms of different coupling schemes. The unification of the binary coupling theory of angular momenta is achieved. There remain, however, unresolved problems such as: A procedure for obtaining paths of minimum length, a counting formula for the number of cubic graphs of order $2(n-1)$ of "angular momentum type," and the physical meaning of the existence of many paths for expressing the same $3(n-1)-j$ coefficient. Chapters 1, 5, and 8 provide more comprehensive details of results presented in this overview.*

We have taken the unusual step of presenting this overview here in the Preface so as to have in one place a reasonable statement of the coherence brought to the subject of angular momentum coupling theory by the methods outlined above, unblurred by the intricate steps needed in its implementation.

The binary coupling theory of angular momenta has relevance to quantum measurement theory. Measurements of the properties of composite systems is only at the present time, in sophisticated experimental set-ups, revealing the behavior of systems prepared in an initial definite state that remains unmeasured (undisturbed) until some later time, when a second measurement is made on the system. It is the prediction of what such a second measurement will yield that is at issue. This problem can be phrased very precisely in terms of doubly stochastic matrices for binary coupled angular momentum states corresponding to complete sets of commuting Hermitian observables; it is so formulated in Chapters 1 and 5, using the property that there is a unique doubly stochastic matrix of order $N_j(\mathbf{j})$ associated with each binary coupled scheme 1 and each binary coupled scheme 2. The probability of a prepared coupled scheme 1 state being in a measured scheme 2 coupled state is just the (row, column) entry in the doubly stochastic matrix corresponding to these respective coupled states. This is the answer given in the context of conventional nonrelativistic quantum theory. We do not speculate on the meaning of this answer to the holistic aspects of complex (or simple) quantal systems. Rather, throughout this volume, we focus on the detailed development of topics and concepts that relate to the binary coupling theory of angular momenta, as developed in [L], [6], and in this volume, leaving their full interpretation for the future.

In many ways, this portion of the monograph can be considered as a mopping-up operation for an accounting of the binary theory of the coupling of arbitrarily many angular momentum systems within the paradigm of conventional nonrelativistic quantum mechanics, in the sense of Kuhn [35, p. 24]. Yet, there is the apt comparison with Complexity Theory as advanced by the Santa Fe Institute — a few simple, algorithmic-like rules that generate an almost unlimited scope of patterns of high informational content. Moreover, binary trees viewed as branching diagrams, are omnipotent as classification schemes for objects of all sorts — their shapes and labelings have many applications going beyond angular momentum systems. The closely related graphs — Cayley's [19] trivalent trees, which originate from a single labeled binary tree — and their joining for pairs of such binary trees define cubic graphs (see [L]). These cubic graphs determine the classification of all binary coupling schemes in angular momentum theory. But they surely extend beyond the context of angular momentum systems in sorting out the diverse patterns of regularity in nature, as discussed briefly in Appendix C.

TOPICAL CONTENTS

We summarize next the principal topics that constitute the present monograph, and the relevant chapters.

Chapter 1, Chapter 5, Chapter 8. Total angular momentum states (reviewed from [L]). The total angular momentum of a physical system is a collective property. The addition of two angular momenta, a problem already solved in the seminal papers in quantum mechanics, is the simplest example, especially for intrinsic internal quantum spaces such as spin space. Here we review the earlier work, emphasizing that the tensor product space in which the angular momentum operators act has the property of all such tensor product spaces: It contains vectors that cannot be obtained as simple products of vectors of the individual parts of the system — tensor product spaces by their very nature are holistic;

that is, are superpositions of the tensor product of the constituent system state vectors. For the most part, our discussions in Chapters 1 and 5 are a summary of material from the first volume needed for this monograph, but now focused more strongly on the properties of the unitary matrices $Z^j(U^{(j)}, V^{(j)})$, called recoupling matrices, where $U^{(j)}$ and $V^{(j)}$ are unitary matrices that given the transformation coefficients from the uncoupled basis to the coupled basis that determines the composite system state vectors. These recoupling matrices satisfy the very important multiplication rule:

$$Z^j(U^{(j)}, V^{(j)}) Z^j(V^{(j)}, W^{(j)}) = Z^j(U^{(j)}, W^{(j)}).$$

In this relation, each of the unitary matrices $U^{(j)}, V^{(j)}, W^{(j)}$ gives the transformation coefficients of a complete coupled set of state vectors that are simultaneous eigenvectors of the mutually commuting set of Hermitian operators given by the squares of the constituent angular momenta $\mathbf{J}^2(i), i = 1, 2, \ldots, n$, the squared total angular momentum $\mathbf{J}^2$ and its 3−component J_3, and an additional set of operators (distinct for each of the three cases), or sets of parameter spaces, that complete the set of state vectors. Each set of coupled state vectors then spans the same tensor product space $\mathcal{H}_\mathbf{j}$, and the elements of the recoupling matrices give the transformation coefficients from one coupled basis set to the other, either for the full tensor product space $\mathcal{H}_\mathbf{j}$, or well-defined subspaces. It is also the case that each of the three recoupling matrices is a doubly stochastic matrix, each of which has a probabilistic interpretation in exactly the same sense as that for von Neumann's density matrices. Thus, the product rule has implications for measurements carried out on systems described by the state vectors corresponding to complete sets of mutually commuting observables.

The above vector space structures are more comprehensive than the notation indicates. This is because we have suppressed the labels in the $SU(2)$ irreducible multiplet ket vector notation $|j_i\, m_i\rangle \in \mathcal{H}_{j_i}$. More generally, these ket vectors are given by $|\alpha_i,\, j_i,\, m_i\rangle$ and constitute a complete set of eigenvectors for the i−th system S_i of the full system S; it is the quantum labels in the sequence α_i that originate from the eigenvalues of a complete set of mutually commuting Hermitian operators (or other complete labeling schemes) that includes $\mathbf{J}^2(i)$ and $J_3(i)$ that give a complete set of eigenvectors of system S_i, which itself could be a composite system with repeated values of the angular momentum quantum numbers j_i, as controlled by the labels α_i. The basic multiplication property of the recoupling matrix still holds under an appropriate adaptation of the notations. *The key concept is always completeness, first in the set of mutually commuting observables, and then of the simultaneous eigenvectors.* Thus, many complex quantal systems come under the purview of the angular momentum structure of composite systems, as we have outlined above.

Chapters 2-7. Permutation matrices and related topics. There is, perhaps, no symmetry group more important for all of quantum physics than the group of permutations of n objects — the permutation or symmetric group S_n. Permutation matrices of order n are the simplest matrix realization of the group S_n by matrices containing a single 1 in each row and column: They consist of the $n!$ rearrangements of the n columns of the identity matrix of order n. But in this work the symmetric group makes its direct appearance in a different context than the Pauli principle; namely, through Birkhoff's [8] theorem that proves the existence of a subset of the set of all $n!$ permutation matrices of order n such that every doubly stochastic matrix of order n can be expanded with positive real coefficients in terms of the subset of permutations matrices. We not only present this aspect of doubly stochastic matrices, but develop a more general theory in Chapter 3 of matrices having the same fixed *line-sum* for all rows and columns. Such matrices include doubly stochastic matrices, magic squares, and alternating sign matrices, which are all of interest in physical theory, as discussed in [L]. Here, additional results of interest are obtained, with Chapters 5, 6, 7 being dedicated to each topic, respectively. We comment further on Chapter 5.

Chapter 5. Doubly stochastic matrices. These matrices are introduced in Chapter 1, and the development of further properties continued here. Recoupling matrices introduced in Chapter 1 are doubly stochastic matrices. Such matrices have a probabilistic interpretation in terms of preparation of states corresponding to complete sets of commuting Hermitian operators. The matrix elements in a given (row, column) of a doubly stochastic matrix is the probability of a prepared eigenstate (labeled by the row) of a complete set of mutually commuting Hermitian operators being the measured eigenstate (labeled by the column) of a second (possibly the same) complete set of mutually commuting Hermitian operators. Here, these eigenstates are taken to be the coupled states corresponding to standard labeled binary trees of order n. This aspect of doubly stochastic matrices is illustrated numerous times. To my knowledge, doubly stochastic matrices were introduced into quantum theory by Alfred Landé [39], the great atomic spectroscopist from whom I had the privilege of hearing first-hand, while a graduate student at The Ohio State University, his thesis that they are fundamental objects underlying the meaning of quantum mechanics.

Chapter 8. Heisenberg's magnetic ring. This physical problem, addressed in depth by Bethe [5] in a famous paper preceding the basic work by Wigner [82] on angular momentum theory and Clebsch-Gordan coefficients is a very nice application of binary coupling theory. The Hamiltonian comes under the full purview of composite systems: Its state vectors can be classified as eigenvectors of the square of the total angular momentum with the usual standard action of the total angular momentum. The exact solutions are given for composite systems containing $n = 2, 3, 4$ constituents, each part with an arbitrary angular momentum. Remarkably, this seems not to have been noticed.

For $n \geq 5$, the magnetic ring problem can be reduced to the calculation of the eigenstates originating from the diagonalization by a real orthogonal matrix of a real symmetric matrix of order equal to the Clebsch-Gordan number $N_j(\mathbf{j})$, which gives the number of occurrences (multiplicity) of a given total angular momentum state in terms of the angular momenta of the individual constituents. This Hamiltonian matrix of order $N_j(\mathbf{j})$ is a real symmetric matrix uniquely determined by certain recoupling matrices originating from binary couplings of the constituent angular momenta. It is an exquisitely complex implementation of the uniform reduction procedure for the calculation of all $3(n-1) - j$ coefficients that arise; this procedure is itself based on paths, shape transformations, recoupling matrices, and their reduction properties. This approach to the Heisenberg ring problem gives a complete and, perhaps, very different viewpoint not present in the Bethe approach, especially, since no complex numbers whatsoever are involved in obtaining the energy spectrum, nor need be in obtaining a complete set of orthonormal eigenvectors. The problem is fully solved in the sense that the rules for computing all $3(n-1) - j$ coefficients that enters into the calculation of the Hamiltonian matrix can be formulated explicitly. Unfortunately, it is almost certain that the real orthogonal matrix required to diagonalize the symmetric Hamilton matrix, with its complicated $3(n-1) - j$ type structure, cannot be determined algebraically. It may also be the case that numerical computations of the elements of the symmetric Hamiltonian matrix are beyond reach, except for simple special cases. Of course, complex phase factors do enter into the classification of the eigenvectors by the cyclic invariance group C_n of the Hamiltonian, but this may not be necessary for many applications.

Finally, there are three Appendices A, B, and C that deal with issues raised in the main text. Appendix C, however, presents natural generalizations of binary tree classifications to other problems, especially, to composite systems where the basic constituents have $U(n)$ symmetry.

MATTERS OF STYLE, READERSHIP, AND RECOGNITION

On matters of readership and style, we repeat portions of the first volume, since they still prevail.

The very detailed Table of Contents serves as a summary of topics covered. The readership is intended to be advanced graduate students and researchers interested in learning of the relation between symmetry and combinatorics and of challenging unsolved problems. The many examples serve partially as exercises, but this monograph is not a textbook. It is hoped that the topics presented promote further and more rigorous developments. While we are unable to attain the eloquence of Mac Lane [51], his book has served as a model.

We mention again, as in [L], some unconventional matters of style. We present significant results in italics, but do not grade and stylize them as lemmas and theorems. Such italicized statements serve as summaries of results, and often do not merit the title as theorems. Diagrams and figures are integrated into the text, and not set aside on nearby pages, so as to have a smooth flow of ideas. Our informality of presentation, including proofs, does not attain the status of rigor demanded in more formal approaches, but our purpose is better served, and our objectives met, by focusing on algorithmic, constructive methods, as illustrated by many examples.

As with the earlier volume, this continuing work is heavily indebted to the two volumes on the quantum theory of angular momentum published with the late Lawrence C. Biedenharn [6]. The present volume is much more limited in scope, addressing special topics left unattended in the earlier work, as well as new problems.

Our motivation and inspiration for working out many details of binary coupling theory originates from the great learning experiences, beginning in the early 1990's, as acquired in combinatorial research papers with William Y. C. Chen, which were strongly encouraged by the late Gian-Carlo Rota, and supplemented by his many informative conversations and inspirational lectures at Los Alamos. We do not reference directly many of the seminal papers by Gian-Carlo Rota, his colleagues, and students in this volume, but these publications were foundational in shaping the earlier volume, and are ever-present here. We acknowledge a few of these again: Désarménien, *et al.* [23]; Kung and Rota [36]; Kung [37]; Roman and Rota [68]; Rota [70]-[71]; Rota, *et al.* [72]; as well as the Handbook by Graham, *et al.* [28]. This knowledge acquisition has continued under the many invitations by William Y. C. Chen, Director, The Center for Combinatorics, Nankai University, PR China, to give lectures on these subjects to students and to participate in small conferences. These opportunities expanded at about the same time in yet another direction through similar activities, organized by Tadeusz and Barbara Lulek, Rzeszòw University of Technology, Poland, on Symmetry and Structural Properties of Condensed Matter, under the purview over the years by Adam Mickiewics University, Poznań, University of Rzeszów, and Rzeszów University of Technology. The interaction with Chinese and Polish students and colleagues has been particularly rewarding. Finally, the constant encouragement by my wife Marge and son Tom provided the friendly environment for bringing both the first and second volumes to completion.

Editors Zhang Ji and Lai Fun Kwong deserve special mention and thanks for their encouragement and support of this project.

James D. Louck

Contents

Preface and Prelude ... vii

Notation ... xxxiii

1 Composite Quantum Systems ... 1
    1.1 Introduction ... 1
    1.2 Angular Momentum State Vectors of a Composite System ... 4
         1.2.1 Group Actions in a Composite System ... 10
    1.3 Standard Form of the Kronecker Direct Sum ... 11
         1.3.1 Reduction of Kronecker Products ... 12
    1.4 Recoupling Matrices ... 14
    1.5 Preliminary Results on Doubly Stochastic Matrices and Permutation Matrices ... 19
    1.6 Relationship between Doubly Stochastic Matrices and Density Matrices in Angular Momentum Theory ... 22

2 Algebra of Permutation Matrices ... 25
    2.1 Introduction ... 25
    2.2 Basis Sets of Permutation Matrices ... 31
         2.2.1 Summary ... 41

3 Coordinates of A in Basis $\mathbb{P}_{\Sigma_n(e,p)}$ ... 43
    3.1 Notations ... 43

	3.2	The A-Expansion Rule in the Basis $\mathbb{P}_{\Sigma_n(e,p)}$	45
	3.3	Dual Matrices in the Basis Set $\Sigma_n(e,p)$	47
	3.3.1	Dual Matrices for $\Sigma_3(e,p)$	48
	3.3.2	Dual Matrices for $\Sigma_4(e,p)$	50
	3.4	The General Dual Matrices in the Basis $\Sigma_n(e,p)$	53
	3.4.1	Relation between the A-Expansion and Dual Matrices	55

4 Further Applications of Permutation Matrices   59

4.1	Introduction		59
4.2	An Algebra of Young Operators		60
4.3	Matrix Schur Functions		63
4.4	Real Orthogonal Irreducible Representations of S_n		67
	4.4.1	Matrix Schur Function Real Orthogonal Irreducible Representations	67
	4.4.2	Jucys-Murphy Real Orthogonal Representations	69
4.5	Left and Right Regular Representations of Finite Groups		72

5 Doubly Stochastic Matrices in Angular Momentum Theory   81

5.1	Introduction		81
5.2	Abstractions and Interpretations		89
5.3	Permutation Matrices as Doubly Stochastic		91
5.4	The Doubly Stochastic Matrix for a Single System with Angular Momentum $\mathbf{J}$		92
	5.4.1	Spin-1/2 System	92
	5.4.2	Angular Momentum-j System	94
5.5	Doubly Stochastic Matrices for Composite Angular Momentum Systems		97
	5.5.1	Pair of Spin-1/2 Systems	97
	5.5.2	Pair of Spin-1/2 Systems as a Composite System	99
5.6	Binary Coupling of Angular Momenta		104

		5.6.1	Complete Sets of Commuting Hermitian Observables 104

- 5.6.2 Domain of Definition $\mathbb{R}_T(\mathbf{j})$ 106
- 5.6.3 Binary Bracketings, Shapes, and Binary Trees . . . 109
- 5.7 State Vectors: Uncoupled and Coupled 115
- 5.8 General Binary Tree Couplings and Doubly Stochastic Matrices 140
  - 5.8.1 Overview 140
  - 5.8.2 Uncoupled States 142
  - 5.8.3 Generalized WCG Coefficients 143
  - 5.8.4 Binary Tree Coupled State Vectors 145
  - 5.8.5 Racah Sum-Rule and Biedenharn-Elliott Identity as Transition Probability Amplitude Relations . . 153
  - 5.8.6 Symmetries of the $6-j$ and $9-j$ Coefficients . . 165
  - 5.8.7 General Binary Tree Shape Transformations 167
  - 5.8.8 Summary 172
  - 5.8.9 Expansion of Doubly Stochastic Matrices into Permutation Matrices 174

6 Magic Squares 177
- 6.1 Review 177
- 6.2 Magic Squares and Addition of Angular Momenta 180
- 6.3 Rational Generating Function of $H_n(r)$ 186

7 Alternating Sign Matrices 195
- 7.1 Introduction 195
- 7.2 Standard Gelfand-Tsetlin Patterns 197
  - 7.2.1 A-Matrix Arrays 199
  - 7.2.2 Strict Gelfand-Tsetlin Patterns 202
- 7.3 Strict Gelfand-Tsetlin Patterns for $\lambda = (n\, n-1 \cdots 2\, 1)$. 202
  - 7.3.1 Symmetries 204

7.4	Sign-Reversal-Shift Invariant Polynomials		206
7.5	The Requirement of Zeros		211
7.6	The Incidence Matrix Formulation		219

8 The Heisenberg Magnetic Ring 223

8.1	Introduction		223
8.2	Matrix Elements of H in the Uncoupled and Coupled Bases		226
8.3	Exact Solution of the Heisenberg Ring Magnet for $n = 2, 3, 4$		230
8.4	The Heisenberg Ring Hamiltonian: Even n		235
	8.4.1	Summary of Properties of Recoupling Matrices	240
	8.4.2	Maximal Angular Momentum Eigenvalues	242
	8.4.3	Shapes and Paths for Coupling Schemes I and II	243
	8.4.4	Determination of the Shape Transformations	245
	8.4.5	The Transformation Method for $n = 4$	249
	8.4.6	The General $3(2f-1) - j$ Coefficients	253
	8.4.7	The General $3(2f-1) - j$ Coefficients Continued	255
8.5	The Heisenberg Ring Hamiltonian: Odd n		261
	8.5.1	Matrix Representations of H	266
	8.5.2	Matrix Elements of $R^{j_2;j_1}$: The $6f - j$ Coefficients	269
	8.5.3	Matrix Elements of $R^{j_3;j_1}$: The $3(f+1) - j$ Coefficients	276
	8.5.4	Properties of Normal Matrices	287
8.6	Recount, Synthesis, and Critique		289
8.7	Action of the Cyclic Group		292
	8.7.1	Representations of the Cyclic Group	295
	8.7.2	The Action of the Cyclic Group on Coupled State Vectors	299
8.8	Concluding Remarks		304

A Counting Formulas for Compositions and Partitions	**305**
A.1 Compositions	305
A.2 Partitions	307
B No Single Coupling Scheme for $n \geq 5$	**313**
B.1 No Single Coupling Scheme Diagonalizing H for $n \geq 5$	313
C Generalization of Binary Coupling Schemes	**317**
C.1 Generalized Systems	317
C.2 The Composite $U(n)$ System Problem	321
Bibliography	**327**
Index	**335**
Errata and Related Notes	**343**

Notation

General symbols

,	comma separator; used non-uniformly
$\mathbb{R}$	real numbers
$\mathbb{C}$	complex numbers
$\mathbb{P}$	positive numbers
$\mathbb{Z}$	integers
$\mathbb{N}$	nonnegative integers
$\mathbb{R}^n$	Cartesian n−space
$\mathbb{C}^n$	complex n−space
$\mathbb{E}^n$	Euclidean n−space
$O(n,R)$	group of real orthogonal matrices of order n
$SO(n,\mathbf{R})$	group of real, proper orthogonal matrices of order n
$U(n)$	group of unitary matrices of order n
$SU(n)$	group of unimodular unitary matrices of order n
$GL(n,\mathbf{C})$	group of complex nonsingular matrices of order n
$\mathbf{M}^p_{n\times n}(\alpha,\alpha')$	set of $n\times n$ matrix arrays with nonnegative elements with row-sum α and column-sum α'
$\times$	ordinary multiplication in split product
$\oplus$	direct sum of matrices
$\otimes$	tensor product of vector spaces, Kronecker (direct) product of matrices
$\delta_{i,j}$	the Kronecker delta for integers i,j
$\delta_{A,B}$	the Kroneker delta for sets A and B
$K(\lambda,\alpha)$	the Kostka numbers
$c^\lambda_{\mu\nu}$	the Littlewood-Richardson numbers
$\mathbb{P}\mathrm{ar}_n$	set of partitions having n parts, including 0 as a part
λ,μ,ν	partitions in the set $\mathbb{P}\mathrm{ar}_n$
$|\mathbb{A}|$	cardinality of a set $\mathbb{A}$
$[n]$	set of integers $\{1,2,\ldots,n\}$

Specialized symbols are introduced as needed in the text;

the list below contains a few of the more general ones:

$\mathbf{J}(i)$ — angular momentem of constituent $i \in [n]$ of a composite system

$\mathbf{K}(i)$ — intermediate angular momentem $i \in [n-2]$ of a composite system

$\mathbf{J}$ — total angular momentem of all constituents of a composite system

$\mathbf{j}$ — sequence $(j_1, j_2, \ldots, j_n)$ of quantum numbers of the constituents of a composite system

$\mathbb{B}_n$ — set of binary bracketings of order n

$\mathbb{T}_n$ — set of binary trees of order n

Sh_T — shape of a binary tree $T \in \mathbb{T}_n$

$Sh_T(\mathbf{j})$ — shape of a standard labeled binary tree $T \in \mathbb{T}_n$

$w(A, C)$ — word in the letters A and C

$|T(\mathbf{j}\,\mathbf{k})_j\,m\rangle$ — simultaneous eigenvector of a complete set of $2n$ angular momentum operators $\mathbf{J}^2(i), i = 1, 2, \ldots, n; \mathbf{J}^2, J_3;$ $\mathbf{K}^2(i), i = 1, 2, \ldots, n-2$ — also called binary coupled state vectors

$\langle T(\mathbf{j}\,\mathbf{k})_j\,m \mid T'(\mathbf{j}'\,\mathbf{k}')_j\,m \rangle$
 — inner product of two binary coupled state vectors

$\mathcal{H}_{\mathbf{j}_i}$ — Hilbert vector space of dimension $2j_i + 1$ that is irreducible under the action of $SU(2)$

$\mathcal{H}_{\mathbf{j}} =$ — $\mathcal{H}_{\mathbf{j}_1} \otimes \mathcal{H}_{\mathbf{j}_2} \otimes \cdots \otimes \mathcal{H}_{\mathbf{j}_n}$: tensor product of the spaces $\mathcal{H}_{\mathbf{j}_i}$ of dimension $(2j_1 + 1) \cdots (2j_n + 1) = N(\mathbf{j})$

$N_j(\mathbf{j})$ — Clebsch-Gordan number

$\mathcal{H}(\mathbf{j}, j, m)$ — subspace of $\mathcal{H}_{\mathbf{j}}$ of order $N_j(\mathbf{j})$

$U^\dagger V$ — Landé form of a doubly stochastic matrix

$R^{S;S'}$ — recoupling matrix for a pair of standard labeled binary trees related by arbitrary shapes S and S'

$R^{S_h;S_{h+1}}$ — recoupling matrix for a pair of standard labeled binary trees of shapes S_h and S_{h+1} related by an elementary shape transformation

$\{\Delta_T(\mathbf{j}\,\mathbf{k})_j \mid \Delta_{T'}(\mathbf{j}'\,\mathbf{k}')_j\}$
 — triangle coefficient that is a $3 \times (n-1)$ matrix array that encodes the structure of the labeled forks of a pair of standard labeled binary trees

NOTATION

$\mathbb{G}_\lambda$	Gelfand-Tsetlin (GT) pattern of shape λ
$\binom{\lambda}{m}$	member of $\mathbb{G}_\lambda$
$E(x)$	linear matrix form
$e_{ij}(x)$	element of $E(x)$
$\mathcal{P}_n$	vector space of linear forms
$\mathbb{P}_{\Sigma_n}$	basis set Σ_n of permutation matrices
$\mathbb{P}_{\Sigma_n(e)}$	basis set $\Sigma_n(e)$ of permutation matrices
$\mathbb{P}_{\Sigma_n(e,p)}$	basis set $\Sigma_n(e,p)$ of permutation matrices
$\mathbb{A}_n$	set of doubly stochastic matrices of order n
$\mathbb{M}_n(r)$	set of magic squares of order n and line-sum r
$\mathbb{A}S_n$	set of alternating sign matrices of order n
l_A	line-sum of a matrix A of fixed line-sum

Chapter 1

Composite Quantum Systems

1.1 Introduction

The group and angular momentum theory of composite quantum systems was initiated by Weyl [80] and Wigner [82]. It is an intricate, but well-developed subject, as reviewed in Biedenharn and van Dam [7], and documented by the many references in [6]. It is synthesized further by the so-called binary coupling theory developed in great detail in [L]. It was not realized at the time that recoupling matrices, the objects that encode the full prescription for relating one coupling scheme to another, are doubly stochastic matrices. This volume develops this aspect of the theory and related topics. We review in this first chapter some of the relevant aspects of the coupling theory of angular momenta for ease of reference. Curiously, these developments relate to the symmetric group S_n, which is a finite subgroup of the general unitary group and which is also considered in considerable detail in the previous volume. But here the symmetric group makes its appearance in the form of one of its simplest matrix (reducible) representations, the so-called permutation matrices. The symmetric group is one of the most important groups in physics (Wybourne [87]), as well as mathematics (Robinson [67]). In physics, this is partly because of the Pauli exclusion principle, which expresses a collective property of the many entities that constitute a composite system; in mathematics, it is partly because every finite group is isomorphic to a symmetric group of some order. While the symmetric group is one of the most studied of all groups, many of its properties that relate to doubly stochastic matrices, and other matrices of physical importance, seem not to have been developed. This review chapter provides the background and motivation for this continued study.

A comprehensive definition of a composite quantal system that is sufficiently broad in scope to capture all possible physical systems is difficult: it will not be attempted here. Instead, we consider some general aspects of complex systems and then restrict our attention to a definition that is sufficient for our needs.

In some instances, a composite quantal system can be built-up by bringing together a collection of known independent quantal systems, initially thought of as being noninteractive, but as parts of a composite system, the subsystems are allowed to be mutually interactive. We call this the *build-up principle* for composite systems. We assume that such a built-up composite system can also be taken apart in the sense that, if the mutual interactions between the known parts are ignored, the subsystems are each described independently and have their separate identities. This is a classical intuitive notion; no attempt is made to place the "putting-together and breaking-apart" process itself in a mathematical framework.

The mathematical model for describing a built-up composite system utilizes the concept of a tensor product of vector spaces. The state space of the $i-$th constituent of such a composite system is given by an inner product vector space $\mathcal{H}_i$, which we take to be a bra-ket vector space in the sense of Dirac [24], and, for definiteness, it is taken to be a separable Hilbert space. Each such Hilbert space then has an orthonormal basis given by

$$\mathbf{B}_i = \{|i, k_i\rangle \mid k_i = 1, 2, \ldots\}, i = 1, 2, \ldots, n. \tag{1.1}$$

The state space of a composite system, built-up from n such independent systems is then the tensor product space $\mathcal{H}$ defined by

$$\mathcal{H} = \mathcal{H}_1 \otimes \mathcal{H}_2 \otimes \cdots \otimes \mathcal{H}_n. \tag{1.2}$$

The orthonormal basis of $\mathcal{H}$ is given in terms of the individual orthonormal bases $\mathbf{B}_i$ of $\mathcal{H}_i$ by

$$\mathbf{B} = \mathbf{B}_1 \otimes \mathbf{B}_2 \otimes \cdots \otimes \mathbf{B}_n. \tag{1.3}$$

A general vector in the linear vector space $\mathcal{H}$ is of the form

$$|\text{ general state}\rangle \tag{1.4}$$

$$= \sum_{k_1 \geq 1} \sum_{k_2 \geq 1} \cdots \sum_{k_n \geq 1} a_{k_1, k_2, \ldots, k_n} \mid 1, k_1\rangle \otimes \mid 2, k_2\rangle \otimes \cdots \otimes \mid n, k_n\rangle,$$

where the coefficients $a_{k_1, k_2, \ldots, k_n}$ are arbitrary complex numbers. Since, in general, $a_{k_1, k_2, \ldots, k_n} \neq a_{1, k_1} a_{2, k_2} \cdots a_{n, k_n}$, it is an evident (and well-known) that a general superposition of state vectors given by (1.4) does not have the form

$$\left(\sum_{k_1 \geq 1} a_{1, k_1} \mid 1, k_1\rangle \right) \otimes \left(\sum_{k_2 \geq 1} a_{1, k_1} \mid 2, k_2\rangle \right) \otimes \cdots \otimes \left(\sum_{k_n \geq 1} a_{n, k_n} \mid n, k_n\rangle \right).$$

$$\tag{1.5}$$

1.1. INTRODUCTION

Thus, there are vectors in a tensor product space that cannot be written as the tensor product of n vectors, each of which belongs to a constituent subspace $\mathcal{H}_i$. This mathematical property already foretells that composite systems have properties that are not consequences of the properties of the individual constituents.

The definition of composite quantal systems and their interactions with measuring devices are fundamental to the interpretation of quantum mechanics. Schrödinger [73] coined the term *entanglement* to describe the property that quantal systems described by the superposition of states in tensor product space are not all realizable as the tensor product of states belonging to the constituent subsystems. Entanglement, in its conceptual basis, is not a mysterious property. It is a natural property of a linear theory based on vector spaces and standard methods for building new vector spaces out of given vector spaces. But the meaning and breadth of such mathematical constructions for the explanation of physical processes can be profound.

We make also the following brief remarks on the methodology of tensor product spaces introduced above. These observations have been made and addressed by many authors; we make no attempt (see Wheeler and Zurek [81]) to cite the literature, our purpose here being simply to place results presented in this volume in the broader context:

Remarks.

1. The build-up principle for composite systems stated above is already to narrow in scope to capture the properties of many physical systems. It does not, for example, include immediately an electron with spin: the spin property cannot be removed from the electron; it is one of its intrinsic properties, along with it mass and charge. Nonetheless, the energy spectrum of a single (nonrelativistic) electron with spin in the presence of a central attractive potential can be described in terms of a tensor product of vector spaces $\mathcal{H}_\psi \otimes \mathcal{H}_l \otimes \mathcal{H}_{1/2}$, where $\mathcal{H}_\psi$, $\mathcal{H}_l$, and $\mathcal{H}_{1/2}$, are, respectively, the space of solutions of the Schrödinger radial equation, the finite-Hilbert space of dimension $2l+1$ of orbital angular momentum states, and the finite-Hilbert space of dimension 2 of spin states (see Ref. [6]). This is indicative of the fact that mathematical techniques developed in specific contexts often have a validity that extends beyond their original intent.

2. It is sometimes helpful to give concrete realizations of tensor product spaces in terms of functions over the real or complex numbers. This is usually possible in ordinary quantum theory, even for spin and other internal symmetries. The abstract tensor product space relation (1.1) is then formulated as

$$\psi(\mathbf{z}_1, \mathbf{z}_2, \ldots, \mathbf{z}_n) \qquad (1.6)$$

$$= \sum_{k_1 \geq 1, k_2 \geq 1, \ldots, k_n \geq 1} a_{k_1, k_2, \ldots, k_n} \psi_{k_1}^{(1)}(\mathbf{z}_1) \psi_{k_2}^{(2)}(\mathbf{z}_2) \cdots \psi_{k_n}^{(n)}(\mathbf{z}_n),$$

where $\psi_{k_i}^{(i)} \in \mathcal{H}_i, i = 1, 2, \ldots, n$, with values $\psi_{k_i}^{(i)}(\mathbf{z}_i)$, where $\mathbf{z}_i$ is a set of real or complex numbers appropriate to the description of the desired property of the i-th part of the system. In the sense of presentation of state vectors in the form (1.6), the tensor product property of state vectors can be referred to as the *factorization assumption* for composite systems.

3. The converse of the build-up principle and its extensions is more difficult to formulate as a general principle, where the first naive question is: Can a composite system be taken apart to reveal its basic constituents? This question is regressive, since it can again be asked of the basic constituents. It does not have an answer without further qualifications.

4. There are fundamental issues associated with the very notion of an isolated quantal system: How does Newton's third law transcribe to quantal systems? Specifically, how does a quantal system interact with its environment such as instruments, classical and quantal, designed to measure certain of its properties. The implication of a measurement performed on a subsystem of a composite system are particularly intriguing, since the subsystems remain as entangled parts of the composite system independent of separation distance, if the whole system is left undisturbed between the time of its preparation and the time of the measurement.

5. Many properties of composite physical systems can be presented exactly by focusing on the properties of tensor products of subspaces of the general state space that can described exactly in terms of separable Hilbert spaces and their tensor products.

We continue now by describing the general setting for composite physical systems from the viewpoint of their angular momentum subspaces.

1.2 Angular Momentum State Vectors of a Composite System

We consider those composite quantal systems such that the state space $\mathcal{H}_i$ of the i-th part of the system contains at least one subspace characterized by the angular momentum $\mathbf{J}(i)$ of the subsystem, each

1.2. ANGULAR MOMENTUM STATE VECTORS

$i = 1, 2, \ldots, n$. The entire composite system is described in terms of a common right-handed inertial reference frame $(\mathbf{e}_1, \mathbf{e}_2, \mathbf{e}_3)$ in Cartesian 3–space $\mathbf{R}^3$, where redescriptions of the system are effected by unitary unimodular $SU(2)$ group transformations of the reference frame as described in detail in [L]. The angular momentum $\mathbf{J}(i)$ of the i-th subsystem is given in terms of its three components relative to the reference frame by $\mathbf{J}(i) = J_1(i)\mathbf{e}_1 + J_2(i)\mathbf{e}_2 + J_3(i)\mathbf{e}_3$, where the components satisfy the commutation relations $[J_1(i), J_2(i)] = iJ_3(i), [J_2(i), J_3(i)] = iJ_1(i), [J_3(i), J_1(i)] = iJ_2(i)$, where the i in the commutator relation is $i = \sqrt{-1}$, and not the subsystem index. The components $J_k(i), k = 1, 2, 3$, of $\mathbf{J}(i)$ have the standard action on each subspace of the states of the subsystem, as characterized by

$$\mathbf{J}^2(i)|j_i\,m_i\rangle = j_i(j_i+1)|j_i\,m_i\rangle,$$
$$J_3(i)|j_i\,m_i\rangle = m_i|j_i\,m_i\rangle,$$
$$J_+(i)|j_i\,m_i\rangle = \sqrt{(j_i-m_i)(j_i+m_i+1)}\,|j_i\,m_i+1\rangle, \quad (1.7)$$
$$J_-(i)|j_i\,m_i\rangle = \sqrt{(j_i+m_i)(j_i-m_i+1)}\,|j_i\,m_i-1\rangle.$$

The notation $\mathcal{H}_{j_i}$ denotes the finite-dimensional Hilbert space of dimension $\dim \mathcal{H}_{j_i} = 2j_i + 1$ with orthonormal basis given by

$$\mathbf{B}_{j_i} = \{|j_i\,m_i\rangle \mid m_i = j_i, j_i - 1, \ldots, -j_i\},$$
$$\langle j_i\,m_i | j_i\,m_i'\rangle = \delta_{m_i, m_i'}, \text{ each pair } m_i, m_i' \in \{j_i, j_i - 1, \ldots, -j_i\}. \quad (1.8)$$

The usual assumptions underlying the derivation of the standard relations (1.7) for the action of the angular momenta components are made; namely, that the linear vector space $\mathcal{H}_i$ over the complex numbers is equipped with an inner product with respect to which the components $J_k(i)$ are Hermitian operators that act linearly on the space $\mathcal{H}_i$ to effect a transformation to a new vector in the space. The operators $J_+(i)$ and $J_-(i)$ are the usual Hermitian conjugate shift operators defined by $J_+(i) = J_1(i) + iJ_2(i)$ and $J_-(i) = J_1(i) - iJ_2(i)$, where the nonindexing i is the complex number $i = \sqrt{-1}$.

The angular momentum components $J_k(i), k = 1, 2, 3$, for distinct subsystems i all mutually commute. It is allowed that each of the vector spaces $\mathcal{H}_i, i = 1, 2, \ldots, n$ can have the same or distinct definitions of inner product, it only being required that the angular momentum components in each subspace are Hermitian with respect to the inner product for that subspace. In the notation for the orthonormal basis vectors $\mathbf{B}_i$ of $\mathcal{H}_{j_i}$ in (1.8), we have suppressed all the extra quantum labels that may be necessary to define a basis for the full space $\mathcal{H}_i$. In applications to specific problems, such labels are to be supplied. It is the properties of the fundamental *standard angular momentum multiplets* defined by (1.7) and (1.8) in the tensor product space

$$\mathcal{H}_{j_1} \otimes \mathcal{H}_{j_2} \otimes \cdots \otimes \mathcal{H}_{j_n} \subset \mathcal{H}_1 \otimes \mathcal{H}_2 \otimes \cdots \otimes \mathcal{H}_n \qquad (1.9)$$

that are the subject of interest here. The analysis concerns only finite-dimensional Hilbert spaces and is fully rigorous. We point out that while we use the term angular momentum to describe the operators with the action (1.7) on the basis (1.8), it would, perhaps, be more appropriate to refer to the space $\mathcal{H}_{j_i}$ as an *irreducible $SU(2)$-multiplet*, since it is not necessary that such operators be interpreted physically as angular momenta. For example, the analysis can be applied to Gell-Mann's eightfold way, since the irreducible $SU(3)$−multiplet is realized in terms of the eight-dimensional Hilbert space $\mathcal{H}_{(2,1,0)}$ can be presented as the direct sum of the angular momentum vector spaces as given by

$$\mathcal{H}_{(2,1,0)} = \mathcal{H}_{(2,1)} \oplus \mathcal{H}_{(2,0)} \oplus \mathcal{H}_{(1,0)} \oplus \mathcal{H}_{(1,1)} = \mathcal{H}_{1/2} \oplus \mathcal{H}_1 \oplus \mathcal{H}'_{1/2} \oplus \mathcal{H}_0. \quad (1.10)$$

The mapping from the subspace $\mathcal{H}_{(a,b)}$ to the angular momentum subspace $\mathcal{H}_j$ is given by $j = (a-b)/2$, which conceals the fact that the two spaces $\mathcal{H}_{(2,1)}$ and $\mathcal{H}_{(1,0)}$, each of which has $j = 1/2$, are, in fact, perpendicular. Thus, while the $SU(2)$−multiplet content is the same, the physical content is quite different, since the two multiplets correspond to particles with different properties within the context of the unitary group $SU(3)$. The space $\mathcal{H}_{(2,1,0)}$ can clearly be incorporated with the framework of relation (1.9) by taking direct sums and paying careful attention to notations.

The concept of the tensor product space of $SU(2)$−multiplets is sufficiently rich in structure to accommodate rather diverse applications and illustrate properties of composite systems. A principal property always to be kept in mind is that this tensor product space is a linear vector space; hence, arbitrary linear combinations of vectors belonging to the space are allowed, and such superpositions show interference in the probabilistic interpretation of measurements.

The total angular momentum operator for the composite system (1.9) is defined by

$$\mathbf{J} = \sum_{i=1}^{n} \oplus \left(\mathbb{I}_{j_1} \otimes \cdots \otimes \mathbf{J}(i) \otimes \cdots \otimes \mathbb{I}_{j_n} \right), \qquad (1.11)$$

where in the direct sum the identity operators $\mathbb{I}_{j_1}, \mathbb{I}_{j_2}, \ldots, \mathbb{I}_{j_n}$ appear in the corresponding positions $1, 2, \ldots, n$, except in position i, where $\mathbf{J}(i)$ stands. This notation and the positioning of $\mathbf{J}(i)$ signify that the angular momentum operator $\mathbf{J}(i)$ acts in the Hilbert space $\mathcal{H}_{j_i}$, and that the unit

1.2. ANGULAR MOMENTUM STATE VECTORS

operators act in all other parts of the tensor product space. We often use the simplified notation

$$\mathbf{J} = \mathbf{J}(1) + \mathbf{J}(2) + \cdots + \mathbf{J}(n) \tag{1.12}$$

for the sum of various angular momentum operators acting in the tensor product space, but such expressions are always to be interpreted in the sense of a direct sum of tensor products of operators of the tensor product form (1.11). (See Sect. 10.5, Compendium A of [L] for a summary of the properties of tensor product spaces in terms of the present notations.)

We introduce the following compact notations to describe the ket-vectors of the tensor product space:

$$\mathbf{j} = (j_1, j_2, \ldots, j_n), \text{ each } j_i \in \{0, 1/2, 1, 3/2, \ldots\}, i = 1, 2, \ldots, n,$$
$$\mathbf{m} = (m_1, m_2, \ldots, m_n), \text{ each } m_i \in \{j_i, j_i - 1, \ldots, -j_i\},$$
$$i = 1, 2, \ldots, n,$$
$$\mathcal{H}_\mathbf{j} = \mathcal{H}_{j_1} \otimes \mathcal{H}_{j_2} \otimes \cdots \otimes \mathcal{H}_{j_n}, \tag{1.13}$$
$$|\mathbf{j}\,\mathbf{m}\rangle = |j_1\,m_1\rangle \otimes |j_2\,m_2\rangle \otimes \cdots \otimes |j_n\,m_n\rangle,$$
$$\mathbb{C}(\mathbf{j}) = \{\mathbf{m} \,|\, m_i = j_i, j_i - 1, \ldots, -j_i;\ i = 1, 2, \ldots, n\}.$$

The set of $2n$ mutually commuting Hermitian operators

$$\mathbf{J}^2(1), J_3(1), \mathbf{J}^2(2), J_3(2), \ldots, \mathbf{J}^2(n), J_3(n) \tag{1.14}$$

is a complete set of operators in the tensor product space $\mathcal{H}_\mathbf{j}$, in that the set of simultaneous eigenvectors $|\mathbf{j}\,\mathbf{m}\rangle, \mathbf{m} \in \mathbb{C}(\mathbf{j})$ is an orthonormal basis; that is, there is no degeneracy left over. The action of the angular momentum operators $\mathbf{J}(i), i = 1, 2, \ldots, n$, is the standard action given by

$$\mathbf{J}^2(i)|\mathbf{j}\,\mathbf{m}\rangle = j_i(j_i + 1)|\mathbf{j}\,\mathbf{m}\rangle,$$
$$J_3(i)|\mathbf{j}\,\mathbf{m}\rangle = m_i|\mathbf{j}\,\mathbf{m}\rangle,$$
$$J_+(i)|\mathbf{j}\,\mathbf{m}\rangle = \sqrt{(j_i - m_i)(j_i + m_i + 1)}\,|\mathbf{j}\,\mathbf{m}_{+1}(i)\rangle, \tag{1.15}$$
$$J_-(i)|\mathbf{j}\,\mathbf{m}\rangle = \sqrt{(j_i + m_i)(j_i - m_i + 1)}\,|\mathbf{j}\,\mathbf{m}_{-1}(i)\rangle,$$
$$\mathbf{m}_{\pm 1}(i) = (m_1, \ldots, m_i \pm 1, \cdots, m_n).$$

The orthonormality of the basis functions is expressed by

$$\langle \mathbf{j}\,\mathbf{m} \,|\, \mathbf{j}\,\mathbf{m}'\rangle = \delta_{\mathbf{m},\mathbf{m}'}, \text{ each pair } \mathbf{m}, \mathbf{m}' \in \mathbb{C}(\mathbf{j}). \tag{1.16}$$

Since the collection of $2n$ commuting Hermitian operators (1.14) refers to the angular momenta of the individual constituents of a physical system, and the action of the angular momentum operators is on the basis vectors of each separate space, the basis $|\mathbf{j}\,m\rangle$, $\mathbf{m} \in \mathbb{C}(\mathbf{j})$, is referred to as the *uncoupled basis* of the space $\mathcal{H}_\mathbf{j}$.

One of the most important observables for a composite system is the total angular momentum defined by (1.11). A set of $n+2$ mutually commuting Hermitian operators, which includes the square of the total angular momentum $\mathbf{J}$ and J_3 is the following:

$$\mathbf{J}^2(1), \mathbf{J}^2(2), \ldots, \mathbf{J}^2(n), \mathbf{J}^2, J_3. \qquad (1.17)$$

This set of $n+2$ commuting Hermitian operators is an incomplete set with respect to the construction of the states of total angular momentum; that is, the simultaneous state vectors of the $n+2$ operators (1.17) do not determine a basis of the space $\mathcal{H}_\mathbf{j}$. There are many ways to complete such an incomplete basis. For example, an additional set of $n-2$ independent $SU(2)$ invariant Hermitian operators, commuting among themselves, as well as with each operator in the set (1.17), can serve this purpose. Other methods of labeling can also be used. For the present discussion, we make the following assumptions:

Assumptions. The incomplete set of simultaneous eigenvectors of the $n+2$ angular momentum operators (1.17) has been extended to a basis of the space $\mathcal{H}_\mathbf{j}$ with properties as follows: A basis set of vectors can be enumerated in terms of an indexing set $\mathbb{R}(\mathbf{j})$ of the form

$$\mathbb{R}(\mathbf{j}) = \left\{ \boldsymbol{\alpha} = (\alpha_1, \alpha_2, \ldots, \alpha_{n-2}), j, m \;\middle|\; \begin{array}{l} j \in \mathbb{D}(\mathbf{j}); \boldsymbol{\alpha} \in \mathbb{A}^{(j)}(\mathbf{j}); \\ m = j, j-1, \ldots, -j \end{array} \right\}, \qquad (1.18)$$

where the domains of definition $\mathbb{D}(\mathbf{j})$ of j and $\mathbb{A}^{(j)}(\mathbf{j})$ of $\boldsymbol{\alpha}$ have the properties as follows. These domains of definition are to be such that for given quantum numbers $\mathbf{j}$ the cardinality of the set $\mathbb{R}(\mathbf{j})$ is given by

$$|\mathbb{R}(\mathbf{j})| = |\mathbb{C}(\mathbf{j})| = \prod_{i=1}^{n}(2j_1 + 1). \qquad (1.19)$$

Moreover, these labels are to be such that the space $\mathcal{H}_\mathbf{j}$ has the orthonormal basis given by the ket-vectors

$$|(\mathbf{j}\,\boldsymbol{\alpha})_{j\,m}\rangle, \; \boldsymbol{\alpha}, j, m \in \mathbb{R}(\mathbf{j}),$$
$$\langle(\mathbf{j}\,\boldsymbol{\alpha})_{j\,m} | (\mathbf{j}\,\boldsymbol{\alpha}')_{j'\,m'}\rangle = \delta_{j,j'}\delta_{m,m'}\delta_{\boldsymbol{\alpha},\boldsymbol{\alpha}'}, \qquad (1.20)$$
$$\boldsymbol{\alpha}, j, m \in \mathbb{R}(\mathbf{j}); \; \boldsymbol{\alpha}', j', m' \in \mathbb{R}(\mathbf{j}).$$

It is always the case that $\mathbb{D}(\mathbf{j})$ is independent of how the extension to a basis through the parameters $\boldsymbol{\alpha}$ is effected and that, for given j, the domain of m is $m = j, j-1, \ldots, -j$. The sequence of quantum labels $\boldsymbol{\alpha}$ also belongs to some domain of definition $\mathbb{A}^{(j)}(\mathbf{j})$ that can depend on j.

1.2. ANGULAR MOMENTUM STATE VECTORS

The actions of the commuting angular momentum operators (1.17) and the total angular momentum $\mathbf{J}$ on the orthonormal basis set (1.20) are given by

$$\mathbf{J}^2(i)|(\mathbf{j}\,\boldsymbol{\alpha})_{jm}\rangle = j_i(j_i+1)|(\mathbf{j}\,\boldsymbol{\alpha})_{jm}\rangle, i = 1, 2, \ldots, n,$$
$$\mathbf{J}^2|(\mathbf{j}\,\boldsymbol{\alpha})_{jm}\rangle = j(j+1)|(\mathbf{j}\,\boldsymbol{\alpha})_{jm}\rangle,$$
$$J_3|(\mathbf{j}\,\boldsymbol{\alpha})_{jm}\rangle = m|(\mathbf{j}\,\boldsymbol{\alpha})_{j\,m+1}\rangle, \qquad (1.21)$$
$$J_+|(\mathbf{j}\,\boldsymbol{\alpha})_{jm}\rangle = \sqrt{(j-m)(j+m+1)}|(\mathbf{j}\,\boldsymbol{\alpha})_{j\,m+1}\rangle,$$
$$J_-|(\mathbf{j}\,\boldsymbol{\alpha})_{jm}\rangle = \sqrt{(j+m)(j-m+1)}|(\mathbf{j}\,\boldsymbol{\alpha})_{j\,m-1}\rangle. \qquad \square$$

The notation for the ket-vectors in (1.20) and (1.21) places the total angular momentum quantum number j and its projection m in the subscript position to accentuate their special role. The set $\mathbb{R}(\mathbf{j})$ enumerates an alternative unique orthonormal basis (1.20) of the space $\mathcal{H}_\mathbf{j}$ that contains the total angular momentum quantum numbers j, m. Any basis set with the properties (1.20)-(1.21) is called a *coupled basis* of $\mathcal{H}_\mathbf{j}$. For $n = 2$, the uncoupled basis set is $\{|j_1\,m_1\rangle\otimes|j_2\,m_2\rangle\,|\,(m_1, m_2) \in \mathbb{C}(j_1, j_2)\}$, where $\mathbb{C}(j_1, j_2) = \{(m_1, m_2) | m_i = j_i, j_i - 1, \ldots, -j_i, i = 1, 2\}$; and the coupled basis set is $\{|(j_1\,j_2)_{jm}\rangle\,|\,(j, m) \in \mathbb{R}(j_1, j_2)\}$, where $\mathbb{R}(j_1, j_2) = \{(j, m) | j = j_1 + j_2, j_1 + j_2 - 1, \ldots, |j_1 - j_2|; m = j, j - 1, \ldots, -j\}$. No extra $\boldsymbol{\alpha}$ labels are required. For $n = 3$, one extra label α_1 is required, and at this point in our discussions, we leave the domain of definition of α_1 unspecified.

Angular momentum coupling theory of composite systems is about the various ways of providing the extra set of $\boldsymbol{\alpha}$ labels and their domains of definition, together with the values of the total angular momentum quantum number j, such that the space $\mathcal{H}_\mathbf{j}$ is spanned by the vectors $|(\mathbf{j}\,\boldsymbol{\alpha})_{jm}\rangle$. It turns out, as shown below, that the set of values that the total angular momentum quantum number j can assume is independent of the α_i; the values of j being $j = j_{\min}, j_{\min} + 1, \ldots, j_{\max}$, for well-defined minimum and maximum values of j that are expressed in terms of $j_1, j_2, \ldots, j_n$. Thus, the burden of completing any basis is placed on assigning the labels α_i in the set

$$\mathbb{R}(\mathbf{j}, j) = \{\boldsymbol{\alpha} = (\alpha_1, \alpha_2, \ldots, \alpha_{n-2})\,|\,\boldsymbol{\alpha} \in \mathbb{A}^{(j)}(\mathbf{j})\}. \qquad (1.22)$$

Such an assignment is called an $\boldsymbol{\alpha}-$*coupling scheme*. Since there are many ways of completing an incomplete basis of a finite vector space, there are also many coupling schemes. In this sense, the structure of the *coupling scheme set* $\mathbb{R}(\mathbf{j}, j)$ is the key object in angular momentum coupling theory; all the details of defining the coupling scheme are to be provided by the domain of definition $\boldsymbol{\alpha} \in \mathbb{A}^{(j)}(\mathbf{j})$.

The cardinality of the sets $\mathbb{R}(\mathbf{j})$ and $\mathbb{C}(\mathbf{j})$ are related by

$$|\mathbb{R}(\mathbf{j})| = \sum_{j=j_{\min}}^{j_{\max}} (2j+1) N_j(\mathbf{j}) = N(\mathbf{j}) = \prod_{i=1}^{n} (2j_i+1) = |\mathbb{C}(\mathbf{j})|, \quad (1.23)$$

where we have defined $N_j(\mathbf{j}) = |\mathbb{R}(\mathbf{j},j)|$. These positive numbers are called Clebsch-Gordan (CG) numbers. They can be generated recursively as discussed in Sect. 2.1.1 of [L].

The orthonormal bases (1.16) and (1.20) of the space $\mathcal{H}_\mathbf{j}$ must be related by a unitary transformation $A^{(\mathbf{j})}$ of order $N(\mathbf{j}) = \prod_{i=1}^{n}(2j_i+1)$ with (row; column) indices enumerated by $(\mathbf{m} \in \mathbb{C}(\mathbf{j}); \alpha, j, m \in \mathbb{R}(\mathbf{j}))$ (see (1.35) below). Thus, we must have the invertible relations:

$$|(\mathbf{j}\,\alpha)_{jm}\rangle = \sum_{\mathbf{m} \in \mathbb{C}(\mathbf{j})} \left(A^{(\mathbf{j})\mathrm{tr}} \right)_{\alpha, j, m;\, \mathbf{m}} |\mathbf{j}\,\mathbf{m}\rangle, \text{ each } \alpha, j, m \in \mathbb{R}(\mathbf{j}), \quad (1.24)$$

$$|\mathbf{j}\,\mathbf{m}\rangle = \sum_{\alpha, j, m \in \mathbb{R}(\mathbf{j})} \left(A^{(\mathbf{j})\dagger} \right)_{\alpha, j, m;\, \mathbf{m}} |(\mathbf{j}\,\alpha)_{jm}\rangle, \text{ each } \mathbf{m} \in \mathbb{C}(\mathbf{j}). \quad (1.25)$$

Note. We have reversed the role of row and column indices here from that used in [L] (see pp. 87, 90, 91, 94, 95), so that the notation accords with that used later in Chapter 5 for coupling schemes associated with binary trees, and the general structure set forth in Sect. 5.1. ☐

The transformation to a coupled basis (1.20) as given by (1.24) effects the full reduction of the n−fold Kronecker product

$$D^{\mathbf{j}}(U) = D^{j_1}(U) \otimes D^{j_2}(U) \otimes \cdots \otimes D^{j_n}(U),\ U \in SU(2), \quad (1.26)$$

of $SU(2)$ unitary irreducible matrix representations. The matrix $D^{\mathbf{j}}(U)$, $U \in SU(2)$, is a reducible unitary representation of $SU(2)$ of dimension $N(\mathbf{j})$, and the transformation (1.25) effects the transformation to a direct sum of irreducible unitary representations $D^j(U)$ (Wigner D−matrices). We next summarize the transformation properties of the coupled and ucoupled bases (1.25) under $SU(2)$ frame rotations.

1.2.1 Group Actions in a Composite System

Under the action of an $SU(2)$ frame rotation of the common frame $(\mathbf{e}_1, \mathbf{e}_2, \mathbf{e}_3)$ used to describe the n constituents of a physical system in Cartesian space $\mathbb{R}^3$, where system i has angular momentum $\mathbf{J}(i) = J_1(i)\mathbf{e}_1 + J_2(i)\mathbf{e}_2 + J_3(i)\mathbf{e}_3$, the orthonormal basis of the subspace

$$\mathcal{H}_{j_i} = \{|j_i\,m_i\rangle\,|\,m_i = j_i, j_i - 1, \ldots, -j_i\} \quad (1.27)$$

of system i undergoes the standard unitary transformation

$$\mathcal{T}_U|j_i\, m'_i\rangle = \sum_{m_i} D^{j_i}_{m_i\, m'_i}(U)|j_i\, m_i\rangle, \text{ each } U \in SU(2). \qquad (1.28)$$

The uncoupled basis $\mathcal{H}_{j_1} \otimes \mathcal{H}_{j_2} \otimes \cdots \otimes \mathcal{H}_{j_n}$ of the angular momentum space $\mathcal{H}_\mathbf{j}$ of the collection of systems undergoes the reducible unitary transformation given by

$$(\mathcal{T}_U \otimes \mathcal{T}_U \otimes \cdots \otimes \mathcal{T}_U)(|j_1\, m'_1\rangle \otimes |j_2\, m'_2\rangle \otimes \cdots \otimes |j_n\, m'_n\rangle)$$
$$= \sum_\mathbf{m} \left(D^{j_1}(U) \otimes D^{j_2}(U) \otimes \cdots \otimes D^{j_n}(U)\right)_{\mathbf{m}\,\mathbf{m}'}$$
$$\times (|j_1\, m_1\rangle \otimes |j_2\, m_2\rangle \otimes \cdots \otimes |j_n\, m_n\rangle), \qquad (1.29)$$

where $\mathbf{m} = (m_1, m_2, \ldots, m_n)$, $\mathbf{m}' = (m'_1, m'_2, \ldots, m'_n)$. This relation is described in the abbreviated notations (1.13) and (1.15)-(1.16) by

$$\mathcal{T}_U|\mathbf{j}\,\mathbf{m}'\rangle = \sum_\mathbf{m} D^\mathbf{j}_{\mathbf{m}\,\mathbf{m}'}(U)|\mathbf{j}\,\mathbf{m}\rangle, \quad D^\mathbf{j}(U) = D^{j_1}(U) \otimes \cdots \otimes D^{j_n}(U), \quad (1.30)$$

for each $U \in SU(2)$. Similarly, the coupled basis (1.20) of $\mathcal{H}_\mathbf{j}$ undergoes the irreducible unitary transformation:

$$\mathcal{T}_U|(\mathbf{j}\,\alpha)_j\, m'\rangle = \sum_m D^j_{m\, m'}(U)|(\mathbf{j}\,\alpha)_j\, m\rangle, \text{ each } U \in SU(2). \qquad (1.31)$$

1.3 Standard Form of the Kronecker Direct Sum

Schur's lemma (see Sect. 10.7.2, Compendium A in [L]) implies that the reducible unitary Kronecker product representation $D^\mathbf{j}(U)$ of $SU(2)$ defined by (1.26) is reducible into a direct sum of irreducible unitary representations $D^j(U)$ by a unitary matrix similarity transformation $U^{(\mathbf{j})}$ of order $N(\mathbf{j}) = \prod_{i=1}^n (2j_i + 1)$:

$$U^{(\mathbf{j})\dagger} D^\mathbf{j}(U) U^{(\mathbf{j})} = \mathbb{D}^\mathbf{j}(U)$$
$$= \begin{pmatrix} \mathbb{D}^{j_{\min}}(U) & 0 & 0 & \cdots & 0 \\ 0 & \mathbb{D}^{j_{\min}+1}(U) & 0 & \cdots & 0 \\ \vdots & \vdots & \vdots & \cdots & \vdots \\ 0 & 0 & 0 & \cdots & \mathbb{D}^{j_{\max}}(U) \end{pmatrix}, \quad (1.32)$$

each $U \in SU(2)$, where the block form on the right defines the matrix $\mathbb{D}^{\mathbf{j}}(U)$ for each $\mathbf{j} = (j_1, j_2, \ldots, j_n)$, which is also of order $N(\mathbf{j})$. Each matrix $\mathbb{D}^j(U), j = j_{\min}, j_{\min}+1, \ldots, j_{\max}$, is itself a matrix direct sum of block form consisting of the same standard irreducible matrix representation $D^j(U)$ of order $2j+1$ of $SU(2)$ repeated $N_j(\mathbf{j})$ times, as given by the Kronecker product

$$\mathbb{D}^j(U) = I_{N_j(\mathbf{j})} \otimes D^j(U)$$

$$= \begin{pmatrix} D^j(U) & 0 & 0 & \cdots & 0 \\ 0 & D^j(U) & 0 & \cdots & 0 \\ \vdots & \vdots & \vdots & \cdots & \vdots \\ 0 & 0 & 0 & \cdots & D^j(U) \end{pmatrix}. \quad (1.33)$$

In this relation, $I_{N_j(\mathbf{j})}$ is the unit matrix of order $N_j(\mathbf{j})$, the Clebsch-Gordan number. The reason for adopting a standard form for the Kronecker direct sum, as given explicitly by (1.32)-(1.33), is so that we can be very specific about the structure of the unitary matrix $U^{(\mathbf{j})}$ that effects the reduction.

1.3.1 Reduction of Kronecker Products

The reduction of the Kronecker product $D^{\mathbf{j}}(U)$ into the standard form of the Kronecker direct sum by the unitary matrix similarity transformation in (1.32) is not unique. There are nondenumerably infinitely many unitary matrices $U^{(\mathbf{j})}$ of order $N(\mathbf{j}) = \prod_{i=1}^{n}(2j_i+1)$ that effect the transformation

$$U^{(\mathbf{j})\dagger} D^{\mathbf{j}}(U) U^{(\mathbf{j})} = \sum_{j=j_{\min}}^{j_{\max}} \oplus \mathbb{D}^j(U). \quad (1.34)$$

The rows and columns of $U^{(\mathbf{j})\dagger}$ are labeled by the indexing sets $\mathbb{R}(\mathbf{j})$ and $\mathbb{C}(\mathbf{j})$ as given by (1.18) and (1.13), respectively:

$$\left(U^{(\mathbf{j})\dagger}\right)_{\alpha, j, m; \mathbf{m}} = \langle (\mathbf{j}\, \alpha)_{jm} \,|\, \mathbf{j}\, \mathbf{m} \rangle = \left(A^{(\mathbf{j})\dagger}\right)_{\alpha, j, m; \mathbf{m}}, \quad (1.35)$$

where we note that these matrix elements are also the transformation coefficients between the coupled and uncoupled basis vectors given by (1.25). The rows and columns can always be ordered such that $U^{(\mathbf{j})}$ effects the standard reduction given by (1.32)-(1.33); that is, given any coupling scheme, the transformation of the Kronecker product to the standard Kronecker direct sum can always be realized.

1.3. STANDARD FORM OF THE KRONECKER DIRECT SUM

There is an intrinsic non-uniqueness in the transformation (1.32) due to the multiplicity structure (1.33) of any standard reduction. Thus, define the matrix $\mathbb{W}^{(\mathbf{j},j)}$ of order $N(\mathbf{j})$ to be the direct product given by

$$\mathbb{W}^{(\mathbf{j},j)} = W^{(\mathbf{j},j)} \otimes I_{2j+1}, \qquad (1.36)$$

where $W^{(\mathbf{j},j)}$ is an **arbitrary complex matrix** of order $N_j(\mathbf{j})$, the CG number. Then, the matrix $\mathbb{W}^{(\mathbf{j},j)}$ commutes with the direct sum matrix $\mathbb{D}^j$ defined by (1.33):

$$\mathbb{W}^{(\mathbf{j},j)} \mathbb{D}^j(U) = \mathbb{D}^j(U) \mathbb{W}^{(\mathbf{j},j)}, \text{ each } U \in SU(2). \qquad (1.37)$$

We may choose $W^{(\mathbf{j},j)}$ in (1.36) to be an arbitrary unitary matrix of order $N_j(\mathbf{j})$; that is, $W^{(\mathbf{j},j)} \in U(N_j(\mathbf{j}))$, the group of unitary matrices of order $N_j(\mathbf{j})$. Then, the direct sum matrix

$$W^{(\mathbf{j})} = \sum_{j=j_{\min}}^{j_{\max}} \oplus \mathbb{W}^{(\mathbf{j},j)} = \sum_{j=j_{\min}}^{j_{\max}} \oplus \left(W^{(\mathbf{j},j)} \otimes I_{2j+1} \right), W^{(\mathbf{j},j)} \in U(N_j(\mathbf{j})),$$
$$(1.38)$$

is a unitary matrix belonging to the unitary group $U(N(\mathbf{j}))$; it has the commuting property given by

$$W^{(\mathbf{j})} D^{\mathbf{j}}(U) = D^{\mathbf{j}}(U) W^{(\mathbf{j})}, \text{ each } U \in SU(2). \qquad (1.39)$$

Thus, if we define the unitary matrix $V^{(\mathbf{j})}$ by $V^{(\mathbf{j})} = W^{(\mathbf{j})} U^{(\mathbf{j})}$, hence,

$$V^{(\mathbf{j})} U^{(\mathbf{j})\dagger} = W^{(\mathbf{j})}, \qquad (1.40)$$

then $V^{(\mathbf{j})}$ also effects, for each $U \in SU(2)$, the transformation:

$$V^{(\mathbf{j})\dagger} D^{\mathbf{j}}(U) V^{(\mathbf{j})} = U^{(\mathbf{j})\dagger} D^{\mathbf{j}}(U) U^{(\mathbf{j})} = \sum_{j=j_{\min}}^{j_{\max}} \oplus D^j(U). \qquad (1.41)$$

Each unitary matrix $V^{(\mathbf{j})}$ effects exactly the same reduction of the Kronecker product representation $D^{\mathbf{j}}(U)$ of $SU(2)$ into standard Kronecker direct sum form as does $U^{(\mathbf{j})}$. We call all unitary similarity transformations with the property (1.41) *standard reductions*.

Summary: Define the subgroup $H(N(\mathbf{j}))$ of the unitary group $U(N(\mathbf{j}))$ of order $N(\mathbf{j}) = \prod_{i=1}^{n}(2j_i+1)$ by

$$H(N(\mathbf{j})) = \left\{ \sum_{j=j_{\min}}^{j_{\max}} \oplus \left(W^{(\mathbf{j},j)} \otimes I_{2j+1} \right) \middle| W^{(\mathbf{j},j)} \in U(N_j(\mathbf{j})) \right\}. \qquad (1.42)$$

Then, if the unitary matrix element $U^{(\mathbf{j})}$ effects the standard reduction, so does every unitary matrix $V^{(\mathbf{j})}$ such that

$$V^{(\mathbf{j})} U^{(\mathbf{j})\dagger} \in H(N(\mathbf{j})). \tag{1.43}$$

1.4 Recoupling Matrices

Let $U^{(\mathbf{j})}$ and $V^{(\mathbf{j})}$ be unitary matrices of order $N(\mathbf{j})) = \prod_{i=1}^{n}(2j_i+1)$ that effect the standard reduction (1.34). The unitary matrix $U^{(\mathbf{j})\dagger}$ corresponds to an $\boldsymbol{\alpha}$-coupling scheme and has its rows enumerated by the elements of a set $\mathbb{R}(\mathbf{j})$ of the form:

$$\mathbb{R}(\mathbf{j}) = \left\{ \boldsymbol{\alpha} \in \mathbb{R}(\mathbf{j},j), j, m \,\middle|\, \begin{array}{l} j = j_{\min}, j_{\min}+1, \ldots, j_{\max}; \\ m = j, j-1, \ldots, -j \end{array} \right\}. \tag{1.44}$$

The domain of definition $\mathbb{R}(\mathbf{j},j)$ of each α_i quantum number in the sequence $\boldsymbol{\alpha}$ is itself a set of the form:

$$\mathbb{R}(\mathbf{j},j) = \{ \boldsymbol{\alpha} = (\alpha_1, \alpha_2, \ldots, \alpha_{n-2}) \,|\, \alpha_i \in \mathbb{A}_i(\mathbf{j},j) \}, \tag{1.45}$$

where each set $\mathbb{A}_i(\mathbf{j},j)$ is uniquely defined in terms of the given angular momenta $\mathbf{j} = (j_1, j_2, \ldots, j_n)$ and j in accordance with the prescribed $\boldsymbol{\alpha}$-coupling scheme. Similarly, the unitary matrix $V^{(\mathbf{j})\dagger}$ corresponds to a $\boldsymbol{\beta}$-coupling scheme and has its rows enumerated by the elements of a set $\mathbb{S}(\mathbf{j})$ of the form:

$$\mathbb{S}(\mathbf{j}) = \left\{ \boldsymbol{\beta} \in \mathbb{S}(\mathbf{j},j), j, m \,\middle|\, \begin{array}{l} j = j_{\min}, j_{\min}+1, \ldots, j_{\max}; \\ m = j, j-1, \ldots, -j \end{array} \right\}, \tag{1.46}$$

$$\mathbb{S}(\mathbf{j},j) = \{ \boldsymbol{\beta} = (\beta_1, \beta_2, \ldots, \beta_{n-2}) \,|\, \beta_i \in \mathbb{B}_i(\mathbf{j},j) \}. \tag{1.47}$$

The column indexing set for each of $U^{(\mathbf{j})}$ and $V^{(\mathbf{j})}$ is the same set of projection quantum numbers $\mathbb{C}(\mathbf{j})$.

There is a set of coupled state vectors associated with each of the unitary matrices $U^{(\mathbf{j})}$ and $V^{(\mathbf{j})}$ given by

$$|(\mathbf{j}\boldsymbol{\alpha})_j m\rangle = \sum_{\mathbf{m} \in \mathbb{C}(\mathbf{j})} \left(U^{(\mathbf{j})tr} \right)_{\boldsymbol{\alpha}, j, m; \mathbf{m}} |\mathbf{j}\,\mathbf{m}\rangle, \quad \boldsymbol{\alpha}, j, m \in \mathbb{R}(\mathbf{j}),$$

$$\tag{1.48}$$

$$|(\mathbf{j}\boldsymbol{\beta})_j m\rangle = \sum_{\mathbf{m} \in \mathbb{C}(\mathbf{j})} \left(V^{(\mathbf{j})tr} \right)_{\boldsymbol{\beta}, j, m; \mathbf{m}} |\mathbf{j}\,\mathbf{m}\rangle, \quad \boldsymbol{\beta}, j, m \in \mathbb{S}(\mathbf{j}).$$

1.4. RECOUPLING MATRICES

The unitary matrices $U^{(\mathbf{j})}$ and $V^{(\mathbf{j})}$ in these transformations are matrices of order $N(\mathbf{j}) = \prod_{i=1}^{n}(2j_i+1)$ in consequence of the equality of cardinality of the sets that enumerate the rows and columns:

$$|\mathbb{R}(\mathbf{j})| = |\mathbb{S}(\mathbf{j})| = |\mathbb{C}(\mathbf{j})| = N(\mathbf{j}). \tag{1.49}$$

Both the α-coupled basis and β-coupled basis are orthonormal basis sets of the same tensor product space $\mathcal{H}_{\mathbf{j}}$ and satisfy all of the standard relations (1.41)-(1.43). Since these orthonormal basis sets span the same vector space, they are related by a unitary transformation of the form:

$$|(\mathbf{j}\,\boldsymbol{\beta})_{jm}\rangle = \sum_{\boldsymbol{\alpha}\in\mathbb{R}(\mathbf{j},j)} Z^{j}_{\boldsymbol{\alpha};\boldsymbol{\beta}}(U^{(\mathbf{j})},V^{(\mathbf{j})})\,|(\mathbf{j}\,\boldsymbol{\alpha})_{jm}\rangle, \tag{1.50}$$

$$Z^{j}_{\boldsymbol{\alpha};\boldsymbol{\beta}}(U^{(\mathbf{j})},V^{(\mathbf{j})}) = \langle(\mathbf{j}\,\boldsymbol{\alpha})_{jm}\,|\,(\mathbf{j}\,\boldsymbol{\beta})_{jm}\rangle = \left(U^{(\mathbf{j})\dagger}V^{(\mathbf{j})}\right)_{\boldsymbol{\alpha},j,m;\boldsymbol{\beta},j,m},$$

$$\boldsymbol{\alpha}\in\mathbb{R}(\mathbf{j},j),\ \boldsymbol{\beta}\in\mathbb{S}(\mathbf{j},j);\ |\mathbb{R}(\mathbf{j},j)| = |\mathbb{S}(\mathbf{j},j)| = N_j(\mathbf{j}).$$

It is the same value of j and m that appear in both sides of the first relation because the vectors in each basis set are eigenvectors of $\mathbf{J}^2$ and J_3. Moreover, the transformation coefficients $Z^{j}_{\boldsymbol{\alpha};\boldsymbol{\beta}}(U^{(\mathbf{j})},V^{(\mathbf{j})})$ are independent of the value $m = j, j-1, \ldots, -j$ of the projection quantum number, as the notation indicates. This is true because the general relation (1.50) can be generated from

$$|(\mathbf{j}\,\boldsymbol{\beta})_{jj}\rangle = \sum_{\boldsymbol{\alpha}\in\mathbb{R}(\mathbf{j},j)} Z^{j}_{\boldsymbol{\alpha};\boldsymbol{\beta}}(U^{(\mathbf{j})},V^{(\mathbf{j})})\,|(\mathbf{j}\,\boldsymbol{\alpha})_{jj}\rangle \tag{1.51}$$

by the standard action of the lowering operator J_-, which does not affect the transformation coefficients.

The unitary transformation coefficients $Z^{j}_{\boldsymbol{\alpha};\boldsymbol{\beta}}(U^{(\mathbf{j})},V^{(\mathbf{j})})$ are called *recoupling coefficients* because they effect the transformation from one set of coupled state vectors to a second set, which here is from the α-coupling scheme to the β-coupling scheme, as given by (1.50).

The matrix $Z^{j}(U^{(\mathbf{j})},V^{(\mathbf{j})})$ with rows elements enumerated by $\boldsymbol{\alpha}\in\mathbb{R}(\mathbf{j},j)$ and column elements enumerated by $\boldsymbol{\beta}\in\mathbb{S}(\mathbf{j},j)$ is defined by

$$\left(Z^{j}(U^{(\mathbf{j})},V^{(\mathbf{j})})\right)_{\boldsymbol{\alpha};\boldsymbol{\beta}} = Z^{j}_{\boldsymbol{\alpha};\boldsymbol{\beta}}(U^{(\mathbf{j})},V^{(\mathbf{j})}) \tag{1.52}$$

is called a *recoupling matrix*. It is assumed that a total order relation can be imposed on the sequences $\boldsymbol{\alpha}\in\mathbb{R}(\mathbf{j},j)$ and $\boldsymbol{\beta}\in\mathbb{S}(\mathbf{j},j)$; hence, the elements of $Z^j(U^{(\mathbf{j})}, V^{(\mathbf{j})})$ can always be arranged in a square matrix of order $N_j(\mathbf{j})$.

But there is still a further direct sum matrix associated with the recoupling matrices $Z^j(U^{(\mathbf{j})}, V^{(\mathbf{j})}), j = j_{\min}, j_{\min}+1, \ldots, j_{\max}$, as expressed by the direct sum relation (1.42). Thus, we define the matrix $Z(U^{(\mathbf{j})}, V^{(\mathbf{j})})$ by

$$Z(U^{(\mathbf{j})}, V^{(\mathbf{j})}) = U^{(\mathbf{j})\dagger} V^{(\mathbf{j})}, \qquad (1.53)$$

with matrix elements given by

$$Z_{\boldsymbol{\alpha},j;\boldsymbol{\beta},j}(U^{(\mathbf{j})}, V^{(\mathbf{j})}) = Z^j_{\boldsymbol{\alpha};\boldsymbol{\beta}}(U^{(\mathbf{j})}, V^{(\mathbf{j})}) = \langle (\mathbf{j}\boldsymbol{\alpha})_{jm} \,|\, (\mathbf{j}\boldsymbol{\beta})_{jm}\rangle. \qquad (1.54)$$

This relation holds for all total angular momentum quantum numbers jm given by all values $j = j_{\min}, j_{\min}+1, \ldots, j_{\max}$, $m = j, j-1, \ldots, -j$. Relation (1.54) is then just the expression of the direct sum relation

$$Z(U^{(\mathbf{j})}, V^{(\mathbf{j})}) = \sum_{j=j_{\min}}^{j_{\max}} \oplus \left(Z^j(U^{(\mathbf{j})}, V^{(\mathbf{j})}) \otimes I_{2j+1} \right). \qquad (1.55)$$

Thus, $Z(U^{(\mathbf{j})}, V^{(\mathbf{j})}) \in H(N(\mathbf{j}))$ is of order $N(\mathbf{j}) = \prod_{i=1}^n (2j_i+1)$, and $Z^j(U^{(\mathbf{j})}, V^{(\mathbf{j})}) \in U(N_j(\mathbf{j}))$ is of order $N_j(\mathbf{j})$, the CG number.

The elements of the unitary matrix $Z(U^{(\mathbf{j})}, V^{(\mathbf{j})})$ are recoupling coefficients for the full basis sets defined by

$$\mathbf{B}^{(1)}_{\mathbf{j}} = \{|(\mathbf{j}\boldsymbol{\alpha})_{jm}\rangle \,|\, \boldsymbol{\alpha}, j, m \in \mathbb{R}(\mathbf{j})\},$$

$$\qquad (1.56)$$

$$\mathbf{B}^{(2)}_{\mathbf{j}} = \{|(\mathbf{j}\boldsymbol{\beta})_{jm}\rangle \,|\, \boldsymbol{\beta}, j, m \in \mathbb{S}(\mathbf{j})\},$$

while the elements of the unitary matrix $Z^j(U^{(\mathbf{j})}, V^{(\mathbf{j})})$ are recoupling coefficients for the sub-basis sets defined by

$$\mathbf{B}^{(1)}_{\mathbf{j},j,m} = \{|(\mathbf{j}\boldsymbol{\alpha})_{jm}\rangle \,|\, \boldsymbol{\alpha} \in \mathbb{R}(\mathbf{j}, j)\},$$

$$\qquad (1.57)$$

$$\mathbf{B}^{(2)}_{\mathbf{j},j,m} = \{|(\mathbf{j}\boldsymbol{\beta})_{jm}\rangle \,|\, \boldsymbol{\alpha} \in \mathbb{S}(\mathbf{j}, j)\}.$$

In [L], we have called the matrix $Z(U^{(\mathbf{j})}, V^{(\mathbf{j})})$ a recoupling matrix and $Z^j(U^{(\mathbf{j})}, V^{(\mathbf{j})})$ a reduced recoupling matrix. As emphasized now, they are both recoupling matrices, but for different basis sets, the first

1.4. RECOUPLING MATRICES

for the tensor product space $\mathcal{H}_\mathbf{j}$ with basis sets (1.56), the second for the tensor product subspace $\mathcal{H}_{\mathbf{j},j}$ with basis sets (1.57):

$$\mathcal{H}_\mathbf{j} = \sum_{j=j_{\min}}^{j_{\max}} \oplus \mathcal{H}_{\mathbf{j},j}. \tag{1.58}$$

The unitary recoupling matrices $Z(U^{(\mathbf{j})}, V^{(\mathbf{j})})$ and $Z^j(U^{(\mathbf{j})}, V^{(\mathbf{j})})$ satisfy the following relations:

$$Z(U^{(\mathbf{j})}, V^{(\mathbf{j})})\, Z(V^{(\mathbf{j})}, W^{(\mathbf{j})}) = Z(U^{(\mathbf{j})}, W^{(\mathbf{j})}),$$

$$Z^j(U^{(\mathbf{j})}, V^{(\mathbf{j})})\, Z^j(V^{(\mathbf{j})}, W^{(\mathbf{j})}) = Z^j(U^{(\mathbf{j})}, W^{(\mathbf{j})}), \tag{1.59}$$

where $U^{(\mathbf{j})}$, $V^{(\mathbf{j})}$, and $W^{(\mathbf{j})}$ are the arbitrary unitary matrices whose elements give the α−coupling scheme (1.48), the β−coupling scheme (1.48), and the γ−coupling scheme, where the latter is of exactly the same form as (1.48):

$$|(\mathbf{j}\,\gamma)_{jm}\rangle = \sum_{\mathbf{m} \in \mathbb{C}(\mathbf{j})} \left(W^{(\mathbf{j})}\right)_{\mathbf{m};\,\gamma,j,m} |\mathbf{j}\,\mathbf{m}\rangle,\quad \gamma, j, m \in \mathbb{T}(\mathbf{j}), \tag{1.60}$$

where the indexing set $\mathbb{T}(\mathbf{j})$ has cardinality $N(\mathbf{j}) = \prod_{i=1}^{n}(2j_i + 1)$. We have analogous to (1.56)-(157) the basis sets defined by

$$\mathbf{B}_\mathbf{j}^{(3)} = \{|(\mathbf{j}\,\gamma)_{jm}\rangle \,|\, \gamma,\, j,\, m \in \mathbb{T}(\mathbf{j})\},$$

$$\mathbf{B}_{\mathbf{j},j,m}^{(3)} = \{|(\mathbf{j}\,\gamma)_{jm}\rangle \,|\, \gamma \in \mathbb{T}(\mathbf{j}, j)\}. \tag{1.61}$$

Thus, in the first relation (1.59), the elements of the unitary matrix $Z(V^{(\mathbf{j})}, W^{(\mathbf{j})})$ are recoupling coefficients for the basis sets $\mathbf{B}_\mathbf{j}^{(2)}$ and $\mathbf{B}_\mathbf{j}^{(3)}$, while the elements of the unitary matrix $Z(U^{(\mathbf{j})}, W^{(\mathbf{j})})$ are recoupling coefficients for the basis sets $\mathbf{B}_\mathbf{j}^{(1)}$ and $\mathbf{B}_\mathbf{j}^{(3)}$. In the second relation (1.59), the elements of the unitary matrix $Z^j(V^{(\mathbf{j})}, W^{(\mathbf{j})})$ are recoupling coefficients for the basis sets $\mathbf{B}_{\mathbf{j},j,m}^{(2)}$ and $\mathbf{B}_{\mathbf{j},j,m}^{(3)}$, while the elements of the unitary matrix $Z^j(U^{(\mathbf{j})}, W^{(\mathbf{j})})$, are recoupling coefficients for the basis sets $\mathbf{B}_{\mathbf{j},j,m}^{(1)}$ and $\mathbf{B}_{\mathbf{j},jm}^{(3)}$.

The multiplication properties (1.59) are very important for the coupling theory of angular momenta of a composite system in which the total angular momentum is taken as an observable. These multiplication properties are intrinsic to the underlying $SU(2)$ multiplet structure of the constituent parts presented above, the multiplicity structure of the Kronecker product, and the assumed completeness that the parameters $\alpha, \beta,$ and γ bring to the set of $n + 2$ commuting Hermitian operators $\mathbf{J}^2(i), i = 1, 2, \ldots, n,$ and $\mathbf{J}^2, J_3$. We will provide many examples of how these extra $n - 2$ parameters are provided by the eigenvalues $k_i(k_i + 1)$ of the squares $\mathbf{K}^2(i), i = 1, 2, \ldots, n-2$ of intermediate angular momenta in the binary coupling theory of angular momenta based on binary trees of order n.

Recoupling matrices for different pairs of coupling schemes are the principal objects of study in the theory of the addition of angular momenta. They bring an unsuspected unity to some of the classical relations in angular momentum theory, as developed in detail in [L]. *Every collection of complete coupling schemes has the structure outlined above.*

We next introduce some additional terminology that pertains to the vector space bases introduced above, as used in quantum theory. Let $\mathcal{H}$ be a finite-dimensional Hilbert space of dimension d (a linear vector space with an inner product) over the complex numbers with orthonormal basis vectors enumerated by $|\,i\,\rangle, i = 1, 2, \ldots, d$. For each pair of sequences $\mathbf{u} = (u_1, u_2, \ldots, u_d)$ and $\mathbf{v} = (v_1, v_2, \ldots, v_d)$ of complex numbers, define the vectors $|\,\mathbf{u}\,\rangle \in \mathcal{H}$ and $|\,\mathbf{v}\,\rangle \in \mathcal{H}$ by

$$|\,\mathbf{u}\,\rangle = \sum_{i=1}^{d} u_i |\,i\,\rangle \text{ and } |\,\mathbf{v}\,\rangle = \sum_{i=1}^{d} v_i |\,i\,\rangle. \tag{1.62}$$

In quantum theory, such vectors are called state vectors or simply states. Then, the inner product

$$\langle\,\mathbf{u}\,|\,\mathbf{v}\,\rangle = \sum_{i=1}^{d} u_i^* v_i \tag{1.63}$$

is called the *probability amplitude* of a transition from the state $|\,\mathbf{u}\,\rangle$ to the state $|\,\mathbf{v}\,\rangle$, or simply the *transition probability amplitude*. The absolute value squared of the inner product

$$|\langle\,\mathbf{u}\,|\,\mathbf{v}\,\rangle|^2 = |\sum_{i=1}^{d} u_i^* v_i|^2 \tag{1.64}$$

is called the probability of a transition from the state $|\,\mathbf{u}\,\rangle$ to the state $|\,\mathbf{v}\,\rangle$, or simply the *transition probability*.

A number of transition probability amplitudes between states and the corresponding transition probabilities can be identified from the Hilbert spaces introduced above. Here we focus on just one such family — the ones that lead in the theory of the binary coupling of angular momenta based on binary trees of order n to the theory of $3(n-1)-j$ coefficients and the many relations between them and lower order such coefficients.

The transformation coefficients of principal interest are those in relations (1.50); that is, the matrix elements of the recoupling matrix $Z^j(U^{(\mathbf{j})}, V^{(\mathbf{j})})$:

$$\left\{ Z^j_{\boldsymbol{\alpha};\boldsymbol{\beta}}(U^{(\mathbf{j})}, V^{(\mathbf{j})}) \,\Big|\, \boldsymbol{\alpha} \in \mathbb{R}(\mathbf{j}, j),\ \boldsymbol{\beta} \in \mathbb{S}(\mathbf{j}, j) \right\}. \tag{1.65}$$

A matrix elements in this set gives the probability amplitude for a transition from the state $|(\mathbf{j}\,\boldsymbol{\alpha})_{j\,m}\rangle$ to the state $|(\mathbf{j}\,\boldsymbol{\beta})_{j\,m}\rangle$. The absolute value squared of this matrix element, as defined by

$$S^j_{\boldsymbol{\alpha};\boldsymbol{\beta}}(U^{(\mathbf{j})}, V^{(\mathbf{j})}) = \left| Z^j_{\boldsymbol{\alpha};\boldsymbol{\beta}}(U^{(\mathbf{j})}, V^{(\mathbf{j})}) \right|^2, \tag{1.66}$$

is, then, the transition probability for a transition from the state $|(\mathbf{j}\,\boldsymbol{\alpha})_{j\,m}\rangle$ to the state $|(\mathbf{j}\,\boldsymbol{\beta})_{j\,m}\rangle$. The matrix $S^j(U^{(\mathbf{j})}, V^{(\mathbf{j})})$ itself is called the *transition probability matrix*.

We now have the quite important result, overlooked in [L]:

The transition probability matrix $S^j(U^{(\mathbf{j})}, V^{(\mathbf{j})})$ is a doubly stochastic matrix of order $N_j(\mathbf{j})$.

1.5 Preliminary Results on Doubly Stochastic Matrices and Permutation Matrices

We next define doubly stochastic matrices and permutation matrices and then state a theorem due to Birkhoff [8] on their relationship (see also Brualdi and Ryser [14]).

A matrix A of order n is a *doubly stochastic matrix* if its elements are nonnegative real numbers such the entries in each row and in each column sum to unity. We introduce the notation

$$\mathbb{A}_n = \{A \mid A \text{ is doubly stochastic}\} \tag{1.67}$$

for the set of all doubly stochastic matrices of order n.

A matrix P of order n is a *permutation matrix* if its elements are all 0 or 1 such that the entries in each row and in each column sum to unity. Each permutation matrix can be denoted by P_π, for some $\pi \in S_n$, the symmetric group of order n. This is because each permutation matrix can be obtained by a permutation π of the columns of the unit matrix I_n of order n. We introduce the notation

$$\mathbb{P}_{S_n} = \{P_\pi \mid \pi \in S_n\} \tag{1.68}$$

for the set of all permutation matrices of order n. Thus, the number of such permutations matrices is $|\mathbb{P}_{S_n}| = |S_n| = n!$. Since the transpose of a permutation matrix is again a permutation matrix, the set of permutation matrices can also be obtained by permutations of the rows of the unit matrix — we will always, unless otherwise stated, describe permutations in terms of their column matrices (vectors).

The symmetric group S_n may be taken to be the set of all $n!$ sequences π in the integers $1, 2, \ldots, n$ with a multiplication rule:

$$S_n = \{\pi = (\pi_1, \pi_2, \ldots, \pi_n) \mid \text{the } n \text{ parts } \pi_i, i = 1, 2, \ldots, n$$
$$\text{of } \pi \text{ are an arrangement of the } n \text{ integers } 1, 2, \ldots, n\}, \tag{1.69}$$
$$\rho\tau = (\rho_{\tau_1}, \rho_{\tau_2}, \ldots, \rho_{\tau_n}), \text{ all } \rho, \tau \in S_n.$$

The following relations all hold for the $n!$ sequences with the multiplication rule (1.69):

1. closure: the product of two elements is again an element.

2. associativity: the product rule $(\pi\rho)\tau = \pi(\rho\tau)$ is true.

3. identity element: $e = (1, 2, \ldots, n)$.

4. inverse: $\pi^{-1} \in S_n$.

Thus, S_n is a group. The (unique) inverse, denoted π^{-1}, of the sequence $\pi = (\pi_1, \pi_2, \ldots, \pi_n) \in S_n$ is obtained from π by writing out the sequence $(\pi^{-1}_{\pi_1} = 1, \pi^{-1}_{\pi_2} = 2, \ldots, \pi^{-1}_{\pi_n} = n)$, and rearranging the entries in this sequence such that the **subscripts** are $1, 2, \ldots, n$ as read left-to-right. As an example, the inverse of $(3, 4, 2, 1)$ is obtained by reading off from the parts that $\pi_3^{-1} = 1, \pi_4^{-1} = 2, \pi_2^{-1} = 3, \pi_1^{-1} = 4$, which upon rearranging gives $(3, 4, 2, 1)^{-1} = (4, 3, 1, 2)$. This presentation of permutations as one-line symbols of sequences with the multiplication rule adapted directly to the sequences themselves is quite useful.

The action of a permutation $\pi = (\pi_1, \pi_2, \ldots, \pi_n) \in S_n$ on an arbitrary sequence of n indeterminates $x = (x_1, x_2, \ldots, x_n)$ is defined by $\pi(x) = (x_{\pi_1}, x_{\pi_2}, \ldots, x_{\pi_n})$. The corresponding product action $\rho\tau$ is

$$(\rho\tau)(x) = \rho(\tau(x)) = \left(x_{(\rho\tau)_1}, x_{(\rho\tau)_2}, \ldots, x_{(\rho\tau)_n}\right), \text{ all pairs } \rho, \tau \in S_n. \tag{1.70}$$

1.5. PRELIMINARY RESULTS

The permutation matrix P_π can be presented as the following matrix:

$$P_\pi = (e_{\pi_1}\, e_{\pi_2} \cdots e_{\pi_n}), \qquad (1.71)$$

where e_i is the unit column matrix with 1 in row i and 0 in all other rows, each $i = 1, 2, \ldots, n$. Thus, the element in row i and column j of P_π is given by

$$(P_\pi)_{ij} = \delta_{i,\pi_j}. \qquad (1.72)$$

A unit row matrix presentation is sometimes used, but we will always use the column presentation, unless otherwise noted.

We can characterize the set of permutation matrices $P_\pi \in \mathbb{P}_n$ in the following way: Each element in the set is a doubly stochastic $(0-1)$ matrix whose elements constitute a group under matrix multiplication, where $P_e = I_n, P_{\pi'} P_\pi = P_{\pi'\pi}, P_\pi^{-1} = P_\pi^{tr}$.

The relation between doubly stochastic matrices and permutation matrices of order n is given by the Birkhoff theorem, which we state as follows:

Birkoff Theorem: *There exist positive real numbers $c_1, c_2, \ldots, c_t$ such that each doubly stochastic matrix $A \in \mathbb{A}_n$ is given by*

$$A = c_1 P_1 + c_2 P_2 + \cdots + c_t P_t, \qquad (1.73)$$

where each P_i is a distinct permutation matrix $P_i \in \mathbb{P}_n$. Conversely, each matrix of the form (1.73), where the c_i are arbitrary positive real numbers is a doubly stochastic matrix.

Thus, the set of permutation matrices which themselves are $(0-1)$ doubly stochastic matrices, is a basis for all doubly stochastic matrices.

Birkoff's theorem can be recast in a geometrical form involving points on the unit sphere. For this description, we introduce the notation $\mathbb{S}^{n!-1} \subset \mathbb{R}^{n!}$ for the unit sphere in Cartesian space of dimension $n!$:

Corollary to Birkoff's Theorem: *There exists a real point $x \in \mathbb{S}^{n!-1}$, with $n!$ coordinates $x_\pi, \pi \in S_n, n \geq 1$, such that each doubly stochastic matrix $A \in \mathbb{A}_n$ is given by*

$$A = \sum_{\pi \in S_n} x_\pi^2 P_\pi, \quad \sum_{\pi \in S_n} x_\pi^2 = 1, \text{ each } x_\pi^2 \geq 0. \qquad (1.74)$$

Conversely, for each $x \in \mathbb{S}^{n!-1}$, the matrix A given by (1.74) is doubly stochastic.

We next prove:

The transition probability matrix $S^j(U^{(j)}, V^{(j)})$ is a doubly stochastic matrix.

Proof. The following row and column sum relations follow directly from the fact that $U^{(j)}$ and $V^{(j)}$ are unitary matrices.

$$\sum_{\boldsymbol{\alpha} \in \mathbb{R}(\mathbf{j},j)} \left(S^j(U^{(j)}, V^{(j)})\right)_{\boldsymbol{\alpha}, \boldsymbol{\beta}} = 1, \text{ each } \boldsymbol{\beta} \in \mathbb{S}(\mathbf{j}, j);$$

$$\sum_{\boldsymbol{\beta} \in \mathbb{S}(\mathbf{j},j)} \left(S^j(U^{(j)}, V^{(j)})\right)_{\boldsymbol{\alpha}; \boldsymbol{\beta}} = 1, \text{ each } \boldsymbol{\alpha} \in \mathbb{R}(\mathbf{j}, j). \qquad \square$$

(1.75)

This result is very important for the interpretation and physical significance of $3n - j$ coefficients, which are the elements of a real orthogonal *recoupling matrix*, also called a *triangle coefficient* in the context of binary tree couplings, developed in detail in [L]. This is because the elements of the doubly stochastic matrices that arise in quantum theory give the probabilities for a transition from one state of a quantum system to another state (see Ref. [44] and Landé [39]). This result can be shown from von Neumann's [78] density matrix formulation of quantum mechanics, as we next discuss.

1.6 Relationship between Doubly Stochastic Matrices and Density Matrices in Angular Momentum Theory

The projection operator $\rho^{\text{op}}_{(\mathbf{j}\boldsymbol{\alpha})_{jm}}$ associated with the normalized total angular momentum state $|(\mathbf{j}\boldsymbol{\alpha})_{jm}\rangle$ is defined in the Dirac bra-ket notation by the relation:

$$\rho^{\text{op}}_{(\mathbf{j}\boldsymbol{\alpha})_{jm}} = |(\mathbf{j}\boldsymbol{\alpha})_{jm}\rangle\langle(\mathbf{j}\boldsymbol{\alpha})_{jm}|. \qquad (1.76)$$

By construction, the projection operator $\rho^{\text{op}}_{(\mathbf{j}\boldsymbol{\alpha})_{jm}}$ is *idempotent*:

$$\left(\rho^{\text{op}}_{(\mathbf{j}\boldsymbol{\alpha})_{jm}}\right)^2 = \rho^{\text{op}}_{(\mathbf{j}\boldsymbol{\alpha})_{jm}}. \qquad (1.77)$$

The action of the projection operator $\rho^{\text{op}}_{(\mathbf{j}\boldsymbol{\alpha})_{jm}}$ on the uncoupled basis

1.6. DOUBLY STOCHASTIC MATRICES AND DENSITY MATRICES

$|\mathbf{j\,m}\rangle, \mathbf{m} \in \mathbb{C}(\mathbf{j})$ (see (1.25) and (1.35)) is given by

$$\rho^{\mathrm{op}}_{(\mathbf{j}\,\boldsymbol{\alpha})_{j\,m}}|\mathbf{j\,m}\rangle = |(\mathbf{j}\,\boldsymbol{\alpha})_{j\,m}\rangle\langle(\mathbf{j}\,\boldsymbol{\alpha})_{j\,m}|\mathbf{j\,m}\rangle = \left(U^{(\mathrm{j})\dagger}\right)_{\boldsymbol{\alpha},j,m;\mathbf{m}} |(\mathbf{j}\,\boldsymbol{\alpha})_{j\,m}\rangle. \tag{1.78}$$

Thus, the matrix elements of the projection operator $\rho^{\mathrm{op}}_{(\mathbf{j}\,\boldsymbol{\alpha})_{j\,m}}$ on the uncoupled basis are given by

$$\langle \mathbf{j\,m'}|\rho^{\mathrm{op}}_{(\mathbf{j}\,\boldsymbol{\alpha})_{j\,m}}|\mathbf{j\,m}\rangle = \left(U^{(\mathrm{j})\,tr}\right)_{\boldsymbol{\alpha},j,m;\mathbf{m'}} \left(U^{(\mathrm{j})\dagger}\right)_{\boldsymbol{\alpha},j,m;\mathbf{m}} \tag{1.79}$$

The matrix $\rho_{(\mathbf{j}\,\boldsymbol{\alpha})_{j\,m}}$ with elements in row $\mathbf{m'}$ and column $\mathbf{m}$ defined by

$$\left(\rho_{(\mathbf{j}\,\boldsymbol{\alpha})_{j\,m}}\right)_{\mathbf{m'},\mathbf{m}} = \langle \mathbf{j\,m'}|\rho^{\mathrm{op}}_{(\mathbf{j}\,\boldsymbol{\alpha})_{j\,m}}|\mathbf{j\,m}\rangle = \left(U^{(\mathrm{j})\,tr}\right)_{\boldsymbol{\alpha},j,m;\mathbf{m'}} \left(U^{(\mathrm{j})\dagger}\right)_{\boldsymbol{\alpha},j,m;\mathbf{m}} \tag{1.80}$$

is called the *density matrix* of the state $|(\mathbf{j}\,\boldsymbol{\alpha})_{j\,m}\rangle$ in the basis $|\mathbf{j\,m}\rangle, \mathbf{m} \in \mathbb{C}(\mathbf{j})$. The matrix $\rho_{(\mathbf{j}\,\boldsymbol{\alpha})_{j\,m}}$ is of order $N(\mathbf{j}) = \prod_{i=1}^{n}(2j_i + 1)$. It is also the case that the trace, denoted Tr, of the density matrix is given by

$$\mathrm{Tr}\rho_{(\mathbf{j}\,\boldsymbol{\alpha})_{j\,m}} = \sum_{\mathbf{m}\in\mathbb{C}(\mathbf{j})} \left(U^{(\mathrm{j})\dagger}\right)_{\boldsymbol{\alpha},j,m;\mathbf{m}} \left(U^{(\mathrm{j})}\right)_{\mathbf{m};\boldsymbol{\alpha},j,m}$$

$$= \left(U^{(\mathrm{j})\dagger}U^{(\mathrm{j})}\right)_{\boldsymbol{\alpha},j,m;\boldsymbol{\alpha},j,m} = 1, \tag{1.81}$$

since $U^{(\mathrm{j})}$ is unitary.

It is a standard result of von Neumann's [78] density matrix formulation of quantum mechanics that the significance of the quantity $\mathrm{Tr}(\rho_{(\mathbf{j}\,\boldsymbol{\alpha})_{j\,m}}\rho_{(\mathbf{j}\,\boldsymbol{\beta})_{j\,m}})$ is the probability that a system prepared in the state $|(\mathbf{j}\,\boldsymbol{\alpha})_{j\,m}\rangle$ will be measured to be in the state $|(\mathbf{j}\,\boldsymbol{\beta})_{j\,m}\rangle$. We have directly from (1.80) that

$$\mathrm{Tr}\left(\rho_{(\mathbf{j}\,\boldsymbol{\alpha})_{j\,m}}\rho_{(\mathbf{j}\,\boldsymbol{\beta})_{j\,m}}\right)$$

$$= \left|\sum_{\mathbf{m}\in\mathbb{C}(\mathbf{j})}\left(U^{(\mathrm{j})\dagger}\right)_{\boldsymbol{\alpha},j,m;\mathbf{m}}\left(V^{(\mathrm{j})}\right)_{\mathbf{m};\boldsymbol{\beta},j,m}\right|^2 = \left|\left(W(U^{(\mathrm{j})},V^{(\mathrm{j})})\right)_{\boldsymbol{\alpha},j,m;\boldsymbol{\beta},j,m}\right|^2$$

$$= \left|Z^j_{\boldsymbol{\alpha};\boldsymbol{\beta}}(U^{(\mathrm{j})},V^{(\mathrm{j})})\right|^2 = S^j_{\boldsymbol{\alpha};\boldsymbol{\beta}}(U^{(\mathrm{j})},V^{(\mathrm{j})}), \tag{1.82}$$

where the *transition probability amplitude matrix* $W(U^{(\mathrm{j})},V^{(\mathrm{j})})$ is defined by

$$W(U^{(\mathrm{j})},V^{(\mathrm{j})}) = U^{(\mathrm{j})\dagger}V^{(\mathrm{j})}. \tag{1.83}$$

We refer to (1.50)-(1.52) and (1.66) for various relations and definitions that enter into the above derivation of (1.82)-(1.83).

In summary, we have the result:

The transition probability matrix $S^j(U^{(j)}, V^{(j)})$ *with (row; column) elements* $(\alpha; \beta)$ *given by*

$$S^j_{\alpha;\beta}(U^{(j)}, V^{(j)}) = \left|\left(W(U^{(j)}, V^{(j)})\right)_{\alpha,j;\beta,j}\right|^2,$$

$$W(U^{(j)}, V^{(j)}) = U^{(j)\dagger} V^{(j)},$$
(1.84)

is a doubly stochastic matrix of order equal to the CG number $N_j(\mathbf{j})$; *the physical significance of these elements is that they give the transition probability from the prepared state* $|(\mathbf{j}\alpha)_{jm}\rangle$ *to the measured state* $|(\mathbf{j}\beta)_{jm}\rangle$, *these probabilities being independent of the projection quantum number* m; *that is, they are invariants under* $SU(2)$ *frame rotations.*

We note again that each pair of binary coupling schemes considered in [L] gives an explicit realization of a doubly stochastic matrix whose elements are transition probabilities from the prepared state corresponding to a given binary coupling scheme to the measured state corresponding to a second binary coupling scheme. The fact that pairs of irreducible binary trees of order n also define $3(n-1) - j$ coefficients opens up a whole new arena for their physical interpretation in terms of measurements of properties of composite systems. The details of this application are developed in Chapters 5 and 8. But first we explore refinements of the Birkhoff theorem.

Birkhoff's theorem does not specify how the expansion coefficients in relation (1.73) are to determined from a given doubly stochastic matrix. There are, in fact, many such expansions for one and the same doubly stochastic matrix. This is because the set of permutation matrices of order $n \geq 3$ is linearly dependent. Thus, in order to have a unique expansion, it is necessary to specify a basis. We consider this in the next Chapter.

Chapter 2

Algebra of Permutation Matrices

2.1 Introduction

The notion of a permutation matrix is quite elementary. The characteristics of their properties in relationship to other matrices would appear, at first, not to be particularly interesting. We demonstrate in this chapter the contrary situation by developing in considerable detail these rather extensive relations.

We find it convenient to consider the set of all complex (real) matrices of order n with the usual rules of addition of matrices and their multiplication by complex numbers to be a finite-dimensional Hilbert space $\mathcal{H}_n$ in which the product of all pairs of "vectors" is defined by the usual product rule for matrices, and, in addition (the Hilbert space property), there is an inner product for each pair of vectors, as defined by

$$(A|B) = \text{Tr}(A^\dagger B) = \sum_{1 \leq i,j \leq n} a^*_{ij} b_{ij}, \text{ each pair } A, B \in \mathcal{H}_n. \tag{2.1}$$

In this definition, we have that $A^\dagger = (A^*)^{tr}$ denotes the complex conjugate transpose (simply transpose for real matrices) of the matrix A with elements $A = (a_{ij})_{1 \leq i,j \leq n}$; the matrix B has elements $B = (b_{ij})_{1 \leq i,j \leq n}$; and Tr denotes the trace of the matrix product.

We next introduce the linear matrix form in the vector space $\mathcal{H}_n$ defined by

$$E(x) = \sum_{\pi \in S_n} x_\pi P_\pi, \tag{2.2}$$

in which $\pi \in S_n$ is a permutation in the symmetric group on the integers

$1, 2, \ldots, n$, the P_π are the permutation matrices of order n defined in Sect. 1.5, and the $x_\pi \in \mathbb{R}$, each $\pi \in S_n$, are arbitrary real variables.

The symbol x that appears in the left-hand side of relation (2.2) denotes the sequence of $n!$ variables $x_\pi, \pi \in S_n$, that occur in the right-hand side in the definition of the linear (matrix) form $E(x)$. We sometimes need to list the variables $x_\pi, \pi \in S_n$, in some sequential order, as read from left-to-right. For this purpose, we use lexicographic (page) order on the sequences for permutations in their one-line presentation $\pi - (\pi_1, \pi_2, \ldots, \pi_n)$: Thus, we write $\pi \prec \pi'$, if the first nonzero difference in the sequence $\pi - \pi' = (\pi_1 - \pi'_1, \pi_2 - \pi'_2, \ldots, \pi_n - \pi'_n)$ is negative. We denote the corresponding ordered sequence x by the notation

$$x = \{x_\pi | \pi \in S_n\}^{\text{ord}} \in \mathbb{R}^{n!}, \tag{2.3}$$

where ord denotes that this is an ordered set. We use a similar notation $x = \{x_\pi | \pi \in S\}^{\text{ord}} \in \mathbb{R}^{|S|}$ to specify the ordered points associated with a subset $S \subset S_n$ of permutations: Thus, x is a point in Cartesian (also called Euclidean) space $\mathbb{R}^{|S|}$ containing $|S|$ points in $\mathbb{R}$, where $|S|$ denotes the cardinality of the set S. In most properties we derive, the ordering is irrelevant.

Two matrices $E(x)$ and $E(y)$ of the form (2.2) satisfy the relation

$$E(\alpha x + \beta y) = \alpha E(x) + \beta E(y), \tag{2.4}$$

for arbitrary real numbers α, β and sequences $x = \{x_\pi | \pi \in S_n\}^{\text{ord}} \in \mathbb{R}^{n!}$ and $y = \{y_\pi | \pi \in S_n\}^{\text{ord}} \in \mathbb{R}^{n!}$. It is because of this property that the matrix $E(x)$ is called a *linear matrix form*.

The element $e_{ij}(x)$ in row i and column j of the linear matrix form $E(x)$ is given by

$$e_{ij}(x) = \sum_{\pi \in S_n} x_\pi \delta_{i, \pi_j} = \sum_{\pi \in S_n^{i,j}} x_\pi, \tag{2.5}$$

where the subset of permutations $S_n^{i,j} \subset S_n$ is defined by

$$S_n^{i,j} = \{\pi \in S_n \mid \pi_j = i\}. \tag{2.6}$$

Thus, there are $(n-1)!$ elements in the set $S_n^{i,j}$ corresponding to all permutations of the integers $(1, 2, \ldots, \hat{i}, \ldots, n)$ in which integer i is missing, and part j of each $\pi \in S_n^{i,j}$ is fixed at value i. The n sets of permutations corresponding to the row-column pairs (i, j) for fixed i and for $j = 1, 2, \ldots, n$ are disjoint, as are those for fixed j and $i = 1, 2, \ldots, n$,

2.1. INTRODUCTION

and their respective unions give the full group S_n of permutations:

$$\bigcup_{j=1}^{n} S_n^{i,j} = S_n, \quad \bigcup_{i=1}^{n} S_n^{i,j} = S_n. \tag{2.7}$$

The $e_{ij}(x)$ are linear forms in the $(n-1)!$ variables $x_\pi, \pi \in S_n^{i,j}$, and these functions satisfy the row and column sum rules

$$\sum_{k=1}^{n} e_{kj}(x) = \sum_{k=1}^{n} e_{ik}(x) = e_1(x), \; i,j = 1,2,\ldots,n, \tag{2.8}$$

where $e_1(x)$ denotes the linear form in all $n!$ variables:

$$e_1(x) = \sum_{\pi \in S_n} x_\pi. \tag{2.9}$$

Thus, the matrix $E(x)$ is a linear matrix form which has each row and column sum equal to $e_1(x)$. We refer to $e_1(x)$ as the *line-sum* of $E(x)$. Thus, we have a principal property of the linear forms $E(x)$:

For every choice of the set of $n!$ real numbers $\{x_\pi \mid \pi \in S_n\}^{\text{ord}}$, the matrix $E(x) = \sum_{\pi \in S_n} x_\pi P_\pi$ has line-sum $e_1(x)$, and the element in row i and column j of $E(x)$ is given by the linear form $e_{ij}(x)$.

The n^2 linear forms $e_{ij}(x), i,j = 1,2,\ldots,n$ are linearly dependent; that is, there exist coefficients $c_{ij}, i,j = 1,2,\ldots,n$, not all zero, such that $\sum_{i,j=1}^{n} c_{ij} e_{ij}(x) = 0$. The next result gives a set of linearly independent linear forms among the $e_{ij}(x)$:

The $b_n = n^2 - 2n + 2$ linear forms given by

$$e_{ij}(x), \; i,j = 1,2,\ldots,n-1, \text{ and } e_{nn}(x) \tag{2.10}$$

are linearly independent, and the remaining $2(n-1)$ linear forms $e_{in}, i = 1,2\ldots,n-1$, and $e_{nj}, j = 1,2,\ldots,n-1$, are linearly dependent on these.

Proof. The method of proof is the standard one for showing linear independence of vectors. Thus, we show that the relation

$$\sum_{i,j=1}^{n} c_{ij} e_{ij}(x) = 0, \tag{2.11}$$

in which

$$c_{1n} = c_{2n} = \cdots = c_{n-1\,n} = c_{n1} = c_{n2} = c_{n\,n-1} = 0 \tag{2.12}$$

implies that the remaining $b_n = (n-1)^2 + 1$ coefficients $c_{ij}, i, j = 1, 2, \ldots, n-1$ and c_{nn} are zero. Substitution of (2.5) into (2.11) and rearrangement of terms gives

$$\sum_{\pi \in S_n} x_\pi \left(\sum_{j=1}^n c_{\pi_j j} \right) = 0, \qquad (2.13)$$

in which the coefficients (2.12) are all to be set to zero. Since relation (2.11) must hold for all $x_n \in \mathbb{R}$, the $e_{ij}(x), i, j = 1, 2, \ldots, n-1$, and $e_{nn}(x)$ are linearly independent if and only if the following $n!$ relations hold:

$$\sum_{j=1}^n c_{\pi_j j} = 0, \; \pi \in S_n. \qquad (2.14)$$

Again, all coefficients (2.12) are to be set to zero in this relation. With these zeros taken into account, we must show that relations (2.14) imply that all the remaining coefficients $c_{ij}, i, j = 1, 2, \ldots, n-1$, and c_{nn} are zero. For example, for $n = 3$, accounting for $c_{13} = c_{23} = c_{31} = c_{32} = 0$, relation (2.14) gives $c_{11} + c_{22} + c_{33} = 0, c_{11} = 0, c_{21} = 0, c_{21} + c_{12} + c_{33} = 0, c_{12} = 0, c_{22} = 0$, from which we conclude all 9 coefficients c_{ij} are zero. To establish this result in general, we first select the two relations corresponding to $\pi = (1, 2, \ldots, n)$ and the transposition $\pi_{k,n}$ that interchanges k and n, which gives the two relations,

$$\sum_{j=1}^n c_{jj} = 0, \quad \sum_{j=1, j \neq k}^{n-1} c_{jj} = 0, \qquad (2.15)$$

where the second relation is valid for each choice $k = 1, 2, \ldots, n-1$. These n relations imply $c_{kk} = 0, k = 1, 2, \ldots, n$. We next select π to be a permutation with $n-3$ cycles of length 1 and a single cycle $\pi_{kln}, k \neq l \neq n$, or length 3, which gives $\sum_{j=1, j \neq k, l, n}^n c_{jj} + c_{lk} = 0$, hence, $c_{lk} = 0$ for all $k \neq l \neq n$.

Thus, the b_n linear forms given by (2.10) are linearly independent. By direct substitution from (2.5), the following three relations are obtained:

$$e_1(x) = \frac{1}{n-2}\left(-e_{nn}(x) + \sum_{i,j=1}^{n-1} e_{ij}(x) \right),$$

$$e_{in}(x) = e_1(x) - \sum_{j=1}^{n-1} e_{ij}(x), i = 1, 2, \ldots, n-1, \qquad (2.16)$$

$$e_{nj}(x) = e_1(x) - \sum_{i=1}^{n-1} e_{ij}(x), j = 1, 2, \ldots, n-1. \quad \square$$

2.1. INTRODUCTION

We next relate the properties of the matrix $E(x)$ to an arbitrary, but given real matrix A of order n having fixed line-sum l_A. Thus, let $A = (a_{ij})_{1 \leq i,j \leq n}$ be a real matrix of order n in which the $(n-1)^2$ elements $a_{ij}, i,j = 1,2,\ldots,n-1$, and a_{nn} are selected arbitrarily. Then, the elements $a_{in}, i = 1,2,\ldots,n-1$, and $a_{nj}, j = 1,2,\ldots,n-1$, of A are given by

$$a_{in} = l_A - \sum_{j=1}^{n-1} a_{ij}, \quad a_{nj} = l_A - \sum_{i=1}^{n-1} a_{ij}. \tag{2.17}$$

It then follows from these relations that

$$(n-2)l_A = -a_{nn} + \sum_{i,j=1}^{n-1} a_{ij}, \tag{2.18}$$

which is in agreement with relations (2.16) with $e_1(x) = l_A$. Moreover, every real matrix A of order n with line-sum l_A has this structure. These properties prove:

Let A be a given real matrix of order n with line-sum $l_A \in \mathbb{R}$. Then, necessary and sufficient conditions that the relation $E(x) = A$ be satisfied are that the system of $n^2 - 2n + 2$ linear equations in the $n!$ variables $x_\pi, \pi \in S_n$ given by

$$e_{ij}(x) = a_{ij}, \; i,j = 1,2,\ldots,n-1, \; e_{nn}(x) = a_{nn} \tag{2.19}$$

be satisfied.

We next derive an important property of line-sums. We formulate the above results directly in terms of the vector space of real matrices $A, B, \ldots$ of order n and given line-sums $l_A, l_B, \ldots$, where we denote this vector space by $\mathcal{P}_n$. Thus, we consider two matrices $A, B \in \mathcal{P}_n$:

$$A = \sum_{\pi \in S_n} a_\pi P_\pi, \; l_A = \sum_{\pi \in S_n} a_\pi, \; a_\pi \in \mathbb{R},$$

$$B = \sum_{\pi \in S_n} b_\pi P_\pi, \; l_B = \sum_{\pi \in S_n} b_\pi, \; b_\pi \in \mathbb{R}. \tag{2.20}$$

We now take the product of the two matrices $A, B \in \mathcal{P}_n$ and obtain:

$$AB = C = \sum_{\pi \in S_n} c_\pi P_\pi \in \mathcal{P}_n, \tag{2.21}$$

where the group multiplication rule for permutation matrices gives the expansion coefficients as

$$c_\pi = \sum_{\rho \in S_n} a_\rho b_{\rho^{-1}\pi}. \tag{2.22}$$

Thus, we find that the line-sum of the product matrix AB satisfies the simple product rule:
$$l_{AB} = l_A l_B, \qquad (2.23)$$
since $l_{AB} = \sum_{\pi \in S_n} c_\pi = (\sum_{\rho \in S_n} a_\rho)(\sum_{\tau \in S_n} b_\tau)$.

A linear vector space that also has a product defined for all pairs of vectors such that the product of each pair is again a vector in the space is called an algebra — here, the *algebra of permutation matrices,* which we denote by $A\mathcal{P}_n$.

It is, of course, the case that there are only $b_n = (n-1)^2 + 1$ linearly independent permutation matrices in the vector space of matrices of order n having fixed line-sum. Thus, let $\Sigma_n \subset S_n$ be an any subset of permutations of S_n of cardinality $|\Sigma_n| = (n-1)^2 + 1$ such that the set of corresponding permutation matrices
$$\mathbb{P}_{\Sigma_n} = \{P_\pi \mid \pi \in \Sigma_n\} \qquad (2.24)$$
is linearly independent. Then, each $A \in A\mathcal{P}_n$ can also be expressed in the form:
$$A = \sum_{\pi \in \Sigma_n} a_\pi P_\pi, \text{ each } a_\pi \in \mathbb{R}. \qquad (2.25)$$

The expansion coefficients, or coordinates a_π relative to the basis $\mathbb{P}_{\Sigma_n}$ are now unique. They can be determined by using the dual basis to $\mathbb{P}_{\Sigma_n}$, a property that we discuss subsequently. The line-sum l_A of A can now be expressed in terms of the expansion coefficients a_π that appear in the form (2.25) by
$$l_A = \sum_{\pi \in \Sigma_n} a_\pi, \qquad (2.26)$$
since $\sum_{k=1}^n \delta_{i,\pi_k} = 1$ and $\sum_{k=1}^n \delta_{k,\pi_j} = 1$, for **each** $\pi \in S_n$ and all $i, j = 1, 2, \ldots, n$. The same numerical value of the line-sum l_A of A is obtained independently of choice of the basis permutations Σ_n; the number l_A is an intrinsic property of the given A.

The study of the linear forms $E(x)$ has been motivated by some earlier results on doubly stochastic matrices obtained by Birkhoff [8], Brualdi and Ryser [14], and [L]. The work in [L] develops the subject within the context of the quantum mechanical state space of many-particle systems, decomposed into angular momentum multiplets. That the theory of fixed-line matrices applies also to magic squares and alternating sign matrices gives the subject the further breadth that elevates the subject to one of general interest. To place the subject in this general context, we define briefly these three classes of matrices:

1. A doubly stochastic matrix A of order n has line-sum $l_A = r = 1$ with elements $a_{ij} \in \mathbb{N}$, the set of nonnegative real numbers. We denote the set of doubly stochastic matrices of order n by $\mathbb{A}_n$.

2. A magic square matrix M of order n has line-sum equal to arbitrary $l_M = r \in \mathbb{N}$, the set of nonnegative integers, with elements $a_{ij} \in \mathbb{N}$ (the zero matrix is included). We denote the set of magic squares of order n with line-sum r by $\mathbb{M}_n(r)$. It is evident that $\mathbb{M}_n(r)/r \subset \mathbb{A}_n$.

3. An alternating sign matrix AS of order n has line-sum $l_{AS} = r = 1$ with elements $a_{ij} \in \mathbb{A} = \{-1, 0, 1\}$, such that when a given row (column) is read from left(top)-to-right(bottom), and all zeros in that row (column) are ignored, the row contains either a single 1, or is an alternating series $1, -1, 1, -1, \ldots, 1, -1, 1$. We denote the set of alternating sign matrices of order n by $\mathbb{A}S_n$.

The principal result for these three classes of matrices is the following:

Let A be a matrix of order n that is doubly stochastic, magic square, or alternating sign. Then, each such matrix can be expressed as a sum over permutation matrices of order n, that is, $A = \sum_{\pi \in S_n} a_\pi P_\pi$, where the expansion coefficients $a_\pi, \pi \in S_n$, belong, respectively, to the sets $\mathbb{N}$, if A is doubly stochastic; to $\mathbb{N}$, if A is magic square; and to $\mathcal{A} = \{-1, 0, 1\}$, if A is alternating sign.

The stated property of doubly stochastic matrices is the Birkhoff [8] result, modified to include 0 coefficients. The analogous property of magic squares and alternating sign matrices are not difficult to prove. The principal focus here will be on doubly stochastic matrices, but we denote short chapters to the other classes of matrices as well.

The permutation matrices $P_\pi, \pi \in S_n$, are simultaneously doubly stochastic, magic square, and alternating sign. Moreover, permutation matrices constitute a basis for all three classes of matrices. As we have proved above, this basis is over-complete, since the set of all $n!$ permutation matrices is linearly dependent ($n \geq 3$). We turn next to the discussion of basis subsets of the set of all permutation matrices of order n, and of the expansion of the matrices of fixed line-sum in terms of such basis sets.

2.2 Basis Sets of Permutation Matrices

It is well-known that the set of real matrices of order n can be considered to be a vector space $\mathcal{V}_n$ of dimension $\dim \mathcal{V}_n = n^2$ over the real numbers (see Gelfand [26], Chapter 1, for a very readable account of Euclidean vector spaces). The coordinates relative to the unit matrix basis $\{e_{ij} \mid i, j = 1, 2, \ldots, n\}$ are the elements of the matrix itself:

$$A = \sum_{i,j=1}^{n} a_{ij} e_{ij}, \qquad (2.27)$$

where the matrix unit e_{ij} is the matrix with 1 in row i and column j and 0 in all other rows and columns. Indeed, as noted at the beginning of this chapter, the vector space $\mathcal{V}_n$ can be taken to be an inner product space with the inner product given by (2.1). With respect to this inner product, the matrix units are then an orthonormal basis: $(e_{ij}|e_{kl}) = \delta_{i,k}\delta_{j,l}$.

Our interest here is in the vector space $\mathcal{P}_n \subset \mathcal{V}_n$ determined by the set $\mathbb{P}_{S_n}$ containing the $n!$ permutation matrices $P_\pi, \pi \in S_n$. But this set of permutation matrices $\mathbb{P}_{S_n}$ is linearly dependent in consequence of the linearly dependence of the linear matrix forms (2.1); indeed, it is evident that the number of linearly independent permutation matrices must equal the number b_n of linearly independent matrix forms. Thus, there must exist a subset $\Sigma_n \subset S_n$ of permutations such that the set of permutation matrices defined by

$$\mathbb{P}_{\Sigma_n} = \{P_\pi \mid \pi \in \Sigma_n\} \tag{2.28}$$

is a basis of the vector space $\mathcal{P}_n$. In such a basis set of permutation matrices, each matrix $A \in \mathcal{P}_n$ having line-sum l_A has the form given by

$$A = \sum_{\pi \in \Sigma_n} a_\pi P_\pi, \tag{2.29}$$

where the $b_n = |\Sigma_n| = (n-1)^2 + 1$ real numbers a_π are now given by the ordered sequence:

$$a = \{a_\pi \mid \pi \in \Sigma_n\}^{\text{ord}}. \tag{2.30}$$

These numbers are called the *coordinates* of A relative to the basis $\mathbb{P}_{\Sigma_n}$. The line-sum l_A of A is given by

$$l_A = \sum_{\pi \in \Sigma_n} a_\pi. \tag{2.31}$$

The coordinates a_π of A in the basis $\mathbb{P}_{\Sigma_n}$ are now uniquely determined by the dual space to $\mathcal{P}_n$, which we denote by $\widehat{\mathcal{P}}_n$. (The dual space to a given linear vector space is an important concept for elucidating the structure of vector spaces (see Gelfand [26]).) The dual linear vector space $\widehat{\mathcal{P}}_n$ has as basis a set of dual permutation matrices given by

$$\widehat{\mathbb{P}}_{\Sigma_n} = \{\widehat{P}_\pi \mid \pi \in \Sigma_n\}, \tag{2.32}$$

where $\widehat{P}_\pi$ is the vector in the dual space $\widehat{\mathcal{P}}_n$ defined by the b_n orthonormality relations:

$$(\widehat{P}_\pi|P_{\pi'}) = \delta_{\pi,\pi'}, \text{ for all } \pi' \in \Sigma_n. \tag{2.33}$$

2.2. BASIS SETS OF PERMUTATION MATRICES

There is, of course, such a set of b_n relations for each $\widehat{P}_\pi, \pi \in \Sigma_n$. The important property is: *The set of b_n relations for each dual permutation matrix uniquely defines the matrix.*

The vector space $\mathcal{P}_n$ and its dual $\widehat{\mathcal{P}}_n$ for the case at hand, are one and the same space of matrices; that is, $\mathcal{P}_n = \widehat{\mathcal{P}}_n$. This is because each dual permutation matrix has fixed line-sum and every fixed-line sum matrix has a unique expansion in terms of the basis set $\mathbb{P}_{\Sigma_n}$, and, conversely, in consequence of the defining property (2.33). The set of permutation matrices $\mathbb{P}_{\Sigma_n}$ and the dual set $\widehat{\mathbb{P}}_{\Sigma_n}$ span the same vector space $\mathcal{P}_n = \widehat{\mathcal{P}}_n$.

Once the set of dual permutation matrices $\widehat{\mathbb{P}}_{\Sigma_n}$ is known, the coordinates $a_\pi, \pi \in \Sigma_n$, of A in the basis $\mathbb{P}_{\Sigma_n}$ in (2.29) are given by

$$a_\pi = (\widehat{P}_\pi \,|\, A) = (A \,|\, \widehat{P}_\pi). \tag{2.34}$$

Explicit expressions for the dual space matrices $\widehat{P}_\pi, \pi \in \Sigma_n$ in terms of the permutation matrices $P_{\pi'}, \pi' \in \Sigma_n$ can be obtained as follows:

$$\widehat{P}_{\pi'} = \sum_{\pi \in \Sigma_n} x_{\pi,\pi'} P_\pi, \tag{2.35}$$

where the matrix X with elements $(X)_{\pi,\pi'} = x_{\pi,\pi'}$ is given as the inverse of the Gram matrix; that is, $X = Z^{-1}$, where the Gram matrix itself is defined as the matrix with elements given by the inner product $(P_\pi | P_{\pi'})$:

$$(Z)_{\pi,\pi'} = z_{\pi,\pi'} = (P_\pi | P_{\pi'}) = \sum_{h=1}^{n} \delta_{\pi_h, \pi'_h}. \tag{2.36}$$

In these relations, the permutations are given by $\pi = (\pi_1, \pi_2, \ldots, \pi_n)$ and $\pi' = (\pi'_1, \pi'_2, \ldots, \pi'_n)$. The matrix Z, with elements $z_{\pi,\pi'}, \pi, \pi' \in \Sigma_n$, is a real, symmetric, nonsingular, positive definite matrix of order $b_n = (n-1)^2 + 1$; that is, it is the Gram matrix of the linearly independent basis set $\mathbb{P}_{\Sigma_n}$.

The summation in the last relation (2.36) is over the **shared parts** of the two permutations π and π'; hence, the elements $z_{\pi,\pi'}$ of the Gram matrix Z are nonnegative integers with the property $z_{\pi,\pi'} \in \{0, 1, \ldots, n\}$. The relation $ZX = I_{b_n}$ is easily derived by direct substitution of $\widehat{P}_{\pi'}$ from (2.35) into the inner product $(\widehat{P}_{\pi'} | P_{\pi''})$, and using the easily proved relations $\delta_{i,k}\delta_{j,k} = \delta_{j,i}\delta_{k,i}$ and $\sum_{k=1}^{n} \delta_{k,\pi_j} = \sum_{k=1}^{n} \delta_{i,\pi_k} = 1$, for an arbitrary permutation $\pi \in S_n$. Also, the line-sum of the dual space matrix $\widehat{P}_{\pi'}$ is

$$l_{\widehat{P}_{\pi'}} = \sum_{\pi \in \Sigma_n} x_{\pi,\pi'}. \tag{2.37}$$

We note that: *The dual matrix of a permutation matrix is not a permutation matrix.* We summarize these results with the statement:

The complete rules for determining the dual space matrices for any basis set $\mathbb{P}_{\Sigma_n}$ of permutation matrices are given by (2.35)-(2.36); the coordinates a_π of A in such a basis are then given by the inner product (2.34) and the line-sum of A by (2.31).

The calculations of inner products with a permutation matrix occur frequently. The following mnemonic is quite useful for the evaluation of the inner product of an arbitrary matrix $A = (a_{ij})_{i \leq 1, j \leq n}$ and a permutation matrix P_π, such as occur in (2.36):

Inner product rule:

$$(A|P_\pi) = \text{sum of the } n \text{ elements of } A \text{ in row } \pi_1 \text{ of column 1,}$$
$$\text{in row } \pi_2 \text{ of column 2, } \ldots, \text{ in row } \pi_n \text{ of column } n. \quad (2.38)$$

There are many ways of selecting a subset of b_n permutations matrices of order n from the full set of $n!$ permutation matrices to obtain a basis set $\mathbb{P}_{\Sigma_n}$. While the number of such basis sets is finite, we have not determined it. Instead, we formulate general criteria for validating when a set of b_n permutation matrices is a basis set and introduce two such sets.

The conditions that a set of permutation matrices $P_\pi, \pi \in \Sigma_n, |\Sigma_n| = b_n = n^2 - 2n + 2$, be linearly independent is the standard one used for any linear vector space, namely, that

$$\sum_{\pi \in \Sigma_n} c_\pi P_\pi = 0 \ (0 \text{ matrix}) \text{ implies } c_\pi = 0, \text{ all } \pi \in \Sigma_n. \quad (2.39)$$

It is useful to formulate this condition as follows. Let $\pi^{(k)} \in \Sigma_n$ denote a permutation with parts given in the one-line notation by

$$\pi^{(k)} = (\pi_1^{(k)}, \pi_2^{(k)}, \ldots, \pi_n^{(k)}), \ k = 1, 2, \ldots, b_n. \quad (2.40)$$

Then, in terms of this notation for the basis set permutations, the conditions for linear independence become

$$\sum_{k=1}^{b_n} c_{\pi^{(k)}} P_{\pi^{(k)}} = 0 \text{ implies } c_{\pi^{(k)}} = 0, k = 1, 2, \ldots, b_n. \quad (2.41)$$

In terms of the matrix elements of the permutation matrices, these conditions are

$$\sum_{k=1}^{b_n} \delta_{i, \pi_j^{(k)}} c_{\pi^{(k)}} = 0, i, j = 1, 2, \ldots, n. \quad (2.42)$$

2.2. BASIS SETS OF PERMUTATION MATRICES

These conditions can be written in the matrix form

$$MC = 0 \text{ (0 column matrix)} \tag{2.43}$$

by defining the $(0-1)$ matrix M with n^2 rows and d_n columns by

$$M^{(k)}_{(i,j)} = \delta_{i,\pi^{(k)}_j}. \tag{2.44}$$

The b_n columns of M are enumerated by $k = 1, 2, \ldots, b_n$; and the n^2 rows by the pairs (i, j), taken, say, in lexicographic order:

$$\Big((1,1),(1,2),\ldots,(1,n);(2,1),(2,2),\ldots,(2,n);$$

$$\ldots;(n,1),(n,2),\ldots,(n,n)\Big). \tag{2.45}$$

We call the $(0-1)$ matrix M the *incidence matrix* of the set of permutation matrices in the basis set $\{P_{\pi^{(1)}}, P_{\pi^{(2)}}, \ldots, P_{\pi^{(b_n)}}\}$.

The general conditions for a basis set of permutation matrices to be linearly independent are:

Necessary and sufficient conditions that any set of b_n permutation matrices $\{P_{\pi^{(k)}} \mid k = 1, 2, \ldots, b_n\}$ be linearly independent are that the row rank (column rank) of the $n^2 \times b_n$ incidence matrix M of this set be b_n.

This result is, of course, fully equivalent to the requirement that the Gram matrix be nonsingular.

The concept of linearly independence (dependence) can be applied directly to the set of permutations $\pi \in S_n$. This is because the inner product of two permutation matrices given by $(P_\pi | P_{\pi'}) = \sum_{h=1}^{n} \delta_{\pi_h, \pi'_h}$ depends solely on the parts of the permutations π and π' themselves. Thus, we can define an inner product $\langle \pi | \pi' \rangle$ for permutations by the rule:

$$\langle \pi | \pi' \rangle = (P_\pi | P_{\pi'}) = \sum_{h=1}^{n} \delta_{\pi_h, \pi'_h}, \text{ all } \pi, \pi' \in S_n. \tag{2.46}$$

Thus, it is quite meaningful to define a set of permutations $\pi, \pi' \in S_n$ to be linearly independent (dependent) if the corresponding Gram matrix is nonsingular (singular). Then, a set of permutation matrices is linearly independent (dependent) if and only if the corresponding set of permutations is linearly independent (dependent). We can now use interchangeably the terminology *linear independent (dependent)* for both permutation matrices and the permutations themselves:

A set of permutations $\Sigma_n \subset S_n$ is a basis set if and only if contains b_n permutations and its incidence matrix M defined by (2.44) has rank b_n.

The inner product (2.46) also applies within the context of the vector space of Young operators; this is outlined briefly in Sect. 4.2, Chapter 4.

We point out that $(0-1)$ matrices are fundamental objects in combinatorics, as is the concept of incidence matrix. The book by Brualdi and Ryser [14] is an excellent source for this subject.

It is not always easy to determine the rank of the incidence matrix, but there is a general case where we can use the incidence matrix to determine sufficiency conditions for a set of permutations Σ_n to be a basis set. We next formulate this result.

Configure the set of permutations $\{\pi^{(1)}, \pi^{(2)}, \ldots, \pi^{(b_n)}\}$ of an arbitrary subset of permutations Σ_n, where $\pi^{(k)} = (\pi_1^{(k)}, \pi_2^{(k)}, \ldots, \pi_n^{(k)})$, into a matrix $\Pi^{(n \times b_n)}$ that has n rows and b_n columns as shown in the following display, where the parts of a given permutation constitute the columns:

$$\Pi^{n \times b_n} = \begin{pmatrix} \pi_1^{(1)} & \pi_1^{(2)} & \cdots & \pi_1^{(b_n)} \\ \pi_2^{(1)} & \pi_2^{(2)} & \cdots & \pi_2^{(b_n)} \\ \vdots & \vdots & & \vdots \\ \pi_n^{(1)} & \pi_n^{(2)} & \cdots & \pi_n^{(b_n)} \end{pmatrix}. \qquad (2.47)$$

We call this matrix the $\Pi-matrix$ of Σ_n.

Then, we have the following result:

Π-matrix rule for linear independence:

The columns of the $\Pi-matrix$ $\Pi^{(n \times b_n)}$ give a basis set of permutations if the columns can be rearranged (if needed) such that the following properties hold: column 1 contains an entry in some row that does not occur in the same row in any of columns 2 through b_n; column 2 contains an entry in some row that does not occur in the same row in any of columns 3 through b_n; ...; column k contains an entry in some row that does not occur in the same row in any of columns $k+1$ through b_n; ...; column $b_n - 1$ contains an entry in some row that does not occur in the same row in column b_n.

Proof. We give two closely related proofs of this result. For the first proof, we consider the incidence matrix M of the permutations in the rearranged $\Pi-matrix$. The permutation in column 1 has a row in the incidence matrix M with 1 as part 1 and 0 in all $b_n - 1$ parts to the right of 1; column 2 has a row in the incidence matrix M with 1 as part 2 and 0 in all $b_n - 2$ parts to the right of 1; ...; column k has a row in the incidence matrix M with 1 as part k and 0 in all $b_n - k$ parts to the right of 1; ...; column $b_n - 1$ has a row in the incidence matrix M with 1 as part $b_n - 1$ and 0 as part b_n; column b_n has a row in the incidence matrix

2.2. BASIS SETS OF PERMUTATION MATRICES

M with 1 as part b_n. The existence of this lower triangle set of rows in the incidence matrix gives the desired proof. For the second proof, the applicable rule is that a permutation that has a part not shared by any other permutation in the collection is linearly independent of those in the collection. Repeated application of this rule to a Π–matrix having the asserted property above, then again proves the linear independence. □

Examples. We illustrate the preceding general results by some examples:

1. $n = 2$: Both permutations $(1,2), (2,1)$ are required for a basis set. The Π–matrix is given by

$$\Pi^{2\times 2} = \begin{pmatrix} 1 & 2 \\ 2 & 1 \end{pmatrix}. \tag{2.48}$$

2. $n = 3$: Any five of the six permutations $(1,2,3), (1,3,2), (2,1,3),$ $(2,3,1), (3,1,2), (3,2,1)$ is a basis set. There is only one relation between the corresponding permutation matrices, which is the sum over even and odd permutations (this relation holds for n) given by

$$P_{1,2,3} + P_{2,3,1} + P_{3,1,2} = P_{1,3,2} + P_{2,1,3} + P_{3,2,1} = \begin{pmatrix} 1 & 1 & 1 \\ 1 & 1 & 1 \\ 1 & 1 & 1 \end{pmatrix}. \tag{2.49}$$

The Π–matrix matrix for the five permutations $(1,2,3), (1,3,2),$ $(2,1,3), (2,3,1), (3,1,2)$ is

$$\Pi^{3\times 5} = \begin{pmatrix} 1 & 1 & 2 & 2 & 3 \\ 2 & 3 & 1 & 3 & 1 \\ 3 & 2 & 3 & 1 & 2 \end{pmatrix}. \tag{2.50}$$

3. $n = 4$: Two examples of basis sets of permutations corresponding to $\Sigma_4(e)$ and $\Sigma_4(e,p)$, which are defined for general n below, have the Π–matrices as follows:

$$\Pi^{3\times 10}(e) = \begin{pmatrix} 1 & 1 & 2 & 2 & 3 & 1 & 1 & 4 & 1 & 2 \\ 2 & 3 & 1 & 3 & 1 & 2 & 4 & 1 & 3 & 3 \\ 3 & 2 & 3 & 1 & 2 & 4 & 2 & 2 & 4 & 4 \\ 4 & 4 & 4 & 4 & 4 & 3 & 3 & 3 & 2 & 1 \end{pmatrix},$$

$$\tag{2.51}$$

$$\Pi^{3\times 10}(e,p) = \begin{pmatrix} 1 & 1 & 2 & 2 & 1 & 1 & 2 & 2 & 3 & 4 \\ 2 & 3 & 1 & 3 & 2 & 3 & 3 & 4 & 2 & 2 \\ 3 & 2 & 3 & 1 & 4 & 4 & 4 & 3 & 4 & 3 \\ 4 & 4 & 4 & 4 & 3 & 2 & 1 & 1 & 1 & 1 \end{pmatrix}.$$

By careful selection of columns, application of the Π-matrix rule above shows that the permutations with parts equal to the columns in each of the cases (2.50)-(2.51) are linearly independent. Thus, one simply identifies, in each example, a column having a part not shared by the other column, casts out that column, and repeats the procedure until all columns have been cast out, except the last.

We introduce two basis sets for permutation matrices of order n that generalize the basis sets for $n = 4$ given by (2.51). For this, we introduce the following definition of a shift operator that acts on sequences.

Definition. The symbol $\langle i, j \rangle$ denotes a *shift operator* whose action on an arbitrary sequence $a = (a_1, a_2, \ldots, a_n), n \geq 2$, is to shift the entry a_i in place i to place j, leaving the relative position of all other entries unchanged (if a_i is removed from a and from $\langle i, j \rangle a$, the two resulting sequences are identical), where this definition is to apply for all $i, j \in \{1, 2, \ldots, n\}$. In this definition of the $\langle i, j \rangle$ shift operator, we make the following identifications:

$$\langle 1, 1 \rangle a = \langle 2, 2 \rangle a = \cdots = \langle n, n \rangle a = a, \tag{2.52}$$

$$\langle i, i+1 \rangle a = \langle i+1, i \rangle a, i = 1, 2, \ldots, n-1.$$

The general action of $\langle i, j \rangle$ on a sequence a for $i < j$ is given by

$$\langle i, j \rangle (a_1, \ldots, a_i, \ldots, a_j, \ldots, a_n)$$
$$= (a_1, \cdots, a_{i-1}, a_{i+1}, \ldots, a_j, a_i, a_{j+1}, \ldots, a_n), \tag{2.53}$$

with obvious adjustments for $i = 1$ and $j = n$, and with i and j interchanged for $i > j$.

Thus, when operating on a given sequence a, there are altogether $n^2 - 2n + 2$ sequences in the set

$$\{\langle i, j \rangle a | i, j = 1, 2, \ldots, n\}. \tag{2.54}$$

An example of the $\langle 3, 5 \rangle$ shift operator action for $n = 5$ is

$$\langle 3, 5 \rangle (a_1, a_2, a_3, a_4, a_5) = (a_1, a_2, a_4, a_5, a_3). \tag{2.55}$$

One of the nice properties possessed by the set of $n^2 - 2n + 2$ permutations obtained from the action of the shift operators on the identity permutation $e = (1, 2, \ldots, n)$ is the inverse property

$$(\langle i, j \rangle e)^{-1} = \langle j, i \rangle e. \tag{2.56}$$

The proof is a straightforward application of the definition of $\langle i, j \rangle e$.

2.2. BASIS SETS OF PERMUTATION MATRICES

The following relations for arbitrary pairs of permutations $\pi, \rho \in S_n$ and permutations given by the action of the shift operator $\langle i, j \rangle$ can be proved directly from the definition of product of permutations and of the action of $\langle i, j \rangle$ on a permutation, where any one of the relations implies the other two:

$$\begin{aligned}
\text{invariant form:} \quad & \pi^{-1}(\langle i, j \rangle \pi) = \rho^{-1}(\langle i, j \rangle \rho), \\
\text{left multiplication:} \quad & \pi(\langle i, j \rangle \rho) = \langle i, j \rangle (\pi \rho), \\
\text{right multiplication:} \quad & (\langle i, j \rangle \rho)\pi = \pi^{-1}(\langle i, j \rangle (\pi \rho))\pi.
\end{aligned} \quad (2.57)$$

We next use the shift operators $\langle i, j \rangle$ to define two bases of the vector space $\mathcal{P}_{\Sigma_n}$, where Σ_n is a specific set of permutations, denoted $\Sigma_n(e)$ and $\Sigma_n(e, p)$, respectively, as we next define.

The basis set $\Sigma_n(e)$:

The union of the following three sets of disjoint permutations is a basis set of permutations:

$$\left\{ (\pi^{(k)}, n) \mid \pi^{(k)} \in \Sigma_{n-1}(e), k = 1, 2, \ldots, b_{n-1} \right\},$$
$$\{\langle i, n \rangle e, \ i = n-1, n-2, \ldots, 1\}, \qquad (2.58)$$
$$\{\langle n, j \rangle e, \ j = n-2, n-3, \ldots, 1\}.$$

Proof. The proof is by induction and use of the Π–matrix rule. First, we note the union of these three sets is the basis set given by (2.50)-(2.51) for $n = 3, 4$. By assumption (induction hypothesis), the first set containing b_{n-1} permutations is linearly independent. The second set has integer n as part $n-1$, part $n-2, \ldots$, part 1, respectively; hence, by the Π–matrix rule, each of these permutations is linearly independent of those in the first set, and, in turn, from one another. The third set has the property that part n of each permutation is distinct from part n of every other permutation in all three sets. Thus, again, by the Π–matrix rule, all $b_{n-1} + 2n - 3 = b_n$ permutations in the three sets (2.58) are linearly independent. $\square$

We write the union of the permutations in the three sets (2.58) as

$$\Sigma_n(e) = \{\langle i, j \rangle e \mid i, j = 1, 2, \ldots, n\}, \qquad (2.59)$$

where only distinct permutations are included in this set (as is the custom for a set versus a multiset). The number $b_n = n^2 - (2n - 2)$ of distinct permutations comes about because of the $2n - 2$ equalities $\langle i, i \rangle e = e, i = 1, 2, \ldots, n$; $\langle i, i+1 \rangle e = \langle i+1, i \rangle e, i = 1, 2, \ldots, n - 1$. Because this basis set is fully generated by application of the shift operators to the identity permutation, it is denoted by $\Sigma_n(e)$. The basis set of permutation matrices is then given by

$$\mathbb{P}_{\Sigma_n(e)} = \{P_\pi \mid \pi \in \Sigma_n(e)\}. \tag{2.60}$$

The basis set $\Sigma_n(e,p)$:

Let $p = (2, 3, \ldots, n, 1) = \langle 1, n \rangle e = \pi^{(1,n)}$. The union of the following three sets of permutations is a basis set:

$$\begin{aligned} S_n^{(1)} &= \{\pi^{(i,j)} = \langle i, j \rangle e \mid 1 \le i < j \le n\} - \{\pi^{(1,n)} = p\}, \\ S_n^{(2)} &= \{\pi^{(i,j)} = \langle i-1, j \rangle p \mid 1 \le j < i-1 \le n-1\}, \\ S_n^{(3)} &= \{\pi^{(1,1)} = e, \pi^{(1,n)} = p\}. \end{aligned} \tag{2.61}$$

Proof. We have written the subsets in the particular form shown because each of these subsets has distinctive properties as illustrated in the next section. These three sets are disjoint and contain a number of permutations given, respectively, by $|S_n^{(1)}| = \binom{n}{2} - 1, |S_n^{(2)}| = \binom{n-1}{2}, |S_n^{(3)}| = 2$; hence, their union contains $n^2 - 2n + 2 = b_n$ permutations. Define the set of permutations analogous to the middle one in (2.61) by $S_n^{(1')} = \{\langle i, j \rangle e \mid 1 \le j < i-1 \le n-1\}$; hence, we have from the left multiplication rule in (2.57) that $S_n^{(2)} = pS_n^{(1')}$. But the set of permutations $S_n^{(1')}$ is a subset of the set $\mathbb{P}_{\Sigma_n(e)}$, which we have proved is linearly independent; hence, the set $S_n^{(1')}$ is linearly independent, and it is an easy result that the set $S_n^{(2)}$ is linearly independent (see the **Remark** below). Since the set of permutations $S_n^{(1)} \cup S_n^{(1')} \cup S_n^{(3)}$ is exactly that given by (2.58)-(2.60), which we have shown to be linearly independent, the set $\Sigma_n(e, p)$ defined by

$$\Sigma_n(e, p) = S_n^{(1)} \cup S_n^{(2)} \cup S_n^{(3)} \tag{2.62}$$

is linearly independent; hence, is a basis set of permutations. The corresponding basis set of permutation matrices is given by

$$\mathbb{P}_{\Sigma_n(e,p)} = \{P_\pi \mid \pi \in \Sigma_n(e,p)\}. \quad \square \tag{2.63}$$

Remark. The multiplication of any basis set Σ_n of permutations $\Sigma_n = \{\pi^{(1)}, \pi^{(2)}, \ldots, \pi^{(b_n)}\}$ of permutations from the left by an arbitrary per-

2.2. BASIS SETS OF PERMUTATION MATRICES

mutation $\pi \in S_n$ again gives a basis set, since

$$\sum_{k=1}^{b_n} a_\pi P_{\pi(k)} = 0 \text{ (0 matrix) implies}$$

$$P_\pi \sum_{k=1}^{b_n} a_\pi P_{\pi(k)} = \sum_{k=1}^{b_n} a_\pi P_\pi P_{\pi(k)} = 0 \text{ (0 matrix)}. \qquad (2.64)$$

The reversal of this relation is also true. But not all basis sets obtained by left multiplication can be distinct. For example, for $n = 3$, there are exactly five distinct basis sets, and the $3! = 6$ sets obtained by left multiplication cannot be distinct. The same result holds for right multiplication, indeed, for simultaneous left and right multiplication by any pair of permutation matrices in S_n. □

We define two sets of permutations Σ_n and Σ'_n to be equivalent or inequivalent, respectively, if the following relations hold:

$$\Sigma'_n = \pi \Sigma_n \pi', \text{ for any pair of permutations } \pi, \pi' \in S_n, \qquad (2.65)$$

$$\Sigma'_n \neq \pi \Sigma_n \pi', \text{ for every pair of permutations } \pi, \pi' \in S_n.$$

These relations are true if and only if the corresponding relations hold for permutation matrices, in which case the sets of permutation matrices $\mathbb{P}_{\Sigma_n}$ and $\mathbb{P}_{\Sigma'_n}$ are also said to be equivalent or inequivalent.

Example. An example of inequivalent sets of permutations is provided by $\Sigma_3(e)$ and $\Sigma_3(e, p)$. To prove this, it is sufficient to show that the assumption (A) that the relation $\pi \rho = (3, 2, 1)\pi'$ holds for all permutations $\pi, \pi' \in S_3$ and all $\rho \in \Sigma_3(e)$ is a contradiction. Since it is true that $\rho \in \Sigma_3(e)$ implies that $\rho^{-1} \in \Sigma_3(e)$ (it is closure that fails for the set $\Sigma_3(e)$ to be a group), the assumption (A) also implies that $\rho\pi = \pi'(3, 2, 1)$ for all permutations $\pi, \pi' \in S_3$ and all $\rho \in \Sigma_3(e)$. But choosing $\pi = (1, 2, 3)$ then implies that $(3, 2, 1)\pi' = \pi'(3, 2, 1)$ for all $\pi' \in S_3$, which is a contradiction. Thus, the sets of permutations $\Sigma_3(e)$ and $\Sigma_3(e, p)$ are inequivalent. The inequivalence of $\Sigma_n(e)$ and $\Sigma_n(e, p)$ for general n can also be proved, but we make no use of this result, and omit the proof.

2.2.1 Summary

Let Σ_n with $|\Sigma_n| = b_n$ be a basis set of permutations. Then, an arbitrary matrix A with fixed line-sum l_A has the expansion given by

$$A = \sum_{\pi \in \Sigma_n} a_\pi P_\pi, \; l_A = \sum_{\pi \in \Sigma_n} a_\pi, \qquad (2.66)$$

where the coordinates a_π of A relative to the basis Σ_n are given by

$$a_\pi = (\widehat{P}_\pi | A) = \sum_{h=1}^{n} \left(\sum_{\pi' \in \Sigma_n} x_{\pi,\pi'} a_{\pi'_h, h} \right). \tag{2.67}$$

We make the following observations regarding this expansion theorem:

1. The matrix X with elements given by $x_{\pi,\pi'}$ is given by the inverse of the Gram matrix: $X = Z^{-1}$. Thus, the dual matrices can be obtained by inverting the Gram matrix Z, whose elements $z_{\pi,\pi'}$ are determined from the basis Σ_n by the relation $z_{\pi,\pi'} = \sum_{h=1}^{n} \delta_{\pi_h, \pi'_h}$, $\pi, \pi' \in \Sigma_n$. But it is not easy to give the explicit inversion for general n.

2. The dual matrices are uniquely determined by the defining relations $(\widehat{P}_\pi | P_{\pi'}) = \delta_{\pi,\pi'}$, $\pi, \pi' \in \Sigma_n$. In this relation, there is no direct reference to the Gram matrix; it is not needed for the determination of the dual matrices, but again this direct procedure is not easily effected for general n.

3. The exact expansion for an arbitrary matrix A of given line-sum in the particular basis $\Sigma_n(e, p)$ can be given without using the dual basis or the Gram matrix. Such an expression could also be developed for other basis sets of permutations, but the basis $\Sigma_n(e, p)$ has special features that allow the development of the result for general n. This expansion has important consequences for the structure of the dual basis and the Gram matrix. It is therefore of considerable interest. The details are presented in the next Chapter.

Chapter 3

Coordinates of A in Basis $\mathbb{P}_{\Sigma_n(e,p)}$

3.1 Notations

The set of permutations $\Sigma_n(e)$ has the unusual property that its elements satisfy all the group postulates except for closure. Despite this appealing property, the set of permutations $\Sigma_n(e,p)$ has features that give it greater simplicity. This is because the set of coordinates $\{a_\pi \,|\, \pi \in \Sigma_n(e,p)\}$ of $A = (a_{ij})_{1 \leq i \leq j \leq n}$ contains more elements that are just single elements a_{ij} of A than does the set of coordinates for the basis set $\Sigma_n(e)$, except for $n=3$, where the number is 3 in both cases (different elements). For $n=4$, the number is 7 for $\Sigma_4(e)$, and 8 for $\Sigma_4(e,p)$. For general $\Sigma_n(e,p)$, the coordinates that are single elements a_{ij} of A are those corresponding to the $\pi^{(i,j)}$ in the set $S_n^{(1)} \cup S_n^{(2)}$, and these are $n(n-2)$ in number (see (2.61)). This means that there are only two coordinates of A that are linear combinations of the elements of A, since $b_n = n(n-2) + 2$. These are the simplifying features that allow the complete determination of all coordinates of A in the expansion of A in terms of the basis set of permutation matrices $P_{\Sigma_n(e,p)}$.

To put into effect the details of this calculation, we require some notations for *indexing sets* adapted to the properties of the basis set $\Sigma_n(e,p)$. We list the new sets and subsets required, and then explain their significance. The sets $S_n^{(t)}$, $t = 1, 2, 3$, in the collection of indexing sets are defined in (2.61). For convenience of reference, we also list the associated sets of permutations and permutation matrices:

Indexing sets:

$$[n] \times [n] = \{(i,j) \mid i \in [n], j \in [n]\},$$
$$\mathbb{I}_n^{(t)} = \{(i,j) \in [n] \times [n] \mid \pi^{(i,j)} \in S_n^{(t)}\}, \ t = 1, 2, 3,$$
$$\mathbb{I}_{(k,l)}^{(n)} = \{(i,j) \in \mathbb{I}_n^{(12)} \mid \pi_l^{(i,j)} = k\}, \text{ for all } (k,l) \in [n] \times [n],$$
$$\mathbb{I}_n^{(12)} = \mathbb{I}_n^{(1)} \cup \mathbb{I}_n^{(2)}, \ \mathbb{I}_n = \mathbb{I}_n^{(1)} \cup \mathbb{I}_n^{(2)} \cup \mathbb{I}_n^{(3)}, \qquad (3.1)$$
$$\mathbb{I}_n^c = [n] \times [n] - \mathbb{I}_n - \{(i,i),(i,i-1) \mid i = 2,3,\ldots,n\}.$$

Basic sets of permutations:

$$\Sigma_n(e,p) = \{\pi^{(i,j)} \mid (i,j) \in \mathbb{I}_n\}, \ \pi^{(i,j)} = (\pi_1^{(i,j)}, \ldots, \pi_n^{(i,j)}),$$
$$\Sigma_n^c(e,p) = \{\pi^{(i,j)} \mid (i,j) \in \mathbb{I}_n^c\},$$
$$S_n^{(1)} = \{\pi^{(i,j)} = \langle i, j \rangle e \mid (i,j) \in \mathbb{I}_n^{(1)}\}, \qquad (3.2)$$
$$S_n^{(2)} = \{\pi^{(i,j)} = \langle i-1, j \rangle p \mid (i,j) \in \mathbb{I}_n^{(2)}\},$$
$$S_n^{(3)} = \{\pi^{(1,1)} = e = (1,2,\ldots,n), \pi^{(1,n)} = p = (2,\ldots,n,1)\}.$$
$$S_{(kl)}^{(n)} = \{\pi^{(i,j)} \mid (i,j) \in \mathbb{I}_{(k,l)}^{(n)}\}.$$

Basic sets of permutation matrices.

$$\mathbb{P}_{\Sigma_n(e,p)} = \{P_{\pi^{(i,j)}} \mid \pi^{(i,j)} \in \Sigma_n(e,p)\},$$
$$\mathbb{P}_{\Sigma_n^c(e,p)} = \{P_{\pi^{(i,j)}} \mid \pi^{(i,j)} \in \Sigma_n^c(e,p)\}, \qquad (3.3)$$
$$\mathbb{P}_n^{(t)} = \{P_{\pi^{(i,j)}} \mid \pi^{(i,j)} \in S_n^{(t)}\}, \ t = 1,2,3,$$
$$\mathbb{P}_{(k,l)}^{(n)} = \{P_{\pi^{(i,j)}} \mid \pi^{(i,j)} \in S_{(k,l)}^{(n)}\}.$$

Set cardinalities:

$$|\Sigma_n(e,p)| = b_n = n^2 - 2n + 2.$$
$$|[n] \times [n]| = n^2, \qquad (3.4)$$
$$|\mathbb{I}_n^{(1)}| = (n^2 - n - 2)/2, \ |\mathbb{I}_n^{(2)}| = (n^2 - 3n + 2)/2, \ |\mathbb{I}_n^{(3)}| = 2,$$
$$|\mathbb{I}_n^{(12)}| = n(n-2), \ |\mathbb{I}_n| = n^2 - 2n + 2 = d_n, \ |\mathbb{I}_n^c| = 2n - 2,$$
$$|\mathbb{I}_{(k,l)}^{(n)}| = |S_{(k,l)}^{(n)}| = |P_{(k,l)}^{(n)}| = \text{not determined}.$$

The set $\mathbb{I}_n^{(1)}$ is the upper triangular set of indices from $[n] \times [n]$, $[n] = \{1, 2, \ldots, n\}$, that are above the principal diagonal with $(1, n)$ removed;

$\mathbb{I}_n^{(2)}$ is the lower triangular set of indices below the principal diagonal with the pairs belonging to the subdiagonal $\{(2,1),(3,2),\ldots,(n-1,n)\}$ removed. The set $\mathbb{I}_{(k,l)}^{(n)} \subset [n] \times [n]$ is the set of permutations $\pi^{(i,j)} = (\pi_1^{(i,j)}, \pi_2^{(i,j)}, \ldots, \pi_n^{(i,j)}) \in \Sigma_n(e,p)$ that have part l equal to k; that is, $\pi_l^{(i,j)} = k$.

The cardinality of a set enumerated by an indexing set is, of course, that of the indexing set itself.

3.2 The A-Expansion Rule in the Basis $\mathbb{P}_{\Sigma_n(e,p)}$

We next give the b_n coordinates $\alpha_{(i,j)}$ of an arbitrary A of fixed line-sum, that is, $A \in \mathcal{P}_n$, in the basis set of permutations $\Sigma_n(e,p)$:

$$A = \sum_{(i,j) \in \mathbb{I}_n} \alpha_{(i,j)} P_{\pi^{(i,j)}}. \tag{3.5}$$

The indexing set $\mathbb{I}_n$ enumerates exactly those permutations in the basis set of permutations $\Sigma_n(e,p)$; hence, $|\mathbb{I}_n| = |\Sigma_n(e,p)| = b_n = (n-1)^2 + 1$. The set $\mathbb{I}_n$ is written as the union of three disjoint subsets $\mathbb{I}_n = \mathbb{I}_n^{(1)} \cup \mathbb{I}_n^{(2)} \cup \mathbb{I}_n^{(3)}$ in order to account for the partitioning $\Sigma_n(e,p) = S_n^{(1)} \cup S_n^{(2)} \cup S_n^{(3)}$ into three disjoint subsets of permutations, as defined in relations (2.61).

The A-expansion rule in the basis $\mathbb{P}_{\Sigma_n(e,p)}$: We first give the results and then their derivation:

The b_n coordinates $\alpha_{(i,j)}, (i,j) \in \mathbb{I}_n$, for the expansion of A are the following:

$$\alpha_{(i,j)} = a_{ij}, \text{ for } (i,j) \in \mathbb{I}_n^{(12)},$$

$$\alpha_{(1,1)} = a_{kk} - \sum_{(i,j) \in \mathbb{I}_{(k,k)}^{(n)}} a_{ij}, \; k = 1, 2, \ldots, n,$$

$$\alpha_{(1,n)} = a_{k\,k-1} - \sum_{(i,j) \in \mathbb{I}_{(k,k-1)}^{(n)}} a_{ij}, \; k = 2, 3, \ldots, n, \tag{3.6}$$

$$\alpha_{(1,n)} = a_{1n} - \sum_{(i,j) \in \mathbb{I}_{(1,n)}^{(n)}} a_{ij}.$$

The indexing sets $\mathbb{I}_{(k,l)}^{(n)}$ for $l = k, k-1$, and $(k,l) = (1,n)$ that appear in the summations in (3.6) are, respectively, the subsets of permutations $\pi^{(i,j)} \in \mathbb{I}_n^{(1)} \cup \mathbb{I}_n^{(2)}$ with parts such that $\pi_k^{(i,j)} = k$, $\pi_{k-1}^{(i,j)} = k$, and $\pi_1^{(i,j)} = n$.

The relations for the coordinates $\alpha_{(1,1)}$ and $\alpha_{(1,n)}$ in (3.6) are each expressed in n different forms. Any of these n relations can be selected for the calculation of the coordinate $\alpha_{(1,1)}$ and $\alpha_{(1,n)}$. This redundancy of relations is a consequence of the fixed line-sum of A that requires the sum of each row and each column to be

$$l_A = \sum_{(i,j) \in \mathbb{I}_n}' \alpha_{(i,j)}. \tag{3.7}$$

Proof of the A−expansion rule. The proof of relations (3.6) uses the following significant property of the basis set $\mathbb{P}_{\Sigma_n(e,p)}$ of permutation matrices:

The permutation matrix $P_{\pi^{(i,j)}}$, $(i,j) \in \mathbb{I}_n^{(12)} = \mathbb{I}_n^{(1)} \cup \mathbb{I}_n^{(2)}$ is the only permutation matrix in the set $\mathbb{P}_n^{(12)}$ that has a 1 in row i and column j.

This is verified by direct examination of the permutations in the set $S_n^{(1)} \cup S_n^{(1)}$ defined by the shift operators in (2.61). This proves the first relation in (3.6), which is needed to carry the proof forward, as next developed. This still leaves open the question of the expansion of the $2n$ elements of A given by the two elements a_{11}, a_{1n} and the $2n - 2$ elements $a_{kl} \in \mathbb{I}_n^c$ in terms of the coordinates $\alpha_{(ij)}$ of A. But again by direct examination of the definition of the permutations $\pi^{(i,j)}$ in terms of shift operators as given by (2.61), the elements of A given by $a_{kk}, k = 1, 2, \ldots, n$ and a_{1n} can be verified directly to satisfy the relations given by (3.6) This derivation yields the relations for the elements a_{ij}, in which the negative terms on the right-hand side of (3.6) occur as positive terms on the left-hand side. An example of such a derivation is the following. From relation (3.5), we have that $a_{11} = \sum_{(i,j) \in \mathbb{I}_n} \alpha_{(i,j)} \delta_{1, \pi_1^{(i,j)}}$. But the permutations in the set (2.61) that have $\pi_1^{(i,j)} = 1$ are just those for which $(i,j) \in \mathbb{I}_{(1,1)}^{(n)}$, together with the identity e. But all $\alpha_{(i,j)}$ for which $(i,j) \in \mathbb{I}_{(1,1)}^{(n)}$ are given by $\alpha_{(i,j)} = a_{ij}$. Thus, we have that $a_{11} = \alpha_{(1,1)} + \sum_{(i,j) \in \mathbb{I}_{(1,1)}^n} a_{ij}$, which gives the $k = 1$ relation in (3.6). The derivation of all other relations in (3.6) proceeds similarly. □

Remarks. The following remarks are intended to complement those made earlier at the end of Chapter 2.

1. The derivation above of the A−expansion rule is based entirely on the structure of the basis set $\mathbb{P}_{\Sigma_n(e,p)}$ itself. The rule shows that $n(n-2)$ of the coordinates are just the elements a_{ij} of A itself; these are all elements

3.3. DUAL MATRICES IN THE BASIS SET $\Sigma_n(e,p)$

of A that do not fall on the principal diagonal or on the down-diagonal diag_{21}, which is defined to be the sequence of length n starting with the element in row 2, column 1 and continuing down the subdiagonal adjacent to the principal diagonal (also called the down-diagonal diag_{11}) and continuing downward to the element in row n, column $n-1$, and then looping upward to capture the element in row 1, column n. Thus, $\text{diag}_{21} = (a_{21}, a_{32}, \ldots, a_{nn-1}, a_{1n})$. The remaining two coordinates $\alpha_{(1,1)}$ and $\alpha_{(1,n)}$ then subsume all the interrelations needed to assure the line-sum constraints on A. A simpler expansion of A in terms of permutation matrices is difficult to imagine. No use is made of the Gram matrix, nor of the dual matrices in obtaining the coordinates of A.

2. If the dual matrices $\widehat{P}_{\pi^{(k,l)}}$, $\pi^{(k,l)} \in \Sigma_n(e,p)$, are considered as known matrices, we can apply the A-expansion rule (3.6) to obtain the coordinates $\alpha_{(i,j),(k,l)}$ of each dual matrix in the basis $\mathbb{P}_{\Sigma_n(e,p)}$ of permutation matrices:

$$\widehat{P}_{\pi^{(k,l)}} = \sum_{(i,j) \in \mathbf{I}_n} \alpha_{(i,j);(k,l)} P_{\pi^{(i,j)}}. \tag{3.8}$$

We can also use the form (3.8) and the defining relations

$$(\widehat{P}_{\pi^{(k,l)}} | P_{\pi^{(i',j')}}) = \delta_{k,i'} \delta_{l,j'} \tag{3.9}$$

to derive the system of equations for the coordinates given by

$$\sum_{(i,j) \in \mathbf{I}_n} \alpha_{(i,j);(k,l)} \left(\sum_{h=1}^{b_n} \delta_{\pi_h^{(i,j)}, \pi_h^{(i',j')}} \right) = \delta_{k,i'} \delta_{l,j'}. \tag{3.10}$$

These equations then have a unique solution for the coordinates. This system of equations can be simplified by using the A-expansion (3.6) as applied to each dual matrix $\widehat{P}_{\pi^{(k,l)}}$.

The coordinates $\alpha_{(i,j);(k,l)}$ are, in turn, the elements of the symmetric matrix X, which is the inverse $X = Z^{-1}$ of the Gram matrix.

We give the explicit dual matrices for the basis $\Sigma_n(e,p)$ in the next section.

3.3 Dual Matrices in the Basis Set $\Sigma_n(e,p)$

It is useful to give the explicit results for the dual matrices for $n = 3, 4$ before giving and proving the general result for $\Sigma_n(e,p)$. Since the general result is quite easily proved, we simply state the $n = 3, 4$ results and comment on some of their features.

3.3.1 Dual Matrices for $\Sigma_3(e,p)$

The basis set of permutations $\Sigma_3(e,p)$ is

$$\Sigma_3(e,p) = \{\pi^{(1,1)} = (1,2,3),\ \pi^{(1,2)} = (2,1,3),\ \pi^{(1,3)} = (2,3,1),$$
$$\pi^{(2,3)} = (1,3,2),\ \pi^{(3,1)} = (3,2,1)\}. \qquad (3.11)$$

We enumerate the rows and columns of matrices of order 5 by the natural page ordering of the pairs (i,j) as given by $(1,1) \mapsto 1, (1,2) \mapsto 2, (1,3) \mapsto 3, (2,3) \mapsto 4, (3,1) \mapsto 5$. (This is different from the page ordering of the permutations $\pi \in S_3$.)

An arbitrary matrix A of fixed line-sum has the expression:

$$A = \alpha_{(1,1)}P_{1,2,3} + \alpha_{(1,2)}P_{2,1,3} + \alpha_{(1,3)}P_{2,3,1} + \alpha_{(2,3)}P_{1,3,2} + \alpha_{(3,1)}P_{3,2,1}$$

$$= \begin{pmatrix} \alpha_{(1,1)} + a_{23} & a_{12} & \alpha_{(1,3)} + a_{31} \\ \alpha_{(1,3)} + a_{12} & \alpha_{(1,1)} + a_{31} & a_{23} \\ a_{31} & \alpha_{(1,3)} + a_{23} & \alpha_{(1,1)} + a_{12} \end{pmatrix}, \qquad (3.12)$$

where $\alpha_{(1,2)} = a_{12}, \alpha_{(2,3)} = a_{23}, \alpha_{(3,1)} = a_{31}$. The coordinates and the line-sum of A are given by

$$\text{coordinates} = (\alpha_{(1,1)}, \alpha_{(1,2)}, \alpha_{(1,3)}, \alpha_{(2,3)}, \alpha_{(3,1)})$$
$$= (a_{11} - a_{23},\ a_{12},\ a_{13} - a_{31},\ a_{23},\ a_{31}), \qquad (3.13)$$
$$l_A = \alpha_{(1,1)} + a_{12} + \alpha_{(1,3)} + a_{23} + a_{31}.$$

It is also the case that the Gram matrix Z, which is easily computed directly from the inner product (2.46), applied to $\Sigma_3(e,p)$, has the expression:

$$Z = \begin{pmatrix} 3 & 1 & 0 & 1 & 1 \\ 1 & 3 & 1 & 0 & 0 \\ 0 & 1 & 3 & 1 & 1 \\ 1 & 0 & 1 & 3 & 0 \\ 1 & 0 & 1 & 0 & 3 \end{pmatrix}. \qquad (3.14)$$

This Gram matrix has the following symmetry properties: Row 1 (column 1) and row 3 (column 3) are permutations (in S_5) of one another, and the remaining three rows (columns) are permutations of row 2 (column 2). This symmetry carries forward, of course, to $X = Z^{-1}$ given by (3.17) below.

Each dual matrix $\widehat{P}_{\pi^{(k,l)}}$, $\pi^{(k,l)} \in \Sigma_3(e,p)$, as given by (3.8), can be calculated by direct solution of the system of relations (3.10), where these relations can be simplified by making use of the A–expansion, applied

3.3. DUAL MATRICES IN THE BASIS SET $\Sigma_n(e,p)$

to $\widehat{P}_{\pi(k,l)}$, and of the symmetries of the Gram matrix Z. The set of dual matrices is found to be the following, where it is relatively simple to check their validity by direct application of the inner product rule (2.38):

$$P_{1,2,3} = \begin{pmatrix} 1 & 0 & 0 \\ 0 & 1 & 0 \\ 0 & 0 & 1 \end{pmatrix}, \quad \widehat{P}_{1,2,3} = \frac{1}{3}\begin{pmatrix} 1 & -1 & 0 \\ 0 & 1 & -1 \\ -1 & 0 & 1 \end{pmatrix},$$

$$P_{2,1,3} = \begin{pmatrix} 0 & 1 & 0 \\ 1 & 0 & 0 \\ 0 & 0 & 1 \end{pmatrix}, \quad \widehat{P}_{2,1,3} = \frac{1}{9}\begin{pmatrix} -1 & 5 & -1 \\ 2 & -1 & 2 \\ 2 & -1 & 2 \end{pmatrix},$$

$$P_{2,3,1} = \begin{pmatrix} 0 & 0 & 1 \\ 1 & 0 & 0 \\ 0 & 1 & 0 \end{pmatrix}, \quad \widehat{P}_{2,3,1} = \frac{1}{3}\begin{pmatrix} 0 & -1 & 1 \\ 1 & 0 & -1 \\ -1 & 1 & 0 \end{pmatrix}, \quad (3.15)$$

$$P_{1,3,2} = \begin{pmatrix} 1 & 0 & 0 \\ 0 & 0 & 1 \\ 0 & 1 & 0 \end{pmatrix}, \quad \widehat{P}_{1,3,2} = \frac{1}{9}\begin{pmatrix} 2 & 2 & -1 \\ -1 & -1 & 5 \\ 2 & 2 & -1 \end{pmatrix},$$

$$P_{3,2,1} = \begin{pmatrix} 0 & 0 & 1 \\ 0 & 1 & 0 \\ 1 & 0 & 0 \end{pmatrix}, \quad \widehat{P}_{3,2,1} = \frac{1}{9}\begin{pmatrix} -1 & 2 & 2 \\ -1 & 2 & 2 \\ 5 & -1 & -1 \end{pmatrix}.$$

We also give the inverse of Z, whose elements are given by the coordinates in the expansion (3.8). Thus, for example, we have that

$$\widehat{P}_{\pi(1,1)} = (2P_{\pi(1,1)} - P_{\pi(1,2)} + P_{\pi(1,3)} - P_{\pi(2,3)} - P_{\pi(3,1)})/3, \quad (3.16)$$

$$\widehat{P}_{\pi(1,2)} = (-3P_{\pi(1,1)} + 5P_{(\pi1,2)} - 3P_{\pi(1,3)} + 2P_{\pi(2,3)} + 2P_{\pi(3,1)})/9.$$

Column 3 of X is then obtained from column 1 by effecting any of the permutations of it that give column 3 of the Gram matrix Z from column 1. Similarly, columns 4, 5 are obtained from column 2 by effecting any of the respective permutations of column 2 of Z that give columns 4, 5:

$$X = Z^{-1} = \frac{1}{9}\begin{pmatrix} 6 & -3 & 3 & -3 & -3 \\ -3 & 5 & -3 & 2 & 2 \\ 3 & -3 & 6 & -3 & -3 \\ -3 & 2 & -3 & 5 & 2 \\ -3 & 2 & -3 & 2 & 5 \end{pmatrix}. \quad (3.17)$$

Of course, given $X = Z^{-1}$, we can also write out all dual matrices using the A−expansion (3.6).

3.3.2 Dual Matrices for $\Sigma_4(e,p)$

The basis set of permutations $\Sigma_4(e,p)$ is

$$\Sigma_4(e,p) = \{\pi^{(1,1)} = (1,2,3,4),\ \pi^{(1,2)} = (2,1,3,4),$$
$$\pi^{(1,3)} = (2,3,1,4),\ \pi^{(1,4)} = (2,3,4,1),\ \pi^{(2,3)} = (1,3,2,4),$$
$$\pi^{(2,4)} = (1,3,4,2),\ \pi^{(3,1)} = (3,2,4,1),\ \pi^{(3,4)} = (1,2,4,3),$$
$$\pi^{(4,1)} = (4,2,3,1),\ \pi^{(4,2)} = (2,4,3,1)\}. \qquad (3.18)$$

Again, rows and columns of matrices are ordered by the rule: $(1,1) \mapsto 1, (1,2) \mapsto 2, (1,3) \mapsto 3, (1,4) \mapsto 4, (2,3) \mapsto 5, (2,4) \mapsto 6, (3,1) \mapsto 7, (3,4) \mapsto 8, (4,1) \mapsto 9, (4,2) \mapsto 10$. (As for $n=3$, this rule is different from the page ordering of the one-line permutations.)

A general $A = (A_1\ A_2\ A_3\ A_4)$ of order 4 of fixed-line sum has the following expression given by (3.6), where, for display purposes, we present A in terms of its columns $A_j, j = 1,2,3,4$:

$$A_1 = \begin{pmatrix} a_{11} \\ a_{21} \\ a_{31} \\ a_{41} \end{pmatrix} = \begin{pmatrix} \alpha_{(1,1)} + a_{23} + a_{24} + a_{34} \\ \alpha_{(1,4)} + a_{12} + a_{13} + a_{42} \\ a_{31} \\ a_{41} \end{pmatrix},$$

$$A_2 = \begin{pmatrix} a_{12} \\ a_{22} \\ a_{32} \\ a_{42} \end{pmatrix} = \begin{pmatrix} a_{12} \\ \alpha_{(1,1)} + a_{31} + a_{34} + a_{41} \\ \alpha_{(1,4)} + a_{13} + a_{23} + a_{24} \\ a_{42} \end{pmatrix},$$

$$A_3 = \begin{pmatrix} a_{13} \\ a_{23} \\ a_{33} \\ a_{43} \end{pmatrix} = \begin{pmatrix} a_{13} \\ a_{23} \\ \alpha_{(1,1)} + a_{12} + a_{41} + a_{42} \\ \alpha_{(1,4)} + a_{24} + a_{31} + a_{34} \end{pmatrix},$$

$$A_4 = \begin{pmatrix} a_{14} \\ a_{24} \\ a_{34} \\ a_{44} \end{pmatrix} = \begin{pmatrix} \alpha_{(1,4)} + a_{31} + a_{41} + a_{42} \\ a_{24} \\ a_{34} \\ \alpha_{(1,1)} + a_{12} + a_{13} + a_{23} \end{pmatrix}. \qquad (3.19)$$

$$\text{line-sum} = \alpha_{(1,1)} + \alpha_{(1,2)} + \alpha_{(1,3)} + \alpha_{(1,4)} + \alpha_{(2,3)} + \alpha_{(2,4)} + \alpha_{(3,1)}$$
$$+ \alpha_{(3,4)} + \alpha_{(4,1)} + \alpha_{(4,2)}. \qquad (3.20)$$

3.3. DUAL MATRICES IN THE BASIS SET $\Sigma_n(e,p)$

$$\begin{aligned}
\text{coordinates} \quad = \quad & (\alpha_{(1,1)}, \alpha_{(1,2)} = a_{12}, \alpha_{(1,3)} = a_{13}, \alpha_{(1,4)}, \alpha_{(2,3)} = a_{23}, \\
& \alpha_{(2,4)} = a_{24}, \alpha_{(3,1)} = a_{31}, \alpha_{(3,4)} = a_{34}, \alpha_{(4,1)} = a_{41}, \\
& \alpha_{(4,2)} = a_{42});
\end{aligned} \quad (3.21)$$

$$\begin{aligned}
\alpha_{(1,1)} &= a_{11} - (a_{23} + a_{24} + a_{34}) = a_{22} - (a_{31} + a_{34} + a_{41}) \\
&= a_{33} - (a_{12} + a_{41} + a_{42}) = a_{44} - (a_{12} + a_{13} + a_{23}), \\
\alpha_{(1,4)} &= a_{14} - (a_{31} + a_{41} + a_{42}) = a_{21} - (a_{12} + a_{13} + a_{42}) \\
&= a_{32} - (a_{13} + a_{23} + a_{24}) = a_{43} - (a_{24} + a_{31} + a_{34}).
\end{aligned}$$

The elements of the Gram matrix Z are easily calculated from $z_{\pi,\pi'} = \sum_{h=1}^{4} \delta_{\pi_h, \pi'_h}$ by counting the number of shared parts in the set (3.18) of permutations $\pi, \pi' \in \Sigma_4(e,p)$:

$$Z = \begin{pmatrix}
4 & 2 & 1 & 0 & 2 & 1 & 1 & 2 & 2 & 1 \\
2 & 4 & 2 & 1 & 1 & 0 & 0 & 0 & 1 & 2 \\
1 & 2 & 4 & 2 & 2 & 1 & 0 & 0 & 0 & 1 \\
0 & 1 & 2 & 4 & 1 & 2 & 2 & 1 & 1 & 2 \\
2 & 1 & 2 & 1 & 4 & 2 & 0 & 1 & 0 & 0 \\
1 & 0 & 1 & 2 & 2 & 4 & 1 & 2 & 0 & 0 \\
1 & 0 & 0 & 2 & 0 & 1 & 4 & 2 & 2 & 1 \\
2 & 0 & 0 & 1 & 1 & 2 & 2 & 4 & 1 & 0 \\
2 & 1 & 0 & 1 & 0 & 0 & 2 & 1 & 4 & 2 \\
1 & 2 & 1 & 2 & 0 & 0 & 1 & 0 & 2 & 4
\end{pmatrix}. \quad (3.22)$$

This Gram matrix has the following symmetry properties: Row 1 (column 1) and row 4 (column 4) are permutations (in S_{10}) of one another, and the remaining rows (columns) are permutations of row 2 (column 2). This symmetry carries forward, of course, to $X = Z^{-1}$ given by (3.25) below.

Each dual matrix $\widehat{P}_{\pi^{(k,l)}}$, $\pi^{(k,l)} \in \Sigma_4(e,p)$, as given by (3.8), can be calculated by direct solution of the system of relations (3.10), where these relations can be simplified by making use of the A-expansion, applied to $\widehat{P}_{\pi^{(k,l)}}$, and of the symmetries of the Gram matrix Z. The set of dual matrices is found to be the following, where it is relatively simple to check their validity by direct application of the inner product rule (2.38):

$$\widehat{P}_{1,2,3,4} = \frac{1}{4} \begin{pmatrix} 1 & -2 & -1 & 0 \\ 0 & 1 & -2 & -1 \\ -1 & 0 & 1 & -2 \\ -2 & -1 & 0 & 1 \end{pmatrix}, \quad \widehat{P}_{2,1,3,4} = \frac{1}{8} \begin{pmatrix} -1 & 5 & -1 & -1 \\ 1 & -1 & 1 & 1 \\ 1 & -1 & 1 & 1 \\ 1 & -1 & 1 & 1 \end{pmatrix},$$

$$\widehat{P}_{2,3,1,4} = \frac{1}{8}\begin{pmatrix} -1 & -1 & 5 & -1 \\ 1 & 1 & -1 & 1 \\ 1 & 1 & -1 & 1 \\ 1 & 1 & -1 & 1 \end{pmatrix}, \quad \widehat{P}_{2,3,4,1} = \frac{1}{4}\begin{pmatrix} 0 & -1 & -2 & 1 \\ 1 & 0 & -1 & -2 \\ -2 & 1 & 0 & -1 \\ -1 & -2 & 1 & 0 \end{pmatrix},$$

$$\widehat{P}_{1,3,2,4} = \frac{1}{8}\begin{pmatrix} 1 & 1 & -1 & 1 \\ -1 & -1 & 5 & -1 \\ 1 & 1 & -1 & 1 \\ 1 & 1 & -1 & 1 \end{pmatrix}, \quad \widehat{P}_{1,3,4,2} = \frac{1}{8}\begin{pmatrix} 1 & 1 & 1 & -1 \\ -1 & -1 & -1 & 5 \\ 1 & 1 & 1 & -1 \\ 1 & 1 & 1 & -1 \end{pmatrix},$$

$$\widehat{P}_{3,2,4,1} = \frac{1}{8}\begin{pmatrix} -1 & 1 & 1 & 1 \\ 1 & 1 & 1 & 1 \\ 5 & -1 & -1 & -1 \\ -1 & 1 & 1 & 1 \end{pmatrix}, \quad \widehat{P}_{1,2,4,3} = \frac{1}{8}\begin{pmatrix} 1 & 1 & 1 & -1 \\ 1 & 1 & 1 & -1 \\ -1 & -1 & -1 & 5 \\ 1 & 1 & 1 & -1 \end{pmatrix},$$

(3.23)

$$\widehat{P}_{4,2,3,1} = \frac{1}{8}\begin{pmatrix} -1 & 1 & 1 & 1 \\ -1 & 1 & 1 & 1 \\ -1 & 1 & 1 & 1 \\ 5 & -1 & -1 & -1 \end{pmatrix}, \quad \widehat{P}_{2,4,3,1} = \frac{1}{8}\begin{pmatrix} 1 & -1 & 1 & 1 \\ 1 & -1 & 1 & 1 \\ 1 & -1 & 1 & 1 \\ -1 & 5 & -1 & -1 \end{pmatrix}.$$

We also give the inverse $X = Z^{-1}$ of Z, whose elements are given by the coordinates in the expansion (3.8). Thus, for example, we have that

$$\widehat{P}_{\pi(1,1)} = (6P_{\pi(1,1)} - 2P_{\pi(1,2)} - P_{\pi(1,3)} + 4P_{\pi(1,4)} - 2P_{\pi(2,3)}$$
$$- P_{\pi(2,4)} - P_{\pi(3,1)} - 2P_{\pi(3,4)} - 2P_{\pi(4,1)} - P_{\pi(4,2)})/4,$$

$$\widehat{P}_{\pi(1,2)} = (-4P_{\pi(1,1)} + 5P_{\pi(1,2)} - P_{\pi(1,3)} - 2P_{\pi(1,4)} + P_{\pi(2,3)}$$
$$+ P_{\pi(2,4)} + P_{\pi(3,1)} + P_{\pi(3,4)} + P_{\pi(4,1)} - P_{\pi(4,2)})/8. \quad (3.24)$$

Column 4 of X is then obtained from column 1 by effecting any of the permutations of it that give column 4 of the Gram matrix Z from column 1. Similarly, columns $3, 5, 6, \cdots, 10$ are obtained from column 2 by effecting any of the respective permutations of column 2 of Z that give columns $3, 5, 6, \cdots, 10$ of Z. Thus, we obtain the matrix $X = Z^{-1}$, as presented in terms of its columns $X = (X_1\, X_2\, \cdots\, X_{b_5})$:

$$\begin{aligned}
X_1 &= (6, -2, -1, 4, -2, -1, -1, -2, -2, -1)/4, \\
X_2 &= (-4, 5, -1, -2, 1, 1, 1, 1, 1, -1)/8, \\
X_3 &= (-2, -1, 5, -4, -1, 1, 1, 1, 1, 1)/8, \\
X_4 &= (4, -1, -2, 6, -1, -2, -2, -1, -1, -2)/4, \\
X_5 &= (-4, 1, -1, -2, 5, -1, 1, 1, 1, 1)/8, \\
X_6 &= (-2, 1, 1, -4, -1, 5, 1, -1, 1, 1)/8, \quad (3.25) \\
X_7 &= (-2, 1, 1, -4, 1, 1, 5, -1, -1, 1)/8, \\
X_8 &= (-4, 1, 1, -2, 1, -1, -1, 5, 1, 1)/8, \\
X_9 &= (-4, 1, 1, -2, 1, 1, -1, 1, 5, -1)/8, \\
X_{10} &= (-2, -1, 1, -4, 1, 1, 1, 1, -1, 5)/8.
\end{aligned}$$

Of course, given Z^{-1}, we can also write out all dual matrices using the A-expansion.

3.4 The General Dual Matrices in the Basis $\Sigma_n(e,p)$

It is not difficult to prove that the general Gram matrix Z has the following symmetry property that generalizes that already noted for $n = 3, 4$:

In the basis $\Sigma_n(e,p)$, the Gram matrix G (also G^{-1}) has the property that row 1 and row n are permutations (in S_{b_n}) of one another and all other $b_n - 2 = n(n\text{-}2)$ rows are permutations of row 2. Since G (also $X = G^{-1}$) is symmetric, the same property is true of columns.

This symmetry of the Gram matrix is a property of the inner product $(P_\pi | P_{\pi'})$, and the fact that only two coordinates, namely, $\alpha_{(1,1)}$ and $\alpha_{(1,n)}$ in the general A–expansion in (3.6) are not elements of A itself. This symmetry subgroup of S_{b_n} of the Gram matrix simplifies substantially the calculation of the dual matrices. The results for $n = 3, 4$ given above illustrate this nicely.

It is convenient to repeat here the definition of the permutations in the basis set $\Sigma_n(e,p)$:

$$\Sigma_n(e,p) = S_n^{(1)} \cup S_n^{(2)} \cup S_n^{(3)}, \qquad (3.26)$$

where each of the subsets $S_n^{(t)}$, $t = 1, 2, 3$, of permutations is defined by

$$\begin{aligned} S_n^{(1)} &= \{\pi^{(i,j)} = \langle i, j\rangle e \mid 1 \leq i < j \leq n\} - \{\pi^{(1,n)} = p\}, \\ S_n^{(2)} &= \{\pi^{(i,j)} = \langle i-1, j\rangle p \mid 1 \leq j < i-1 \leq n-1\}, \quad (3.27) \\ S_n^{(3)} &= \{\pi^{(1,1)} = e, \pi^{(1,n)} = p\}, \end{aligned}$$

where $e = (1, 2, \ldots, n)$ and $p = (2, 3, \ldots, n, 1) = \langle 1, n\rangle e = \pi^{(1,n)}$. The symbol $\langle i, j\rangle$ denotes the shift operator defined by (2.52)-(2.53), whose action on any sequence $(a_1, a_2, \ldots, a_n)$ is to shift part i to part j and leave the relative positions of all other $n - 1$ parts unchanged.

Guided by the explicit calculations for $n = 3, 4$, we can conjecture, and then prove, the following expression for the dual matrix $\widehat{P}_{\pi^{(1,2)}} = \widehat{P}_{2,1,3,4,\ldots,n}$ for all $n \geq 2$:

$$\widehat{P}_{\pi^{(1,2)}} = \frac{1}{n^2} \begin{pmatrix} -n+2 & b_n & -n+2 & -n+2 & \cdots & -n+2 \\ 2 & -n+2 & 2 & 2 & \cdots & 2 \\ 2 & -n+2 & 2 & 2 & \cdots & 2 \\ \vdots & \vdots & \vdots & \vdots & \cdots & \vdots \\ 2 & -n+2 & 2 & 2 & \cdots & 2 \end{pmatrix}.$$

$$(3.28)$$

54 CHAPTER 3. COORDINATES OF A IN BASIS $\mathbb{P}_{\Sigma_n(e,p)}$

Proof. The defining relations for the dual permutation matrix $\widehat{P}_{\pi(1,2)}$ are

$$(\widehat{P}_{\pi(1,2)} \mid P_\pi) = \text{Tr}(\widehat{P}^{tr}_{\pi(1,2)} P_\pi) = \sum_{j=1}^{n} (\widehat{P}_{\pi(1,2)})_{\pi_j, j}$$

$$= \delta_{\pi(1,2), \pi}, \text{ each } \pi \in \Sigma_n(e, p). \qquad (3.29)$$

The inner product (3.29) is the sum of the elements in the (row, column) positions in the matrix (3.28) given by (row π_1, column 1) + (row π_2, column 2) + $\cdots$ + (row π_n, column n). For $\pi = (1, 2, \ldots, n)$, this sum is that of the diagonal terms, which is 0. Since all rows of $\widehat{P}_{\pi(1,2)}$ are equal, excepting row 1, and all columns of $\widehat{P}_{\pi(1,2)}$ are equal, excepting column 2, all permutations $\pi \in \Sigma_n(e, p)$ **not** of the form $(1, 2, \pi_3, \ldots, \pi_n)$ must also give a sum that is 0. This includes all permutations $\pi \in \Sigma_n(e, p)$, excepting $\pi = \pi^{(1,2)}$, for which the sum is (row 2, column 1) + (row 2, column 1) + (row 1, column 2) + (row 3, column 3) + $\cdots$ + (row n, column n) = $(2 + b_n + 2(n-2))/n^2 = 1$. Thus, the dual matrix $\widehat{P}_{\pi(1,2)} = \widehat{P}_{2,1,3,4,\ldots,n}$ satisfies the defining conditions (2.33). $\square$

The $n(n-2) - 1$ dual matrices $\widehat{P}_{\pi(i,j)}, (i,j) \in \mathbb{I}_n^{(1)} \cup \mathbb{I}_n^{(2)}$ are now obtained from (3.28) as follows: Row i and column j of $\widehat{P}_{\pi(i,j)}$ are exactly row 1 and column 2 of $\widehat{P}_{\pi(1,2)}$; the other columns are the $n-1$ identical columns of (3.28) in which now the entry $-n+2$ occurs in row i. The number $b_n = n(n-2) + 2$ stands at the intersection of row i and column j, and the line-sum of each of these matrices is $1/n$. The proof that each of the dual permutation matrices defined by (3.30) below satisfies the defining relation $(\widehat{P}_{\pi(i,j)} \mid P_\pi) = \delta_{\pi(i,j), \pi}$, each $\pi \in \Sigma_n(e, p)$, is just a variation of the one given above for $\widehat{P}_{\pi(1,2)}$. Thus, we have that the dual matrix $\widehat{P}_{\pi(i,j)}, (i,j) \in \mathbb{I}_n^{(1)} \cup \mathbb{I}_n^{(2)}$ is the following simple matrix:

$$\widehat{P}_{\pi(i,j)} = \qquad (3.30)$$

$$= \frac{1}{n^2} \begin{pmatrix} & & & (\text{col } j) & & & \\ 2 & \cdots & 2 & -n+2 & 2 & \cdots & 2 \\ \vdots & \cdots & \vdots & \vdots & \vdots & \cdots & \vdots \\ 2 & \cdots & 2 & -n+2 & 2 & \cdots & 2 \\ -n+2 & \cdots & -n+2 & b_n & -n+2 & \cdots & -n+2 \\ 2 & \cdots & 2 & -n+2 & 2 & \cdots & 2 \\ \vdots & \cdots & \vdots & \vdots & \vdots & \cdots & \vdots \\ 2 & \cdots & 2 & -n+2 & 2 & \cdots & 2 \end{pmatrix} (\text{row } i).$$

3.4. THE GENERAL DUAL MATRICES IN THE BASIS $\Sigma_n(e,p)$

We have now obtained the dual permutation matrices in the basis $\Sigma_n(e,p)$ for all but two cases, $\widehat{P}_e$ and $\widehat{P}_p$. The feature of these two remaining dual matrices that stands out in the examples given for $n = 3, 4$ is that row 1, row 2, ..., row n are related by the group of n cyclic permutations $(1, 2, \ldots, n), (n, 1, 2, \ldots, n-1), (n-1, n, 1, \ldots, n-2), \ldots, (2, 3, \ldots, n, 1)$. This property generalizes to give the following two matrices:

$$\widehat{P}_e = \frac{1}{n}\begin{pmatrix} 1 & -n+2 & \cdots & & 0 \\ 0 & 1 & -n+2 & \cdots & -1 \\ -1 & 0 & 1 & \cdots & -n+2 \\ \vdots & \vdots & & \cdots & \vdots \\ -n+2 & -n+3 & \cdots & & 1 \end{pmatrix}, \quad (3.31)$$

$$\widehat{P}_p = \frac{1}{n}\begin{pmatrix} 0 & -1 & \cdots & -n+2 & 1 \\ 1 & 0 & \cdots & -n+3 & -n+2 \\ -n+2 & 1 & \cdots & -n+4 & -n+3 \\ \vdots & \vdots & & \vdots & \vdots \\ -1 & -2 & \cdots & 1 & 0 \end{pmatrix}. \quad (3.32)$$

In the definition of $\widehat{P}_e$, the first row is the sequence of length n given by $(1, -n+2, \ldots, 0)$ that begins with $1, -n+2$ and then the successive parts increase by 1 until 0 is reached. Row 2 is then the cyclic permutation of row 1 that begins with 0; row 3 the cyclic permutation of row 1 that begins with -1; ...; row $n-1$ the cyclic permutation of row 1 that begins with $-n+3$; row n the cyclic permutation of row 1 that begins with $-n+2$. Similarly, in the definition of $\widehat{P}_p$, the first row is the sequence of length given by $(0, -1, \ldots, -n+2, 1), n \geq 3$. Row 2 is then the cyclic permutation of row 1 that begins with 1; row 3 the cyclic permutation of row 1 that begins with $-n+2$; ...; row $n-1$ the cyclic permutation of row 1 that begins with -2; row n the cyclic permutation of row 1 that begins with -1.

The line-sum of $\widehat{P}_e$ and of $\widehat{P}_p$ is $-(n-3)/2$. The proof that these are, indeed, the correct dual matrices for general n is by direct verification of the defining conditions (2.33) in the style carried out above.

3.4.1 Relation between the A-Expansion and the Dual Matrices

We have written out fully the dual matrices above in terms of their elements. These matrices can be expressed very concisely in terms of the 1–matrix J_n of order n whose elements are all equal to 1, and in terms of the n^2 matrix units $e_{ij}, 1 \leq i, j \leq n$, of order n, where e_{ij} has 1 in row i and column j, and 0 in all other rows and columns. Thus, we have the following expressions for the dual permutation matrices in the basis $\Sigma_n(e, p)$:

$$\widehat{P}_{\pi(i,j)} = \frac{2}{n^2} J_n - \frac{1}{n}\left(\sum_{k=1}^{n}(e_{ik} + e_{kj})\right) + e_{ij}, \ (i,j) \in \mathbb{I}_n^{(12)},$$

$$\widehat{P}_e = \frac{1}{n}\sum_{i=1}^{n}\sum_{j=i+1}^{i+n}(j-i-n+1)e_{ij}, \qquad (3.33)$$

$$\widehat{P}_p = \frac{1}{n}\sum_{j=1}^{n}\sum_{i=j+2}^{j+1+n}(i-j-n)e_{ij}.$$

In the expressions for $\widehat{P}_e$ and $\widehat{P}_p$, the following cyclic (congruent modulo n) identifications are to be made:

$$e_{i,rn+j} \equiv e_{ij}, \ r \in \mathbb{P}, \ j = 1, 2, \ldots, n;$$
(3.34)
$$e_{ri+n,j} \equiv e_{ij}, \ r \in \mathbb{P}, \ i = 1, 2, \ldots, n.$$

Using the easily proved relations $(A^{tr}|e_{ij}) = (e_{ij}|A^{tr}) = a_{ij}$ for arbitrary real A and $(A|J_n) = (J_n|A) = nl_A$ for an arbitrary A of fixed line-sum l_A, it is straightforward to prove from (3.33) that

$$(\widehat{P}_{\pi(i,j)}|A^{tr}) = a_{ij}, \ (i,j) \in \mathbb{I}_n^{(12)},$$

$$(\widehat{P}_e|A^{tr}) = \frac{1}{n}\sum_{i=1}^{n}\sum_{j=i+1}^{i+n}(j-i-n+1)a_{ij} = \alpha_{(1,1)}, \quad (3.35)$$

$$(\widehat{P}_p|A^{tr}) = \frac{1}{n}\sum_{j=1}^{n}\sum_{i=j+2}^{j+1+n}(i-j-n)a_{ij} = \alpha_{(1,n)}.$$

(In these relations, the same cyclic identifications for the elements a_{ij} of A are made as for the e_{ij} in (3.34).) The coordinates $\alpha_{(1,1)}$ and $\alpha_{(1,n)}$ given by relations (3.35) are related to those given by (3.6) by addition of the respective n relations in (3.6) and division by n. Thus, the expansion of an arbitrary matrix A of fixed line-sum obtained by the A–expansion method and the dual matrix method are in agreement.

3.4. THE GENERAL DUAL MATRICES IN THE BASIS $\Sigma_n(e,p)$ 57

Concluding Remarks.

1. The contents of Chapters 2 and 3 are conceptually quite elementary, and while rather intricate in their details, these subjects do not seem to be in the literature. This material is included here to show the role of the binary coupling theory of angular momentum in leading naturally to doubly stochastic matrices and their relationship to measurement theory of composite systems, as well as to a generalization of the Regge magic square characterization of the Racah coefficients, presented in Chapter 6. This all seems particularly useful in view of the present day focus on quantum theory from the viewpoint of information theory, as presented, for example by Nielsen and Chuang [57]. In this direction, we also acknowledge helpful discussions with K. Życzkowski and D. Chriściński relating to the geometry of quantum mechanics and bring attention to their work: I. Bengtsson and K. Życzkowski [4] and D. Chriściński [21].

2. The expansion of matrices of fixed line-sum into a sum of linearly independent permutation matrices fills-out a topic on the theory of special classes of matrices. But, while doubly stochastic matrices, magic squares, and alternating sign matrices themselves find applications to physical problems, there seems to be no corresponding physical interpretation for the expansion coordinates of such matrices in terms of a basis set of permutation matrices. We address this briefly for doubly stochastic matrices in Sect. 5.8.9, with disappointing results.

3. We would like also to acknowledge the major contributions of Dr. John H. Carter to the material presented in Chapters 2 and 3. Most of this work was carried out during the period July 2003-August 2004, when he worked as a postdoctoral student at Los Alamos. He deserves full credit for the discovery and proof of the number $b_n = (n-1)^2 + 1$ of linearly independent matrices of fixed line-sum, together with the introduction of the shift operators $\langle i, j \rangle$, and the two basis sets $\Sigma_n(e)$ and $\Sigma_n(e,p)$ that they generate. This research was presented as a preprint in Ref. [16]. The paper was submitted for publication, and provisionally accepted provided revisions be carried out. The required revisions were never properly completed; it remains unpublished, through no fault of Carter, although the number b_n is cited in Ref. [17].

4. It came to our attention recently that the number b_n was published in a related context by Jakubczyk *et al.* [32]. No doubt it appears elsewhere in the published literature. We are also indebted to Dr. Aram Harrow, University of Bristol, for pointing out in a private communication on 12/03/2010 that the reducible representation $\{P_\pi \,|\, \pi \in S_n\}$ of order n of the symmetric group S_n is reducible

by a real orthogonal similarity transformation to the direct sum of the identity representation and a real orthogonal irreducible representation $\{R_\pi \,|\, \pi \in S_n\}$ of order $n-1$, and that this property can be used to prove $b_n = (n-1)^2 + 1$. The irreducible representation $\{R_\pi\}$ then has the simple character set given by

$$\operatorname{Tr} R_\pi = -1 + \operatorname{Tr} P_\pi = -1 + \sum_{i=1}^{n} \delta_{i,\pi_i}, \text{ each } \pi \in S_n. \qquad (3.36)$$

Thus, the number of linearly independent permutation matrices has a direct group theoretical origin, as well as that derived by Carter by purely algebraic means.

Chapter 4

Further Applications of Permutation Matrices

4.1 Introduction

There is, perhaps, no group that has more universal applications in quantum physics than the symmetric group because of the extraordinary restrictions put on the quantum states of composite systems by the Pauli principle for such applications. Hundreds of papers and many books have been written about the symmetric group. We address this subject again because the representations of the symmetric group S_n in irreducible form are obtained naturally, directly from the irreducible representations of the unitary group $U(n)$. The setting for the irreducible representations of $U(n)$ themselves is within the more general framework of the matrix Schur functions, whose properties are the subject of many pages in [L]: It is the matrix Schur functions evaluated on the numerical permutation matrices of order n that establish the relationship.

There is another less well-known representation of the symmetric group that is determined by the complete set of mutually commuting Hermitian operators known as the Jucys-Murphy operators. Again, the permutation matrices play a crucial role for obtaining the irreducible representations.

The algebra of permutation matrices has an isomorphic realization as an algebra of Young operators, which are of great importance in physical problems (see Hamermesh [30] and Wybourne [87] for many applications). This structure illustrates yet another aspect of the role of permutation matrices in physical theory. Indeed, from the viewpoint of operators acting in a vector space, it is the algebra of Young operators that underlies the occurrence of the symmetric group as a subgroup of the general unitary group, when the latter is realized as Lie algebraic operators acting on the irreducible Gelfand-Tsetlin basis vectors (see [L]). Thus, the group structure relating to the matrix Schur functions in the first paragraph could also be presented from the viewpoint of Young operators for the unitary group, while the Jucys-Murphy operator approach could be presented as a specialization of the matrix Schur functions. The main point is: *Each property of the algebra of permutation matrices (over $\mathbb{R}$ or $\mathbb{C}$, as specified) can be expressed as an identical property of the algebra of Young operators (over $\mathbb{R}$ or $\mathbb{C}$) simply by mapping a given permutation matrix to the corresponding Young operator.*

This chapter addresses each of the above topics, since each is a continuing application or aspect of the subject matter developed in [L]. We begin with an algebra of Young operators, since it is relevant to each of the other topics. We give a derivation of a set of all irreducible representations of S_n directly from the matrix Schur functions, although we could have used the Young operator approach. The derivation of a second set of all irreducible representations of S_n uses the Young operator approach, although we could have used the matrix Schur functions directly. The presentation given here is adapted from Ref. [47].

4.2 An Algebra of Young Operators

The viewpoint can be adopted that the permutations in the group S_n are operators that act in the set of all sequences of length n. The component parts or "variables" of such a "point" in a general theory can be taken to be indeterminates that can be chosen in the context of a given application. The classical work of Young [88], Weyl [80], and Robinson [67], establishing the foundations of this subject, is well-known. Our brief discussion here is to relate the linear vector space of Young operators to the linear vector space of permutation matrices. These two spaces are one-to-one (isomorphic) under the mapping $P_\pi \mapsto \pi$, where we now regard a permutation π as an operator. In order to keep clear the viewpoint between permutations simply as elements of the group S_n and the more general viewpoint that they are operators acting in some vector space, we denote the operator associated with each $\pi \in S_n$ by Y_π, which we call a *Young operator*. Of course, the concept of a linear operator requires that we specify the linear vector space on which the operator acts, and that we define the rule for its action in that vector space. In this section, we develop a very special algebra of Young operators by choosing a very simple vector space in which the operators act. Our interest here is in showing that this theory is isomorphic to the algebra of permutation matrices, hence, may be regarded as an application of permutation matrices (or conversely).

We recall that a linear operator is a rule for mapping vectors in a linear vector space $\mathcal{V}$ into new vectors in the same space. Let $\mathcal{V}_n$ denote a finite-

4.2. AN ALGEBRA OF YOUNG OPERATORS

dimensional Hilbert space of dimension $\dim \mathcal{V}_n = n$ with an orthonormal basis set of vectors $\mathbf{B}_n = \{|1\rangle, |2\rangle, \ldots, |n\rangle\}$ (Dirac bra-ket notation). Then, a linear operator Y_π, each $\pi = (\pi_1, \pi_2, \ldots, \pi_n) \in S_n$, is defined by

$$Y_\pi |i\rangle = |\pi_i\rangle, \quad \text{each } i = 1, 2, \ldots, n. \tag{4.1}$$

The linear property then means that the action of $Y_\pi, \pi \in S_n$, on each vector $|\alpha\rangle = \sum_{i=1}^n \alpha_i |i\rangle \in \mathcal{V}_n$ is given by

$$Y_\pi \sum_{i=1}^n \alpha_i |i\rangle = \sum_{i=1}^n \alpha_i Y_\pi |i\rangle$$

$$= \sum_{i=1}^n \alpha_i |\pi_i\rangle = \sum_{i=1}^n \alpha_{\pi_i^{-1}} |i\rangle = |\alpha_{\pi^{-1}}\rangle, \tag{4.2}$$

$$Y_\pi |\alpha\rangle = |\alpha_{\pi^{-1}}\rangle.$$

The second relation summarizes the first one, and the sequence $\alpha_{\pi^{-1}}$ is the rearrangement of the parts of $\alpha = (\alpha_1, \alpha_2, \ldots, \alpha_n)$ defined by $\alpha_{\pi^{-1}} = (\alpha_{\pi_1^{-1}}, \alpha_{\pi_2^{-1}}, \ldots, \alpha_{\pi_n^{-1}})$ for $\pi^{-1} = (\pi_1^{-1}, \pi_2^{-1}, \ldots, \pi_2^{-1})$. It is sufficient for our present purpose to take the parts of α to be real: $\alpha_i \in \mathbb{R}$. The linear operator Y_π is the simplest example of a *Young operator* associated with the symmetric group S_n. The product $Y_{\pi'} Y_\pi$ of two such Young operators is then defined by $(Y_{\pi'} Y_\pi)|\alpha\rangle = Y_{\pi'}(Y_\pi |\alpha,\rangle)$, which must hold for each vector $|\alpha\rangle \in \mathcal{V}_n$. This relation gives the operator identity on $\mathcal{V}_n$ expressed by $Y_{\pi'} Y_\pi = Y_{\pi'\pi}$, as required of a group action.

We can now extend the definition of a Young operator to arbitrary sums (over $\mathbb{R}$) of the *elementary* Young operators Y_π, as given by

$$Y(x) = \sum_{\pi \in S_n} x_\pi Y_\pi, \quad \text{each } x_\pi \in \mathbb{R}. \tag{4.3}$$

It is a straightforward exercise to prove the well-known result: *The set of all linear operators* $\{Y(x) \,|\, x \in \mathbb{R}^n\}$ *is itself a linear vector space* $\mathcal{V}_n^{op}$ *over the real numbers.* We also refer to $Y(x)$ as a Young operator. But now we have the additional property that, for each pair of Young operators $Y(x) \in \mathcal{V}_n^{op}$ and $Y(y) \in \mathcal{V}_n^{op}$, it also the case that the product $Y(x)Y(y)$ is a Young operator; that is, we have

$$Y(x)Y(y) = Y(z) \in \mathcal{V}_n^{op}; \quad z_\pi = \sum_{\rho \in S_n} x_\rho y_{\rho^{-1}\pi}. \tag{4.4}$$

Thus, we have an *algebra of Young operators*. See Hamermesh [30, pp. 239-246] for a more general discussion of this algebra from a physical point of view.

The matrix elements of the elementary Young operator Y_π on the orthonormal basis $\mathbf{B}_n$ of the vector space $\mathcal{V}_n$ are given by

$$\langle i | Y_\pi | j \rangle = \langle i | \pi_j \rangle = \delta_{i,\pi_j} = (P_\pi)_{ij}. \tag{4.5}$$

This result implies:

Every relation in Chapter 2 and Chapter 3 that holds between permutation matrices P_π and their linear combinations $A, B, \ldots$ over $\mathbb{R}$ also hold between the elementary Young operators Y_π and their linear combinations $Y(a), Y(b), \ldots$ over $\mathbb{R}$.

We repeat several of the significant relations:

1. Dimension of the vector space $\mathcal{V}_n^{\mathrm{op}}$ of Young operators: $\dim \mathcal{V}_n^{\mathrm{op}} = b_n = (n-1)^2 + 1$.

2. Inner product of elementary Young operators: $\langle Y_\pi | Y_{\pi'} \rangle = (P_\pi | P_{\pi'}) = \sum_{i=1}^n \delta_{\pi_i, \pi_i'}$.

3. Inner product of Young operators $Y(a)$ and $Y(b)$: $\langle Y_a | Y_b \rangle = \sum_{i=1}^n \sum_{\pi, \pi' \in \Sigma_n} \delta_{\pi_i, \pi_i'} a_\pi b_{\pi'}$ for $A = \sum_{\pi \in \Sigma_n} a_\pi P_\pi$, $B = \sum_{\pi \in \Sigma_n} b_\pi P_\pi$, where Σ_n is a linearly independent basis set of permutations of S_n.

4. Elementary dual Young operators defined by the orthogonality requirements: $\langle \widehat{Y}_\pi | Y_{\pi'} \rangle = (\widehat{P}_\pi | P_{\pi'}) = \delta_{\pi, \pi'}$, each pair $\pi, \pi' \in \Sigma_n$.

5. Coordinates in a basis: $Y(a) = \sum_{\pi \in \Sigma_n} a_\pi Y_\pi$, each $a_\pi \in \mathbb{R}$, where $a_\pi = \langle \widehat{Y}_\pi | Y(a) \rangle = (\widehat{P}_\pi | P(a))$.

Example 1.

$$P_{1,2,3} + P_{2,3,1} + P_{3,1,2} - P_{1,3,2} - P_{3,2,1} - P_{2,1,3} = 0 \text{ (zero matrix)},$$
$$\text{if and only if} \tag{4.6}$$
$$Y_{1,2,3} + Y_{2,3,1} + Y_{3,1,2} - Y_{1,3,2} - Y_{3,2,1} - Y_{2,1,3} = 0^{\mathrm{op}} \text{ (zero operator)},$$

is an operator identity on the space $\mathcal{V}_n$.

Example 2. Relations for the dual permutation matrices in the basis $\Sigma_3(e, p)$ given in Chapter 3 carry over exactly to the elementary Young operators and their duals simply by making the substitutions $P_\pi \mapsto Y_\pi$ and $\widehat{P}_\pi \mapsto \widehat{Y}_\pi$. Thus, the following five dual Young operators are obtained in terms of the elementary Young operators $Y_\pi, \pi \in \Sigma_3(e, p)$ from the columns of the inverse $X = Z^{-1}$ of the Gram matrix given by (3.17):

$$\widehat{Y}_{1,2,3} = (2Y_{1,2,3} - Y_{2,1,3} + Y_{2,3,1} - Y_{1,3,2} - Y_{3,2,1})/3,$$
$$\widehat{Y}_{2,1,3} = (-3Y_{1,2,3} + 5Y_{2,1,3} - 3Y_{2,3,1} + 2Y_{1,3,2} + 2Y_{3,2,1})/9,$$
$$\widehat{Y}_{2,3,1} = (Y_{1,2,3} - Y_{2,1,3} + 2Y_{2,3,1} - Y_{1,3,2} - Y_{3,2,1})/3, \quad (4.7)$$
$$\widehat{Y}_{1,3,2} = (-3Y_{1,2,3} + 2Y_{2,1,3} - 3Y_{2,3,1} + 5Y_{1,3,2} + 2Y_{3,2,1})/9,$$
$$\widehat{Y}_{3,2,1} = (-3Y_{1,2,3} + 2Y_{2,1,3} - 3Y_{2,3,1} + 2Y_{1,3,2} + 5Y_{3,2,1})/9.$$

All the general relations between the dual Young operators $\widehat{Y}_{\pi^{(i,j)}}, (i,j) \in \mathbb{I}_n^{(1)} \cup \mathbb{I}_n^{(1)}, \widehat{Y}_e$, and $\widehat{Y}_p$ and the Young operators $Y_{\pi^{(i,j)}}, (i,j) \in \mathbb{I}_n^{(1)} \cup \mathbb{I}_n^{(1)}, Y_e$, and Y_p for the dual permutation matrices by relations (3.28)-(3.35) also hold simply by making the mapping $P_\pi \mapsto Y_\pi$ and the mapping $\widehat{P}_\pi \mapsto \widehat{Y}_\pi$. Indeed, had the properties of Young operators been developed along these lines first, we would regard the permutation matrices as a (faithful) representation of this operator algebra on the basis $\mathcal{V}_n$.

We hasten to point out that Young operators can be defined on other vectors spaces that give rise to a more general theory of their algebra (see Hamermesh [30]). But it is always the case that the inner product $\langle Y_\pi | Y_{\pi'} \rangle = \sum_{i=1}^n \delta_{\pi_i, \pi'_i}$ can always be defined, since it depends only on the properties of the permutations $\pi, \pi' \in S_n$.

4.3 Matrix Schur Functions

A real, orthogonal, irreducible representation of the symmetric group is obtained directly from the specialization of a class of many-variable polynomials known as *matrix Schur functions*. Whereas an ordinary Schur function is based on a single partition, a matrix Schur function is based on a number of stacked partitions that satisfy certain "betweenness" relations. The definition and properties of the matrix Schur functions is the subject of considerable length in [L]. Here, we list in Items 1-14 some of the properties of matrix Schur functions for the present application to the irreducible representations of S_n. The matrix Schur functions are fully defined by various of these relations, as discussed and proved in [L], where these functions are also referred to as D^λ-polynomials and Gelfand-Tsetlin (GT) polynomials. The matrix Schur functions are homogeneous polynomials in the indeterminates (variables) $Z^A = \prod_{1 \leq i,j \leq n} z_{ij}^{a_{ij}}$, each $a_{ij} \in \mathbb{N}$, as given by

$$D \begin{pmatrix} m' \\ \lambda \\ m \end{pmatrix}(Z) = \sum_{A \in \mathbb{M}_{n \times n}^p(\alpha, \alpha')} C \begin{pmatrix} m' \\ \lambda \\ m \end{pmatrix}(A) \frac{Z^A}{A!}, \quad (4.8)$$

CHAPTER 4. APPLICATIONS OF PERMUTATION MATRICES

where $\alpha \in w\binom{\lambda}{m}$ and $\alpha' \in w\binom{\lambda}{m'}$ are weights of the respective lower and upper GT patterns. We define the various symbols and list important properties of these polynomials in the fourteen items below:

1. Gelfand-Tsetlin patterns:

    (a) λ is a partition with n parts: $\lambda = (\lambda_1, \lambda_2, \ldots, \lambda_n)$, each $\lambda_i \in \mathbb{N}$, with $\lambda_1 \geq \lambda_2 \geq \cdots \geq \lambda_n \geq 0$.

    (b) $\binom{\lambda}{m}$ is a stacked array of partitions:

    $$\binom{\lambda}{m} = \begin{pmatrix} \lambda_1 & \lambda_2 & \cdots & \lambda_j & \cdots & & \lambda_{n-1} & \lambda_n \\ m_{1,n-1} & m_{2,n-1} & \cdots & m_{j,n-1} & \cdots & & m_{n-1,n-1} & \\ & & & \vdots & & & & \\ & & m_{1,j} & m_{2,j} & \cdots & m_{j,j} & & \\ & & & \vdots & & & & \\ & & & m_{1,2} & m_{2,2} & & & \\ & & & & m_{1,1} & & & \end{pmatrix}, \quad (4.9)$$

    in which row j is a partition with j parts, and adjacent rows $(\mathbf{m}_{j-1}, \mathbf{m}_j), j = 2, 3, \ldots, n$, satisfy the "betweenness conditions:"

    $$\mathbf{m}_j = (m_{1,j}, m_{2,j}, \ldots, m_{j,j}), j = 1, 2, \ldots, n; \; \mathbf{m}_n = \lambda,$$
    $$m_{i,j-1} \in [m_{i,j}, m_{i+1,j}], \text{ each } i = 1, 2, \ldots, j-1, \quad (4.10)$$
    $[a, b], a \leq b$, denotes the closed interval of nonnegative integers $a, a+1, \ldots, b$.

    (c) Lexical patterns (patterns that satisfy betweenness rule (4.10)):

    $$\mathbb{G}_\lambda = \left\{ \binom{\lambda}{m} \;\middle|\; m \text{ is a lexical pattern} \right\}. \quad (4.11)$$

2. The notation $\begin{pmatrix} m' \\ \lambda \\ m \end{pmatrix}$ denotes a pair of lexical GT patterns that share the same partition λ, where $\binom{m'}{\lambda}$ denotes the inverted pattern $\binom{\lambda}{m'}$, and the shared partition λ is written only once. Thus, it is these *double GT patterns* that enumerate the matrix Schur functions $D\begin{pmatrix} m' \\ \lambda \\ m \end{pmatrix}(Z)$ in the n^2 indeterminates $Z = (z_{ij})_{1 \leq i,j \leq n}$.

3. The C-coefficients in the right-hand side in the definition of the matrix Schur functions that multiply the polynomial factor $Z^A/A!$ are, up to multiplying factors, the elements of a real orthogonal matrix. The details are not important here (see [L]).

4.3. MATRIX SCHUR FUNCTIONS

4. The set $\mathbb{M}^p_{n\times n}(\alpha,\alpha')$:

   (a) The weight $w\binom{\lambda}{m}$ of a GT pattern $\binom{\lambda}{m}$ is defined to be the sequence $w\binom{\lambda}{m}$ with n parts given by the sum of the parts in row j minus the sum of the parts in row $j-1$:

   $$w\binom{\lambda}{m} = (m_{1,1}, m_{1,2}+m_{2,2}-m_{1,1}, \ldots, \sum_{i=1}^{n} m_{i,n} - \sum_{i=1}^{n-1} m_{i,n-1}). \quad (4.12)$$

   The weights of the lower and upper patterns are denoted, respectively, by

   $$\begin{aligned}\alpha &= (\alpha_1,\alpha_2,\ldots,\alpha_n) = w\binom{\lambda}{m}, \\ \alpha' &= (\alpha'_1,\alpha'_2,\ldots,\alpha'_n) = w\binom{\lambda}{m'}.\end{aligned} \quad (4.13)$$

   (b) Homogeneity properties of the polynomials:

   degree α_i in row $z_i = (z_{i1}, z_{i2}, \ldots, z_{in})$ of Z,
   degree α'_j in column $z^j = (z_{1j}, z_{2j}, \ldots, z_{nj})$ of Z, $\quad (4.14)$
   $p = \sum_i \lambda_i = \sum_i \alpha_i = \sum_i \alpha'_i.$

   (c) $\mathbb{M}^p_{n\times n}(\alpha,\alpha')$ denotes the set of all matrix arrays A of order n such that the sum of all elements in row i of A is α_i, and the sum of all elements in column j of A is α'_j, where $\lambda \vdash p, \alpha \vdash p, \alpha' \vdash p$.

5. The summation in (4.8) is over all $A \in \mathbb{M}^p_{n\times n}(\alpha,\alpha')$, where $\alpha = w\binom{\lambda}{m}$ and $\alpha' = w\binom{\lambda}{m'}$: *The matrix Schur functions are homogeneous polynomials of total degree p in the indeterminates Z.*

6. Orthogonality relations in an inner product denoted $(,)$ (isomorphic to the boson inner product often used in physics):

   $$\left(Z^{A'}, Z^A\right) = A!\delta_{A',A},$$

   $$\left(D\begin{pmatrix}m'\\ \lambda\\ m\end{pmatrix}(Z), D\begin{pmatrix}m'''\\ \lambda'\\ m''\end{pmatrix}(Z)\right) = \delta_{m,m''}\delta_{m',m'''}\delta_{\lambda,\lambda'} M(\lambda),$$

   $$M(\lambda) = \prod_{i=1}^{n}(\lambda_i + n - i)! \Big/ 1!2!\cdots(n-1)!\mathrm{Dim}\lambda, \quad (4.15)$$

$$\mathrm{Dim}\lambda = \prod_{1\leq i<j\leq n}(\lambda_i - \lambda_j + j - i)\Big/ 1!2!\cdots(n-1)! \quad \text{(Weyl)},$$

$$D^\lambda(I_n) = I_{\mathrm{Dim}\lambda}.$$

7. Invertibility:

$$\frac{Z^A}{A!} = \sum_{\lambda \vdash p} \sum_{m,m' \in \mathbb{G}_\lambda(\alpha,\alpha')} \frac{1}{M(\lambda)A!} C\begin{pmatrix} m' \\ \lambda \\ m \end{pmatrix}(A) D\begin{pmatrix} m' \\ \lambda \\ m \end{pmatrix}(Z), \tag{4.16}$$

each $A \in \mathbb{M}^p_{n\times n}(\alpha,\alpha')$, where $\mathbb{G}_\lambda(\alpha,\alpha')$ denotes the set of double GT patterns of weights $\alpha = w\binom{\lambda}{m}$ and $\alpha' = \binom{\lambda}{m'}$.

8. Robinson, Schensted, Knuth (RSK) identity (expresses invertibility):

$$\sum_{\lambda \vdash p} K(\lambda,\alpha) K(\lambda,\alpha') = |\mathbb{M}^p_{n\times n}(\alpha,\alpha')|, \tag{4.17}$$

where $K(\lambda,\alpha)$ denotes the multiplicity of weight α in shape λ; it is known as the *Kostka number*.

9. Multiplication property: Fundamental multiplication rule:

$$D^\lambda(X) D^\lambda(Y) = D^\lambda(XY), \text{ for arbitrary } X, Y. \tag{4.18}$$

10. Transpositional symmetry: Z replaced by the transpose Z^{tr}:

$$D^\lambda(Z^{tr}) = (D^\lambda(Z))^{tr}. \tag{4.19}$$

11. Group property: If Z is specialized to be a member of any matrix group, continuous or finite, representations of that group are obtained. Thus, the irreducible representations of the general linear group $GL(n,\mathbb{C})$ and the unitary group $U(n)$ are so obtained. This applies as well to any multiplicative matrix algebra.

12. Diagonal property: Z restricted to the diagonal matrix denoted $\mathrm{diag} Z = \mathrm{diag}(z_1, z_2, \ldots, z_n)$:

$$D\begin{pmatrix} m' \\ \lambda \\ m \end{pmatrix}(\mathrm{diag} Z) = \delta_{m,m'} z_1^{\alpha_1} z_2^{\alpha_2} \cdots z_n^{\alpha_n}, \tag{4.20}$$

where $\alpha = (\alpha_1, \alpha_2, \ldots, \alpha_n)$ is the weight of the GT pattern $\binom{\lambda}{m}$.

13. Schur functions. The trace of the above relation gives the ordinary Schur functions:
$$s_\lambda(z) = \mathrm{Tr} D^\lambda(\mathrm{diag} Z) = \sum_\alpha K(\lambda, \alpha) x^\alpha, \qquad (4.21)$$

where the sum is over the distinct weights α of $\binom{\lambda}{m}$.

14. Generalized MacMahon's master theorem:
$$\frac{1}{\det(I_{n^2} - tX \otimes Y)} = \sum_{k \geq 0} t^k \sum_{\lambda \vdash k} \mathrm{Tr} D^\lambda(X) \mathrm{Tr} D^\lambda(Y), \qquad (4.22)$$

where X and Y are arbitrary matrices of order n of commuting indeterminates, and $X \otimes Y$ is their Kronecker product (see Méndez [53, 54], Schwinger [74], Louck [44], Michel and Zhilinskii [55], [L]).

4.4 Real Orthogonal Irreducible Representations of S_n

4.4.1 Matrix Schur Function Real Orthogonal Irreducible Representations

The irreducible repesentations of the symmetric group are obtained from the matrix Schur functions defined by (4.8) in three steps:

(i). Choose $\lambda \vdash n$ and patterns m and m' to have weights $\alpha = w(\genfrac{}{}{0pt}{}{\lambda}{m}) = \alpha' = w(\genfrac{}{}{0pt}{}{\lambda}{m'}) = (1^n)$ (1 repeated n times) in (4.8) to obtain:

$$D\begin{pmatrix} m' \\ \lambda \\ m \end{pmatrix}(Z) = \sum_{\pi' \in S_n} C\begin{pmatrix} m' \\ \lambda \\ m \end{pmatrix} (P_{\pi'}) z_{\pi'_1,1} z_{\pi'_2,2} \cdots z_{\pi'_n,n}. \qquad (4.23)$$

This result is correct because each $A \in \mathrm{M}^p_{n \times n}(1^n, 1^n)$ must be a permutation matrix $P_{\pi'} \in S_n$; hence, $a_{ij} = (P_{\pi'})_{ij} = \delta_{i,\pi'_j}$, which gives $(z_{ij})^{\delta_{i,\pi'_j}} = z_{\pi'_j,j}$.

(ii). Choose $Z = P_\pi, \pi \in S_n$, in (4.23) and use $z_{ij} = (P_\pi)_{ij} = \delta_{i,\pi_j}$:

$$D\begin{pmatrix} m' \\ \lambda \\ m \end{pmatrix}(P_\pi) = C\begin{pmatrix} m' \\ \lambda \\ m \end{pmatrix}(P_\pi). \qquad (4.24)$$

This result follows in consequence of $z_{ij} = \delta_{i,\pi_j}$ in (4.23); hence, $z_{\pi'_1,1} z_{\pi'_2,2} \cdots z_{\pi'_n,n} = \delta_{\pi'_1,\pi_1} \delta_{\pi'_2,\pi_2} \cdots \delta_{\pi'_n,\pi_n}$.

(iii). Choose $\lambda \vdash n$ and $Z = P_\pi, \pi \in S_n$ in (4.8) to obtain:

$$D\begin{pmatrix} m' \\ \lambda \\ m \end{pmatrix}(P_\pi) = 0, \text{ unless } \alpha = w\begin{pmatrix} \lambda \\ m \end{pmatrix} = \alpha' = w\begin{pmatrix} \lambda \\ m' \end{pmatrix} = (1^n). \quad (4.25)$$

This result follows because of $\prod_{i,j=1}^n (\delta_{i,\pi_j})^{a_{ij}} = \prod_{i,j=1}^n \delta_{i,\pi_j} = 0$, $\pi \in S_n$, independently of $A \in \mathbb{M}_{n \times n}^p(\alpha, \alpha')$, where we always take $(\delta_{i,j})^0 = \delta_{i,j}$.

We now define the matrix $S^\lambda(P_\pi)$, each $\pi \in S_n$, to be the matrix with elements in row m and column m' given by

$$S\begin{pmatrix} m' \\ \lambda \\ m \end{pmatrix}(P_\pi) = D\begin{pmatrix} m' \\ \lambda \\ m \end{pmatrix}(P_\pi),$$

for all weights $w\begin{pmatrix} \lambda \\ m \end{pmatrix} = w\begin{pmatrix} \lambda \\ m' \end{pmatrix} = (1^n).$ \quad (4.26)

It follows from properties (i)-(iii) that the matrices in the set $\{S^\lambda(P_\pi) \mid \pi \in S_n\}$ constitute a real orthogonal representation of S_n for each partition $\lambda = (\lambda_1, \lambda_2, \ldots, \lambda_n) \vdash n$:

$$S^\lambda(P_\pi) S^\lambda(P_{\pi'}) = S^\lambda(P_\pi P_{\pi'}) = S^\lambda(P_{\pi\pi'}). \quad (4.27)$$

The order of these matrix representations is obtained by counting the number of GT patterns in the set $\mathbf{G}_\lambda$, $\lambda \vdash n$, with weights $= w\binom{\lambda}{m} = w\binom{\lambda}{m'} = (1^n)$. But this number is just the Kostka number $K(\lambda, 1^n)$. We conclude that the dimension of the representation $S^\lambda(P_\pi)$ is given by

$$\dim \lambda = K(\lambda, 1^n). \quad (4.28)$$

The RSK relation (4.17) reduces now to the sum-of-squares identity:

$$n! = \sum_{\lambda \vdash n} K(\lambda, 1^n) K(\lambda, 1^n). \quad (4.29)$$

This is, of course, also just the expression of the Frobenius character formula, applied to the identity class (see Hamermesh [30, relation (7-25)).

It is also the case that the Frobenius dimension is generated by the general Frobenius formula for generating all characters. The formula for generating the Kostka numbers is obtained from the generating formula (7-24) in Hamermesh [30] by choosing the class to be (1^n) and bringing

4.4. ORTHOGONAL IRREDUCIBLE REPRESENTATIONS

the relation to the notational form used in [L] and the present volume:

$$\left(p_1(x)\right)^n \prod_{1 \leq i < j \leq n} (x_i - x_j) = \sum_{\lambda \vdash n} \dim \lambda \sum_{\pi \in S_n} \varepsilon_\pi \, x_{\pi_1}^{\lambda_1 + n - 1} x_{\pi_1}^{\lambda_2 + n - 2} \cdots x_{\pi_1}^{\lambda_n},$$

$$p_1(x) = x_1 + x_2 + \cdots + x_n, \tag{4.30}$$

$\varepsilon_\pi = 1$, even permutations; $\varepsilon_\pi = -1$, odd permutations.

If the bijection of GT patterns to standard Young tableaux is used (see [L]), then one obtains the result that the GT pattern enumeration of the elements of S^λ is just that given by the standard Young tableau enumeration presented in Hamermesh [30, pp.199–200] for S_4 and S_5. Moreover explicit character tables are listed for $S_1 - S_5$, demonstrating explicitly the sum-of-squares formula (4.29).The irreducible representations of S_n obtained from $U(n)$, as described above, are orthogonally equivalent to the Yamanouchi representations.

It is also of interest that relation (4.23) gives a whole class of matrices $D^\lambda(Z), \lambda \vdash n$, such that the determinantal multiplication rule holds; that is, gives a generalization of determinant (see Littlewood [42, Chapter VI]):

$$D^\lambda(X) D^\lambda(Y) = D^\lambda(XY). \tag{4.31}$$

4.4.2 Jucys-Murphy Real Orthogonal Representations

The permutation $(i,j) \in S_n$ is defined to be $\pi = (\pi_1, \pi_2, \ldots, \pi_n)$, in which all parts $\pi_k = k$, except for parts i and j, for which $\pi_i = j$ and $\pi_j = i$ for $i \neq j$. The permutation (i,j) is called a *transposition*, since it interchanges i and j in its action, and leaves all other integers unchanged. It has the properties $(i,j) = (j,i) = (i,j)^{-1}$. Thus, there are in all $n(n-1)/2$ distinct transpositions in S_n, which can be chosen to be $(i,j), 1 \leq i < j \leq n$.

It is well-known from the work of Lulek [48], and others (see, for example, Lulek et al. [49], Jakubczyk et al. [32]), that the $n-1$ Jucys-Murphy operators (Young operators) defined by $M_2 = Y_{(1,2)}, M_3 = Y_{(1,3)} + Y_{(2,3)}, M_4 = Y_{(1,4)} + Y_{(2,4)} + Y_{(3,4)}, \ldots, M_n = Y_{(1,n)} + Y_{(2,n)} + \cdots + Y_{(n-1,n)}$ mutually commute. In our presentation, we choose to replace the Jucys-Murphy (JM) operators by the modified set $M_1^{(n)} = M_1, M_2^{(n)} = M_1 + M_2, M_3^{(n)} = M_1 + M_2 + M_3, \cdots$; that is, we use the modified Jucys-Murphy operators $M_k^{(n)}$ defined by

$$M_k^{(n)} = \sum_{1 \leq i < j \leq k} Y_{(i,j)}, \quad k = 2, 3, \ldots, n. \tag{4.32}$$

This set is, of course, still an independent set of $n-1$ mutually commuting Young operators belonging to the group algebra of S_n.

We now use the vector space $\mathcal{V}_n$ of dimension $\dim \mathcal{V}_n = n$ with basis $\mathbf{B}_n$ defined at the beginning of Sect. 4.2 (see (4.1)-(4.4)) to transform the Young operator relation (4.32) into matrix form. The action of each Young operator $Y_\pi, \pi \in S_n$ on the basis $\mathbf{B}_n$ is defined by relation (4.1). Since each elementary Young operator $Y_{(i,j)}$ is represented by the permutation matrix $P_{(i,j)}$, the matrix realization of the Young operator (4.32) on the basis $\mathcal{V}_n$ is given by the following matrix relation of order n:

$$P_k^{(n)} = \sum_{1 \leq i < j \leq n} P_{(i,j)} = \begin{pmatrix} J_k + \frac{1}{2}k(k-3)I_k & 0_{n-k} \\ & \\ 0_{n-k} & \binom{k}{2}I_{n-k} \end{pmatrix}. \quad (4.33)$$

Here J_k denotes the matrix of order k with elements all equal to 1, and 0_{n-k} denotes the zero matrix of order $n-k$. For $k = n$, the last row and column are to be omitted in this relation; that is, $P_n^{(n)} = J_n + \frac{1}{2}n(n-3)I_n$. The spectrum of the JM operators $M_k^{(n)}$ is exactly that obtained by diagonalizing this real symmetric matrix.

The characteristic equation of the matrix $P_k^{(n)}$ is given by

$$\left(\lambda - \left(\binom{k-1}{2} + 1\right)\right)^{k-1} \left(\lambda - \binom{k}{2}\right)^{n-k+1} = 0, \quad (4.34)$$

where $\binom{a}{m}$ denotes a binomial coefficient. Thus, we find that the real symmetric matrix $P_k^{(n)}$ of order n has two distinct eigenvalues with multiplicities $k-1$ and $n-k+1$, respectively:

$$\lambda_k^{(n)}(1) = \binom{k-1}{2} - 1 \text{ and } \lambda_k^{(n)}(2) = \binom{k}{2}, \quad (4.35)$$

where this result holds for each of the matrices for $k = 2, 3, \ldots, n$, and the binomial coefficient $\binom{1}{2}$ is defined to be 0.

There always exists a real orthogonal matrix R of order n that simultaneously diagonalizes a set of commuting symmetric matrices. Thus, there exists an R such that

$$R^T P_k^{(n)} R = D_k^{(n)} = \begin{pmatrix} \left(\binom{k-1}{2} - 1\right) I_{k-1} & 0_{n-k+1} \\ & \\ 0_{n-k+1} & \binom{k}{2} I_{n-k+1} \end{pmatrix}, \quad (4.36)$$

4.4. ORTHOGONAL IRREDUCIBLE REPRESENTATIONS

where R is the same for all $k = 2, 3, \ldots, n$. For the case at hand, the matrix $R = (R_1 \, R_2 \, \cdots \, R_n)$ is unique up to $\pm$ sign of its columns $R_i, i = 1, 2, \ldots, n$, a result that can be proved as follows: Relation (4.36) implies $DR = RD$, where $D = \sum_{k=2}^{n} D_k^{(n)}$ is a diagonal matrix with **distinct** elements. But then it follows trivially that R is unique up to choice of signs of its columns. This is just the expression in matrix form that the set of $n-1$ mutually commuting symmetric matrices $P_k^{(n)}, k = 2, 3, \ldots, n$ is a complete set. Equivalently, this is the expression of the fact (Lulek [48]) that the set of $n-1$ JM operators (4.32) is a complete set of mutually commuting Hermitian operators on the space $\mathcal{V}_n$ over $\mathbb{C}$. (The Hermitian property is expressed by $\langle a | M_k^{(n)} | b \rangle = \left(\langle b | M_k^{(n)} | a \rangle \right)^* = \sum_{i,j=1}^{n} a_i b_j^* (P_k^{(n)})_{ij}$ for each pair of vectors $|a\rangle, |b\rangle \in \mathcal{V}_n$.)

We thus obtain the JM **reducible representation** J_π given by orthogonal similarity to a permutation matrix:

$$J_\pi = R^{tr} P_\pi R, \; \pi \in S_n, \quad (4.37)$$

where this representation is characterized as the diagonalization of the complete set of $n-1$ real symmetric matrices in relation (4.33) by a real, orthogonal matrix R, which is unique up to $\pm$ signs of its columns.

The irreducible real orthogonal JM representations can now be obtained from the matrix Schur functions in exactly the same manner as the GT representation $S^\lambda(P_\pi), \pi \in S_n$, given by (4.26)-(4.27). We replace P_π in relation (4.26) by $P_\pi = R J_\pi R^{tr}$ from (4.37), and use the general multiplication property (4.18) to obtain $S^\lambda(P_\pi) = S^\lambda(R) S^\lambda(J_\pi) S^\lambda(R^{tr})$. Thus, we obtain the following Jucys-Murphy irreducible representations of S_n :

$$S^\lambda(J_\pi) = (S^\lambda(R))^{tr} S^\lambda(P_\pi) S^\lambda(R), \text{ each } \pi \in S_n. \quad (4.38)$$

The matrix $S^\lambda(R)$ of order $\dim\lambda = K(\lambda, 1^n)$ is the real orthogonal matrix obtained from the matrix Schur function by setting $Z = R$, choosing $\lambda \vdash n$, and restricting the weights of the patterns to be $\binom{\lambda}{m} = \binom{\lambda}{m'} = (1^n)$.

As expected, each real orthogonal irreducible JM representation of S_n is orthogonally similar to the real orthogonal irreducible GT representation corresponding to the same partition $\lambda \vdash n$.

4.5 Left and Right Regular Representations of Finite Groups

Permutation matrices have a much broader role in the representation of groups than that presented for the symmetric group S_n, as we next review.

Let G_n be any finite group of order n with its n elements given by

$$G_n = \{g_1 = e, g_2, \ldots, g_n\}. \tag{4.39}$$

Select any set of n orthonormal vectors from $\mathbb{R}^n$, and label the orthonormal vectors in this basis, denoted $\mathbf{B}_n$, by the elements of the group itself:

$$\mathbf{B}_n = \{|g_1\rangle, |g_2\rangle, \ldots, |g_n\rangle\}, \langle g_i | g_j \rangle = \delta_{ij}, \; i,j = 1, 2, \ldots, n. \tag{4.40}$$

For each $g_k \in G_n$, define the left and right actions, respectively, of G_n on $\mathbf{B}_n$ by

$$L_{g_k}|g_j\rangle = |g_k g_j\rangle, \; R_{g_k}|g_j\rangle = |g_j g_k^{-1}\rangle, j = 1, 2, \ldots, n. \tag{4.41}$$

The following properties of the left and right action on the basis $\mathbf{B}_n$ are now verified directly from these definitions:

$$L_{g_{k'}} L_{g_k} = L_{g_{k'} g_k}, \; R_{g_{k'}} R_{g_k} = R_{g_{k'} g_k}, \; L_{g_k} R_{g_{k'}} = R_{g_{k'}} L_{g_k}, \tag{4.42}$$

for all pairs $g_{k'}, g_k \in G_n$.

Proof. Let $g_i = g_k g_j$ and $g_{i'} = g_j g_k^{-1}$. Then, we have the identities:

$$\begin{aligned}
L_{g_{k'}} L_{g_k}|g_j\rangle &= L_{g_{k'}}\left(L_{g_k}|g_j\rangle\right) = L_{g_{k'}}|g_i\rangle = |g_{k'} g_i\rangle \\
&= |(g_{k'} g_k) g_j\rangle = L_{g_{k'} g_k}|g_j\rangle, \\
R_{g_{k'}} R_{g_k}|g_j\rangle &= R_{g_{k'}}\left(R_{g_k}|g_j\rangle\right) = R_{g_{k'}}|g_{i'}\rangle = |g_{i'} g_{k'}^{-1}\rangle \\
&= |g_j (g_{k'} g_k)^{-1}\rangle = R_{g_{k'} g_k}|g_j\rangle, \\
R_{g_{k'}} L_{g_k}|g_j\rangle &= R_{g_{k'}}\left(L_{g_k}|g_j\rangle\right) = R_{g_{k'}}|g_i\rangle = |g_i g_{k'}^{-1}\rangle = |g_k g_j g_{k'}^{-1}\rangle, \\
L_{g_k} R_{g_{k'}}|g_j\rangle &= L_{g_k}\left(R_{g_{k'}}|g_j\rangle\right) = L_{g_k}|g_j g_{k'}^{-1}\rangle = |g_k g_j g_{k'}^{-1}\rangle,
\end{aligned} \tag{4.43}$$

where these relations hold, first of all, for all $g_j \in G_n$; then, for arbitrary selection of $g_k \in G_n$ and $g_{k'} \in G_n$. $\square$

The action of each of the operators L_{g_k} and R_{g_k} on the orthonormal basis $\mathbf{B}_n$ is represented by a matrix of order n, the order of the group

4.5. REGULAR REPRESENTATIONS OF FINITE GROUPS

G_n, by a $0-1$ matrix; that is, a matrix containing only 0 and 1, with matrix elements as follows:

$$\left(P_{g_k}^{(l)}\right)_{ij} = \langle g_i | L_{g_k} | g_j \rangle = \delta_{g_i, g_k g_j}, \quad i, j = 1, 2, \ldots, n, \tag{4.44}$$

$$\left(P_{g_k}^{(r)}\right)_{ij} = \langle g_i | R_{g_k} | g_j \rangle = \delta_{g_i, g_j g_k^{-1}}, \quad i, j = 1, 2, \ldots, n.$$

The sets of matrices with the matrix elements so defined are called *left* and *right regular matrix representations*, respectively, of the group G_n:

$$\mathbb{P}^{(l)} = \{P_{g_k}^{(l)} \mid k = 1, 2, \cdots, n\} \text{ and } \mathbb{P}^{(r)} = \{P_{g_k}^{(r)} \mid k = 1, 2, \cdots, n\}. \tag{4.45}$$

These representations of G_n then satisfy the same rules of multiplication as the abstract group itself; that is, the following relations are satisfied for all $i, j = 1, 2, \ldots, n$:

$$P_{g_i}^{(l)} P_{g_j}^{(l)} = P_{g_i g_j}^{(l)}, \quad P_{g_i}^{(r)} P_{g_j}^{(r)} = P_{g_i g_j}^{(r)}, \quad P_{g_i}^{(l)} P_{g_j}^{(r)} = P_{g_j}^{(r)} P_{g_i}^{(l)}. \tag{4.46}$$

The sets of left and right regular matrix representations (4.46) are permutation matrices, as is easily shown. What is important is:

The distribution of 0 and 1 into elements of the matrices is intrinsic to the group structure itself, as determined by the Kronecker delta functions in (4.44). Given either the left regular matrix representation or the right regular representation the abstract group multiplication of group elements can be reconstructed.

It is useful to see directly from the group multiplication table the origin of the left and right regular matrix representations.

Group multiplication table for an abstract group G_n:

	g_1	$\cdots$	g_j	$\cdots$	g_n
g_1	$g_1 g_1$	$\cdots$	$g_1 g_j$	$\cdots$	$g_1 g_n$
$\vdots$	$\vdots$		$\vdots$		$\vdots$
g_i	$g_i g_1$	$\cdots$	$g_i g_j$	$\cdots$	$g_i g_n$
$\vdots$	$\vdots$		$\vdots$		$\vdots$
g_n	$g_n g_1$	$\cdots$	$g_n g_j$	$\cdots$	$g_n g_n$

(4.47)

The row and column indices $(i, j) \mapsto (g_i, g_j)$ are read in one-to-one correspondence with group labels. We also recall that each group element

$g_k \in G_n$ occurs exactly once in each row and in each column. For given ordering of the group elements $g_1, g_2, \ldots, g_n$ in the first row and first column of the multiplication table, the distribution of the element $g_k \in G_n$ among the rows and columns of the multiplication table is uniquely determined by the abstract group multiplication rules.

There are various ways to associate a set of n permutation matrices of order n with the group multiplication table (4.47), each of which utilizes the distribution of a given element g_k among the rows and columns, which, as pointed out above, is fixed by the multiplication rules for the abstract group itself. What differs among the various sets of permutation matrices is the rule for associating their (i, j) matrix elements with the positions of the g_k. We give three such examples:

$$\left(P_{g_k}\right)_{i,j} = \begin{cases} 1, & \text{for all } i, j \text{ for which } g_i g_j = g_k, \\ 0, & \text{for all } i, j \text{ for which } g_i g_j \neq g_k. \end{cases} \quad (4.48)$$

$$\left(P_{g_k}^{(l)}\right)_{i,j} = \begin{cases} 1, & \text{for all } i, j \text{ for which } g_i g_j^{-1} = g_k, \\ 0, & \text{for all } i, j \text{ for which } g_i g_j^{-1} \neq g_k. \end{cases} \quad (4.49)$$

$$\left(P_{g_k}^{(r)}\right)_{i,j} = \begin{cases} 1, & \text{for all } i, j \text{ for which } g_i^{-1} g_j = g_k, \\ 0, & \text{for all } i, j \text{ for which } g_i^{-1} g_j \neq g_k. \end{cases} \quad (4.50)$$

Of interest here, of course, are the two sets of permutation matrices defined by relations (4.49) and (4.50):

$$\mathbb{P}_n^{(l)} = \{P_{g_k}^{(l)} \mid g_k \in G_n\} \text{ and } \mathbb{P}_n^{(r)} = \{P_{g_k}^{(r)} \mid g_k \in G_n\}. \quad (4.51)$$

These sets of permutation matrices then have the property that the permutation matrices in one set commute with those in the other set.

The row-column indices that specify the matrix elements of the left and right permutation matrices $\mathbb{P}_n^{(l)}$ and $\mathbb{P}_n^{(r)}$ can be put in somewhat better form by defining the index $\bar{i}$ by

$$g_{\bar{i}} = g_i^{-1}, \text{ each } i = 1, 2, \ldots, n. \quad (4.52)$$

In terms of this notation, we have: The elements of the left permutation matrix $P_{g_k}^{(l)}$ and the right permutation matrix $P_{g_k}^{(l)}$ are given by

$$\left(P_{g_k}^{(l)}\right)_{i\bar{j}} = \delta_{i,\bar{j}}, \ g_{\bar{j}} = g_i^{-1} g_k, \ i = 1, 2, \ldots, n,$$

$$\left(P_{g_k}^{(r)}\right)_{\bar{i}j} = \delta_{\bar{i},j}, \ g_{\bar{i}} = g_k g_j^{-1}, \ j = 1, 2, \ldots, n. \quad (4.53)$$

4.5. REGULAR REPRESENTATIONS OF FINITE GROUPS

These relations have the following interpretation in terms of the group multiplication table (4.47): To obtain the elements of $P_{g_k}^{(l)}$ keep the order $g_1, g_2, \ldots, g_n$ of the rows, but rearrange the elements of the columns to the order $g_{\bar{1}} = g_1^{-1} = g_1, g_{\bar{2}} = g_2^{-1}, \ldots, g_{\bar{n}} = g_n^{-1}$. The element g_k then stands in row i and column $\bar{j}$ defined by $g_{\bar{j}} = g_i^{-1} g_k$, for each $i = 1, 2, \ldots, n$. To obtain the elements of $P_{g_k}^{(r)}$ rearrange the elements of the rows to the order $g_{\bar{1}} = g_1^{-1} = g_1, g_{\bar{2}} = g_2^{-1}, \ldots, g_{\bar{n}} = g_n^{-1}$, but keep the order $g_1, g_2, \ldots, g_n$ of the columns. The element g_k then stands in row $\bar{i}$ defined by $g_{\bar{i}} = g_k g_j^{-1}$ and column j, for each $j = 1, 2, \ldots, n$.

The subsets of permutations of S_n that enter into the enumeration of the left and right regular permutation matrices with elements given by (4.53) are read off the permutation matrices themselves by using the rule $(P_\pi)_{i,j} = \delta_{i,\pi_j}$ for each permutations $\pi = (\pi_1, \pi_2, \ldots, \pi_n) \in S_n$ (see (1.72)). This rule applied to the left permutation matrix $P_{g_k}^{(l)}$ and the right permutation matrix $P_{g_k}^{(r)}$ in (4.53) gives the permutations corresponding to the group elements g_k, respectively:

$$g_k \mapsto \pi^{(k)} = \left(\pi_1^{(k)}, \pi_2^{(k)}, \ldots, \pi_n^{(k)}\right), \quad \pi_{\bar{j}}^{(k)} = i,$$

for $\bar{j}$ given by $g_{\bar{j}} = g_i^{-1} g_k$ for given k and i;

$$g_k \mapsto \pi^{(k)} = \left(\pi_1^{(k)}, \pi_2^{(k)}, \ldots, \pi_n^{(k)}\right), \quad \pi_{\bar{i}}^{(k)} = j, \tag{4.54}$$

for $\bar{i}$ given by $g_{\bar{i}} = g_k g_j^{-1}$ for given k and j.

The simplest way to implement this mapping rule is to follow those given by (4.53), which we now illustrate explicitly:

Modified group multiplication table for an abstract group:

	$g_{\bar{1}}$	$\cdots$	$g_{\bar{j}}$	$\cdots$	$g_{\bar{n}}$
g_1	$g_1 g_{\bar{1}}$	$\cdots$	$g_1 g_{\bar{j}}$	$\cdots$	$g_1 g_{\bar{n}}$
$\vdots$	$\vdots$		$\vdots$		$\vdots$
g_i	$g_i g_{\bar{1}}$	$\cdots$	$g_i g_{\bar{j}}$	$\cdots$	$g_i g_{\bar{n}}$
$\vdots$	$\vdots$		$\vdots$		$\vdots$
g_n	$g_n g_{\bar{1}}$	$\cdots$	$g_n g_{\bar{j}}$	$\cdots$	$g_n g_{\bar{n}}$

(4.55)

It is now the case that the element of the group in position $(i, \bar{j})$ is given by $g_i g_{\bar{j}} = g_k$, this holding for all places $i, \bar{j} \in \{1, 2, \ldots, n\}$. It is from this structure that one reads off the permutation $\pi^{(k)}$ corresponding to the group element g_k and the left permutation matrix in (4.54), each $k = 1, 2, \ldots, n$. A similar situation holds for the right permutation matrices, where it is now the rows from (4.47) that are relabeled by the $g_{\bar{i}}$.

It is easy to show that these permutation matrices satisfy the important "completeness" relations

$$\sum_{k=1}^{n} P^{(l)}_{\pi^{(k)}} = \sum_{k=1}^{n} P^{(r)}_{\pi^{(k)}} = J_n,$$

$$J_n P^{(l)}_{\pi^{(k)}} = P^{(l)}_{\pi^{(k)}} J_n = J_n; \quad J_n P^{(r)}_{\pi^{(k)}} = P^{(r)}_{\pi^{(k)}} J_n = J_n.$$

(4.56)

where J_n is the matrix of order n with entry 1 in each row and column (see the remark below (4.67)).

This construction in terms of the group multiplication table shows vividly how either the set of left or right regular representations by permutation matrices encode exactly the abstract group multiplication rules themselves. Thus, if one is given such a group $\mathbb{P}_n^{(l)}$ of left regular permutation matrices for some group G_n, but not the group itself, then row k of the multiplication table is given by $\left(g_{\pi_1^{(k)}}, g_{\pi_2^{(k)}}, \ldots, g_{\pi_n^{(k)}}\right)$; hence, all rows can be written out, and therefore the columns, from which the right regular permutation matrices can be read off. Similarly, starting from a known right regular representation of some group G_n, the group multiplication table is constructed from its columns $\left(g_{\rho_1^{(k)}}, g_{\rho_2^{(k)}}, \ldots, g_{\rho_n^{(k)}}\right), k = 1, 2, \ldots, n$, followed by the construction of the left regular representation.

This raises the question as to when a subgroup of permutation matrices $\mathbb{P}_n \subset \mathbb{P}_{S_n}$ of order n, selected from the subset of all $n!$ permutation matrices of order n of the full symmetric group, constitutes a regular matrix representation of some finite group G_n of order n. The answer is quite simple:

A subgroup $\mathbb{P}_n \subset \mathbb{P}_{S_n}$ of order n of the full permutation group is a regular representation of some group G_n if and only if the first relation (4.54) holds, in which case this subgroup of permutation matrices can be taken to be the left regular representation of the group G_n, and the accompanying right regular representation of G_n can then constructed from the second relation in (4.54).

Proof. Since the first relation (4.54) holds, we can choose, without qualifying the proof, the permutation matrix $P_{\pi^{(k)}} \in \mathbb{P}_n$ to be the unique permutation having first part $\pi_1^{(k)} = k$. (This choice only determines the ordering of the rows in the multiplication table and not its abstract multiplication properties.) Then, row k of the multiplication table of G_n

4.5. REGULAR REPRESENTATIONS OF FINITE GROUPS

is given by $\left(k, g_{\pi_2^{(k)}}, \ldots, g_{\pi_n^{(k)}}\right)$, each $k = 1, 2, \ldots, n$. No two entries in any given column are the same, since this would violate condition (4.54). Thus, we obtain the group multiplication table of some group G_n, and write out its right regular representation by permutation matrices by use of the second relation in (4.54). □

We can go much further. The following three results are classical for a (left or right) regular representation $P_{g_k}, g_k \in G_n$, of a finite group G_n over the field of complex numbers: (i) $\mathrm{Tr} P_{g_k} = n\delta_{1,k}$; (ii) the n matrices in the regular representation are linearly independent; and (iii) there exists a unitary similarity transformation that completely reduces the regular representation to a direct sum of all of the irreducible representations of G_n, where each such irreducible representation is repeated a number of times equal to its dimension (order of the irreducible matrices), which is also the number of classes. The existence of two commuting regular representations — left and right, as discussed above — has rather profound consequences for the determination of the irreducible representations of every finite group. We do not pursue this further here. We note, however, that these results carry to its natural conclusion the well-known result that every finite group G_n is isomorphic to a subgroup of order n of the set of permutations S_n. Our purpose here is simply to place in evidence the role of permutation matrices, since they enter in a great variety of contexts in this volume. An added incentive is to place the determination of all groups of permutation matrices of order n that qualify as representations of finite groups in the context of the combinatorial theory of $(0-1)$–matrices as developed in Brualdi and Ryser [14].

We emphasize again that the commuting left and right regular permutation matrices representing a group G_n are matrices of order equal to that of the group. For example, if the group is S_n, where the order is $n!$, and, by convention, the subscript n in S_n is the length of the permutations, not the order of the group, the left and right regular permutation matrices are of order $n!$. For example, for $n = 3$, the left and right regular representation matrices of S_3 are of order 6, and every group of order six is characterized, up to isomorphism (ordering of group elements), by its pair of left and right regular permutation matrices of order 6.

Summary. We summarize the principal results above on left and right permutation matrices of any abstract group $G_n = \{g_1, g_2, \cdots, g_n\}$ of order n. First, define the two subsets of permutations in S_n in terms of the elements of G_n as follows:

$$\Sigma_n^{(l)} = \left\{ \pi^{(k)} = (\pi_1^{(k)}, \pi_2^{(k)}, \ldots, \pi_n^{(k)}) \;\middle|\; \begin{array}{l} \pi_{\bar{j}}^{(k)} = i, \text{all } \bar{j}, k, i \in \{1, \ldots, n\} \\ \text{such that } g_{\bar{j}} = g_i^{-1} g_k \end{array} \right\} \quad (4.57)$$

$$\Sigma_n^{(r)} = \left\{ \pi^{(k)} = (\pi_1^{(k)}, \pi_2^{(k)}, \ldots, \pi_n^{(k)}) \;\middle|\; \begin{array}{l} \pi_{\bar{i}}^{(k)} = j, \text{all } \bar{i}, k, j \in \{1, \ldots, n\} \\ \text{such that } g_{\bar{i}} = g_k g_j^{-1} \end{array} \right\}.$$

Then, each permutation in one set commutes with each permutation in the other set. Second, define two sets of corresponding permutation matrices as follows:

$$\mathbb{P}_{\Sigma_n^{(l)}} = \left\{ P_{\pi^{(k)}} \,\middle|\, \pi^{(k)} \in \Sigma_n^{(l)} \right\}, \quad \mathbb{P}_{\Sigma_n^{(r)}} = \left\{ P_{\pi^{(k)}} \,\middle|\, \pi^{(k)} \in \Sigma_n^{(r)} \right\}. \qquad (4.58)$$

Then, each permutation matrix in one set commutes with each permutation matrix in the other set. Indeed, the permutations and the permutations matrices obey identical multiplication rules. But these sets are not unique because suitable sets are also obtained by similarity transformations:

$$\pi \, \Sigma_n^{(l)} \, \pi^{-1}, \; \pi \, \Sigma_n^{(r)} \, \pi^{-1}, \; P_\pi \, \Sigma_n^{(l)} \, P_{\pi^{-1}}, \; P_\pi \, \Sigma_n^{(r)} \, P_{\pi^{-1}}. \qquad (4.59)$$

where $\pi \in S_n$ is an arbitrary permutation. □

We conclude this chapter by giving the example of the above results for groups of order four:

Examples.

$$\begin{array}{c|cccc}
\text{cyclic group } C_4 & & & & \\
\hline
 & g_1 & g_2 & g_3 & g_4 \\
\hline
g_1 & g_1 & g_2 & g_3 & g_4 \\
g_2 & g_2 & g_3 & g_4 & g_1 \\
g_3 & g_3 & g_4 & g_1 & g_2 \\
g_4 & g_4 & g_1 & g_2 & g_3
\end{array}
\qquad
\begin{array}{c|cccc}
\text{Klein's four-group} & & & & \\
\hline
 & g_1 & g_2 & g_3 & g_4 \\
\hline
g_1 & g_1 & g_2 & g_3 & g_4 \\
g_2 & g_2 & g_1 & g_4 & g_3 \\
g_3 & g_3 & g_4 & g_1 & g_2 \\
g_4 & g_4 & g_3 & g_2 & g_1
\end{array}
\qquad (4.60)$$

Both of these groups are abelian; hence, the group tables are symmetric in their rows and columns, and the left and right regular permutation matrices coincide. We accordingly drop the superscript (l) and (r) from the notation. The groups of permutation matrices are easily read off the multiplication tables:

Cyclic group C_4:

$$P_{1,2,3,4} = \begin{pmatrix} 1 & 0 & 0 & 0 \\ 0 & 1 & 0 & 0 \\ 0 & 0 & 1 & 0 \\ 0 & 0 & 0 & 1 \end{pmatrix}, \; P_{2,3,4,1} = \begin{pmatrix} 0 & 0 & 0 & 1 \\ 1 & 0 & 0 & 0 \\ 0 & 1 & 0 & 0 \\ 0 & 0 & 1 & 0 \end{pmatrix},$$

$$P_{3,4,1,2} = \begin{pmatrix} 0 & 0 & 1 & 0 \\ 0 & 0 & 0 & 1 \\ 1 & 0 & 0 & 0 \\ 0 & 1 & 0 & 0 \end{pmatrix}, \; P_{4,1,2,3} = \begin{pmatrix} 0 & 1 & 0 & 0 \\ 0 & 0 & 1 & 0 \\ 0 & 0 & 0 & 1 \\ 1 & 0 & 0 & 0 \end{pmatrix}.$$

$$(4.61)$$

4.5. REGULAR REPRESENTATIONS OF FINITE GROUPS

Klein's four-group:

$$P_{1,2,3,4} = \begin{pmatrix} 1 & 0 & 0 & 0 \\ 0 & 1 & 0 & 0 \\ 0 & 0 & 1 & 0 \\ 0 & 0 & 0 & 1 \end{pmatrix}, \quad P_{2,1,4,3} = \begin{pmatrix} 0 & 1 & 0 & 0 \\ 1 & 0 & 0 & 0 \\ 0 & 0 & 0 & 1 \\ 0 & 0 & 1 & 0 \end{pmatrix},$$

(4.62)

$$P_{3,4,1,2} = \begin{pmatrix} 0 & 0 & 1 & 0 \\ 0 & 0 & 0 & 1 \\ 1 & 0 & 0 & 0 \\ 0 & 1 & 0 & 0 \end{pmatrix}, \quad P_{4,3,2,1} = \begin{pmatrix} 0 & 0 & 0 & 1 \\ 0 & 0 & 1 & 0 \\ 0 & 1 & 0 & 0 \\ 1 & 0 & 0 & 0 \end{pmatrix}.$$

First, we consider the cyclic group C_4. The permutation matrix $P_{g_2} = P_{2,3,4,1}$ generates the full set of permutations matrices; that is, $P_{g_k} = \left(P_{g_2}\right)^{k-1}$, $k = 1, 2, 3, 4$, and the group element g_2 is of order four; that is, $g_2^4 = g_1$, which implies that the characteristic roots of P_{g_2} are the fourth roots of unity. Thus, to find the unitary matrix that diagonalizes all four of the commuting, real orthogonal permutation matrices, it is sufficient to diagonalize P_{g_2}. By elementary means, we find that the unitary matrix given by

$$U = \frac{1}{2} \begin{pmatrix} 1 & 1 & 1 & 1 \\ 1 & -i & -1 & i \\ 1 & -1 & 1 & i \\ 1 & i & -1 & -i \end{pmatrix}$$

(4.63)

effects the desired transformation to diagonal form given by

$$U^\dagger L_{g_2} U = \begin{pmatrix} 1 & 0 & 0 & 0 \\ 0 & i & 0 & 0 \\ 0 & 0 & -1 & 0 \\ 0 & 0 & 0 & -i \end{pmatrix} = \begin{pmatrix} 1 & 0 & 0 & 0 \\ 0 & \omega_1 & 0 & 0 \\ 0 & 0 & \omega_2 & 0 \\ 0 & 0 & 0 & \omega_3 \end{pmatrix},$$

(4.64)

where $\omega_k = e^{i\pi(k-1)/2}$, $k = 1, 2, 3, 4$, are the fourth roots of unity. By taking the second and third powers of this relation, we obtain the four one-dimensional representations of the abelian group C_4, which is the same as its character table, and can be displayed just as for the multiplication table, since each element is in a class by itself:

$$\chi(C_4) = \begin{array}{c|cccc} & g_1 & g_2 & g_3 & g_4 \\ \hline g_1 & 1 & 1 & 1 & 1 \\ g_2 & 1 & \omega_2 & \omega_3 & \omega_4 \\ g_3 & 1 & \omega_2^2 & \omega_3^2 & \omega_4^2 \\ g_4 & 1 & \omega_2^3 & \omega_3^3 & \omega_4^3 \end{array} \qquad (4.65)$$

The determination of the irreducible representations of the Klein four group proceeds similarly. The group elements g_2, g_3, g_4 are all of order two, and therefore each of the three symmetric permutation matrices $P_{g_2} = P_{2,1,4,3}, P_{g_3} = P_{3,4,1,3}, P_{g_4} = P_{4,3,2,1}$ has eigenvalues ± 1, and there exists a unique (up to signs of its columns) real orthogonal matrix R that effects their simultaneous diagonalization. Elementary calculations quickly give the following results:

$$\begin{aligned} R^{tr} P_{g_1} R &= \mathrm{diag}(1,1,1,1), & R^{tr} P_{g_2} R &= \mathrm{diag}(1,1,-1,-1), \\ R^{tr} P_{g_3} R &= \mathrm{diag}(1,-1,1,-1), & R^{tr} P_{g_4} R &= \mathrm{diag}(1,-1,-1,1), \end{aligned}$$
(4.66)

$$R = \frac{1}{2} \begin{pmatrix} 1 & 1 & 1 & 1 \\ 1 & 1 & -1 & -1 \\ 1 & -1 & 1 & -1 \\ 1 & -1 & -1 & 1 \end{pmatrix},$$

$$\chi = \begin{array}{c|cccc} & g_1 & g_2 & g_3 & g_4 \\ \hline g_1 & 1 & 1 & 1 & 1 \\ g_2 & 1 & 1 & -1 & -1 \\ g_3 & 1 & -1 & 1 & -1 \\ g_4 & 1 & -1 & -1 & 1 \end{array} \qquad (4.67)$$

The Jucys-Murphy transpositional permutation matrices have no role in this construction, since there are no transposition matrices in the regular representations, a result that is true for general n (see (4.56)). The above case for groups of order four, for which both possible groups are abelian, does not exhibit the full structure of the two commuting regular representations, the left and right representations by permutation matrices. It is the commuting property of these two sets of real orthogonal matrices that allows for the complete reduction for general finite groups, in which the property (4.56) has a central role; it assures, for example, that the vector $(1, 1, \ldots, 1)/\sqrt{n}$ is always a simultaneous eigenvector; that is, the identity representation is always present.

Chapter 5

Doubly Stochastic Matrices in Angular Momentum Theory

5.1 Introduction

A doubly stochastic matrix A of order n is, by definition, a matrix of nonnegative elements such that the elements in each row and column sum to unity. The occurrence of doubly stochastic matrices in conventional quantum mechanics is pervasive, since there is a doubly stochastic matrix $S(U)$ of order n associated with every unitary matrix U of order n as defined by its elements in row i and column j by $S_{ij}(U) = |U_{ij}|^2$, $i, j = 1, 2, \ldots, n$. But this general observation is too generic to lead to detailed consequences.

Our purpose here is to identify classes of doubly stochastic matrices associated with the binary coupling theory of angular momenta and the complete sets of commuting Hermitian observables that characterize such composite angular momentum systems. Thus, for the most part, the focus is on very special sets of unitary matrices that give doubly stochastic matrices. The key property we use is the notion of a complete set of mutually commuting Hermitian observables, indeed, of several such sets whose simultaneous eigenvectors define different orthonormal bases of the same finite-dimensional Hilbert space. We define these concepts more carefully below, but note at the outset that we are then led naturally to general sets of doubly stochastic matrices that we denote by QA_n, as defined below in (5.4); their definition includes the double stochastic matrices that occur in the binary coupling theory of angular momenta, and leads, in particular, to a probabilistic interpretation of all $3(n-1) - j$ coefficients.

Motivated by the applications mentioned above and by the work of Alfred Landé [39], we introduce the following class of doubly stochastic matrices. Let $U(n)$ denote the set of all unitary matrices of order n, and define the unitary matrix $W(U,V)$ by

$$W(U,V) = U^\dagger V, \text{ each pair } U, V \in U(n). \tag{5.1}$$

This definition then has the property $W(U,U) = W(V,V) = I_n$, which is very useful for our applications. Then, the matrix $S(U,V)$ with element in row i and column j given by

$$S_{ij}(U,V) = |W_{ij}(U,V)|^2 \tag{5.2}$$

is doubly stochastic; that is,

$$S(U,V)J = JS(U,V) = J, \text{ each pair } (U,V) \in U(n), \tag{5.3}$$

where $J = J_n$ is the matrix of order n with element 1 in each row and column. We define the set QA_n of doubly stochastic matrices by

$$\mathrm{QA}_n = \{U^\dagger V \mid U, V \in U(n)\}. \tag{5.4}$$

Choosing $U = I_n$ in (5.1) and (5.2) shows that the unitary group $U(n)$ itself defines a set of doubly stochastic matrices, as noted above; we can also choose $V = I_n$. The simultaneous replacements $U \mapsto U^\dagger$ and $V \mapsto V^\dagger$ gives $W(U^\dagger, V^\dagger) = UV^\dagger$, which preserves the property $W(U,U) = W(V,V) = I_n$; that is, $W(U^\dagger, U^\dagger) = W(V^\dagger, V^\dagger) = I_n$. Further properties are noted below.

Doubly stochastic matrices of the type QA_n arise in a reasonable general setting of conventional quantum mechanics as to justify their development. A key role is played by complete sets of commuting Hermitian observables, especially for physical systems possessing properties that can be described or modeled by a separable Hilbert spaces $\mathcal{H}$. We can then in many cases focus attention on the finite-dimensional inner product spaces whose direct sum constitutes the full separable Hilbert space $\mathcal{H}$. Even if such a separable Hilbert space for the whole physical system cannot be identified, it can still be very useful to identify finite-dimensional inner product spaces that give information about special features of the given physical system. (We refer to both finite vector spaces and denumerably infinite separable vector spaces with an inner product as Hilbert spaces.) The action of some symmetry group of the physical system is often the setting that provides such finite-dimensional Hilbert spaces as the carrier spaces of unitary irreducible representations.

The bound states of a quantum-mechanical system can often be characterized by a set of mutually commuting Hermitian operators, say, a set $\mathbb{H}_N^{op} = \{H_1^{op}, H_2^{op}, \ldots, H_N^{op}\}$. We adapt a famous definition from Dirac [24] on complete sets of such observables to our present needs:

5.1. INTRODUCTION

Let $\mathcal{H}_n$ be a finite-dimensional Hilbert space of dimension $\dim \mathcal{H}_n$. Let $\mathbb{H}_N^{op} = \{H_1^{op}, H_2^{op}, \ldots, H_N^{op}\}$ denote a set of mutually commuting Hermitian operators on the space $\mathcal{H}_n$ whose action on the space $\mathcal{H}_n$ maps the space into itself. Then, the set $\mathbb{H}_N^{op}$ of mutually commuting Hermitian operators is complete with respect to the Hilbert space $\mathcal{H}_n$ if and only if the set of normalized simultaneous eigenvectors of the N operators in the set $\mathbb{H}_N^{op}$ is an orthonormal basis of $\mathcal{H}_n$.

The importance of complete sets of mutually commuting Hermitian operators with respect to a Hilbert space on which their action is defined is to give a unique orthonormal basis of the space itself, up to arbitrary multiplicative phase factors of unit modulus. While this restriction to finite-dimensional Hilbert spaces may seem overly severe, we will show that it still allows a very rich structure of doubly stochastic matrices to be identified in quantum theory, especially in the context of the binary coupling theory of composite angular momentum systems.

Unitary matrices now makes their natural appearance, as follows: We have the following system of eigenvalue relations on each such Hilbert space $\mathcal{H}_n$ described above:

$$H_m^{op}|\mathbf{h}\rangle = h_m|\mathbf{h}\rangle, \text{ each } H_m^{op} \in \mathbb{H}_N^{op}, \qquad (5.5)$$

where $\mathbf{h}$ denotes the sequence $\mathbf{h} = (h_1, h_2, \ldots, h_N)$ of (necessarily real) eigenvalues of the complete set $\mathbb{H}_N^{op}$ of Hermitian operators. Each eigenvalue h_m itself belongs to some real domain of definition A_m, which is a finite set that depends on the problem at hand. The domains of definition in the sequence $\mathbb{A}_N = (A_1, A_2, \ldots, A_N)$ can relate to one another in rather intricate ways. We write $\mathbf{h} \in \mathbb{A}_N$ to denote that each $h_m \in A_m, m = 1, 2, \ldots, N$. An essential requirement is that the set of allowed values that each $h_m, m = 1, 2, \ldots, N$ can assume must enumerate exactly $\dim \mathcal{H}_n$ orthonormal vectors so as to give an orthonormal basis of the space $\mathcal{H}_n$. The orthonormality comes from the fact that all the operators are Hermitian and mutually commute; hence, there exists a unitary transformation of the space $\mathcal{H}_n$ that defines the basis on which all the operators have the diagonal action given by (5.5). Completeness assures that the normalized simultaneous eigenvectors are unique, up to arbitrary multiplicative complex factors of unit modulus and ordering.

It is useful to formulate all of the above in terms of commuting Hermitian matrices and the unitary matrix that effects their diagonalization. We have assumed the existence of a finite-dimensional Hilbert space such that the action of each $H_m^{op} \in \mathbb{H}_N^{op}$ maps the vector space into itself. Each such Hilbert space also possesses an orthonormal basis, say, a set of vectors denoted in the Dirac ket notation and defined by

$$\mathcal{H}_n - basis = \{|k\rangle \,|\, k = 1, 2, \ldots, d_n\}, \; d_n = \dim \mathcal{H}_n. \qquad (5.6)$$

The space $\mathcal{H}_n$ is also a linear vector space with an inner product, say,

⟨ | ⟩; hence, we have the orthonormality of basis vectors expressed by

$$\langle k \,|\, k' \rangle = \delta_{k,k'}, \text{ all pairs } k, k' = 1, 2, \ldots, d_n. \tag{5.7}$$

The choice of the orthonormal $\mathcal{H}_n$–basis (5.6) is fully arbitrary.

The action of each $H_m^{op} \in \mathbb{H}_N^{op}$ is then given by

$$H_m^{op} |k'\rangle = \sum_{k=1}^{d_n} (H_m)_{k,k'} |k\rangle, \tag{5.8}$$

where the matrix H_m is a Hermitian matrix of order $d_n = \dim\mathcal{H}_n$ with elements in row k and column k' given by

$$(H_m)_{k,k'} = \langle k \,|H_m^{op}|\, k' \rangle. \tag{5.9}$$

The Hermitian matrix H_m is called the *representation* of H_m^{op} on the orthonormal $\mathcal{H}_n$–basis.

It is well-known (see, for example, Perlis [60]), that there exists a unitary matrix U of order n that simultaneously diagonalizes a set of mutually commuting Hermitian matrices $\mathbb{H}_N = (H_1, H_2, \ldots, H_N)$; that is, there exists a unitary matrix U such that

$$U^\dagger H_m U = D_m, \; m = 1, 2, \ldots, N. \tag{5.10}$$

This relation can also be written in terms of the d_n columns $U_\mathbf{h}$, $\mathbf{h} \in \mathbb{A}_n$ as

$$H_m U_\mathbf{h} = d_{m,\mathbf{h}} U_\mathbf{h}, \; m = 1, 2, \ldots, N, \tag{5.11}$$

where $d_{m,\mathbf{h}} = \left(D_m\right)_{\mathbf{h},\mathbf{h}}$ is the eigenvalue in column $\mathbf{h}$ = row $\mathbf{h}$ of the diagonal matrix D_m. It is the columns of the unitary matrix U that constitute the eigenvectors, and it is the custom in physics to enumerate the eigenvectors by the eigenvalues themselves.

The whole point of the completeness condition of the set of mutually commuting Hermitian matrices $\mathbb{H}_N$ is that then the unitary matrix U that effects the simultaneous diagonalization has the following properties:

The matrix U in relation (5.10) is unique up to the choice of complex factors of unit modulus that multiply each of its columns and to the ordering of its columns; the ordering of columns fixes the order in which the eigenvalues $d_{m,\mathbf{h}}$ are entered along the diagonal in each of the diagonal matrices D_m in (5.10).

An alternative way of describing the non-uniqueness of U is: If the matrix U diagonalizes a complete set of mutually commuting Hermitian matrices, then so does the unitary matrix W defined by

$$W = U \, \text{diag}\left(e^{i\phi_1}, e^{i\phi_2}, \ldots, e^{i\phi_{d_n}}\right) P_\pi, \; \pi \in S_{d_n}, \; d_n = \dim\mathcal{H}_n, \tag{5.12}$$

5.1. INTRODUCTION

where P_π is an arbitrary permutation matrix of the full symmetric group S_{d_n}. This non-uniqueness can be avoided only by phase factor and ordering **conventions**.

The unitary matrix U that brings all of the commuting Hermitian matrices to diagonal form gives the transformation of the selected orthonormal $\mathcal{H}_n$−basis to the new orthonormal H_n−basis in which the Hermitian matrices are all diagonal:

$$H_n - basis = \{|\mathbf{h}\rangle \mid \mathbf{h} \in \mathbb{A}_N\}. \tag{5.13}$$

Thus, we have the invertible basis relations:

$$|\mathbf{h}\rangle = \sum_{k=1}^{d_n} U_{k,\mathbf{h}} |k\rangle, \text{ each } \mathbf{h} \in \mathbb{A}_N, |\mathbb{A}_N| = d_n,$$

$$|k\rangle = \sum_{\mathbf{h}\in\mathbb{A}_N} U_{k,\mathbf{h}}^* |\mathbf{h}\rangle, \text{ each } k = 1, 2, \ldots, d_n. \tag{5.14}$$

The basis transformation given by the first of relations (5.14) is the matrix element form of the unitary matrix similarity transformation (5.10):

$$\sum_{k',k=1}^{d_n} U_{\mathbf{h}',k'}^\dagger \langle k'|H_m|k\rangle U_{k,\mathbf{h}} = \delta_{\mathbf{h}',\mathbf{h}} \left(D_m\right)_{\mathbf{h}',\mathbf{h}}, \text{ all } \mathbf{h}', \mathbf{h} \in \mathbb{A}_N. \tag{5.15}$$

It is typical in physical applications that the enumeration of different orthonormal sets of basis vectors of one and the same vector space requires the use of distinct indexing sets, which often indicate their different physical origins. This phenomenon is illustrated by the $\mathcal{H}_n$−basis and the H_n−basis, where the physics comes from the H_n−basis and the context of the application. Caution must be exercised or invalid relations can result from misplacement of row and column indices.

In summary, we conclude from the above results:

To each complete set of mutually commuting Hermitian operators with respect to a finite-dimensional Hilbert space $\mathcal{H}_n$ of dimension $dim\mathcal{H}_n$ there corresponds a unitary matrix $U_{k,\mathbf{h}}$ with rows and columns enumerated by $k = 1, 2, \ldots, dim\mathcal{H}_n$ and $\mathbf{h} \in \mathbb{A}_N$. The matrix U of order $dim\mathcal{H}_n$ is unique up to multiplication of its columns by arbitrary complex numbers of modulus 1 and arbitrary permutations of its columns. The matrix $S(U)$ with elements defined by

$$S_{k,\mathbf{h}}(U) = |U_{k,\mathbf{h}}|^2 \tag{5.16}$$

is a doubly stochastic matrix of order $dim\mathcal{H}_n$.

We point out that the above results for mutually commuting Hermitian matrices generalize to normal matrices, where a normal matrix is one that commutes with its Hermitian conjugate, but we make no use of this result, and refer to Perlis [60] for this generalization.

We next adapt the above results to include a second complete set $\mathbb{K}_N^{op} = \{K_1^{op}, K_2^{op}, \ldots, K_N^{op}\}$ of mutually commuting Hermitian operators on the same space $\mathcal{H}_n$. Then, all the results given above apply directly simply by making the appropriate changes in notations, listed as follows:

Simultaneous eigenvalue equations:
$$K_m^{op} |\mathbf{k}\rangle = k_m |\mathbf{k}\rangle, \text{ each } K_m^{op} \in \mathbb{K}_N^{op}. \tag{5.17}$$

Domain of definition of eigenvalues:
$$\mathbf{k} = (k_1, k_2, \ldots, k_N), \text{ each } k_m \in B_m, \tag{5.18}$$
$$\mathbf{k} \in \mathbf{B}_N = (B_1, B_2, \ldots, B_N).$$

Operator action in the $\mathcal{H}_n$–basis:
$$K_m^{op} |k'\rangle = \sum_{k=1}^{d_n} (K_m)_{k,k'} |k\rangle; \quad \left(K_m\right)_{k,k'} = \langle k| K_m^{op} |k'\rangle. \tag{5.19}$$

Simultaneous diagonalization of set of Hermitian matrices
$\mathbb{K}_N = (K_1, K_2, \ldots, K_N)$ by unitary matrix V:
$$V^\dagger K_m V = D'_m, \ m = 1, 2, \ldots, N,$$
$$K_m V_\mathbf{k} = d'_{m,\mathbf{k}} V_\mathbf{k}, \ m = 1, 2, \ldots, N, \tag{5.20}$$
$$d'_{m,\mathbf{k}} = \left(D'_m\right)_{\mathbf{k},\mathbf{k}}.$$

Orthonormal basis:
$$K_n - basis = \{|\mathbf{k}\rangle \mid \mathbf{k} \in \mathbb{B}_N\}. \tag{5.21}$$

Invertible basis relations:
$$|\mathbf{k}\rangle = \sum_{k=1}^{d_n} V_{k,\mathbf{k}} |k\rangle, \text{ each } \mathbf{k} \in \mathbb{B}_N, |\mathbb{B}_N| = d_n,$$
$$\tag{5.22}$$
$$|k\rangle = \sum_{\mathbf{k} \in \mathbb{B}_N} V^*_{k,\mathbf{k}} |\mathbf{h}\rangle, \text{ each } k = 1, 2, \ldots, d_n.$$

We use the same complete basis $\{|k\rangle \mid k = 1, 2, \ldots, d_n\}$ of the space $\mathcal{H}_n$ given by (5.6) in formulating the above H_n–basis and K_n–basis

5.1. INTRODUCTION

representation of the complete set of Hermitian commuting matrices $\mathbb{H}_N$ and $\mathbb{K}_N$, respectively.

The H_n-basis and K_n-basis are orthonormal basis of the same vector space $\mathcal{H}_n$ of dimension $d_n = \dim \mathcal{H}_n$; hence, the two bases are related by a unitary transformation $W(U,V)$, as defined by

$$|\mathbf{k}\rangle = \sum_{\mathbf{h} \in \mathbb{A}_n} W_{\mathbf{h},\mathbf{k}}(U,V)|\mathbf{h}\rangle, \quad \langle \mathbf{h} | \mathbf{k}\rangle = W_{\mathbf{h},\mathbf{k}}(U,V). \qquad (5.23)$$

The unitary transformations (5.14) and (5.22) now give the elements of the matrix $W(U,V) = \langle \mathbf{h} | \mathbf{k} \rangle$ of order d_n with rows enumerated by $\mathbf{h} \in \mathbb{A}_N$ and columns enumerated by $\mathbf{k} \in \mathbb{B}_N$ as

$$W(U,V) = U^\dagger V. \qquad (5.24)$$

The elements of $W(U,V)$ are also expressed in terms of the column vectors $U_\mathbf{h}$ of H_m in (5.11) and the column vectors $V_\mathbf{k}$ of H_k in (5.20) by the row on column relationship

$$W_{\mathbf{h},\mathbf{k}}(U,V) = U_\mathbf{h}^\dagger V_\mathbf{k}. \qquad (5.25)$$

Thus, we obtain the double stochastic matrix $S(U,V)$ with elements given by

$$S_{\mathbf{h},\mathbf{k}}(U,V) = |U_\mathbf{h}^\dagger V_\mathbf{k}|^2. \qquad (5.26)$$

It is also useful to retain the factorization rule (5.24) for transition probability amplitudes directly in terms of the inner products of orthonormal basis vectors that define the transition probability amplitudes, $U_{k,\mathbf{h}} = \langle k | \mathbf{h}\rangle$ from (5.14), $\langle k | \mathbf{k}\rangle$ from (5.22), and $\langle \mathbf{h} | \mathbf{k}\rangle = W_{\mathbf{h},\mathbf{k}}(U,V)$ from (5.23):

$$\langle \mathbf{h} | \mathbf{k}\rangle = \sum_{k=1}^{d_n} \langle \mathbf{h} | k\rangle \langle k | \mathbf{k}\rangle. \qquad (5.27)$$

This relation is just the expression of the completeness of the orthonormal basis $\{|k\rangle \,|\, k = 1, 2, \ldots, d_n\}$ of the originally selected space $\mathcal{H}_n$, which we did not in our presentation above associate with still another complete set of mutually commuting Hermitian operators, although in the application to the theory of binary coupling of angular momenta this is the case (see also Chapters 1 and 8) — the so-called uncoupled basis.

We point out that the terminology "recoupling matrix," used frequently in [L], and also in this volume, in the context of the pairwise coupling theory of angular momenta arises quite naturally from the above construction of two new orthonormal bases of one and the same given vector space $\mathcal{H}_n$; namely, the H_n-basis and the K_n-basis, which can be called coupled bases. The transformation coefficients between the two coupled bases are then aptly called recoupling coefficients, and the matrix of the transformation a recoupling matrix.

It is also worth noting that the factored form $W(U,V) = U^\dagger V$ of the transition probability amplitudes has the effect of transforming the problem of solving the nonlinear quantum equations of motion for the transition probability amplitudes into the linear problem of diagonalizing various complete sets of commuting Hermitian matrices that correspond to complete sets of commuting Hermitian observables of a physical system.

Remarks.

(1). From the general viewpoint of an inner product vector space $\mathcal{V}$ over the complex numbers, the above results are quite trivial, and the reason for introducing different bases of one and the same vector space can be asked. The general answer, although necessarily somewhat vague, is that in physical problems one often seeks a sub-basis of the general state space of the quantum mechanical system in question such that the behavior of the system in the energy region of interest can be modeled in a reasonably accurate manner — this often requires introducing different orthonormal bases of one and the same subspace.

(2). It is the Dirac notion of complete sets of commuting Hermitian observables that brings physical content to a model — it is for this reason that the eigenvalues of such complete sets are used in enumerating various vector spaces and their bases. Also, one has the added significance of the unitary matrix $W(U,V)$ in the sense of von Neumann's density matrix interpretation: The absolute value squared of the elements of this matrix give the transition probability for a transition from the prepared state $|\mathbf{h}\rangle$ to the measured state $|\mathbf{k}\rangle$. We always follow the practice of calling the state vector $|\mathbf{h}\rangle$ associated with the transformation U the prepared state and that of the state vector $|\mathbf{k}\rangle$ associated with the transformation V the measured state. Moreover, the general structure exhibited above and leading to the transition amplitude probability $W(U,V) = U^\dagger V$ is exactly what is needed for the probabilistic interpretation of the recoupling coefficients that arise in the binary coupling of the angular momenta associated with complex composite systems.

(3). We remark further that "prepared state" means that the given quantum system has by some means been placed or is known to be in the quantum state described by all the eigenvalues corresponding to the complete set of Hermitian operators $\mathbb{H}_N$, and "measured state" means the system has been measured to be in the quantum state described by all the eigenvalues corresponding to the complete set of Hermitian operators $\mathbb{K}_N$. Most significantly, between the determination of the prepared state and the measured state, the system must remained in complete isolation; that is, there must be no intermediate measurements of any kind or any other disruptions of its deterministic time evolution. It is then the case that the transition probability matrix $W(U,V)$ is constant in time between the time t_1 of the prepared state and the time t_2 of the measured state: $W(U,V) = W(U_t, V_t)$, $t_1 \leq t \leq t_2$. Here it is assumed that the time evolution is governed by a time-independent Hamiltonian operator H^{op} that is modeled by a Hermitian matrix H of order n on a subspace $\mathcal{H}_n$ of the system of interest, and, moreover, that H *is a common member of, or fully determined by, the Hermitian operators in each of the complete commuting sets* $\mathbb{H}_n$ *and* $\mathbb{K}_n$. In this case, the transition probability amplitude $W(U,V)$ is constant in time, as is also the transition probability $S(U,V)$ itself; that is, the constant-in-time relations

$$W(U_t, V_t) = W(U,V), \quad S(U_t, V_t) = S(U,V), \text{ all } t_1 \leq t \leq t_2 \quad (5.28)$$

hold for all times t for which the deterministic time evolution relations of quantum mechanics hold.

We fill in the steps of Item 3. Under the stated conditions, the time evolution U_t of the unitary matrix U and the time evolution V_t of the unitary matrix V are governed by the relations:

$$i\hbar \frac{\partial U_t}{\partial t} = HU_t, \quad i\hbar \frac{\partial V_t}{\partial t} = HV_t, \, t_1 \leq t \leq t_2. \quad (5.29)$$

Here we are regarding each column of the unitary matrix U_t (resp., V_t) as a state vector in the sense of the matrix relations (5.11), resp., (5.20), so that H is a Hermitian matrix realization of the Hamiltonian H^{op} on the space $\mathcal{H}_n$. Then, we have that

$$U_t = e^{-i(t-t')H/\hbar} U_{t'}, \; V_t = e^{-i(t-t')H/\hbar} V_{t'}, t \geq t', \quad (5.30)$$

which gives

$$U_t^\dagger V_t = (U_{t'}^\dagger e^{i(t-t')H/\hbar})(e^{-i(t-t')H/\hbar} V_{t'}) = U_{t'}^\dagger V_{t'}. \quad (5.31)$$

To the extent that relations (5.29) can be maintained, there are no constraints on the time between a prepared state and a measured state.

5.2 Abstractions and Interpretations

Our presentation of doubly stochastic matrices above is based on the writings of Alfred Landé, as acknowledged in the Preface. We have adapted his results to finite doubly stochastic matrices as needed for application to the binary coupling theory of angular momenta of composite systems. The set of unitary matrices of the Landé form (see (1.84))

$$W(U,V) = U^\dagger V, \quad (5.32)$$

and the associated doubly stochastic matrix $S(U,V)$ with elements in row i and column j defined by

$$S_{ij}(U,V) = |W_{ij}(U,V)|^2, \qquad (5.33)$$

are, of course, well-defined for all unitary matrices $U, V \in U(n)$. It is useful to summarize their basic properties and state their physical significance, as enumerated by Landé [39].

The first three properties are stated directly in terms of the doubly stochastic matrices $S(U,V)$ themselves, and are easily verified:

$$\begin{array}{rll} (i) & S(U,U) & = I_n, \\ (ii) & S(U,V) & = S^T(V,U), \\ (iii) & S(U,V) & = S(U^*,V^*). \end{array} \qquad (5.34)$$

The transition probability amplitude matrix $W(U,V) = U^\dagger V$ itself has the multiplication property:

$$(iv) \quad W(U,V) = W(U,V')W(V',V). \qquad (5.35)$$

Relations (i)-(iii) hold for all pairs $U, V \in U(n)$, and relation (iv) for all triples $U, V, V' \in U(n)$. Relation (iv) is the basic multiplication rule that applies to all transition probability amplitude matrices of the product form (5.32). *Its validity does not enter into the verification of properties (i)-(iii) of the doubly stochastic matrices themselves.* It is an additional property, as discussed further below.

Each of the properties (i)-(iii) has a direct physical interpretation, as described briefly by the following statements:

(i). If a physical system has been prepared in a state corresponding to the unitary symmetry U, then the probability that the measured state will be U is unity.

(ii). This symmetry is referred to as the two-way symmetry by Landé. The probability of a system prepared in a state corresponding to the unitary symmetry U and being measured in a state corresponding to the unitary symmetry V equals the probability of a state corresponding to the unitary symmetry V and being measured in a state corresponding to the unitary symmetry U. Landé interprets this result as the quantum analogue of the classical deterministic mechanics property that the time development of a system is symmetric with respect to the interchange of initial and final states.

(iii). The transformation $U \mapsto U^*, V \mapsto V^*$ expresses the invariance of the transition probability matrix $S(U,V)$ under time reversal.

Property (iv) is the most profound:

(iv). The multiplication property of the underlying transition probability amplitudes $W(U, V)$ expresses the precise form of the interference of the transition probabilities $S(U, V)$ for quantal systems by imposing this multiplication rule on the transition probability amplitudes themselves. Its validity is what accounts for many of the nonintuitive features of the quantal domain such as entanglement.

5.3 Permutation Matrices as Doubly Stochastic

Permutation matrices give the simplest class of doubly stochastic matrices. The question arises as to their physical significance. To address this, we set $U^\dagger V = P_\pi$, $\pi = (\pi_1, \pi_2, \ldots, \pi_n) \in S_n$, in relation (5.32). This gives $V = UP_\pi = (U_{\pi_1} U_{\pi_2}, \ldots, U_{\pi_n})$; that is, $V_i = U_{\pi_i}$, so that the columns of V are given by the permutation π of the columns of $U = (U_1 U_2 \cdots U_n)$. We adapt this result to the general setting of complete sets of commuting Hermitian operators given in Sect. 5.1.

We require in relation (5.25) that

$$W(U, V) = U^\dagger V = P_\pi, \quad \pi \in S_{d_n}. \tag{5.36}$$

This relation, in turn, requires that the unitary matrix V be given in terms of U by $V = UP_\pi$; that is, V has elements given in terms of the elements of U by

$$V_{k,\mathbf{k}} = \sum_{\mathbf{h} \in \mathbb{A}_n} U_{k,\mathbf{h}} \left(P_\pi\right)_{\mathbf{h},\mathbf{k}} = U_{k,\mathbf{h}_\pi}, \tag{5.37}$$

since $\left(P_\pi\right)_{\mathbf{h},\mathbf{k}} = \delta_{\mathbf{h}, \pi_\mathbf{k}}$, where we have that $\mathbf{h} = (h_1, h_2, \ldots, h_N)$ and $\pi_\mathbf{k} = (\pi_{k_1}, \pi_{k_2}, \ldots, \pi_{k_N})$.

We conclude:

If a system is prepared in the simultaneous eigenvector state $|\mathbf{h}\rangle$ of a complete set of commuting Hermitian operators $\mathbb{H}_n$, where the state $|\mathbf{h}\rangle$ is expressed in terms of the orthonormal basis set $|k\rangle, k = 1, 2, \ldots, d_n$ of the given Hilbert space $\mathcal{H}_n$ by the unitary matrix U, then it will be measured with certainty 1 to be in the simultaneous eigenvector state $|\mathbf{h}\rangle$ of the complete set of commuting Hermitian operators $\mathbb{K}_n$, where the state $|\mathbf{h}\rangle$ is expressed in terms of the orthonormal basis set $|k\rangle, k = 1, 2, \ldots, d_n$ of the given Hilbert space $\mathcal{H}_n$ by the unitary matrix $V = UP_\pi$, where π is an arbitrary permutation $\pi \in S_{d_n}$.

The case where the two sets of complete commuting Hermitian operators are the same, $\mathbb{K}_n = \mathbb{H}_n$, is trivial, since then all the simultaneous eigenvectors are the same, and the column labeling of the transition probability matrix is just the permutation π of the row labeling, so that the prepared state and the measured state are the same state. The case where the two sets of mutually commuting Hermitian operators are distinct would appear to suggest a deeper meaning — but this is questionable. This is because the construction of the transition probability matrix is unique up to ordering of its rows and columns (including the case of same sets): The imposition of the extra condition (5.36) may, or may not, be compatible with this construction, as it always is for the same sets. Since we have found no cases for distinct complete sets of mutually commuting Hermitian matrices for which a permutation matrix obtains, it may well be that only the trivial case of same sets can be realized in angular momentum systems. There are, however, cases where a special choice of parameters can give a single 1 and all 0's in some rows and columns, which itself implies that some prepared states and some distinct observed states occur with certainty. For example, choose $a_1 = a_2 = b_1 = b_2 = 0$ and $|a_3| = |b_3| = 1$ in (5.74) below.

We next give a number of examples of doubly stochastic matrices that occur in quantum theory. In the present approach to the probabilities of quantum theory, through doubly stochastic matrices, it is the unitary matrices corresponding to complete sets of mutually commuting Hermitian observable that play the primary role. Bohr and Ulfbeck [11] have argued that the whole of (nonrelativistic) quantal physics is the manifestation of unitary symmetries. Such unitary symmetries lead directly to the doubly stochastic matrices that Alfred Landé regarded as the fundamental entities of quantum mechanics. We next present some explicit examples of doubly stochastic matrices originating from physical theory.

5.4 The Doubly Stochastic Matrix for a Single System with Angular Momentum J

5.4.1 Spin-1/2 System

We first introduce and interpret the unitary Hermitian matrix $U_\mathbf{a} = U_\mathbf{a}^\dagger$ defined by

$$\begin{aligned} U_\mathbf{a} &= \frac{1}{\sqrt{2(1+a_3)}} \begin{pmatrix} 1+a_3 & a_1 - ia_2 \\ a_1 + ia_2 & -1 - a_3 \end{pmatrix} \\ &= \frac{1}{\sqrt{2(1+a_3)}} (a_1\sigma_1 + a_2\sigma_2 + (1+a_3)\sigma_3), \end{aligned} \qquad (5.38)$$

5.4. SINGLE SYSTEM DOUBLY STOCHASTIC MATRICES

where $\mathbf{a} = a_1\mathbf{e}_1 + a_2\mathbf{e}_2 + a_3\mathbf{e}_3$ denotes a real unit vector ($\mathbf{a}\cdot\mathbf{a}=1$) in Cartesian 3–space $\mathbb{R}^3$, and the σ_i are the Pauli matrices defined by

$$\sigma_1 = \begin{pmatrix} 0 & 1 \\ 1 & 0 \end{pmatrix},\ \sigma_2 = \begin{pmatrix} 0 & -i \\ i & 0 \end{pmatrix},\ \sigma_3 = \begin{pmatrix} 1 & 0 \\ 0 & -1 \end{pmatrix}. \tag{5.39}$$

The triad of unit vectors $(\mathbf{e}_1, \mathbf{e}_2, \mathbf{e}_3)$ defines a right-handed inertial reference frame in $\mathbb{R}^3$.

We next give an identity between the Pauli matrices and arbitrary unit vectors $\mathbf{n} = n_1\mathbf{e}_1 + n_2\mathbf{e}_2 + n_3\mathbf{e}_3$ and $\mathbf{a} = a_1\mathbf{e}_1 + a_2\mathbf{e}_2 + a_3\mathbf{e}_3$, where the components of these unit vectors are related by

$$n_1 = \frac{a_1}{\sqrt{2(1+a_3)}},\ n_2 = \frac{a_2}{\sqrt{2(1+a_3)}},\ n_3 = \frac{1+a_3}{\sqrt{2(1+a_3)}}. \tag{5.40}$$

It is easily verified that $\mathbf{a}\cdot\mathbf{a} = \mathbf{n}\cdot\mathbf{n} = 1$. Then, the following matrix identity of order 2 holds, as verified by direct matrix multiplication:

$$U_\mathbf{a}^\dagger \sigma_3 U_\mathbf{a} = \mathbf{a}\cdot\boldsymbol{\sigma}. \tag{5.41}$$

We next interpret the above relations in terms of angular momentum, where it is convenient to use the same notation for operators and matrices. Thus, we put $\mathbf{J} = \boldsymbol{\sigma}/2$, where the total angular momentum quantum number j is now identified to be $j = 1/2$. Each set of Hermitian matrices $\{\mathbf{J}^2 = \frac{3}{4}I_2,\ J_3 = \frac{1}{2}\sigma_3\}$ and $\{\mathbf{J}^2 = \frac{3}{4}I_2,\ J_\mathbf{a} = \frac{1}{2}\mathbf{a}\cdot\boldsymbol{\sigma}\}$ is a complete commuting set of Hermitian angular momentum matrices for $j = 1/2$; that is, $n = 2j+1 = 2$, so that we have a finite-Hilbert space of dimension 2. The two sets of eigenvalue relations read:

$$\mathbf{J}^2 I_2 = \tfrac{3}{4}I_2,\ J_3 I_2 = I_2(\tfrac{1}{2}\sigma_3),$$

$$\mathbf{J}^2 U_\mathbf{a} = U_\mathbf{a}(\tfrac{3}{4}I_2),\ J_\mathbf{a} U_\mathbf{a} = U_\mathbf{a}(\tfrac{1}{2}\sigma_3). \tag{5.42}$$

Thus, we have that $U = I_2$ and $V = U_\mathbf{a}$, which is exactly relations (5.11) and (5.20) for $n = 2$ and the angular momentum matrices at hand. Thus, the transition probability amplitude matrix $W(U,V) = U^\dagger V$ has $U = I_2$ and $V = U_\mathbf{a} = U_\mathbf{a}^\dagger$; hence, $W(I_2, U_\mathbf{a}) = U_\mathbf{a}$. Thus, the doubly stochastic matrix $S(U,V)$ is given by

$$S(I_2, U_\mathbf{a}) = \frac{1}{2}\begin{pmatrix} 1+a_3 & 1-a_3 \\ 1-a_3 & 1+a_3 \end{pmatrix}. \tag{5.43}$$

The rows and columns in the doubly stochastic matrix (5.43) can be labeled by the projection quantum numbers $(m, m_\mathbf{a})$, since $j = 1/2$

is fixed. The standard convention in angular momentum theory is to label from the greatest to the least value of m as read top-to-bottom down the rows with the same rule for $m_\mathbf{a}$ as read left-to-right across the columns. Thus, the interpretation of the elements of the transition probability matrix $S(I_2, U_\mathbf{a})$ is: A spin $j = 1/2$ system that has been prepared in the state with spin component $m = 1/2$ in direction $\mathbf{e_3}$ will be measured with probability $(1 + a_3)/2$ to have spin component $m_\mathbf{a} = 1/2$ in direction $\mathbf{a}$ and with probability $(1 - a_3)/2$ to have spin component $m_\mathbf{a} = -1/2$ in direction $\mathbf{a}$; if the system has been prepared in the state with spin component $m = -1/2$ in direction $\mathbf{e_3}$, it will be measured with probability $(1 - a_3)/2$ to have spin component $m_\mathbf{a} = 1/2$ in direction $\mathbf{a}$ and with probability $(1 + a_3)/2$ to have spin component $m_\mathbf{a} = -1/2$ in direction $\mathbf{a}$. (We note that in relations (31) and (32) in Ref. [44], we have chosen $U = \sigma_3$, which is allowed, but not required, since it effects a column sign change in I_2 used in the first relation in (5.42). The choice of $W(\sigma_3, U_\mathbf{a})$ does give $W(\sigma_3, \sigma_3)$ as a multiple of the unit matrix, if desired.)

5.4.2 Angular Momentum–j System

The preceeding results are a special case of a general result that applies to a single system with an arbitrary angular momentum $\mathbf{J}$. The pair of complete sets of mutually commuting Hermitian operators is $\mathbb{H}_N^{op} = \{H_1^{op}, H_2^{op}, \ldots, H_{N-2}^{op}, \mathbf{J}^2, J_3 = \mathbf{e_3} \cdot \mathbf{J}\}$ and $\mathbb{K}_N^{op} = \{H_1^{op}, H_2^{op}, \ldots, H_{N-2}^{op}, \mathbf{J}^2, J_\mathbf{a} = \mathbf{a} \cdot \mathbf{J}\}$, where the two sets differ only by the choice of the direction of quantization of the component of angular momenta, $J_3 = \mathbf{e_3} \cdot \mathbf{J}$ versus $J_\mathbf{a} = \mathbf{a} \cdot \mathbf{J}$, where $\mathbf{a} = a_1\mathbf{e_1} + a_2\mathbf{e_2} + a_3\mathbf{e_3}$ is a unit vector specifying an arbitrary direction in Cartesian 3-space $\mathbb{R}^3$. Also, all the Hermitian observables $\mathbb{H}_{N-2}^{op} = \{H_1^{op}, H_2^{op}, \ldots, H_{N-2}^{op}\}$ are taken to commute with the total angular momentum $\mathbf{J}$; hence, we always have the freedom to choose the direction of quantization. This property is true of all isolated physical systems — the assumption of rotational quantum symmetry under all $SU(2)$ frame rotations.

The action of the angular momentum components (J_1, J_2, J_3) on any irreducible subspace $\mathcal{H}_j$, $j \in \{0, 1/2, 1, 3/2, 2, \ldots\}$ is the standard action:

$$\mathbf{J}^2|jm\rangle = j(j+1)|jm\rangle, \quad J_3|jm\rangle = m|jm\rangle,$$
$$J_\pm|jm\rangle = \sqrt{(j \mp m)(j \pm m + 1)}\,|j\,m \pm 1\rangle;$$

(5.44)

$$\mathbf{J}^2|j\,m_\mathbf{a}\rangle = j(j+1)|j\,m_\mathbf{a}\rangle, \quad J_\mathbf{a}|j\,m_\mathbf{a}\rangle = m_\mathbf{a}|j\,m_\mathbf{a}\rangle,$$
$$J_\pm|j\,m_\mathbf{a}\rangle = \sqrt{(j \mp m_\mathbf{a})(j \pm m_\mathbf{a} + 1)}\,|j\,m_\mathbf{a} \pm 1\rangle.$$

In both sets of these relations, we have suppressed the labels $\mathbf{h} =$

5.4. SINGLE SYSTEM DOUBLY STOCHASTIC MATRICES

$(h_1, h_2, \ldots, h_{N-2})$ that should be included for a complete labeling of the state vectors; that is, $|(h) jm\rangle$ and $|(h) jm_a\rangle$. It is important to recognize, however, that the domains of definition of the h−labels can at most restrict the domain of definition of the total angular momentum j to be a subset of $\{0, 1/2, 1, 3/2, 2, \ldots\}$, possibly with repeated values. The standard irreducible (j, m)−structure as presented by relations (5.44) of the angular momentum multiplets cannot be altered for a system with the quantum rotation group $SU(2)$ symmetry: Either pair of mutually commuting set of Hermitian observables, $\{\mathbf{J}^2, J_3\}$ or $\{\mathbf{J}^2, J_\mathbf{a}\}$ is a complete set of operators for characterizing every such finite-dimensional Hilbert space $\mathcal{H}_j$ of dimension given by $\dim \mathcal{H}_j = 2j + 1$.

The simultaneous eigenvectors on the space $\mathcal{H}_j$ of the two commuting Hermitian observables in the set $\{\mathbf{J}^2, J_3\}$ and the set $\{\mathbf{J}^2, J_\mathbf{a}\}$ are given, respectively, by relations (5.44). The two orthonormal bases of the space $\mathcal{H}_j$ are related by

$$|j\, m_a\rangle = \sum_{m\in\{j, j-1, \ldots, -j\}} D^j_{m\, m_\mathbf{a}}(U_\mathbf{a}) |j\, m\rangle, \quad m_a \in \{j, j-1, \ldots, -j\}, \quad (5.45)$$

where the transformation coefficients are the famous Wigner D^j−functions. For the problem at hand (see Ref. [6] and [L]), these coefficients can also be presented in terms of the matrix Schur functions, as given by

$$D^j_{m\, m_\mathbf{a}}(U_\mathbf{a}) = \sqrt{(j+m)!(j-m)!(j+m_\mathbf{a})!(j-m_\mathbf{a})!}$$

$$\times \sum_{A \in M^{2j}_{2\times 2}(\alpha, \alpha_\mathbf{a})} \frac{u_{11}^{a_{11}} u_{12}^{a_{12}} u_{21}^{a_{21}} u_{22}^{a_{22}}}{a_{11}!a_{12}!a_{21}!a_{22}!}, \quad (5.46)$$

where $U_\mathbf{a}$ is the unitary Hermitian matrix defined by (5.38), which we have written in the following form:

$$U_\mathbf{a} = \begin{pmatrix} u_{11} & u_{12} \\ u_{21} & u_{22} \end{pmatrix} = \begin{pmatrix} n_3 & n_1 - in_2 \\ n_1 + in_2 & -n_3 \end{pmatrix}, \quad (5.47)$$

$$(a_1, a_2, a_3) = (n_1 n_2, n_2 n_3, 2n_3^2 - 1), \quad \mathbf{n} \cdot \mathbf{n} = 1.$$

The row and column sums in the matrix of exponents A given by

$$M^{2j}_{2\times 2}(\alpha, \alpha_\mathbf{a}) = \begin{pmatrix} a_{11} & a_{12} \\ a_{21} & a_{22} \end{pmatrix} \quad (5.48)$$

are the weights $\alpha = (\alpha_1, \alpha_2) = (j+m, j-m)$ and $\alpha_\mathbf{a} = (\alpha_{\mathbf{a}1}, \alpha_{\mathbf{a}2}) = (j+m_\mathbf{a}, j-m_\mathbf{a})$. The result $D^{1/2}(U_\mathbf{a}) = U_\mathbf{a}$ is recovered for $j = 1/2$.

The inner product of state vectors corresponding to the two sets of operators $\{\mathbf{J}^2, J_3\}$ and $\{\mathbf{J}^2, J_\mathbf{a}\}$ is

$$\langle j\,m\,|\,j\,m_\mathbf{a}\rangle = D^j_{m\,m_\mathbf{a}}(U_\mathbf{a}). \qquad (5.49)$$

Thus, from the general theory in Sect. 5.1, we have that $U = I_{2j+1}$ and $V = V^\dagger = D^j(U_\mathbf{a})$; hence, the transition probability amplitude matrix is

$$W(I_{2j+1}, V) = D^j(U_\mathbf{a}). \qquad (5.50)$$

Thus, we find that the transition probability matrix $S(U, V)$ is the matrix of order $2j + 1$ with elements:

$$S_{m\,m_\mathbf{a}}(I_{2j+1}, D^j(U_\mathbf{a})) = \left| D^j_{m\,m_\mathbf{a}}(U_\mathbf{a}) \right|^2. \qquad (5.51)$$

The proof of the above relations for the pair of commuting Hermitian observables $\{\mathbf{J}^2, J_3 = \mathbf{e}_3 \cdot \mathbf{J}\}, \{\mathbf{J}^2, J_\mathbf{a} = \mathbf{a} \cdot \mathbf{J}\}$, and the Wigner D^j–functions is given in many contexts in [6] and [L]; in particular, the connections with rotations in physical 3–space $\mathbb{R}^3$ are the relations (see [6], Eqs.(2.6),(3.42), (3.45), (3.46), (3.86)-(3.89), (3.89)):

$$U^\dagger(\phi, \mathbf{n})(\mathbf{b} \cdot \mathbf{J}) U(\phi, \mathbf{n}) = \mathbf{a} \cdot \mathbf{J}, \; U = e^{i\phi \mathbf{n} \cdot \mathbf{J}},$$

$$\begin{pmatrix} a_1 \\ a_2 \\ a_3 \end{pmatrix} = R(\phi, \mathbf{n}) \begin{pmatrix} b_1 \\ b_2 \\ b_3 \end{pmatrix},$$

$$R(\phi, \mathbf{n}) = I_3 + N \sin\phi + N^2 (1 - \cos\phi), \qquad (5.52)$$

$$N = \begin{pmatrix} 0 & -n_3 & n_2 \\ n_3 & 0 & -n_1 \\ -n_2 & n_1 & 0 \end{pmatrix},$$

where $\mathbf{a}$ and $\mathbf{b}$ are unit vectors, and we are here using the notation $\mathbf{J} = (J_1, J_2, J_3)$ to denote the standard matrix representations of order $2j + 1$ of the angular components; that is, the first relation above is the matrix realization of the operator relation, which is identical in form. The special matrix relations given in (5.38)-(5.43) are the above set of relations applied to $j = 1/2$ and $\mathbf{b} = \mathbf{e}_3$. Thus, the directional vector $\mathbf{a}$ is obtained from the directional frame vector $\mathbf{e}_3$ by an active positive rotation (right-hand rule) by angle $\phi = \pi$ about the vector $\mathbf{n}$.

The Hermitian property of the unitary matrix $V = V^\dagger = D^j(U_\mathbf{a})$ is a consequence of the general transpositional property $(D^j(U_\mathbf{a}))^{tr} = D^j(U_\mathbf{a}^{tr})$ of the general matrix Schur functions, followed by complex conjugation:

$$D^{j^\dagger}(U_\mathbf{a}) = D^j(U_\mathbf{a}^\dagger) = D^j(U_\mathbf{a}). \qquad (5.53)$$

5.5. COUPLED ANGULAR MOMENTUM SYSTEMS

The probabilistic meaning of the state vector inner product relation

$$|\langle j\, m\, |\, j\, m_\mathbf{a}\rangle|^2 = |D^j_{m\, m_\mathbf{a}}(U_\mathbf{a})|^2 \tag{5.54}$$

is: *The prepared state $|j\,m\rangle$ will be observed with probability $\left|D^j_{m\,m_\mathbf{a}}(U_\mathbf{a})\right|^2$ to be the measured state $|j\,m_\mathbf{a}\rangle$*; this, then, is the meaning of element in row m and column $m_\mathbf{a}$ of the doubly stochastic matrix $S(I_{2j+1}, D^j(U_\mathbf{a}))$, where $m, m_\mathbf{a} \in \{j, j-1, \ldots, -j\}$.

5.5 Doubly Stochastic Matrices for Composite Angular Momentum Systems

The Kronecker direct product or tensor product $U \otimes V$ of two unitary matrices of order n and m is a unitary matrix of order nm, where the tensor product is defined by

$$U \otimes V = \begin{pmatrix} u_{11}V & u_{12}V & \cdots & u_{1n}V \\ u_{21}V & u_{22}V & \cdots & u_{2n}V \\ \vdots & \vdots & \vdots & \vdots \\ u_{n1}V & u_{n2}V & \cdots & u_{nn}V \end{pmatrix}. \tag{5.55}$$

The element in row k and column l of the (i,j)–block of this matrix is denoted by

$$(U \otimes V)_{ik;jl} = u_{ij}v_{kl}, \tag{5.56}$$

where the rows and columns of $U \otimes V$ are enumerated by the pairs (i, k) and (j, l), where $i, k = 1, 2, \ldots, nm$; $j, l = 1, 2, \ldots, nm$. This tensor product then satisfies the multiplication rule:

$$(U' \otimes V')(U \otimes V) = U'U \otimes V'V, \text{ for } U', U \in U(n); \; V', V \in U(m). \tag{5.57}$$

The tensor product of vector spaces and the associated tensor product of state vectors, operators, and matrices is the underlying mathematics for forming composite physical systems from simpler individual systems in quantum mechanics. The structure of doubly stochastic matrices given in the preceding section extends to the quantum theory of composite systems through this tensor or Kronecker product of matrices.

5.5.1 Pair of Spin-1/2 Systems

We begin by describing the simplest physical system for which the above tensor product structures apply, a pair of spin-1/2 systems described by

CHAPTER 5. DOUBLY STOCHASTIC MATRICES IN AMT

(5.38)-(5.43), the first for direction **a** and the second for direction **b**, each referred to a common right-handed inertial reference frame $(\mathbf{e}_1, \mathbf{e}_2, \mathbf{e}_3)$. This construction is also given in detail in Ref. [44], but we repeat it here in summary format, since it serves as a the model for the general binary coupling schemes given in the next section. All angular momentum operators are given in their matrix realizations (same symbols as for operators). The two copies of the spin$-1/2$ system given by (5.38)-(5.43) are:

$$\mathbf{S}(1) = \tfrac{1}{2}\boldsymbol{\sigma} \otimes I_2, \ S_\mathbf{a}(1) = \tfrac{1}{2}(\mathbf{u} \cdot \boldsymbol{\sigma}) \otimes I_2, \tag{5.58}$$

$$\mathbf{S}(2) = I_2 \otimes \tfrac{1}{2}\boldsymbol{\sigma}, \ S_\mathbf{b}(2) = I_2 \otimes \tfrac{1}{2}(\mathbf{b} \cdot \boldsymbol{\sigma}).$$

The two complete sets of four commuting Hermitian matrices of interest are:

$$\{\mathbf{S}^2(1) = \tfrac{3}{4}I_4, \ S_3(1) = \tfrac{1}{2}\sigma_3 \otimes I_2, \ \mathbf{S}^2(2) = \tfrac{3}{4}I_4, \ S_3(2) = I_2 \otimes \tfrac{1}{2}\sigma_3\}, \tag{5.59}$$

$$\{\mathbf{S}^2(1) = \tfrac{3}{4}I_4, \ S_\mathbf{a}(1) = \tfrac{1}{2}(\mathbf{a} \cdot \boldsymbol{\sigma}) \otimes I_2, \ \mathbf{S}^2(2) = \tfrac{3}{4}I_4, \ S_\mathbf{b} = I_2 \otimes \tfrac{1}{2}(\mathbf{b} \cdot \boldsymbol{\sigma})\}.$$

Each of these sets of Hermitian matrices is complete with respect to $\mathcal{H}_4$: The unitary matrices $U = I_4$ and $V = U_\mathbf{a} \otimes U_\mathbf{b}$ diagonalize the respective sets (5.59), where the tensor product is obtained from (5.38):

$$U_{(\mathbf{a},\mathbf{b})} = U_\mathbf{a} \otimes U_\mathbf{b} = \frac{1}{\sqrt{4(1+a_3)(1+b_3)}} \times \tag{5.60}$$

$$\begin{pmatrix} (1+a_3)(1+b_3) & (1+a_3)(b_1-ib_2) & (a_1-ia_2)(1+b_3) & (a_1-ia_2)(b_1-ib_2) \\ (1+a_3)(b_1+ib_2) & -(1+a_3)(1+b_3) & (a_1-ia_2)(b_1+ib_2) & -(a_1-ia_2)(1+b_3) \\ (a_1+ia_2)(1+b_3) & (a_1+ia_2)(b_1-ib_2) & -(1+a_3)(1+b_3) & -(1+a_3)(b_1-ib_2) \\ (a_1+ia_2)(b_1+ib_2) & -(a_1+ia_2)(1+b_3) & -(1+a_3)(b_1+ib_2) & (1+a_3)(1+b_3) \end{pmatrix}.$$

Thus, we have the diagonalization relations for the matrices (5.58):

$$U^\dagger_{(\mathbf{a},\mathbf{b})} \mathbf{S}^2(1) U_{(\mathbf{a},\mathbf{b})} = \tfrac{3}{4}I_4, \ U^\dagger_{(\mathbf{a},\mathbf{b})} S_\mathbf{a}(1) U_{(\mathbf{a},\mathbf{b})} = \tfrac{1}{2}\sigma_3 \otimes I_2, \tag{5.61}$$

$$U^\dagger_{(\mathbf{a},\mathbf{b})} \mathbf{S}^2(2) U_{(\mathbf{a},\mathbf{b})} = \tfrac{3}{4}I_4, \ U^\dagger_{(\mathbf{a},\mathbf{b})} S_\mathbf{b}(2) U_{(\mathbf{a},\mathbf{b})} = I_2 \otimes \tfrac{1}{2}\sigma_3.$$

Thus, the transition probability amplitude matrix for this system is given by $W(U,V)$, where $U = I_4$ and $V = U_{(\mathbf{a},\mathbf{b})}$, which gives

$$W(I_4, U_\mathbf{a} \otimes U_\mathbf{b}) = U_\mathbf{a} \otimes U_\mathbf{b}, \tag{5.62}$$

$$S(I_4, U_\mathbf{a} \otimes U_\mathbf{b}) = S(I_2,, U_\mathbf{a}) \otimes S(I_2,, U_\mathbf{b}),$$

5.5. COUPLED ANGULAR MOMENTUM SYSTEMS

where the $S(I_2,,U_\mathbf{a})$ is the transition probability for spin$-1/2$ system 1 and $S(I_2,,U_\mathbf{b})$ that for spin$-1/2$ system 2, as given by (5.43). For an uncoupled system of two spin$-1/2$, the transition probability matrix factors into the tensor product of the transition probability matrices for the separate systems; in the language of density matrices, all states are *pure states;* there is no interference of probabilities between independent systems. Interference produces transition probability matrices that cannot be written as such products, the so-called *mixed states.*

The (row, column) enumeration of the elements in the Kronecker product transition probability matrix (5.60) is the following:

$$\Big(S(I_2,,U_\mathbf{a}) \otimes S(I_2,,U_\mathbf{b})\Big)_{(m_\mathbf{a},m_\mathbf{b});(m'_\mathbf{a},m'_\mathbf{b})} = \Big|\big(U_\mathbf{a} \otimes U_\mathbf{b}\big)_{(m_\mathbf{a},m_\mathbf{b});(m'_\mathbf{a},m'_\mathbf{b})}\Big|^2$$

$$= \Big(S(I_2,,U_\mathbf{a})\Big)_{m_\mathbf{a},m'_\mathbf{a}} \Big(S(I_2,,U_\mathbf{b})\Big)_{m_\mathbf{b},m'_\mathbf{b}}, \qquad (5.63)$$

where the pairs $(m_\mathbf{a}, m_\mathbf{b})$ and $(m'_\mathbf{a}, m'_\mathbf{b})$ take on the following values as read down the columns and across the rows in (5.60):

$$(m_\mathbf{a}, m_\mathbf{b}) = (1/2, 1/2), (1/2, -1/2), (-1/2, 1/2), (-1/2, -1/2), \qquad (5.64)$$

$$(m'_\mathbf{a}, m'_\mathbf{b}) = (1/2, 1/2), (1/2, -1/2), (-1/2, 1/2), (-1/2, -1/2).$$

For example, if the system is prepared in the state $(m_\mathbf{a}, m_\mathbf{b}) = (-1/2, 1/2)$ and measured in the state $(m'_\mathbf{a}, m'_\mathbf{b}) = (-1/2, -1/2)$, which is the element in row 3 and column 4 of (5.60), the probability of the measured state is $(1 + a_3)(1 - b_3)/4$, as also read off from relation (5.43) or (5.63).

5.5.2 Pair of Spin-1/2 Systems as a Composite System

A pair of spin-1/2 systems can also be described in terms of the total angular momentum of the constituents. Since the total angular momentum is a collective property of the whole system, the system is termed a composite system; nonintuitive properties of the transition probabilities emerge, as encoded in the associated doubly stochastic matrix.

We summarize the construction of the doubly stochastic matrix in question, as given in Ref. [44]:

1. The relevant two complete sets of mutually commuting Hermitian matrices are given by

$$\{\mathbf{S}^2(1), S_3(1), \mathbf{S}^2(2), S_3(2)\} \text{ and } \{\mathbf{S}^2(1), \mathbf{S}^2(2), \mathbf{S}^2, S_3\}, \qquad (5.65)$$

where $\mathbf{S}$ is obtained from $\mathbf{S}(1)$ and $\mathbf{S}(2)$ by addition of angular momenta:

$$\mathbf{S} = \mathbf{S}(1) + \mathbf{S}(2). \qquad (5.66)$$

2. Explicit angular momentum matrices:

$$\mathbf{S}^2(1) = \mathbf{S}^2(2) = \tfrac{3}{4}I_4,$$

$$S_3(1) = \frac{1}{2}\begin{pmatrix} 1 & 0 & 0 & 0 \\ 0 & 1 & 0 & 0 \\ 0 & 0 & -1 & 0 \\ 0 & 0 & 0 & -1 \end{pmatrix}, \quad S_3(2) = \frac{1}{2}\begin{pmatrix} 1 & 0 & 0 & 0 \\ 0 & -1 & 0 & 0 \\ 0 & 0 & 1 & 0 \\ 0 & 0 & 0 & -1 \end{pmatrix},$$

$$\mathbf{S}^2 = \begin{pmatrix} 2 & 0 & 0 & 0 \\ 0 & 1 & 1 & 0 \\ 0 & 1 & 1 & 0 \\ 0 & 0 & 0 & 2 \end{pmatrix}, \quad S_3 = \begin{pmatrix} 1 & 0 & 0 & 0 \\ 0 & 0 & 0 & 0 \\ 0 & 0 & 0 & 0 \\ 0 & 0 & 0 & -1 \end{pmatrix}. \tag{5.67}$$

3. Diagonalization of $\{\mathbf{S}^2(1), \mathbf{S}^2(2), \mathbf{S}^2, S_3\}$:

$$C\,\mathbf{S}^2(1)\,C^{tr} = \tfrac{3}{4}I_4, \quad C\,\mathbf{S}^2(2)\,C^{tr} = \tfrac{3}{4}I_4,$$

$$C\,\mathbf{S}^2\,C^{tr} = \begin{pmatrix} 2 & 0 & 0 & 0 \\ 0 & 2 & 0 & 0 \\ 0 & 0 & 2 & 0 \\ 0 & 0 & 0 & 0 \end{pmatrix}, \quad C\,S_3\,C^{tr} = \begin{pmatrix} 1 & 0 & 0 & 0 \\ 0 & 0 & 0 & 0 \\ 0 & 0 & -1 & 0 \\ 0 & 0 & 0 & 0 \end{pmatrix}, \tag{5.68}$$

$$C = \begin{pmatrix} 1 & 0 & 0 & 0 \\ 0 & 1/\sqrt{2} & 1/\sqrt{2} & 0 \\ 0 & 0 & 0 & 1 \\ 0 & 1/\sqrt{2} & -1/\sqrt{2} & 0 \end{pmatrix}.$$

The elements of matrix C are the Clebsch-Gordan coefficients for $j_1 = j_2 = 1/2$ and $j = 0, 1$ that effect the transformation from the so-called uncoupled basis (the first set in (5.65) to the coupled basis (the second set in (5.65), where the coefficients are ordered, by custom in the various tables (see [6]), into the real orthogonal matrix C, with (row, column) indices as follows:

$$\begin{aligned} \text{rows; top-to-bottom}: (s,m) &= (1,1),(1,0),(1,-1),(0,0), \\ \text{columns; left-to-right}: (m_1, m_2) &= (1/2,1/2),(1/2,-1/2), \\ &\quad (-1/2,1/2),(-1/2,-1/2). \end{aligned} \tag{5.69}$$

Thus, the four simultaneous eigenvectors of the complete set of mutually commuting Hermitian matrices $\mathbf{S}^2(1), \mathbf{S}^2(2), \mathbf{S}^2, S_3$ are the columns of the matrix C, as exhibited in relations (5.68). This system has $U = C^{tr}$ and $V = I_4$. Thus, the transition probability amplitude matrix is

5.5. COUPLED ANGULAR MOMENTUM SYSTEMS

$W(U, V) = U^\dagger V = C$, which gives the transition probability matrix as the following doubly stochastic matrix:

$$S(C^{tr}, I_4) = \begin{pmatrix} 1 & 0 & 0 & 0 \\ 0 & 1/2 & 1/2 & 0 \\ 0 & 0 & 0 & 1 \\ 0 & 1/2 & 1/2 & 0 \end{pmatrix}. \qquad (5.70)$$

Thus, with rows and columns labeled by (5.69), the meaning of the entry in row (j, m) and column (m_1, m_2) is that the prepared spin state (j, m) of the composite system, which is constituted of particle 1 and particle 2, each of spin $s_1 = s_2 = 1/2$, will be measured to be in the spin projection state $(m_1,, m_2)$ of particle 1 and particle 2, respectively, with probability equal to the entry in row (j, m) and column (m_1, m_2) of the doubly stochastic matrix $S(C^{tr}, I_4)$. There are, of course, four possible prepared states (rows) and, for each such prepared state, four possible measured states (columns).

We next consider the following generalization of the above results (5.65)-(5.70) for a pair of spin-1/2 systems by taking the first spin-1/2 system to be quantized with respect to an arbitrary direction **a** and that of the second spin-1/2 system with respect to an arbitrary direction **b**. Thus, we take the two complete systems of mutually commuting Hermitian matrices to be

$$\{\mathbf{S}^2(1), \mathbf{S}^2(2), \mathbf{S}^2, S_3 = \mathbf{e_3} \cdot \mathbf{S}\}, \quad \{\mathbf{S}^2(1), S_\mathbf{a}(1), \mathbf{S}^2(2), S_\mathbf{b}(2)\}, \quad (5.71)$$

where the explicit matrices of order four in these sets are given as follows:

$$\mathbf{S}(1) = \tfrac{1}{2}\boldsymbol{\sigma} \otimes I_2, \; \mathbf{S}(2) = I_2 \otimes \tfrac{1}{2}\boldsymbol{\sigma}, \; \mathbf{S} = \mathbf{S}(1) + \mathbf{S}(2),$$

$$S_\mathbf{a}(1) = \tfrac{1}{2}(\mathbf{a} \cdot \boldsymbol{\sigma}) \otimes I_2, \; S_\mathbf{b}(2) = I_2 \otimes \tfrac{1}{2}(\mathbf{b} \cdot \boldsymbol{\sigma}). \qquad (5.72)$$

Thus, the direction $\mathbf{e_3}$ is that of the 3-axis of a right-handed triad of unit vectors, and **a** and **b** are arbitrary directions specified by these unit vectors relative to this frame. We take the first set in (5.71) as characterizing the prepared state, and the second set as characterizing the measured state.

The simultaneous diagonalization of the commuting Hermitian matrices $\{\mathbf{S}^2(1), \mathbf{S}^2(2), \mathbf{S}^2, S_3 = \mathbf{e_3} \cdot \mathbf{S}\}$ is that already effected in relations (5.67)-(5.68) by the unitary matrix $U = C^{tr}$. The simultaneous diagonalization of the commuting Hermitian matrices $\{\mathbf{S}^2(1), S_\mathbf{a}(1), \mathbf{S}^2(2), S_\mathbf{b}(2)\}$ is that already effected in relations (5.60)-(5.61) by the unitary matrix $V = U_\mathbf{a} \otimes U_\mathbf{b}$. We thus arrive at the following result: The transition probability amplitude matrix for which the prepared states are the eigenvectors of the complete set of mutually commuting Hermitian matrices of order four given by $\{\mathbf{S}^2(1), \mathbf{S}^2(2), \mathbf{S}^2, S_3\}$ and for which the measured states are the eigenvectors of the complete set of mutually commuting Hermitian matrices of order four given by $\{\mathbf{S}^2(1), S_\mathbf{a}(1), \mathbf{S}^2(2), S_\mathbf{b}(2)\}$ is the matrix $W(U,V)$ as follows:

$$W(U,V) = W(C^{tr}, U_\mathbf{a} \otimes U_\mathbf{b}) = C(U_\mathbf{a} \otimes U_\mathbf{b}). \qquad (5.73)$$

The direct calculation of the corresponding transition probability matrix $S(C^{tr}, U_\mathbf{a} \otimes U_\mathbf{b})$ gives the following doubly stochastic matrix:

$$S(C^{tr}, U_\mathbf{a} \otimes U_\mathbf{b}) = \qquad (5.74)$$

$$\frac{1}{4}\begin{pmatrix} (1+a_3)(1+b_3) & (1+a_3)(1-b_3) & (1-a_3)(1+b_3) & (1-a_3)(1-b_3) \\ 1+\mathbf{a}\cdot\mathbf{b}-2a_3b_3 & 1-\mathbf{a}\cdot\mathbf{b}+2a_3b_3 & 1-\mathbf{a}\cdot\mathbf{b}+2a_3b_3 & 1+\mathbf{a}\cdot\mathbf{b}-2a_3b_3 \\ (1-a_3)(1-b_3) & (1-a_3)(1+b_3) & (1+a_3)(1-b_3) & (1+a_3)(1+b_3) \\ 1-\mathbf{a}\cdot\mathbf{b} & 1+\mathbf{a}\cdot\mathbf{b} & 1+\mathbf{a}\cdot\mathbf{b} & 1-\mathbf{a}\cdot\mathbf{b} \end{pmatrix}.$$

(We have interchanged the (row, column) labeling of the elements of C in this result from that used in [44] to accord with customary labeling. The calculation of the full matrix (5.74) is, however, in agreement with that Eq. (52).)

The rows and columns of the doubly stochastic matrix $S(C, U_\mathbf{a} \otimes U_\mathbf{b})$ given by (5.74) are labeled top-to-bottom down the rows and left-to-right across the columns as follows:

$$\begin{aligned} \text{rows}: \quad & (s,m) = (1,1), (1,0), (1,-1), (0,0); \\ \text{columns}: \quad & (m_\mathbf{a}, m_\mathbf{b}) = (1/2, 1/2), (1/2, -1/2), \\ & (-1/2, 1/2), (-1/2, -1/2). \end{aligned} \qquad (5.75)$$

We next state the meaning of the entries in the doubly stochastic matrix $S(C^{tr}, U_\mathbf{a} \otimes U_\mathbf{b})$ with rows and columns labeled as described in (5.75). The following properties of the quantum states are to hold:

1. The prepared state is the total spin state: (s,m) with projection m in direction $\mathbf{e}_3$, where the unit vectors $(\mathbf{e}_1, \mathbf{e}_2, \mathbf{e}_3)$ constitute a right-handed system of axes in physical 3-space $\mathbb{R}^3$.

2. The properties of the measured state comes with many subtleties: The measurement must be effected in such a manner that it is consistent with the notion of one measurement, since the entire composite system must remain undisturbed during the time interval between the prepared state and the measured state. Accordingly, it must be possible in the experimental arrangement to make separate but simultaneous measurements on part 1 with spin state given by $(s_1 = 1/2, m_\mathbf{a})$ with projection $m_\mathbf{a}$ in direction $\mathbf{a} = a_1\mathbf{e}_1 + a_2\mathbf{e}_2 + a_3\mathbf{e}_3$ and on part 2 with spin state given by $(s_2 = 1/2, m_\mathbf{b})$ with projection $m_\mathbf{b}$ in direction $\mathbf{b} = b_1\mathbf{e}_1 + b_2\mathbf{e}_2 + b_3\mathbf{e}_3$, since otherwise the composite system has been disturbed prior to the measurement.

5.5. COUPLED ANGULAR MOMENTUM SYSTEMS

Under these assumptions of prepared and measured states, we conclude:

The probability that the prepared state (s,m) will be measured in the state $(m_\mathbf{a}, m_\mathbf{b})$ is equal to the entry in row (s,m) and column $(m_\mathbf{a}, m_\mathbf{b})$ of the doubly stochastic matrix $S(C^{tr}, U_\mathbf{a} \otimes U_\mathbf{b})$.

It will be noted that the probabilities associated with a prepared singlet state $(s,m) = (0,0)$ are rotational invariants. Thus, for a prepared singlet state, we have the following interpretations of the transition probabilities in row 4 of $S(C^{tr}, U_\mathbf{a} \otimes U_\mathbf{b})$ given by

$$(1 - \mathbf{a} \cdot \mathbf{b})/4,\ (1 + \mathbf{a} \cdot \mathbf{b})/4,\ (1 + \mathbf{a} \cdot \mathbf{b})/4,\ (1 - \mathbf{a} \cdot \mathbf{b})/4. \qquad (5.76)$$

The measured state with projection quantum number $m_\mathbf{a}$ of system 1 in direction $\mathbf{a}$ and of the projection quantum number $m_\mathbf{b}$ of system 2 in direction $\mathbf{b}$ has probability $(1 - \mathbf{a} \cdot \mathbf{b})/4$ for state $(m_\mathbf{a}, m_\mathbf{a}) = (1/2, 1/2)$; probability $(1 + \mathbf{a} \cdot \mathbf{b})/4$ for state $(m_\mathbf{a}, m_\mathbf{a}) = (1/2, -1/2)$; probability $(1 + \mathbf{a} \cdot \mathbf{b})/4$ for state $(m_\mathbf{a}, m_\mathbf{a}) = (-1/2, 1/2)$; probability $(1 - \mathbf{a} \cdot \mathbf{b})/4$ for state $(m_\mathbf{a}, m_\mathbf{a}) = (-1/2, -1/2)$. It is important to notice that *it is not true that $m = m_\mathbf{a} + m_\mathbf{b}$, because the directions $\mathbf{e}_3, \mathbf{a}, \mathbf{b}$ are, in general, not equal.*

The directions $\mathbf{a}$ and $\mathbf{b}$ can be chosen arbitrarily. For simplicity of example, we choose $\mathbf{b} = \mathbf{a}$. Then, the singlet state will be measured with probability 0 for part 1 to be in the projected quantum number state $m_\mathbf{a} = 1/2$ in direction $\mathbf{a}$ and for part 2 to be in the projected quantum number state $m_\mathbf{a} = 1/2$ in the same direction $\mathbf{a}$. Thus, the final measured state is not that of two independent parts: the singlet state remains composite, no matter what the separation of its parts, this holding valid for all such times for which no part of the composite system is measured (or disturbed by unknown influences). This result is at odds with the idea that, when the two parts of the composite system are far removed from one another, a measurement on system 1 cannot influence the measurement on system 2. But the *two-part harmony* of the composite system (to paraphrase a concept accredited to structural biologist Claude Lévi-Strauss [41]) requires the probabilities of interference to be exactly those given by the doubly stochastic matrix $S(C^{tr}, U_\mathbf{a} \otimes U_\mathbf{b})$.

It will be noticed that the probabilities in the doubly stochastic matrix $S(C^{tr}, U_\mathbf{a} \otimes U_\mathbf{b})$ come in both forms: pure states for the total angular momentum states $(s,m) = (1, \pm 1)$; mixed states for the total angular momentum states $(s,m) = (1,0)$ or $(0,0)$.

Wigner [84-85] has used the rotational invariants (5.76) to argue the validity of the Bell [3] inequalities, which imply that a local, hidden-variable theory of quantum mechanics cannot be constructed that is consistent with the predictions of conventional quantum theory.

It is not our purpose here to engage in the several viewpoints of the interpretation of quantum theory presented by various authors, including the early considerations of Einstein, *et al.* [25], Schrödinger [73], Bohm [9], Bohm and Hiley [10], and Bell [3], and the more recent ideas advanced, for example, by Griffiths [29], Omnès [58], Gell-Mann [27], Penrose [59], and Leggett [40] (see the comprehensive overview Wheeler and Zurek [81] for viewpoints advanced before 1983). Rather, we continue with the construction of doubly stochastic matrices within the special context of angular momentum theory and their conventional quantum-mechanical interpretations, since they illustrate vividly many subtleties of the interference of probabilities inherent in the quantum states of composite systems.

5.6 Binary Coupling of Angular Momenta

The purpose of this section is to place the concept of a recoupling matrix in the binary theory of coupling of n angular momenta within the context of doubly stochastic matrices. The reading of this section requires familiarity with the notion of a recoupling matrix as developed in great detail in [L]. However, we do summarize here the principal results.

5.6.1 Complete Sets of Commuting Hermitian Observables

We have already given in Chapter 1 the general setting for the coupling of n angular momenta. In particular, relations (1.14)-(1.16) detail a complete set of mutually commuting Hermitian operators and their simultaneous eigenvector-eigenvalue relations, this structure defining what is called the uncoupled scheme. Relations (1.17)-(1.25) similarly detail a second complete set of mutually commuting Hermitian operators and their simultaneous eigenvector-eigenvalue relations, this structure defining what is called the $\alpha-$coupling scheme. In this Chapter, we supply the set of all possible $\alpha-$coupling schemes associated with the intermediate angular momenta for all possible pairwise schemes for adding n angular momenta, as enumerated by the labeling of all binary trees of order n.

The term "pairwise" means that any two distinct angular momenta, say, $\mathbf{J}(h)$ and $\mathbf{J}(l)$, are selected from the set $\{\mathbf{J}(1), \mathbf{J}(2), \ldots, \mathbf{J}(n)\}$; the selected pair is then coupled according to the standard rules for the addition of two angular momenta: $\mathbf{K}(1) = \mathbf{J}(h) + \mathbf{J}(l)$. Then, two angular momenta are selected from the set of $n - 1$ angular momenta $\left\{\mathbf{K}(1), \left\{\{\mathbf{J}(1), \mathbf{J}(2), \ldots, \mathbf{J}(n)\} - \{\mathbf{J}(h), \mathbf{J}(l)\}\right\}\right\}$ and added according to the standard rules for the addition of two angular momenta. This process is continued until all of the original n angular momenta have been included, which requires $n - 2$ such arbitrary selections and gives rise to

5.6. BINARY COUPLING OF ANGULAR MOMENTA

$n-2$ *intermediate angular momenta* $\mathbf{K}(1), \mathbf{K}(2), \ldots, \mathbf{K}(n-2)$ that we denote by the order of their selection. The set of angular momentum operators

$$\mathbf{J}^2(1), \mathbf{J}^2(2), \ldots, \mathbf{J}^2(n), \mathbf{J}^2, J_3, \mathbf{K}^2(1), \mathbf{K}^2(2), \ldots, \mathbf{K}^2(n-2) \quad (5.77)$$

is then a second complete set of $2n$ mutually commuting Hermitian operators for the coupling of n angular momenta.

Care must be exercised in interpreting the $n-2$ squared intermediate angular momenta in (5.77): *Each selection process gives a distinct set of intermediate angular momenta* $\{\mathbf{K}(1), \mathbf{K}(2), \ldots, \mathbf{K}(n-2)\}$ *in that each such set presents each* $\mathbf{K}(i)$ *as a sum of two distinct angular momenta that includes intermediate angular momenta previously selected and those from the set* $\{\mathbf{J}(1), \mathbf{J}(2), \ldots, \mathbf{J}(n)\}$. We will review subsequently, through examples, the result proved in [L] that each set of intermediate angular momenta is one-to-one with the members $T \in \mathbb{T}_n$ of a collection of graphs known as *binary trees*. We adopt at the outset a notation for the intermediate angular momenta operators that encodes this information: $\{\mathbf{K}_T(1), \mathbf{K}_T(2), \ldots, \mathbf{K}_T(n-2)\}$, each $T \in \mathbb{T}_n$.

We have the following set of $2n$ simultaneous eigenvalue-eigenvector relations for the selection of each such T-coupling scheme, as obtained from a standard labeled binary tree $T(\mathbf{j}\,\mathbf{k})_{jm}$, each binary tree $T \in \mathbb{T}_n$:

$$\mathbf{J}^2(i)|T(\mathbf{j}\,\mathbf{k})_{jm}\rangle = j_i(j_i+1)|T(\mathbf{j}\,\mathbf{k})_{jm}\rangle, i = 1, 2, \ldots, n,$$

$$\mathbf{J}^2|T(\mathbf{j}\,\mathbf{k})_{jm}\rangle = j(j+1)|T(\mathbf{j}\,\mathbf{k})_{jm}\rangle,$$

$$J_3|T(\mathbf{j}\,\mathbf{k})_{jm}\rangle = m|T(\mathbf{j}\,\mathbf{k})_{jm+1}\rangle,$$

$$J_+|T(\mathbf{j}\,\mathbf{k})_{jm}\rangle = \sqrt{(j-m)(j+m+1)}|T(\mathbf{j}\,\mathbf{k})_{jm+1}\rangle, \quad (5.78)$$

$$J_-|T(\mathbf{j}\,\mathbf{k})_{jm}\rangle = \sqrt{(j+m)(j-m+1)}|T(\mathbf{j}\,\mathbf{k})_{jm-1}\rangle,$$

$$\mathbf{K}_T^2(i)|T(\mathbf{j}\,\mathbf{k})_{jm}\rangle = k_i(k_i+1)|T(\mathbf{j}\,\mathbf{k})_{jm}\rangle, i = 1, 2, \ldots, n-2.$$

For economy of notation, we use the same k_i notation for the enumeration of the eigenvalues of $\mathbf{K}_T^2(i)$, where the distinction between these eigenvalues is made through the domain of definition of $\mathbf{k} \in \mathbb{K}_T^{(j)}(\mathbf{j})$ defined below. For each binary tree $T \in \mathbb{T}_n$, the simultaneous eigenvectors in (5.78) constitute an orthonormal basis of the vector space $\mathcal{H}_\mathbf{j} = \mathcal{H}_{j_1} \otimes \mathcal{H}_{j_2} \otimes \cdots \otimes \mathcal{H}_{j_n}$ of dimension $\dim \mathcal{H}_\mathbf{j} = \prod_{i=1}^n (2j_i+1)$, where $\mathbf{k}, j, m$ take on all possible values in their domains of definition for specified angular momenta $\mathbf{j}$.

We recall that $\mathbf{j} = (j_1, j_2, \ldots, j_n)$, where each j_i can be arbitrarily selected from $\{0, 1/2, 1, 3/2, \ldots\}$, but once selected are fixed in the definition of the Hilbert space. The domains of the intermediate quantum labels $\mathbf{k} = (k_1, k_2, \ldots, k_{n-2})$ and the total angular momentum quantum number j are then uniquely determined by the Clebsch-Gordan rule for the addition of two angular momenta and the coupling scheme for the pairwise addition, while the projection quantum number m takes on the values $m = j, j-1, \ldots, -j$, for each allowed value of j. For each set of prescribed $\mathbf{j}$, we denote this unique domain of definition of $\mathbf{k}, j, m$ by $\mathbb{R}_T(\mathbf{j})$; its general structure is:

$$\mathbb{R}_T(\mathbf{j}) = \left\{ \mathbf{k} = (k_1, k_2, \ldots, k_{n-2}), j, m \; \middle| \; \begin{array}{l} j \in \mathbb{J}(\mathbf{j}); \mathbf{k} \in \mathbb{K}_T^{(j)}(\mathbf{j}); \\ m = j, j-1, \ldots, -j \end{array} \right\}, \quad (5.79)$$

where the domains of definition $\mathbb{J}(\mathbf{j})$ of j and $\mathbb{K}_T^{(j)}(\mathbf{j})$ of the $\mathbf{k}$ are defined in detail below. First, we enumerate the orthonormality properties of the coupled basis (5.78) of the Hilbert space $\mathcal{H}_{\mathbf{j}}$. Thus, we define the set of orthonormal basis vectors $\mathbf{B}_T(\mathbf{j})$, each $T \in \mathbb{T}_n$:

$$\mathbf{B}_T(\mathbf{j}) = \{ |T(\mathbf{j}\,\mathbf{k})_{j\,m}\rangle \mid \mathbf{k}, j, m \in \mathbb{R}_T(\mathbf{j}) \},$$
$$\langle (T(\mathbf{j}\,\mathbf{k})_{j\,m} \mid T(\mathbf{j}\,\mathbf{k}')_{j'\,m'} \rangle = \delta_{j,j'} \delta_{m,m'} \delta_{\mathbf{k},\mathbf{k}'}, \quad (5.80)$$
$$\text{for all } \mathbf{k}, j, m \in \mathbb{R}_T(\mathbf{j}); \; \mathbf{k}', j', m' \in \mathbb{R}_T(\mathbf{j}).$$

5.6.2 Domain of Definition $\mathbb{R}_T(\mathbf{j})$

We continue the discussion of the structure of the indexing set $\mathbb{R}_T(\mathbf{j})$ whose elements $\mathbf{k}, j, m$ enumerate the orthonormal basis vectors defined by the complete set of commuting Hermitian operators (5.78) for specified $\mathbf{j}$ and $T \in \mathbb{T}_n$.

1. The indexing set $\mathbb{J}(\mathbf{j})$ is the set of values that the total angular momentum quantum number j can assume in the coupling of n angular momenta:

$$\mathbb{J}(\mathbf{j}) = \{j_{\min}, j_{\min} + 1, \ldots, j_{\max}\}. \quad (5.81)$$

The minimum and maximum values of j are given in terms of $j_1, j_2, \ldots, j_n$ by

$$j_{\min} = \min\{j_1 \pm j_2 \pm \cdots \pm j_n\}, \quad j_{\max} = j_1 + j_2 + \cdots + j_n, \quad (5.82)$$

where $j_{\min}$ is the least nonnegative integer or half-integer in the set obtained from the sum $j_1 \pm j_2 \pm \cdots \pm j_n$ by choosing all possible 2^{n-1} signs $\pm$. Each such value of j occurs with the multiplicity $N_j(\mathbf{j})$, which is called the Clebsch-Gordan (CG) number for the addition of n angular momenta $\mathbf{j}$.

The CG numbers $N_j(\mathbf{j})$ are generated in the following manner. We introduce the notation $\langle a, b \rangle$ for the set of angular momenta that occur

5.6. BINARY COUPLING OF ANGULAR MOMENTA

in the addition of two angular momenta with quantum numbers $a, b \in \{0, 1/2, 1, 3/2, \ldots\}$:

$$\langle a, b \rangle = \{a+b, a+b-1, \ldots, |a-b|\}. \quad \text{(triangle rule)} \quad (5.83)$$

Thus, $\langle a, b \rangle$ is the set of CG numbers for $n = 2$. The rule (5.83) of adding two angular momenta is often called *the triangle rule*. These CG numbers satisfy the following relations:

symmetry rule: $\langle a, b \rangle = \langle b, a \rangle$;

(5.84)

distribution rule: $c \in \langle a, b \rangle$ implies $a \in \langle b, c \rangle$ and $b \in \langle c, a \rangle$.

We now define the multiset of CG numbers for the addition of n angular momenta by the recursion relation:

$$\langle j_1, j_2, \ldots, j_n \rangle = \left\langle \langle j_1, j_2, \ldots, j_{n-1} \rangle, j_n \right\rangle = \{ \langle a, j_n \rangle \mid a \in \langle j_1, j_2, \ldots, j_{n-1} \rangle \}. \quad (5.85)$$

Example:

$$\langle a, b, c \rangle = \langle \langle a, b \rangle, c \rangle = \{ \langle a+b, c \rangle, \langle a+b-1, c \rangle, \ldots, \langle |a-b|, c \rangle \}, \quad (5.86)$$

where, of course, this result is to be expanded further by application of (5.83) to obtain the numerical entries in the multiset $\langle a, b, c \rangle$. We note also the properties:

permutational symmetry: $\langle a, b, c \rangle = \langle a, c, b \rangle = \langle b, c, a \rangle$
$\langle b, a, c \rangle = \langle c, a, b \rangle = \langle c, b, a \rangle$;

(5.87)

distributional symmetry: $d \in \langle a, b, c \rangle$, $a \in \langle b, c, d \rangle$,
$b \in \langle c, d, a \rangle$, $c \in \langle d, a, b \rangle$. □

An essential feature of the above generation of the set $\langle \mathbf{j} \rangle$ is its independence of the coupling scheme. Also, the permutational and distributional symmetries generalize to the general set $\langle \mathbf{j} \rangle$. (See [L] for proof.)

The main point is:

The set $\mathbb{J}(\mathbf{j})$ is the collection of distinct elements in the multiset $\langle \mathbf{j} \rangle$, which includes each $j \in \mathbb{J}(\mathbf{j})$ a number of times equal to the CG number $N_j(\mathbf{j})$.

2. The indexing set $\mathbb{K}_T^{(j)}(\mathbf{j})$ is the set of values that the intermediate angular momentum quantum numbers $\mathbf{k} = (k_1, k_2, \ldots, k_{n-2})$ can assume for each $j \in \mathbb{J}(\mathbf{j})$; it depends not only on the values j_i of the n angular

momenta in **j** and on the total angular momentum j, but very essentially on the pairing of the angular momenta in the given coupling scheme, which, we have noted, is determined by a binary tree T. It is useful to see this interplay by an example before giving the description of the general case.

Example. A single example will make clear how the elements of the sets $\mathbf{k} \in \mathbb{K}_T^{(j)}(\mathbf{j})$ are found. We select $n = 4$ and choose the coupling scheme to be that corresponding to the placement of parenthesis pairs ( ) as follows, where here we lower the index i that enumerates angular momenta to the subscript position to avoid confusion with parenthesis pairs that encode coupling instructions to the total angular momentum $\mathbf{J}$ (the binary tree T is the first one in (5.95), as presented in the next section):

$$\mathbf{J} = \Big(\big((\mathbf{J}_1 + \mathbf{J}_2) + \mathbf{J}_3\big) + \mathbf{J}_4\Big). \tag{5.88}$$

$$\mathbf{K}_1 = \mathbf{J}_1 + \mathbf{J}_2, \ \mathbf{K}_2 = \mathbf{K}_1 + \mathbf{J}_3, \ \mathbf{J} = \mathbf{K}_2 + \mathbf{J}_4.$$

To systematize such pairwise coupling schemes, we associate with (5.88) a *triangle pattern* given by

$$\begin{pmatrix} j_1 & k_1 & k_2 \\ j_2 & j_3 & j_4 \\ k_1 & k_2 & j \end{pmatrix}, \tag{5.89}$$

in which each column is a triangle of angular momentum quantum numbers; that is, the entries (a, b, c) in each column satisfy the relations given by (5.83)-(5.84). Thus, for given $\mathbf{j} = (j_1, j_2, j_3, j_4)$ and $j \in \mathbb{J}(\mathbf{j})$, the values of the intermediate quantum labels k_1 and k_2 are uniquely determined by

$$k_1 \in \langle j_1, j_2 \rangle, \ k_2 \in \langle k_1, j_3 \rangle, \ k_2 \in \langle j_4, j \rangle. \tag{5.90}$$

If we were to leave the last condition in the form $j \in \langle k_2, j_4 \rangle$ with j unspecified, then we would simple generate the multiset $\langle \mathbf{j} \rangle$, which, we have noted, is independent of the coupling scheme. The simple rule of requiring $k_2 \in \langle j_4, j \rangle$ for each selected $j \in \langle \mathbf{j} \rangle$ has the effect of restricting the vales of k_1 and k_2 to exactly those subsets that can give rise to the selected value of j.

Example. $j_1 = j_2 = 1/2, j_3 = j_4 = 1$. Using the recursion relation described above, we easily calculate the multiset of values $\langle 1/2, 1/2, 1, 1 \rangle = \{0, 0, 1, 1, 1, 1, 2, 2, 2, 3\}$, which gives the values of j and their multiplicities, the CG numbers. For each value of j, the projection quantum number m takes on the vales $m = j, j-1, \ldots, -j$. Thus, altogether,

5.6. BINARY COUPLING OF ANGULAR MOMENTA

there are $\prod_{i=1}^{4}(2j_i + 1) = 36$ distinct total angular momentum states $|(1/2, 1/2, 1, 1; k_1, k_2)_{jm}\rangle$. The unique enumeration of the simultaneous eigenvectors of the eight mutually commuting Hermitian operators in (5.78) is then obtained simply by determining the pairs (k_1, k_2) such that the three columns of the triangle pattern

$$\begin{pmatrix} 1/2 & k_1 & k_2 \\ 1/2 & 1 & 1 \\ k_1 & k_2 & j \end{pmatrix} \tag{5.91}$$

are all angular momentum triangles (a, b, c), where we take $j = 0, 1, 2, 3$, in turn. Thus, for the specified $\mathbf{j} = (1/2, 1/2, 1, 1)$, we find:

$$j = 3: \ (k_1, k_2) \in \{(1, 2)\} = \mathbb{K}_T^{(3)}(\mathbf{j});$$

$$j = 2: \ (k_1, k_2) \in \{(0, 1), (1, 1), (2, 1)\} = \mathbb{K}_T^{(2)}(\mathbf{j});$$

$$j = 1: \ (k_1, k_2) \in \{(0, 1), (1, 0), (1, 1), (1, 2)\} = \mathbb{K}_T^{(1)}(\mathbf{j}); \tag{5.92}$$

$$j = 0: \ (k_1, k_2) \in \{(0, 1), (1, 1)\} = \mathbb{K}_T^{(0)}(\mathbf{j}).$$

Each of the thirty-six $|(1/2, 1/2, 1, 1; k_1, k_2)_{jm}\rangle$ orthonormal vectors is labeled by a unique sequence $(1/2, 1/2, 1, 1; k_1, k_2; j, m)$. □

The general relation between the cardinality of the set $\mathbb{R}(\mathbf{j})$, the cardinality of the set $|\mathbb{K}_T^{(j)}(\mathbf{j})| = N_j(\mathbf{j})$ (CG numbers), the allowed values of jm, and the dimensionality $N(\mathbf{j})$ of the uncoupled basis, as specified by the set of values $\mathbf{j}$, is given by

$$|\mathbb{R}_T(\mathbf{j})| = \sum_{j=j_{\min}}^{j_{\max}} (2j + 1) N_j(\mathbf{j}) = \prod_{i=1}^{n} (2j_i + 1) = N(\mathbf{j}). \tag{5.93}$$

The domain of definition $\mathbf{k} \in \mathbb{K}_T^{(j)}(\mathbf{j})$ is uniquely encoded by a *triangle pattern* associated with a standard labeled binary tree $T(\mathbf{j}\,\mathbf{k})_{jm}$: *It is the theory of binary trees and their standard labeling that synthesizes the structure of the simultaneous eigenvector-eigenvalue relations (5.78) of each complete set of commuting Hermitian operators (5.77).*

5.6.3 Binary Bracketings, Shapes, and Binary Trees

There are, as remarked earlier, many ways of selecting a system of $n-2$ intermediate angular momenta to complete the set (5.77). The counting and classification of all such schemes constitutes the subject of the binary coupling theory of angular momenta. The underlying and unifying combinatorial concepts of binary trees, trivalent trees, and cubic graphs play basic roles in the development of this subject as set forth in considerable detail in [L]. In this section, we review results from [L], presenting those that have particular relevance for the present volume.

The enumeration of all binary coupling schemes is the problem of inserting $n-1$ parenthesis pairs ( ) into a sequence of n symbols, in all possible ways such that certain conditions are satisfied. To describe this, we introduce the concept of a *binary bracketing* of a sequence of n objects:

A binary bracketing of order 1 is a symbol X_1 with no parenthesis pair enclosing it. A binary bracketing of order $n \geq 2$ of the sequence $X_1 X_2 \cdots X_n$ of n symbols is a sequence containing these n symbols in the designated order and $n-1$ parenthesis pairs ( ) such that: (i) each parenthesis pair contains at least two X_i–symbols; (ii) the replacement of any parenthesis pair $(\cdots)$ and all its enclosed symbols, denoted $\cdots$, which consists of X_i–symbols and possibly other parenthesis-pair symbols, by a single symbol X_1 leaves behind a binary bracketing of lower order of the remaining X_i symbols, not originally enclosed by ( ).

This is a nonconstructive definition, but one that provides criteria as to whether a sequence consisting of n symbols $X_1 X_2 \cdots X_n$ and $n-1$ symbols ( ) is a binary bracketing, or not. Indeed, this process can be continued to reduce every binary bracketing to a single symbol with no parenthesis pair.

A closely related, but somewhat more abstract definition can be stated. The primitive objects out of which every binary bracketing is built are a symbol ∘ and a parenthesis pair ( ), where the ∘ itself is a binary bracket of order 1 (no parenthesis pair):

A binary bracketing of order $n \geq 2$ of the sequence ∘ ∘⋯∘ of n symbols, called elements or points, is a sequence containing these n symbols and $n-1$ parenthesis pairs ( ) such that: (i) each parenthesis pair contains at least two ∘ points; (ii) the replacement of any parenthesis pair and all its enclosed $\cdots$ points and enclosed parenthesis-pair symbols by a single ∘ point leaves behind a binary bracketing of lower order.

Examples. The binary bracketings of orders $1, 2, 3, 4$ are given by

$$n=1: \circ; \quad n=2: (\circ\circ); \quad n=3: \big((\circ\circ)\circ\big), \big(\circ(\circ\circ)\big);$$

$$n=4: \big((\circ\circ)\circ)\circ\big), \big(\circ\big(\circ(\circ\circ)\big)\big), \big((\circ\circ)(\circ\circ)\big),$$

$$\big(((\circ\circ)\circ)\circ\big), \big(\circ\big((\circ\circ)\circ\big)\big). \qquad (5.94)$$

When convenient we use parenthesis pairs of different sizes, so that the object "parenthesis pair" can be easily identified, but, of course, for

5.6. BINARY COUPLING OF ANGULAR MOMENTA

larger n this is not possible, but the pairing can always be identified starting from the outside and working inward, or inversely. Here, as a mathematical object, the parenthesis pair is a primitive symbol, just as is o. In the second definition above, using only o points and parenthesis pairs ( ), it is usually the intent in applications to fill-in the o points with distinct symbols that fit the context of the application.

We have dwelt somewhat overlong on the concept of a bracketing as described above because it will come to play a major role is our subsequent developments of coupling schemes for angular momenta in consequence of the notion of transformations between the various bracketings of order n.

The above definition, as noted, is nonconstructive, although a constructive definition can be described (see the Preface). Instead, we use the diagrammatic presentation afforded by the notion of a binary tree of order n, which places the notion of a binary bracketing on a fully combinatoric and constructive footing. We have described in detail in [L] how each binary tree is constructed from the root of the tree by the process of bifurcation of a point or by termination in an endpoint. The n o points of a binary tree are the endpoints, and all internal points including the root of the tree are denoted by •, as illustrated in many binary tree diagrams to follow. This division into n o endpoints and $n-1$ internal points is very useful for our purposes — it is one-to-one with the notion of a bracketing of the n points o o $\cdots$ o by $n-1$ parenthesis pairs ( ), as discussed further below.

The construction of binary trees by the bifurcation-termination procedure and the level enumeration is described in detail in [L]. We next recall the diagrams for $n = 4$ for comparison with the bracketing concept.

Example. Binary trees of order $n = 4$:

$(((\text{o o}) \text{o}) \text{o}) \quad \rightarrow$

$(\text{o} (\text{o} (\text{o o}))) \quad \rightarrow$

$$((\circ\,\circ)(\circ\,\circ)) \quad \rightarrow \quad \text{[tree]} \quad \rightarrow \quad \begin{pmatrix} \circ & \circ & \bullet \\ \circ & \circ & \bullet \\ \bullet & \bullet & \bullet \end{pmatrix}$$

$$((\circ\,(\circ\,\circ))\,\circ) \quad \rightarrow \quad \text{[tree]} \quad \rightarrow \quad \begin{pmatrix} \circ & \circ & \bullet \\ \circ & \bullet & \circ \\ \bullet & \bullet & \bullet \end{pmatrix} \qquad (5.95)$$

$$(\circ\,((\circ\,\circ)\,\circ)) \quad \rightarrow \quad \text{[tree]} \quad \rightarrow \quad \begin{pmatrix} \circ & \bullet & \circ \\ \circ & \circ & \bullet \\ \bullet & \bullet & \bullet \end{pmatrix}$$

The sequences of parentheses and $\circ$'s on the left are just the five binary bracketings of order 4 of the four symbols $\circ\,\circ\,\circ\,\circ$ in the form (5.94). The binary bracketings are then mapped to the binary tree shown, which, in turn, are mapped to the 3×3 arrays of the symbols $\circ$ and $\bullet$ shown. This 3×3 array of $\circ$ and $\bullet$ points on the right of each binary tree is a unique array that we next describe.

A binary trees of arbitrary order may be considered to consist of *forks*, which we next define. There are four types of forks:

(5.96)

The concept of a fork, which is defined to be the triangle of three points and two lines depicted in (5.96) with a line incident on the *root of the fork*, which is always a $\bullet$ point, and two endpoints. The occurrence of four types, which accommodates all four possible ways of assigning $\circ$ and *bullet* to the endpoints, is basic to the quantum theory of angular momentum of composite systems. It is for this reason that we have described in the Preface how an arbitrary bracketing of $n\,\circ$ points can be viewed as the "pasting" of $n-1$ forks. Each of the binary trees of order four in (5.56) has three forks. The array on the right in (5.95) encodes the information as to which types of forks make up the binary tree, and is called a *fork array*. The columns of the fork array are read off the binary tree by starting at the top level and reading left-to-right across the successive levels, where the assignment of $\circ$ and $\bullet$ to the endpoints of each fork is in the clockwise sense about the $\bullet$ root of each fork.

5.6. BINARY COUPLING OF ANGULAR MOMENTA

We further make the following definition:

Definition: The shape of a binary tree of order n is defined to be the binary bracketing of order n to which it corresponds.

The shape of a binary tree is read off in the obvious way directly from the binary tree, or given the shape — the binary bracketing — the diagram of the binary tree can be drawn. The number of binary trees of shape n is given by the Catalan number a_n:

$$a_n = |\mathbb{T}_n| = \frac{1}{n}\binom{2n-2}{n-1}, \quad n \geq 1. \tag{5.97}$$

The positive integer n is the number of $\circ$ points, $n-1$ is the number of $\bullet$ points (also the number of forks), and $2n-2$ is the total number of points, not counting the root of the entire tree. Endpoints are of degree 1, bullet points of degree 3, except for the tree root, which is always of degree 2.

In general, a binary tree of order n has $n-1$ forks selected from the set of four types (5.95) that are pasted together in various ways to construct the full set $\mathbb{T}_n$ of binary trees of order n, the number of which is the Catalan number a_n. Each a_n distinct fork array then gives a corresponding $3 \times (n-1)$ triangle pattern; there are five such fork matrices shown in (5.95). All of this is set forth in great detail in [L] in its application to the binary coupling theory of angular momenta, with many more examples being exhibited in this volume. Here, however, we develop much further the notion of *transformation of shapes* and its relationship to the enumeration of all complete sets of commuting Hermitian operators (5.77)-(5.78), the systematic calculation of all $3(n-1) - j$ coefficients as doubly stochastic matrices, and their probabilistic interpretations, and more:

All of these complex structures emerge from the simple notion of shapes composed from the primitive elements $\circ$ and ( ).

It is useful to illustrate again these concepts for the addition of four angular momenta, since we have at our disposal in (5.95) the full set of shapes, the level diagram of the binary tree, and the occurrence of each type of fork:

Example. $n = 4$: We have the following labeled binary tree corresponding to the last shape in (5.95):

$$(j_2 ((j_4\, j_1)\, j_3)) \;\rightarrow\; \begin{array}{c} j_4\, \circ \quad \circ\, j_1 \\ k_1 \bullet \quad \circ\, j_3 \\ j_2\, \circ \quad \bullet\, k_2 \\ \bullet \\ j \end{array} \;\rightarrow\; \begin{pmatrix} j_4 & k_1 & j_2 \\ j_1 & j_3 & k_2 \\ k_1 & k_2 & j \end{pmatrix} \tag{5.98}$$

This diagram corresponds to the binary bracketing of the total angular momentum **J** as encoded exactly by the shape of the binary tree:

$$\mathbf{J} = \Big(\mathbf{J}(2) + \big((\mathbf{J}(4) + \mathbf{J}(1)) + \mathbf{J}(3)\big)\Big). \qquad \Box \qquad (5.99)$$

While we could have labeled the ○ points by these four angular momenta, we follow the practice (used subsequently) of labeling these points by the j_i eigenvalue quantum numbers of the respective squared angular momenta. Thus, the four ○ points are labeled by the permutation of the angular momentum labels in the set $\{j_1, j_2, j_3, j_4\}$ that corresponds to the placement of the angular momentum operators in the shape in (5.99). The • points are labeled by the coupled angular momentum quantum numbers k_1, k_2, and j, where the root of the entire tree is always to be labeled by j. It will be noticed that each parenthesis pair in the shape symbol determines a coupled angular momentum : $(j_4\,j_1)$ determines k_1; $((j_4\,j_1)j_3)$ determines k_2; $(j_2((j_4\,j_1)j_3)$ determines j. The triangle pattern at the right in (5.98) then records the angular momenta associated with each fork, as read clockwise around the fork as illustrated. As explained earlier, each column of the triangle pattern constitutes an elementary triangle of quantum labels subject to the CG rule for the addition of pairs of angular momenta as expressed in relations (5.83)-(5.84). $\Box$

We will be making many assignments of labels to the points of a binary tree of order n, which always has n ○ points as endpoints, $n-1$ • points as roots of forks, with the • point at the lowest level called the root of the binary tree. We recall from [L] what we call the standard assignment of quantum numbers for the complete set of commuting Hermitian operators in (5.77)-(5.78):

Standard rule: There are three parts to the rule, the first two referring to the shape of the binary tree, the third to the triangle pattern (at the right in (5.98)):

(i) Assign any one-line permutation $j_\pi = (j_{\pi_1}, j_{\pi_2}, \ldots, j_{\pi_n})$, $\pi \in S_n$, of $(j_1, j_2, \ldots, j_n)$ to the sequence ○ ○ $\cdots$ ○ of n endpoints in the shape symbol, as read left-to-right, ignoring the parenthesis pairs;

(ii) assign j to the root of the entire binary tree, and then the k_i by the level rule: start at the top level of the binary tree and assign $k_1, k_2, \ldots, k_{n-2}$ successively to the root • point of each fork as read from left-to-right across each level, starting at the top level and proceeding down the levels to level 1 at the bottom (the root of the entire tree is level 0); and

(iii) assign the entries in the $3 \times (n-1)$ triangle pattern one-to-one with the k_i root label of each fork, so that column i of the triangle pattern is col(a_i, b_i, k_i), where (a_i, b_i) are the left-right pair of labels of the fork k_i, there being in general four such types (5.96).

5.7 State Vectors: Uncoupled and Coupled

We first illustrate the preceding results in detail for the first few cases:

Examples. Uncoupled and coupled state vectors for $n = 2, 3, 4$:

1. $n = 2$. Standard addition of two angular momenta. Wigner-Clebsch-Gordon coefficients. There is only one binary tree of order two:

$$(\circ \, \circ) \mapsto T = \vee \mapsto \begin{pmatrix} \circ \\ \circ \\ \bullet \end{pmatrix}. \tag{5.100}$$

(a) Uncoupled basis of $\mathcal{H}_{j_1} \otimes \mathcal{H}_{j_2}$:

$$\{|j_1\, m_1\rangle \otimes |j_2\, m_2\rangle \,|\, m_1, m_2 \in \mathbb{C}(j_1, j_2)\}, \tag{5.101}$$

$$\mathbb{C}(j_1, j_2) = \left\{ m_1, m_2 \,\middle|\, \begin{array}{l} m_1 = j_1, j_1 - 1, \ldots, -j_1; \\ m_2 = j_2, j_2 - 1, \ldots, -j_2 \end{array} \right\}, \tag{5.102}$$

(b) Coupled basis of $\mathcal{H}_{j_1} \otimes \mathcal{H}_{j_2}$ (no intermediate angular momenta):

$$|T(j_1, j_2)_{j\,m}\rangle = \sum_{m_1, m_2} C^{j_1\, j_2\, j}_{m_1\, m_2\, m} |j_1\, m_1\rangle \otimes |j_2\, m_2\rangle, \tag{5.103}$$

$$\mathbb{R}_T(j_1, j_2) = \left\{ j, m \,\middle|\, \begin{array}{l} j = j_1 + j_2, j_1 + j_2 - 1, \ldots, |j_1 - j_2|; \\ m = j, j - 1, \ldots, -j \end{array} \right\}. \tag{5.104}$$

The coefficients $C^{j_1\, j_2\, j}_{m_1\, m_2\, m}$ in the linear transformation (5.103) of the uncoupled basis of the space $\mathcal{H}_{j_1} \otimes \mathcal{H}_{j_2}$ are called Wigner-Clebsch-Gordan (WCG) coefficients. These coefficients effect a real orthogonal transformation of the uncoupled basis to the coupled basis on which the total angular momentum has the standard action.

The two complete sets of commuting Hermitian operators for the uncoupled and coupled bases are

$$\{\mathbf{J}_1^2, J_3(1), \mathbf{J}^2(1), J_3(2)\} \text{ and } \{\mathbf{J}^2(1), \mathbf{J}^2(2), \mathbf{J}^2, J_3\}, \tag{5.105}$$

which are diagonal, respectively, on the state vectors (5.101) with eigenvalues given by (1.7) and on the state vectors (5.103) with eigenvalues given by (5.78) in which no intermediate angular momenta appear.

For specified $j_1, j_2 \in \{0, 1/2, 1, \ldots\}$, the WCG coefficient $C^{j_1\, j_2\, j}_{m_1\, m_2\, m}$ is defined for all values $m_1, m_2 \in \mathbb{C}(j_1, j_2)$ defined by (5.102) and all values $j, m \in \mathbb{R}_T(j_1, j_2)$ defined by (5.104); its value is 0, unless $m_1 + m_2 = m$. This value 0 is required to accommodate the operator identity $J_3 =$

$J_3(1) + J_3(2)$. For all other values of m_1, m_2, j, m the WCG coefficient $C_{m_1 m_2 m}^{j_1 j_2 j}$ is undefined. (This way of defining the WCG coefficient has many advantages.)

The conditions

$$j \in \{j_1 + j_2, j_1 + j_2 - 1, \ldots, |j_1 - j_2|\}, \quad m = m_1 + m_2 \qquad (5.106)$$

are called the *triangle rule* for the addition of two angular momenta, and the projection quantum number *sum rule*. The fact that the two sets of orthonormal vectors, (5.101) and (5.103), span the same vector space $\mathcal{H}_{j_1} \otimes \mathcal{H}_{j_2}$ implies the dimensionality relationship:

$$\sum_{j=|j_1-j_2|}^{j_1+j_2} (2j+1) = (2j_1+1)(2j_2+1), \qquad (5.107)$$

which, of course, can be verified directly.

The orthonormality of the set of coupled state vectors follows from their being simultaneous eigenvectors of the two sets of commuting Hermitian operators (5.105). We write these relation in the following forms:

$$\sum_{m_1, m_2} C_{m_1 m_2 m}^{j_1 j_2 j} C_{m_1 m_2 m}^{j_1 j_2 j'} = \delta_{j,j'}, \qquad (5.108)$$

$$\sum_{j,m} C_{m_1 m_2 m}^{j_1 j_2 j} C_{m'_1 m'_2 m}^{j_1 j_2 j} = \delta_{m_1, m'_1} \delta_{m_2, m'_2}, \qquad (5.109)$$

where the summations in each of these relations is to be carried out over all values for which the coefficients are defined.

The WCG coefficients for the addition of two angular momentum can be represented in terms of a binary tree on three points: two external points presented as ○ and one root point presented as ● (single fork):

$$C\begin{pmatrix} j_1\, m_1 \quad j_2\, m_2 \\ \diagdown\!\diagup \\ j\, m \end{pmatrix} = C_{m_1 m_2 m}^{j_1 j_2 j}, \qquad (5.110)$$

for all quantum labels such that the WCG coefficients are defined. The eigenvectors (5.103), presented in terms of the binary tree notation, are:

$$\left| \begin{matrix} j_1 \quad j_2 \\ \diagdown\!\diagup \\ j\, m \end{matrix} \right\rangle = \sum_{m_1, m_2} C_{m_1 m_2 m}^{j_1 j_2 j} |j_1\, m_1\rangle \otimes |j_2\, m_2\rangle. \qquad (5.111)$$

5.7. STATE VECTORS: UNCOUPLED AND COUPLED

There is only one shape (∘∘) and two ways of assigning (j_1, j_2) to the endpoints of the corresponding binary tree: The one given by (5.111), and the second one defined by

$$\left| \begin{array}{cc} j_2 & j_1 \\ & \vee \\ & jm \end{array} \right\rangle = \sum_{m_1, m_2} C^{j_2 \, j_1 \, j}_{m_1 \, m_2 \, m} \, |j_1 \, m_1\rangle \otimes |j_2 \, m_2\rangle$$

$$= (-1)^{j_1+j_2-j} \left| \begin{array}{cc} j_2 & j_1 \\ & \vee \\ & jm \end{array} \right\rangle. \qquad (5.112)$$

This identity of state vectors is a consequence of the well-known symmetry relation of a WCG coefficient given by (see [6] for a list of such symmetries in the notations used here):

$$C^{j_2 \, j_1 \, j}_{m_2 \, m_1 \, m} = (-1)^{j_1+j_2-j} C^{j_1 \, j_2 \, j}_{m_1 \, m_2 \, m}. \qquad (5.113)$$

The sets of vectors (5.111)-(5.112) satisfy the following orthonormality relations:

$$\left\langle \begin{array}{cc} j_1 & j_2 \\ \vee \\ jm \end{array} \middle| \begin{array}{cc} j_1 & j_2 \\ \vee \\ j'm' \end{array} \right\rangle = \delta_{j,j'} \delta_{m,m'},$$

$$\left\langle \begin{array}{cc} j_2 & j_1 \\ \vee \\ jm \end{array} \middle| \begin{array}{cc} j_2 & j_1 \\ \vee \\ j'm' \end{array} \right\rangle = \delta_{j,j'} \delta_{m,m'}, \qquad (5.114)$$

$$\left\langle \begin{array}{cc} j_1 & j_2 \\ \vee \\ jm \end{array} \middle| \begin{array}{cc} j_2 & j_1 \\ \vee \\ j'm' \end{array} \right\rangle = (-1)^{j_1+j_2-j} \delta_{j,j'} \delta_{m,m'}.$$

In particular, the orthonormal basis set of vectors given by

$$\mathbf{B}_{j_1,j_2} = \left\{ \left| \begin{array}{cc} j_1 & j_2 \\ \vee \\ jm \end{array} \right\rangle \,\middle|\, \begin{array}{l} j = |j_1 - j_2|, |j_1 - j_2| + 1, \ldots, j_1 + j_2; \\ m = j, j-1, \ldots, -j \end{array} \right\} \qquad (5.115)$$

defines a new basis of the vector space $\mathcal{H}_{j_1} \otimes \mathcal{H}_{j_2}$ on which the set of operators $\mathbf{J}^2(1), \mathbf{J}^2(2), \mathbf{J}^2, J_3$ is diagonal. Moreover, the subspace with basis $\mathbf{B}_{j_1,j_2,j}$ obtained from (5.115) for prescribed j, but with $m = j, j-1, \ldots, -j$ in (5.115) is an $SU(2)$–multiplet of dimension $2j+1$ on which the total angular momentum $\mathbf{J}$ has the standard action.

It is convenient already at this initial stage to introduce the notion of a one-column triangle pattern and a triangle coefficient of order two.

Definitions. A one-column triangle pattern is any triple (a, b, c) of angular momentum quantum numbers, where $a, b, c \in \{0, 1/2, 1, 3/2, \ldots\}$, such that $c \in \langle a, b \rangle$, now presented as a column, in the triangle coefficient of order two defined by

$$\left\{ \begin{array}{c|c} a & a' \\ b & b' \\ c & c' \end{array} \right\} = \sum_{\alpha,\beta} C^{a\ b\ c}_{\alpha\ \beta\ \gamma} C^{a'\ b'\ c'}_{\alpha\ \beta\ \gamma'}, \qquad (5.116)$$

in which (a', b') is a permutation of (a, b). Thus, a triangle coefficient of order 2 is a 3×2 array consisting of a left triangle pattern and a right triangle pattern that encodes the orthonormality of the state vectors (5.111) and (5.112) in the quantum numbers $a = j_1, b = j_2, c = j$, where we have relabeled the angular momentum quantum numbers in a more generic notation, as often done. Thus, relations (5.114) give

$$\left\{ \begin{array}{c|c} a & a \\ b & b \\ c & c' \end{array} \right\} = \left\{ \begin{array}{c|c} b & b \\ a & a \\ c & c' \end{array} \right\} = \delta_{c,c'},$$

$$\qquad (5.117)$$

$$\left\{ \begin{array}{c|c} a & b \\ b & a \\ c & c' \end{array} \right\} = \left\{ \begin{array}{c|c} b & a \\ a & b \\ c & c' \end{array} \right\} = (-1)^{a+b-c} \delta_{c,c'}.$$

While this all seems quite trivial, it is the starting point of the theory of general triangle coefficients that arise naturally in our subsequent generalization to standard labeled binary trees and their associated state vectors. Relations (5.117) play an important role in the classification of all possible binary coupled states for the addition of n angular momenta. They are not arbitrary — they originate from the properties of the underlying WCG coefficients and the *basic symmetry* (5.113); they are essential structural elements of the general theory of triangle coefficients of order $2(n-1)$, as shown in [L] and emphasized even greater here.

2. $n = 3$. Addition of three angular momenta. Racah coefficients: There are two binary trees of order three:

$$((\circ \circ) \circ) \mapsto T = \quad \vee \quad \mapsto \quad \begin{pmatrix} \circ & \bullet \\ \circ & \circ \\ \bullet & \bullet \end{pmatrix},$$

$$\qquad (5.118)$$

$$((\circ \circ) \circ) \mapsto T' = \quad \vee \quad \mapsto \quad \begin{pmatrix} \circ & \bullet \\ \circ & \circ \\ \bullet & \bullet \end{pmatrix}.$$

5.7. STATE VECTORS: UNCOUPLED AND COUPLED

(a) Uncoupled basis of $\mathcal{H}_{j_1} \otimes \mathcal{H}_{j_2} \otimes \mathcal{H}_{j_3}$:

$$\{|j_1\,m_1\rangle \otimes |j_2\,m_2\rangle \otimes |j_3\,m_3\rangle|\, m_1, m_2, m_3 \in \mathbb{C}(j_1,j_2,j_2)\}, \tag{5.119}$$

$$\mathbb{C}(j_1,j_2,j_2) = \left\{ m_1, m_2, m_3 \,\middle|\, \begin{array}{l} m_1 = j_1, j_1-1, \ldots, -j_1; \\ m_2 = j_2, j_2-1, \ldots, -j_2; \\ m_3 = j_3, j_3-1, \ldots, -j_3 \end{array} \right\}.$$

(b) First coupled basis of $\mathcal{H}_{j_1} \otimes \mathcal{H}_{j_2} \otimes \mathcal{H}_{j_3}$:

$$|T(j_1,j_2,j_3;k_1)_{j\,m}\rangle$$

$$= \sum_{m_1,m_2,m_3} C_{T(\mathbf{j}\,\mathbf{m};k_1)_{j\,m}} |j_1\,m_1\rangle \otimes |j_2\,m_2\rangle \otimes |j_3\,m_3\rangle, \tag{5.120}$$

where the domain of definition of the quantum numbers in this coupled state vector are given for specified $\mathbf{j} = (j_1, j_2, j_3)$ by

$$\mathbf{R}_T(\mathbf{j}) = \left\{ k_1, j, m \,\middle|\, \begin{array}{l} j = j_{\min}, \ldots, j_{\max}; \\ m = j, j-1, \ldots, -j; \\ k_1 \in \mathbb{K}_T^{(j)}(\mathbf{j}) \end{array} \right\}. \tag{5.121}$$

The domain of definition $\mathbb{K}_T^{(j)}(\mathbf{j})$ of k_1 is given explictly in (5.126) below. The transformation coefficients $C_{T(\mathbf{j}\,\mathbf{m};k_1)_{j\,m}}$ give a real orthogonal transformation from the orthonormal bases of the space $\mathcal{H}_{j_1} \otimes \mathcal{H}_{j_2} \otimes \mathcal{H}_{j_3}$ of dimension $(2j_1+1)(2j_2+1)(2j_3+1)$ to the coupled basis defined by (5.120)-(5.121). Here we have chosen the binary tree $T \in \mathbb{T}_3$ to be the one of shape $((\circ\,\circ)\,\circ)$; we also write relation (5.120) as

$$\left|\begin{array}{c} j_1 \quad\quad j_2 \\ \diagdown\diagup\quad j_3 \\ k_1 \quad\diagup \\ jm \end{array}\right\rangle = \sum_{m_1,m_2,m_3} C_{T(\mathbf{j}\,\mathbf{m};k_1)_{j\,m}} |j_1\,m_1\rangle \otimes |j_2\,m_2\rangle \otimes |j_3\,m_3\rangle. \tag{5.122}$$

The binary bracketing of three angular momentum corresponding to the shape $((\circ\,\circ)\circ)$ is

$$\mathbf{J} = \Big(\big(\mathbf{J}(1)+\mathbf{J}(2)\big)+\mathbf{J}(3)\Big) = \mathbf{K}(1)+\mathbf{J}(3),\ \mathbf{K}(1) = \mathbf{J}(1)+\mathbf{J}(2). \tag{5.123}$$

The transformation coefficients in (5.122) are exactly the product of WCG coefficients read off the pair of forks in the binary tree at the left in the manner of (5.110) (with proper projection quantum numbers adjoined):

$$C_{T(\mathbf{j}\,\mathbf{m};k_1)_{j\,m}} = C^{j_1\ j_2\ k_1}_{m_1,m_2,\mu_1} C^{k_1\ j_3\ j}_{\mu_1,m_3,m}. \tag{5.124}$$

We note, in particular, that this *generalized WCG coefficient* is zero unless $\mu_1 = m_1 + m_2$ and $m = m_1 + m_2 + m_3$; the summation in (5.120) and (5.122) is over all values for which the product of coefficients (5.124) is defined.

The triangle pattern associated with the standard labeled binary tree appearing in the state vector (5.122) is

$$\begin{pmatrix} j_1 & k_1 \\ j_2 & j_3 \\ k_1 & j \end{pmatrix}. \tag{5.125}$$

Thus, for prescribed values of $j_1, j_2, j_3 \in \{0, 1/2, 1, 3/2, \cdots\}$ and $j \in \{j_{\min}, j_{\min} + 1, \ldots, j_{\max} = j_1 + j_2 + j_3\}$, the domain of definition of $k_1 \in \mathbb{K}_T^{(j)}(j_1, j_2, j_3)$ is the intersection of the sets $\langle j_1, j_2 \rangle$ and $\langle j_3, j \rangle$:

$$\mathbb{K}_T^{(j)}(j_1, j_2, j_3) = \langle j_1, j_2 \rangle \cap \langle j_3, j \rangle. \tag{5.126}$$

We can, of course, repeat the above construction for the second shape $(\circ (\circ \circ))$ with the assignment $(j_1 (j_2 j_3))$ to obtain the following relation:

$$\left| \begin{matrix} & j_2 & j_3 \\ j_1 & & k'_1 \\ & j\,m & \end{matrix} \right\rangle = \sum_{m_1, m_2, m_3} C_{T'(\mathbf{j}\,\mathbf{m}; k'_1)_{j\,m}} |j_1\,m_1\rangle \otimes |j_2\,m_2\rangle \otimes |j_3\,m_3\rangle, \tag{5.127}$$

in which $C_{T'(\mathbf{j}\,\mathbf{m}; k'_1)_{j\,m}}$ denotes the product of WCG coefficients given by

$$C_{T'(\mathbf{j}\,\mathbf{m}; k'_1)_{j\,m}} = C^{j_2\ j_3\ k'_1}_{m_2, m_3, \mu'_1} C^{j_1\ k'_1\ j}_{m_1, \mu'_1, m}. \tag{5.128}$$

This binary tree with shape $(\circ (\circ \circ))$ corresponds to the binary bracketing of three angular momentum given by

$$\mathbf{J} = \Big(\mathbf{J}(1) + \big(\mathbf{J}(2) + \mathbf{J}(3) \big) \Big) = \mathbf{J}(1) + \mathbf{K}'(1),\ \mathbf{K}'(1) = \mathbf{J}(2) + \mathbf{J}(3). \tag{5.129}$$

The triangle pattern associated with this binary coupling is

$$\begin{pmatrix} j_2 & j_1 \\ j_3 & k'_1 \\ k'_1 & j \end{pmatrix}. \tag{5.130}$$

Thus, for prescribed values of $j_1, j_2, j_3 \in \{0, 1/2, 1, 3/2, \cdots\}$ and $j \in \{j_{\min}, j_{\min} + 1, \ldots, j_{\max} = j_1 + j_2 + j_3\}$, the domain of definition of $k'_1 \in \mathbb{K}_{T'}^{(j)}(j_1, j_2, j_3)$ is the intersection of the sets $\langle j_2, j_3 \rangle$ and $\langle j_1, j \rangle$:

$$\mathbb{K}_{T'}^{(j)}(j_1, j_2, j_3) = \langle j_2, j_3 \rangle \cap \langle j_1, j \rangle. \tag{5.131}$$

5.7. STATE VECTORS: UNCOUPLED AND COUPLED

The two binary trees in state vectors (5.122) and (5.127), respectively, are those with labeled shapes $((j_1\,j_2)\,j_3)$ and $(j_1\,(j_2\,j_3))$. But there are $3! = 6$ ways to assign j_1, j_2, j, to each shape. Thus, there is a total of twelve state vectors, six of unlabeled shape form $((\circ\circ)\circ)$ and six of unlabeled shape form $(\circ(\circ\circ))$. This, in turn, gives 144 inner products $\langle\cdots|\cdots\rangle$, each of which, in turn, defines a triangle coefficient with a left and right triangle pattern. But we have, symbolically, for each pair T, T' of such labeled binary trees, the identity between inner products and triangle coefficients given by

$$\langle T | T' \rangle = \{L \,|\, R\} = \langle T' | T \rangle = \{R \,|\, L\}. \tag{5.132}$$

The symmetry under interchange of bra- and ket-vectors and, correspondingly, between left and right triangle patterns is because the inner product is real.

The key concept for understanding the structure of triangle patterns and the associated triangle coefficient defined on a left and right triangle pattern is the notion of a *common fork* between two standard labeled binary trees $T, T' \in \mathbb{T}_n$. Thus, let $f(a, b, k_i)$ denote a labeled fork:

$$f(a, b, k_i) \;=\; \underset{k_i}{\overset{a\quad b}{\diagdown\!\diagup}}\;,\quad \diamond = \circ \text{ or } \bullet. \tag{5.133}$$

Definition. Let the fork $f(a, b, k_i)$ be the fork constituent of the standard labeled binary tree $T \in \mathbb{T}_n$, located in the position labeled by the fork root quantum number k_i. Let either the fork $f(a, b, k'_i)$ or the fork $f(b, a, k'_i)$ (a and b interchanged) be the fork constituent of the standard labeled binary tree $T' \in \mathbb{T}_n$ located in the position labeled by the fork root quantum number k'_i. Then, the pair of forks $f(a, b, k_i), f(a, b, k'_i)$ is called a *common fork*, as is also the pair $f(a, b, k_i), f(b, a, k'_i)$.

The concept of a common fork, possessed by two standard labeled binary trees in the set $\mathbb{T}_n$ of all binary trees of order n is an elementary notion, but it has far-reaching consequences in its application to the classification of all binary couplings of n angular momenta. Because of this importance, we formulate this property in full generality here, referring to [L] or to Sect. 5.8 for more detailed defintions. It is an invaluable tool for the calculation of the numerical values of triangle coefficients, as will be demonstrated numerous times in the application to the Heisenberg ring carried out in Chapter 8. We formulate this property of triangle coefficients, and a second equally elementary property for general n, since together they synthesize the method of classification of all binary couplings of n angular momenta:

1. **Reduction property.** The triangle coefficient of a pair of labeled binary trees of order n that possess a common fork is *reduced* from

order $2(n-1)$ to order $2(n-2)$, as expressed by

$$\left\{\cdots \begin{array}{c} a \\ b \\ k_i \end{array} \cdots \middle| \cdots \begin{array}{c} a \\ b \\ k'_{i'} \end{array} \cdots \right\} = \delta_{k_i, k'_{i'}} \left\{ \cdots \middle| \cdots \right\},$$
(5.134)

$$\left\{\cdots \begin{array}{c} a \\ b \\ k_i \end{array} \cdots \middle| \cdots \begin{array}{c} b \\ a \\ k'_{i'} \end{array} \cdots \right\} = \delta_{k_i, k'_{i'}} (-1)^{a+b-k_i} \left\{ \cdots \middle| \cdots \right\},$$

where the triangle coefficient on the right-hand side of this relation is obtained from the one on the left-hand side simply by deleting column i and columns i' from the left and right patterns of the left-hand triangle coefficient. The $\cdots$ denote the remaining unchanged columns. These relations are the generalization of (5.117). A triangle coefficient with no common forks is called *irreducible*.

2. **Phase-transformation property.** All triangle coefficients of order $2(n-1)$ satisfy the phase-transformation property:

$$\left\{\cdots \begin{array}{c} a \\ b \\ k_i \end{array} \cdots \middle| \cdots \right\} = (-1)^{a+b-k_i} \left\{\cdots \begin{array}{c} b \\ a \\ k_i \end{array} \cdots \middle| \cdots \right\}.$$
(5.135)

This relation holds also for the right-most column in the left pattern, where $k_{n-1} = j$, as well as for columns of the right pattern.

Properties (5.134) and (5.135) are easily proved properties of the explicit expressions of triangle coefficients in terms of WCG coefficients (see [L]); they depend only on the symmetry properly (5.113) of the WCG coefficient. Because of this pair of relations, phase transformations of triangle coefficients, which includes those in (5.134), always enter in the form of products of one or more factors of the form

$$\omega_{a,b,k_i} = (-1)^{a+b-k_i}, \text{ all } a, b \in \{0, 1/2, 1, \ldots\}, \ k_i \in \langle a, b \rangle, \quad (5.136)$$

where this factor originates from the WCG symmetry property (5.113). This is the only type of phase factor that occurs in our use of triangle coefficients and their transformation properties. Products, of course, can be simplified in various ways, but it is often advisable to keep the *primitive phase factors* (5.136) in products to avoid errors. (Results equivalent to (5.134)-(5.135) are stated in [L].)

The above properties of triangle coefficients, applied to $n = 3$, give the complete classification of all binary coupling schemes of order $n = 3$, either as products of primitive phase factors or as products of primitive

5.7. STATE VECTORS: UNCOUPLED AND COUPLED

phase factor times a Racah coefficient, multiplied always by a product of $\sqrt{(2k_1+1)(2k_2+1)}$-like dimension factors associated with a pair of internal quantum numbers.

Classification of binary couplings of three angular momenta

The set of $(n!a_n)^2 = ((6)(2))^2 = 144$ triangle coefficients of order four corresponding to all possible assignments of labels to the left and right triangle patterns reduces to 48 that are equal to phase factors or to zero in consequence of a common fork, and 96 that can be obtained from the following one by primitive phase transformations and re-identification of the arguments of the Racah coefficient:

$$\left\{ \begin{array}{c} a \\ k \\ j \end{array} \bigvee \begin{array}{c} b \\ c \end{array} \middle| \begin{array}{c} b \\ a \\ j \end{array} \bigvee \begin{array}{c} c \\ k' \end{array} \right\}$$

$$= \left\{ \begin{array}{cc|cc} a & k & b & a \\ b & c & c & k' \\ k & j & k' & j \end{array} \right\} = \sqrt{(2k+1)(2k'+1)}\, W(abjc; kk'). \quad (5.137)$$

The ninety-six relations between Racah coefficients and triangle coefficients are obtained by operations as follows on relation (5.137): First, we obtain six relations by permuting a, b, c in both sides; then, in the left-hand side of each of the six relations, we perform phase-transformations (5.135) separately on each of the four columns of the triangle coefficient, which gives forty-eight relations; and, finally, we interchange left and right patterns in each of these forth-eight, as well as the interchange of k and k' in the left-hand side, so as to have k in the left pattern and k' in the right pattern. This gives exactly $(6)(8)(2) = 96$ distinct relations between triangle coefficients of order four and Racah coefficients in the standard ordering of patterns:

There is only one coefficient, the Racah coefficient with permutations of its parameters, always multiplied by $\sqrt{(2k+1)(2k'+1)}$ and a primitive phase factor depending on the column involved. This is the complete classification of all binary couplings of three angular momenta.

This result itself is quite striking, but still would be quite a complex result, since a triangle coefficient of order four with no common forks is a summation over four WCG coefficients, each of which itself entails a summation of factorial terms over a single summation parameter. The fact that this complex object can be brought to a comprehensible form — a single summation of factorial terms with all summations over projection quantum numbers effected — is due to Racah [62-63]. The Racah coefficient, in turn, is related to a standard object in mathematics, the $_4F_3$ hypergeometric function of unit argument. We refer to [6] and [L] for a great deal more discussion of this relationship, but note again here, that to our knowledge, *the full implications for hypergeometric function theory have not been set forth, expecially in the context that Racah coefficients are the basic entities out of which triangle coefficients of all orders are built.*

The triangle coefficient of a pair of binary coupling schemes that has no common forks between its left and right triangle patterns does not reduce to a lower order triangle coefficient; the pair of coupling schemes is therefore called *irreducible*, as also is the triangle coefficient. Then, the Racah coefficients relate all other irreducible binary coupling schemes to the one given by (5.137),as described above. We rewrite (5.137) in the standard angular momentum notations:

$$\left| \begin{array}{c} j_2 \diagdown \diagup j_3 \\ j_1 \diagdown k'_1 \\ jm \end{array} \right\rangle = \sum_{k_1 \in \mathbb{K}_T^{(j)}(\mathbf{j})} \left\{ \begin{array}{cc|cc} j_1 & k_1 & j_2 & j_1 \\ j_2 & j_3 & j_3 & k'_1 \\ k_1 & j & k'_1 & j \end{array} \right\} \left| \begin{array}{c} j_1 \diagdown \diagup j_2 \\ k_1 \diagdown j_3 \\ jm \end{array} \right\rangle, \quad (5.138)$$

each $k'_1 \in \mathbb{K}_{T'}^{(j)}(\mathbf{j})$, where the transformation coefficients are given by

$$\left\{ \begin{array}{cc|cc} j_1 & k_1 & j_2 & j_1 \\ j_2 & j_3 & j_3 & k'_1 \\ k_1 & j & k'_1 & j \end{array} \right\} = \sqrt{(2k_1+1)(2k'_1+1)} W(j_1 j_2 j j_3; k_1 k'_1). \quad (5.139)$$

The orthornormal basis vectors given above that span the tensor product space $\mathcal{H}_{j_1} \otimes \mathcal{H}_{j_2} \otimes \mathcal{H}_{j_3}$ can be partitioned into various subsets that span sub vector spaces with properties of particular interest. Two such subsets of basis vectors are defined from (5.122) by

$$\mathbf{B}_{\mathbf{j},j;k_1} = \left\{ \left| \begin{array}{c} j_1 \diagdown \diagup j_2 \\ k_1 \diagdown j_3 \\ jm \end{array} \right\rangle \middle| m = j, j-1, \ldots, -j \right\},$$

$$k_1 \in \mathbb{K}_T^{(j)}(\mathbf{j}), \quad \mathbf{j} = (j_1, j_2, j_3); \quad (5.140)$$

$$\mathbf{B}_{\mathbf{j},j,m} = \left\{ \left| \begin{array}{c} j_1 \diagdown \diagup j_2 \\ k_1 \diagdown j_3 \\ jm \end{array} \right\rangle \middle| k_1 \in \mathbb{K}_T^{(j)}(\mathbf{j}) \right\}, \quad \mathbf{j} = (j_1, j_2, j_3).$$

The first basis $\mathbf{B}_{\mathbf{j},j;k_1}$ defines a vector subspace of dimension $2j+1$ of the tensor product space $\mathcal{H}_{j_1} \otimes \mathcal{H}_{j_2} \otimes \mathcal{H}_{j_3}$, there being $|\mathbb{K}_T^{(j)}(\mathbf{j})| = N_j(\mathbf{j})$ such

5.7. STATE VECTORS: UNCOUPLED AND COUPLED

perpendicular spaces, each of which is the carrier space of the irreducible representation $D^j(U)$ of $SU(2)$ under $SU(2)$ frame rotations, since the total angular momentum $\mathbf{J}$ has the standard action on each of these subspaces. The second basis $\mathbf{B_{j,}}_{j,m}$ defines a vector subspace $\mathcal{H}_{\mathbf{j},j,m}$ of dimension $N_j(\mathbf{j})$, there being such a perpendicular vector subspace of $\mathcal{H}_{j_1} \otimes \mathcal{H}_{j_2} \otimes \mathcal{H}_{j_3}$ for each $j = j_{\min}, j_{\min}+1, \ldots, j_{\max}$ and each $m = j, j-1, \ldots, -j$. Thus, the counting of all the vectors in either of the two subspace decompositions gives the formula:

$$(2j_1+1)(2j_2+1)(2j_3+1) = \sum_{j=j_{\min}}^{j_{\max}} N_j(\mathbf{j})(2j+1). \tag{5.141}$$

The transformation (5.138) is between different basis sets of the same vector subspace $\mathcal{H}_{\mathbf{j},j,m}$. This is just one example of many such transformations, as we next describe. The following result holds for $n = 3$:

Let $|T(\mathbf{j}_\pi k_1)_{jm}\rangle$, each $T \in \mathbb{T}_3, \pi \in S_3$, denote any of the $3!(2) = 12$ standard labeled binary coupling schemes. Then, each irreducible pair (no common fork) of coupled state vectors in this set contains a basis of $N_j(\mathbf{j})$ orthonormal vectors that span the same space $\mathcal{H}_{\mathbf{j},j,m}$. The inner product of basis vectors gives the irreducible triangle coefficients of order six that determine the real linear transformation between basis vectors, as well as the domains of definition of the intermediate quantum numbers that enumerate the orthonormal basis vectors.

It is convenient to note here that the so-called $6-j$ coefficient is often introduced in place of the Racah W-coefficient because of the enhanced symmetries that the $6-j$ coefficient exhibits:

$$\left\{ \begin{array}{ccc} a & b & k \\ c & j & k' \end{array} \right\} = (-1)^{a+b+c+j} W(abjc; kk'). \tag{5.142}$$

We prefer to use the triangle pattern notation for two reasons: (i) It exhibits exactly through its columns the triangles that enter into the definition of the coefficient, as well as their domains of definition by application of the simple CG rule; and (ii) it generalizes to a universal notation for the binary coupling of arbitrary many angular momenta in which the accounting for all elementary triangles and phase factors is manageable.

We have reviewed in some detail the coupling of three angular momenta because it indicates clearly the structures that will be encountered in the general case of n angular momenta. The relation to recoupling matrices and to cubic graphs is still to be supplied — we defer this after considering $n = 4$ and the appearance of the Wigner $9-j$ coefficient.

3. $n = 4$. Addition of four angular momenta. The Wigner $9-j$ coefficients (Wigner [83]). The five binary trees of order four are listed in

(5.95), together with their shapes and fork matrices.

(a) Uncoupled basis of $\mathcal{H}_{j_1} \otimes \mathcal{H}_{j_2} \otimes \mathcal{H}_{j_3} \otimes \mathcal{H}_{j_4}$:

$$\left\{ \begin{array}{c} |j_1\, m_1\rangle \otimes |j_2\, m_2\rangle \\ \otimes |j_3\, m_3\rangle \otimes |j_4\, m_4\rangle \end{array} \,\bigg|\, m_1, m_2, m_3, m_4 \in \mathbb{C}(j_1, j_2, j_3, j_4) \right\}, \tag{5.143}$$

$$\mathbb{C}(j_1, j_2, j_3, j_4) = \left\{ m_1, m_2, m_3, m_4 \,\bigg|\, \begin{array}{l} m_1 = j_1, j_1 - 1, \ldots, \ -j_1; \\ m_2 = j_2, j_2 - 1, \ldots, -j_2; \\ m_3 = j_3, j_3 - 1, \ldots, -j_3; \\ m_4 = j_4, j_4 - 1, \ldots, -j_4 \end{array} \right\}.$$

(b) First coupled basis of $\mathcal{H}_{j_1} \otimes \mathcal{H}_{j_2} \otimes \mathcal{H}_{j_3} \otimes \mathcal{H}_{j_4}$: To illustrate such a basis, we use the binary tree from the list (5.95) that has shape (binary bracketing) given by $\bigl((\circ\circ)(\circ\circ)\bigr) \mapsto T$, which we label in accordance with the angular momentum pairwise additions given by

$$\mathbf{K}(1) = \mathbf{J}(1) + \mathbf{J}(2), \quad \mathbf{K}(2) = \mathbf{J}(3) + \mathbf{J}(4), \quad \mathbf{J} = \mathbf{K}(1) + \mathbf{K}(2). \tag{5.144}$$

Then, we have the following definition of state vectors:

$$|T(\mathbf{j}\,\mathbf{k})_{j\,m}\rangle = \left| \begin{array}{c} j_1 \quad j_2 \quad j_3 \quad j_4 \\ k_1 \diagdown\diagup \diagdown\diagup k_2 \\ j\,m \end{array} \right\rangle = \sum_{m_1, m_2, m_3, m_4} C_{T(\mathbf{j}\,\mathbf{m};\mathbf{k})_{j\,m}}$$

$$\times |j_1\, m_1\rangle \otimes |j_2\, m_2\rangle \otimes |j_3\, m_3\rangle \otimes |j_4\, m_4\rangle, \tag{5.145}$$

where the real orthogonal transformation coefficients between the two orthonormal bases of the space $\mathcal{H}_{j_1} \otimes \mathcal{H}_{j_2} \otimes \mathcal{H}_{j_3} \otimes \mathcal{H}_{j_4}$ are given by

$$C_{T(\mathbf{j}\,\mathbf{m};\mathbf{k})_{j\,m}} = C^{j_1\ j_2\ \ k_1}_{m_1, m_2, m_1+m_2}\, C^{j_3\ j_4\ \ k_2}_{m_3, m_4, m_3+m_4}\, C^{k_1\ \ k_2\ \ j}_{m_1+m_2, m_3+m_4, m}, \tag{5.146}$$

in which $\mathbf{j}\,\mathbf{m} = (j_1, j_2, j_3, j_4, m_1, m_2, m_3, m_4)$ and $\mathbf{k} = (k_1, k_2)$. Again, the generalized WCG coefficient $C_{T(\mathbf{j}\,\mathbf{m};\mathbf{k})_{j\,m}}$ is the product of ordinary WCG coefficients as read-off the three forks of the binary tree on the left, and the summation in (5.145) is over all values of the projection quantum numbers m_i for which the generalized WCG coefficient is defined. The pair of complete sets of mutually commuting Hermitian operators that define the uncoupled basis (5.143) and coupled basis (5.145) are

$$\mathbf{J}^2(1), J_3(1), \mathbf{J}^2(2), J_3(2), \mathbf{J}^2(3), J_3(3), \mathbf{J}^2(4), J_3(4); \tag{5.147}$$

$$\mathbf{J}^2, J_3, \mathbf{J}^2(1), \mathbf{J}^2(2), \mathbf{J}^2(3), \mathbf{J}^2(4), \mathbf{K}_T^2(1), \mathbf{K}_T^2(2).$$

5.7. STATE VECTORS: UNCOUPLED AND COUPLED

The first set of operators has the simultaneous eigenvectors and eigenvalues given by (1.7), while the second set those given by (5.78). We repeat the latter for use in the Heisenberg ring problem in Chapter 8:

$$\mathbf{J}^2|T(\mathbf{jk})_{jm}\rangle = j(j+1)|T(\mathbf{jk})_{jm}\rangle,\ J_3|T(\mathbf{jk})_{jm}\rangle = m|T(\mathbf{jk})_{jm}\rangle;$$
$$\mathbf{J}^2(i)|T(\mathbf{jk})_{jm}\rangle = j_i(j_i+1)|T(\mathbf{jk})_{jm}\rangle,\ i=1,2,3,4; \quad (5.148)$$
$$\mathbf{K}_T^2(i)|T(\mathbf{jk})_{jm}\rangle = k_i(k_i+1)|T(\mathbf{jk})_{jm}\rangle,\ i=1,2.$$

There are $n!a_n = 4!(5) = 120$ ways of defining a binary coupled state vector, corresponding to the five binary trees (5.95) and the twenty-four ways of assigning j_1, j_2, j_3, j_4 to each shape. For each such assignment the domain of definition of the intermediate angular momenta (k_1, k_2) is uniquely determined by the shape and the assignment of the j_1, j_2, j_3, j_4 to the shape. This is illustrated fully by the simultaneous eigenvectors (5.145) of the complete set of commuting Hermitian operators in relations (5.147). Thus, the triangle pattern for the coupling scheme (5.145) is given by

$$T(\mathbf{jk})_j = \begin{matrix} j_1 & j_2 & j_3 & j_4 \\ k_1 & & & k_2 \\ & & j & \end{matrix} \mapsto \begin{pmatrix} j_1 & j_3 & k_1 \\ j_2 & j_4 & k_2 \\ k_1 & k_2 & j \end{pmatrix}. \quad (5.149)$$

Thus, for specified $j \in \{j_{\min}, j_{\min}+1, \ldots, j_{\max}\}$, the domains of definition of k_1 and k_2 are just the set of values for which all three columns of the triangle pattern (5.149) satisfy the triangle rules; that is,

$$(k_1, k_2) \in \mathbb{K}_T^{(j)}(\mathbf{j}) = \left\{ (k_1, k_2) \left| \begin{matrix} k_1 \in \langle j_1, j_2 \rangle, k_1 \in \langle k_2, j \rangle, \\ k_2 \in \langle j_3, j_4 \rangle, k_2 \in \langle k_1, j \rangle \end{matrix} \right. \right\}. \quad (5.150)$$

We recall that the cardinality of this set is $|\mathbb{K}_T^{(j)}(\mathbf{j})| = N_j(\mathbf{j})$, independently of the binary tree T and its labeling.

But now there are $(120)^2 = 14,400$ inner product corresponding to all ways of assigning the $4!= 24$ permutations of (j_1, j_2, j_3, j_4) to the ∘ points of the 5 unlabeled binary trees (5.95), or $7,200$, accounting for the symmetry (5.132)) of the real inner product. The problem is to classify all these inner products, or, equivalently, the associated triangle coefficients of order six, which consist of a left pattern and a right pattern, such a pattern being illustrated by (5.149). If the inner product is for a pair of labeled binary trees that share a common fork, then a lower order triangle coefficient obtains in consequence of the reduction property (5.34); if the inner product is for an irreducible pair of labeled binary trees (no common fork), then it will have an irreducible triangle coefficient of order six. Thus, the determination of the structure of all such irreducible triangle coefficients of order six is the problem that is posed. Every such problem is well-defined, since such inner products are fully determined in terms of unique generalized WCG coefficients associated with the pair of binary trees that define the triangle coefficient. Relations (5.134)-(5.135) are crucial to the classification problem. This is the subject dealt with at great length in [L], including the role of cubic graphs in classifying $3(n - 1) - j$ coefficients into "types." The main result we need to know here for the $9 - j$ coefficient is that there is only one $9 - j$ coefficient, the Wigner $9 - j$ coefficient. It can, however, arise from various pairs of irreducible binary trees of order four and their irreducible triangle coefficients of order six. It is also the case that certain irreducible triangle coefficients that arise from irreducible pairs of binary trees of order 4 are, first of all, expressible as a sum over three irreducible triangle coefficients of order four, but the summation can be effected to give a simple product of two irreducible triangle coefficients of order four. In this case, the irreducible triangle coefficient does not lead to a new object — it is not called a $9-j$ coefficient. This happens for the important Biedenharn-Elliott (B-E) identity (see [L]). This depends of the concept of a path and the property that there can be more than one path that transforms one labeled binary tree of order four into a second one of order four. In the general case, we have the following result:

Irreducible triangle coefficients always determine a unique cubic graph, but the details of their structure depends crucially on the structure of the cubic graph itself through a result known as the Yutsis factor theorem. Certain types of cubic graphs lead to $3(n - 1) - j$ coefficients that can be expressed as summations over the basic irreducible triangle coefficients of order four, but which are also summable into a product of lower-order $3r - j$ coefficients, or as a summation over such products. This is an important and general aspect of the binary coupling theory of angular momenta, which produces countably infinite many relations between irreducible triangle coefficients of order four.

This role of cubic graphs is significant — they distinguish between the irreducible triangle coefficients that factor into lower order triangle coefficients, and those that do not. Thus, cubic graphs have an important role in determining the properties of irreducible triangle coefficients, although, as will be demonstrated many times in this volume, they have no direct role in our method of expressing the triangle coefficients in terms of the irreducible triangle coefficients of order four.

We repeat: The problem of determining which irreducible triangle coefficients of order 6 factor and those that do not, that is, those that give the $9 - j$ coefficient is solved completely by writing out (uniquely) the cubic graph from the six triangles corresponding to the columns of all irreducible pairs of labeled binary trees — those with isomorphic

5.7. STATE VECTORS: UNCOUPLED AND COUPLED

cubic graphs give the same Wigner $9-j$ coefficient, usually in different forms. The classification is a finite problem, of the same sort as done in Example 3 for Racah coefficients. We have not carried this out. Instead, we present here, and subsequently in Chapter 9, several such versions of the $9-j$ coefficient.

The $9-j$ coefficient can be obtained as the irreducible triangle coefficient that gives the transformation coefficient between the orthonormal basis (5.145) and a second such basis of the same unlabeled shape, now filled-in with $((j_1\,j_3)(j_2\,j_4))$. The binary bracketing of angular momenta is given by

$$\mathbf{K}'(1) = \mathbf{J}(1) + \mathbf{J}(3), \quad \mathbf{K}'(2) = \mathbf{J}(2) + \mathbf{J}(4), \quad \mathbf{J} = \mathbf{K}'(1) + \mathbf{K}'(2). \quad (5.151)$$

Because of the importance of these relations in the Heisenberg ring problem in Chapter 8, we present this second basis in full detail:

(c) Second coupled basis of $\mathcal{H}_{j_1} \otimes \mathcal{H}_{j_2} \otimes \mathcal{H}_{j_3} \otimes \mathcal{H}_{j_4}$: We have the following definition of state vectors:

$$|T(\mathbf{j}'\,\mathbf{k}')_{j\,m}\rangle = \left| \begin{array}{c} j_1 \quad j_3 \quad j_2 \quad j_4 \\ k'_1 \quad\quad\quad k'_2 \\ j\,m \end{array} \right\rangle = \sum_{m_1,m_2,m_3,m_4} C_{T(\mathbf{j}'\,\mathbf{m}';\mathbf{k}')_{j\,m}}$$

$$\times |j_1\,m_1\rangle \otimes |j_2\,m_2\rangle \otimes |j_3\,m_3\rangle \otimes |j_4\,m_4\rangle, \quad (5.152)$$

where the real orthogonal transformation between the two orthonormal bases of the space $\mathcal{H}_{j_1} \otimes \mathcal{H}_{j_2} \otimes \mathcal{H}_{j_3} \otimes \mathcal{H}_{j_4}$ is given by

$$C_{T(\mathbf{j}'\,\mathbf{m}';\mathbf{k}')_{j\,m}} = C^{j_1\ \ j_3\ \ k'_1}_{m_1,m_3,m_1+m_3} C^{j_2\ \ j_4\ \ k'_2}_{m_2,m_4,m_2+m_4} C^{k'_1\ \ k'_2\ \ j}_{m_1+m_3,m_2+m_4,m}, \quad (5.153)$$

in which $\mathbf{j}'\,\mathbf{m}' = (j_1, j_3, j_2, j_4, m_1, m_3, m_2, m_4)$ and $\mathbf{k}' = (k'_1, k'_2)$. Again, the generalized WCG coefficient $C_{T(\mathbf{j}'\,\mathbf{m}';\mathbf{k}')_{j\,m}}$ is the product of ordinary WCG coefficients as read-off the three forks of the binary tree on the left, and the summation in (5.152) is over all values of the projection quantum numbers m_i for which the generalized WCG coefficient is defined. We note, in particular, that the summation over $\mathbf{m} = (m_1, m_2, m_3, m_4)$ is not an error, since each m_i in $\mathbf{m} = (m_1, m_3, m_2, m_4)$ is matched with j_i in $\mathbf{j}' = (j_1, j_3, j_2, j_4)$.

The complete set of mutually commuting Hermitian operators that define the coupled basis (5.152) is

$$\mathbf{J}^2, J_3, \mathbf{J}^2(1), \mathbf{J}^2(2), \mathbf{J}^2(3), \mathbf{J}^2(4), \mathbf{K}'^2_T(1), \mathbf{K}'^2_T(2). \quad (5.154)$$

This set of operators has the simultaneous eigenvectors and eigenvalues as follows:

$$\mathbf{J}^2|T(\mathbf{j'}\,\mathbf{k'})_{jm}\rangle = j(j+1)|T(\mathbf{j'}\,\mathbf{k'})_{jm}\rangle,\ J_3|T(\mathbf{j'}\,\mathbf{k'})_{jm}\rangle = m|T(\mathbf{j'}\,\mathbf{k'})_{jm}\rangle;$$
$$\mathbf{J}^2(i)|T(\mathbf{j'}\,\mathbf{k'})_{jm}\rangle = j_i(j_i+1)|T(\mathbf{j'}\,\mathbf{k'})_{jm}\rangle,\ i=1,2,3,4; \qquad (5.155)$$
$$\mathbf{K'}^2_T(i)|T(\mathbf{j'}\,\mathbf{k'})_{jm}\rangle = k'_i(k'_i+1)|T(\mathbf{j'}\,\mathbf{k'})_{jm}\rangle,\ i=1,2.$$

The triangle pattern for the coupling scheme (5.151) is given by

$$T(\mathbf{j'}\,\mathbf{k'})_j = \begin{array}{c} j_1 \quad j_3 \quad j_2 \quad j_4 \\ k'_1 \diagdown\!\!\diagup \;\; \diagdown\!\!\diagup k'_2 \\ j \end{array} \mapsto \begin{pmatrix} j_1 & j_2 & k'_1 \\ j_3 & j_4 & k'_2 \\ k'_1 & k'_2 & j \end{pmatrix}. \qquad (5.156)$$

Thus, for specified $j \in \{j_{\min}, j_{\min}+1, \ldots, j_{\max}\}$, the domains of definition of k'_1 and k'_2 are just the set of values for which all three columns of the triangle pattern (5.156) satisfy the triangle rules; that is,

$$(k'_1, k'_2) \in \mathbb{K}^{(j)}_T(\mathbf{j'}) = \left\{ (k'_1, k'_2) \ \Big|\ \begin{array}{l} k'_1 \in \langle j_1, j_3 \rangle, k'_1 \in \langle k'_2, j \rangle, \\ k'_2 \in \langle j_2, j_4 \rangle, k'_2 \in \langle k'_1, j \rangle \end{array} \right\}. \qquad (5.157)$$

We recall that the cardinality of this set is $|\mathbb{K}^{(j)}_T(\mathbf{j'})| = N_j(\mathbf{j})$, independently of the binary tree labeling (see (5.150)).

We now obtain the following state vector transformation, with the linear transformation coefficients given by an irreducible triangle coefficient of order six:

$$\left| \begin{array}{c} j_1 \quad j_3 \quad j_2 \quad j_4 \\ k'_1 \diagdown\!\!\diagup \;\; \diagdown\!\!\diagup k'_2 \\ j\, m \end{array} \right\rangle \qquad (5.158)$$

$$= \sum_{(k_1,k_2)\in \mathbb{K}^{(j)}_T(\mathbf{j})} \left\{ \begin{array}{ccc} j_1 & j_3 & k_1 \\ j_2 & j_4 & k_2 \\ k_1 & k_2 & j \end{array} \Bigg| \begin{array}{ccc} j_1 & j_2 & k'_1 \\ j_3 & j_4 & k'_2 \\ k'_1 & k'_2 & j \end{array} \right\} \left| \begin{array}{c} j_1 \quad j_2 \quad j_3 \quad j_4 \\ k_1 \diagdown\!\!\diagup \;\; \diagdown\!\!\diagup k_2 \\ j\, m \end{array} \right\rangle.$$

The Wigner $9-j$ coefficient is now defined in terms of the inner product of the two sets of orthonormal basis vectors in (5.158) by

$$\left\langle \begin{array}{c} j_1 \quad j_2 \quad j_3 \quad j_4 \\ k_1 \diagdown\!\!\diagup \;\; \diagdown\!\!\diagup k_2 \\ j \end{array} \Bigg| \begin{array}{c} j_1 \quad j_3 \quad j_2 \quad j_4 \\ k'_1 \diagdown\!\!\diagup \;\; \diagdown\!\!\diagup k'_2 \\ j \end{array} \right\rangle$$

5.7. STATE VECTORS: UNCOUPLED AND COUPLED

$$= \left\{\begin{matrix} j_1 & j_3 & k_1 \\ j_2 & j_4 & k_2 \\ k_1 & k_2 & j \end{matrix}\right\} \left\{\begin{matrix} j_1 & j_2 & k'_1 \\ j_3 & j_4 & k'_2 \\ k'_1 & k'_2 & j \end{matrix}\right\} = \sum_k \left\{\begin{matrix} j_1 & k_1 \\ j_2 & j_3 \\ k_1 & k \end{matrix} \middle| \begin{matrix} j_1 & j_2 \\ j_3 & k'_1 \\ k'_1 & k \end{matrix}\right\}$$

$$\times \left\{\begin{matrix} j_3 & k_1 \\ j_4 & k_2 \\ k_2 & j \end{matrix} \middle| \begin{matrix} k_1 & j_4 \\ j_3 & k \\ k & j \end{matrix}\right\} \left\{\begin{matrix} j_2 & k'_1 \\ j_4 & k'_2 \\ k'_2 & j \end{matrix} \middle| \begin{matrix} j_2 & j_4 \\ k'_1 & k \\ k & j \end{matrix}\right\} \quad (5.159)$$

$$= \sqrt{(2k_1+1)(2k'_1+1)(2k_2+1)(2k'_2+1)} \left\{\begin{matrix} j_1 & j_3 & k'_1 \\ j_2 & j_4 & k'_2 \\ k_1 & k_2 & j \end{matrix}\right\}$$

$$= \sqrt{(2k_1+1)(2k'_1+1)(2k_2+1)(2k'_2+1)} \sum_k (-1)^{j_1+k-k_1-k'_1}$$

$$\times (2k+1) W(j_4 j_3 j k_1; k_2 k) W(j_3 j_1 k j_2; k'_1 k_1) W(k'_1 j_2 j j_4; k k'_2).$$

The summation is over all k−values for which the triangle relations on the columns of the triangle coefficients in which k appears are satisfied. The three-column bracket expression in line four is Wigner's symbol for the $9-j$ coefficient. This expression is derived in [L, p.147]; it is also present here in terms of triangle coefficients of order four (the summation contains no phase factors, which is often the case). Alternative forms for the $9-j$ coefficient are given in Chapter 8, each of which arises naturally in the context of that application; these forms give different summation expressions for the same $9-j$ coefficient, as verified by the fact that they all define isomorphic cubic graphs.

Here we can follow the rules given for $n = 3$ for identifying two relevant subsets of orthonormal basis vectors that define important subspaces of the tensor product space $\mathcal{H}_{j_1} \otimes \mathcal{H}_{j_2} \otimes \mathcal{H}_{j_3} \otimes \mathcal{H}_{j_4}$. First of all, the extension of the basis set given there to $\mathbf{j} = (j_1, j_2, j_3, j_4)$ thus obtaining the orthonormal basis $\mathbf{B}_{\mathbf{j}, jm}$ now with $(k_1, k_2) \in \mathbb{K}_T^{(j)}(\mathbf{j})$, each $T \in \mathbb{T}_4$. This then gives the basis set $\mathbf{B}_{\mathbf{j}, j; k_1, k_2}$ of dimension $2j+1$ of the tensor product space $\mathcal{H}_{j_1} \otimes \mathcal{H}_{j_2} \otimes \mathcal{H}_{j_3} \otimes \mathcal{H}_{j_3}$, there being $|\mathbb{K}^{(j)}(\mathbf{j})| = N_j(\mathbf{j})$ such perpendicular spaces, each of which is the carrier space of the irreducible representation $D^j(U)$ of $SU(2)$ under $SU(2)$ frame rotations, since the total angular momentum $\mathbf{J}$ has the standard action on each of these subspaces.

The orthonormal basis transformation illustrated by relation (5.158) is between different basis sets of the same vector subspace $\mathcal{H}_{\mathbf{j}, j, m}$. This is just one example of many such transformations, as we next describe. The following result holds for $n = 4$. We state the result for general n, from which the case $n = 4$ is easily recovered:

Let $|T(\mathbf{j}_\pi \mathbf{k})_{jm}\rangle$, each $T \in \mathbb{T}_n, \pi \in S_n$, denote any of the $n! a_n$ standard labeled binary coupling schemes. Then, each irreducible pair (no common fork) of coupled state vectors in this set contains a basis of $N_j(\mathbf{j})$ orthonormal vectors that span the same space $\mathcal{H}_{\mathbf{j},j,m}$. The inner product of basis vectors gives the irreducible triangle coefficients of order $2(n-1)$ that determine the real linear transformation between basis vectors, as well as the domains of definition of the intermediate quantum numbers that enumerate the orthonormal basis vectors, as given by $\mathbf{k} = \mathbb{A}_T^{(j)}(\mathbf{j})$, and obtained by applying the simple addition of angular momentum rule to the $2(n-1)$ columns of the triangle coefficient.

The two subspace structures described above are embodied in the dimensionality relation

$$\prod_{i=1}^{n}(2j_i+1) = \sum_{j=j_{\min}}^{j_{\max}} N_j(\mathbf{j})(2j+1). \tag{5.160}$$

We have not yet given the method of the calculation of triangle coefficients. This is, of course, included in the expansive discussion in [L]. It suffices for our present purposes to note that these methods rely on five concepts: the notion of the shape (already defined) of a binary tree, a recoupling matrix, a path between the labeled shapes of a pair of coupled binary tree (equivalently, between the corresponding state vectors), the notion of a shape transformation between such shapes, and the notion of a cubic graph that serves to distinguish between irreducible triangle coefficients that do or do not give $3(n-1)-j$ coefficients and gives the type of the $3(n-1)-j$ coefficients, when several types are admitted. These operations, together with the properties expressed by (5.134)-(5.135) allow for the calculation of every $3(n-1)-j$ coefficient in terms of triangle coefficients of order four (Racah coefficients). We return in Sect. 5.8 to a full review of this method; it is implemented in great detail in Chapter 8 in the application to the Heisenberg ring problem. Next, we show how doubly stochastic matrices enter into the examples above.

Examples. Doubly stochastic matrices for binary coupling of angular momenta for $n = 2, 3, 4$:

Example 1. $n = 2$. The inner product of state vectors

$$\langle T(j_1, j_2)_{jm} | j_1\, m_1; j_2\, m_2 \rangle = C^{j_1\ j_2\ j}_{m_1\, m_2\, m} \tag{5.161}$$

gives the elements of the transition probability amplitude matrix for a transition from the prepared composite state $|T(j_1, j_2)_{jm}\rangle$ to the measured uncoupled state $|j_1\, m_1; j_2\, m_2\rangle$. The real orthogonal matrix $C^{(j_1 j_2)}$ with elements given by the WCG coefficients (5.161) is of order $N_{j_1, j_2} = (2j_1+1)(2j_2+1)$; the matrix is defined to have (row, column) elements

5.7. STATE VECTORS: UNCOUPLED AND COUPLED

defined by
$$\left(C^{(j_1 j_2)}\right)_{(j,m);(m_1,m_2)} = C^{j_1\ j_2\ j}_{m_1\ m_2\ m}. \tag{5.162}$$

This enumeration of the elements of the matrix $C^{(j_1 j_2)}$ is given in *standard form* as follows:

rows; top-to-bottom :
$$(j,m) : j \in \{j_1 + j_2, j_1 + j_2 - 1, \ldots, |j_1 - j_2|\},\ m \in \{j, j-1, \ldots, -j\}; \tag{5.163}$$

columns; left-to-right :
$$(m_1, m_2) : m_1 \in \{j_1, j_1 - 1, \ldots, -j_1\},\ m_2 \in \{j_2, j_2 - 1, \ldots, -j_2\}.$$

The pairs themselves are ordered by the rule that $(a, b) > (a', b')$, if $a > a'$, and $(a, b) > (a, b')$, if $b > b'$. Moreover, the elements of $C^{(j_1 j_2)}$ are defined to be zero unless $m_1 + m_2 = m$. The orthogonality relations $C^{(j_1 j_2)tr} C^{(j_1 j_2)} = C^{(j_1 j_2)} C^{(j_1 j_2)tr} = I_{N_{(j_1,j_2)}}$ are just the matrix form of the orthogonality relations (5.108)-(5.109). It is customary to order the rows and columns from the greatest pair to the least pair as read top-to-bottom down the rows and left-to-right across the columns.

The matrix $S(C^{(j_1 j_2)tr}, I_{N_{(j_1,j_2)}})$ with elements given by

$$S_{(j,m);(m_1,m_2)}(C^{(j_1 j_2)tr}, I_{N_{(j_1,j_2)}}) = \left(C^{j_1\ j_2\ j}_{m_1\ m_2\ m}\right)^2 \tag{5.164}$$

is doubly stochastic. The meaning of the elements of this matrix is: The prepared coupled state $|T(j_1, j_2)_j m\rangle$ is the measured uncoupled state $|j_1 m_1\rangle \otimes |j_2 m_2\rangle$ with probability $\left(C^{j_1\ j_2\ j}_{m_1\ m_2\ m}\right)^2$. For $j_1 = j_2 = 1/2$, this is just the example given earlier by relations (5.67)-(5.70).

Example 2. $n = 3$. The structural relations are similar to those for $n = 2$. We list the corresponding results with very brief descriptions. The elements of the transition probability amplitude matrix are the following for the coupled state vector (5.120)-(5.124):

$$\langle T(j_1, j_2, j_3; k_1)_j m \mid j_1 m_1; j_2 m_2; j_3 m_3 \rangle$$
$$= C^{j_1\ j_2\ k_1}_{m_1\ m_2\ m_1+m_2}\ C^{k_1\ j_3\ j}_{m_1+m_2\ m_3\ m} = \left(C^{(j_1 j_2 j_3)}_T\right)_{(k_1,j,m);(m_1,m_2,m_3)} \tag{5.165}$$

$$C^{(j_1 j_2 j_3) tr}_T C^{(j_1 j_2 j_3)}_T = C^{(j_1 j_2 j_3)}_T C^{(j_1 j_2 j_3) tr}_T = I_{N(j_1,j_2,j_3)}.$$

The generalized real orthogonal WCG matrix $C^{(j_1 j_2 j_3)}_T$ is of order $N(\mathbf{j}) = (2j_1 + 1)(2j_2 + 1)(2j_3 + 1)$, and the element in row (k, j, m) and column

(m_1, m_2, m_3) is zero unless $m_1+m_2+m_3 = m$. The orthogonality relation follows from the product form of the generalized WCG coefficients. The subscript T denotes that the binary tree is of shape $((\circ\circ)\circ) \mapsto T \in \mathbb{T}_3$.

The elements of the transition probability matrix are as follows:

$$S_{k_1,j,m;\,m_1,m_2,m_3}\left(C_T^{(j_1 j_2 j_3)\,tr}, I_{N(j_1,j_2,j_3)}\right) \tag{5.166}$$

$$= \left(\left(C_T^{(j_1 j_2 j_3)}\right)_{k_1,j,m;\,m_1,m_2,m_3}\right)^2 = \left(C_{m_1\ m_2\ m_1+m_2}^{j_1\ j_2\ k_1}\ C_{m_1+m_2\ m_3\ m}^{k_1\ j_3\ j}\right)^2.$$

The prepared coupled state $|T(j_1, j_2, j_3; k_1)_{jm}\rangle$ is the measured uncoupled state $|j_1\, m_1\rangle \otimes |j_2\, m_2\rangle \times |j_3\, m_3\rangle$ with probability given by the element (5.166) in row (k_1, j, m) and column (m_1, m_2, m_3). Rows and columns are ordered by the rule for sequences (a, b, c) given by $(a, b, c) > (a', b', c')$, if $a > a'$; $(a, b, c) > (a, b', c')$, if $b > b'$; $(a, b, c) > (a, b, c')$, if $c > c'$.

But now there is extra information that comes into play through the transformation (5.138) between the coupled states (5.120)-(5.126) and the coupled states (5.127)-(5.131), which we now rewrite in terms of the recoupling matrix in the following form :

$$|T'(\mathbf{j}\, k_1')\rangle_{jm} = \sum_{k_1 \in \mathbb{K}_T^{(j)}(\mathbf{j})} \left(R_{T,T'}^{\mathbf{j};\mathbf{j}}\right)_{k_1,j;\,k_1',j} |T(\mathbf{j}\, k_1)\rangle_{jm}, \tag{5.167}$$

where $\mathbf{j} = (j_1, j_2, j_3)$ are the labels assigned to the shape $((\circ\circ)\circ) \mapsto T \in \mathbb{T}_3$ (in (5.122)) and the shape $(\circ(\circ\circ)) \mapsto T' \in \mathbb{T}_3$ (in (5.127)).

The transformation coefficients in (5.167) have the following expression in terms of inner product and triangle coefficients:

$$\left(R_{T,T'}^{\mathbf{j};\mathbf{j}}\right)_{k_1,j;\,k_1',j} = \langle T(\mathbf{j}\, k_1)_{jm} | T'(\mathbf{j}\, k_1')_{jm}\rangle$$

$$= \left(C_T^{(\mathbf{j})} C_{T'}^{(\mathbf{j})tr}\right)_{k_1,j;\,k_1',j} = \left\{\begin{array}{cc|cc} j_1 & k_1 & j_2 & j_1 \\ j_2 & j_3 & j_3 & k_1' \\ k_1 & j & k_1' & j \end{array}\right\}. \tag{5.168}$$

These are the matrix elements of the *recoupling matrix* for the coupled states shown (see (5.137)). These matrix elements are independent of m (see relations (1.50)-(1.51) for the proof of this in general). The (row, column) enumeration of the elements of $R_{T,T'}^{\mathbf{j};\mathbf{j}}$ is given by

$$k_1 \in \mathbb{K}_T^{(j)}(\mathbf{j}) \text{ and } k_1' \in \mathbb{K}_{T'}^{(j)}(\mathbf{j}). \tag{5.169}$$

Because the cardinality of these two domains of definition are equal; that is, $|\mathbb{K}_T^{(j)}(\mathbf{j})| = |\mathbb{K}_{T'}^{(j)}(\mathbf{j})| = N_j(\mathbf{j})$, the recoupling matrix is, for specified

5.7. STATE VECTORS: UNCOUPLED AND COUPLED

$\mathbf{j} = (j_1, j_2, j_3)$ and $j \in \{j_{\min}, j_{\min} + 1, \ldots, j_{\max}\}$ (and any allowed m), a real orthogonal matrix of order equal to the CG number:

$$R^{\mathbf{j};\mathbf{j}\,tr}_{T,T'} R^{\mathbf{j};\mathbf{j}}_{T,T'} = R^{\mathbf{j};\mathbf{j}}_{T,T'} R^{\mathbf{j};\mathbf{j}\,tr}_{T,T'} = I_{N_j(\mathbf{j})}. \qquad (5.170)$$

This relation also expresses the orthogonality relations of the Racah coefficients, as stated in terms of its matrix elements:

$$\left(R^{\mathbf{j};\mathbf{j}}_{T,T'}\right)_{k_1,j;k_1',j} = \sqrt{(2k_1 + 1)(2k_1' + 1)}\, W(j_1 j_2 j j_3; k_1 k_1'),$$

$$\sum_{k_1}(2k_1 + 1)(2k_1' + 1) W(j_1 j_2 j j_3; k_1 k_1') W(j_1 j_2 j j_3; k_1 k_1'') = \delta_{k_1', k_1''}, \quad (5.171)$$

$$\sum_{k_1'}(2k_1 + 1)(2k_1' + 1) W(j_1 j_2 j j_3; k_1 k_1') W(j_1 j_2 j j_3; k_1''' k_1') = \delta_{k_1, k_1'''}.$$

We also state these orthogonality relations in terms of the triangle coefficients of order four:

$$\sum_{k_1} \left\{ \begin{array}{cc|cc} j_1 & k_1 & j_2 & j_1 \\ j_2 & j_3 & j_3 & k_1' \\ k_1 & j & k_1' & j \end{array} \right\} \left\{ \begin{array}{cc|cc} j_1 & k_1 & j_2 & j_1 \\ j_2 & j_3 & j_3 & k_1'' \\ k_1 & j & k_1'' & j \end{array} \right\} = \delta_{k_1', k_1''},$$

(5.172)

$$\sum_{k_1'} \left\{ \begin{array}{cc|cc} j_1 & k_1 & j_2 & j_1 \\ j_2 & j_3 & j_3 & k_1' \\ k_1 & j & k_1' & j \end{array} \right\} \left\{ \begin{array}{cc|cc} j_1 & k_1''' & j_2 & j_1 \\ j_2 & j_3 & j_3 & k_1' \\ k_1''' & j & k_1' & j \end{array} \right\} = \delta_{k_1, k_1'''}.$$

The summation in the first expression is over all $k_1 \in \mathbb{K}_T^{(j)}(\mathbf{j})$, while $k_1', k_1'' \in \mathbb{K}_{T'}^{(j)}(\mathbf{j})$. Similarly, the summation in the second expression is over all $k_1' \in \mathbb{K}_{T'}^{(j)}(\mathbf{j})$, while $k_1, k_1''' \in \mathbb{K}_T^{(j)}(\mathbf{j})$. Thus, the choice of notation in relations (5.171) and (5.172) reflects these domains of definition. It is the property of having rows and columns of the recoupling matrix (5.167) labeled by quantum numbers having distinct domains of definition that leads to such distinctions in the orthogonality relations.

We have the following probabilistic interpretation of the above results: The transition probability amplitude matrix $W(U, V)$ has $U = C_T^{\mathbf{j}\,tr}, V = C_{T'}^{\mathbf{j}\,tr}$; hence, it is the recoupling matrix above of order $N_j(\mathbf{j})$ with elements defined by (5.169):

$$W\left(C_T^{\mathbf{j}\,tr}, C_{T'}^{\mathbf{j}\,tr}\right) = C_T^{\mathbf{j}} C_{T'}^{\mathbf{j}\,tr} = R_{T,T'}^{\mathbf{j};\mathbf{j}}. \qquad (5.173)$$

The transition probability matrix $S(C_T^{\mathbf{j}\,tr}, C_{T'}^{\mathbf{j}\,tr})$ is the matrix with elements given by

$$S_{k_1,j;k'_1,j}\left(C_T^{\mathbf{j}\,tr}, C_{T'}^{\mathbf{j}\,tr}\right) = \left(\left\{\begin{matrix} j_1 & k_1 \\ j_2 & j_3 \\ k_1 & j \end{matrix}\,\middle|\,\begin{matrix} j_2 & j_1 \\ j_3 & k'_1 \\ k'_1 & j \end{matrix}\right\}\right)^2. \qquad (5.174)$$

Thus, the prepared coupled state

$$|T(\mathbf{j}\,k_1)\rangle_{jm} = \left|\begin{matrix} j_1 \circ \quad \circ j_2 \\ k_1 \bullet\!\!\!\diagup\!\!\!\diagdown\!\bullet\, j_3 \\ jm \end{matrix}\right\rangle \qquad (5.175)$$

is the measured coupled state

$$|T'(\mathbf{j}\,k'_1)\rangle_{jm} = \left|\begin{matrix} j_2 \circ \quad \circ j_3 \\ j_1 \circ \bullet\!\!\!\diagdown\! k'_1 \\ jm \end{matrix}\right\rangle \qquad (5.176)$$

with probability given by the square of the (real) triangle coefficient with left pattern determined by its corresponding labeled binary tree and similarly for the right pattern, which, in turn, is expressed in terms of the Racah coefficient by (5.171). This probability is independent of m.

There are, of course, all of the other slightly varied expressions of relations (5.168)-(5.176) resulting from all assignments of permutations of j_1, j_2, j_3 to either shape in (5.118) with no common forks, but they all involve similar right and left triangle patterns and the Racah coefficients in the manner described in this example.

Example 3. $n = 4$. We list results for $n = 4$ corresponding to $n = 2, 3$ with minimal description. Also, we use the short boldface notations to conserve space: $|\mathbf{j}\,\mathbf{m}\rangle = |j_1\,m_1; j_2\,m_2; j_3\,m_3; j_4\,m_4\rangle = |j_1\,m_1\rangle \otimes |j_2\,m_2\rangle \otimes |j_3\,m_3\rangle \otimes |j_4\,m_4\rangle$, where $\mathbf{j} = (j_1, j_2, j_3, j_4)$, $\mathbf{m} = (m_1, m_2, m_3, m_4)$, $\mathbf{k} = (k_1, k_2)$. The inner product of state vectors

$$\langle T(\mathbf{j}\,\mathbf{k})_{jm} | \mathbf{j}\,\mathbf{m}\rangle = C_{T(\mathbf{j}\,\mathbf{m};\mathbf{k})_{jm}} = \left(C_T^{(\mathbf{j})}\right)_{\mathbf{k},j,m;\mathbf{m}}$$

$$= C^{j_1\;j_2\;k_1}_{m_1,m_2,m_1+m_2}\,C^{j_3\;j_4\;k_2}_{m_3,m_4,m_3+m_4}\,C^{k_1\;k_2\;j}_{m_1+m_2,m_3+m_4,m} \qquad (5.177)$$

gives the (row;column) $= (\mathbf{k}, j, m; \mathbf{m})$ elements of the transition probability amplitude matrix $W(U, V) = W(C_T^{(\mathbf{j})\,tr}, I_{N(\mathbf{j})})$ for the binary coupling described by relations (5.143)-(5.146), where T is the binary tree

5.7. STATE VECTORS: UNCOUPLED AND COUPLED

of shape $((\circ\circ)(\circ\circ))$ filled-in with $\mathbf{j} = (j_1, j_2, j_3, j_4)$. As always, the generalized WCG coefficient is zero unless $m = m_1 + m_2 + m_3 + m_4$. This matrix is of order $N(\mathbf{j}) = \prod_{i=1}^{4}(2j_i + 1)$. The elements of the transition probability matrix $S(U,V) = S(C_T^{(\mathbf{j})\,tr}, I_{N(\mathbf{j})})$ are given by

$$S_{\mathbf{k},j,m;\mathbf{m}}\left(C^{(\mathbf{j})\,tr}, I_{N(\mathbf{j})}\right) = \left(\left(C_T^{(\mathbf{j})}\right)_{\mathbf{k},j,m;\mathbf{m}}\right)^2,$$
$$\mathbf{k}, j, m \in \mathbb{R}_T(\mathbf{j}),\ \mathbf{m} \in \mathbb{C}_T(\mathbf{j}). \qquad (5.178)$$

Thus, the prepared binary coupled state $|T(\mathbf{j\,k})_{j\,m}\rangle$ is the measured state $|\mathbf{j\,m}\rangle$ with transition probability given by the matrix element (5.178) in row $\mathbf{k}, j, m$ and column $\mathbf{m}$ of the doubly stochastic matrix $S(C^{(\mathbf{j})\,tr}, I_{N(\mathbf{j})})$.

We point out that we have used a somewhat more refined notation $\mathbf{m} \in \mathbb{C}_T(\mathbf{j})$ for the set of values that the projection quantum numbers can assume; this set was previously defined as being are over all values for which the generalized WCG coefficients are defined, but now the amended notation reflects that this set depends on the labeled binary tree T from which the domain can be read off exactly.

A similar description applies to the second set of binary coupled state vectors described in relations (5.151)-(5.157), where now we have $\mathbf{j}' = (j_1, j_3, j_2, j_4)$, $\mathbf{m}' = (m_1, m_3, m_2, m_4)$, $\mathbf{k}' = (k_1', k_2')$. The inner product of state vectors is:

$$\langle T(\mathbf{j}'\,\mathbf{k}')_{j\,m} | \mathbf{j\,m} \rangle = C_{T(\mathbf{j}'\,\mathbf{m}';\mathbf{k}')_{j\,m}} = \left(C_T^{(\mathbf{j}')}\right)_{\mathbf{k}',j,m;\mathbf{m}}$$
$$= C^{j_1\ j_3\ k_1'}_{m_1,m_3,m_1+m_3}\, C^{j_2\ j_4\ k_2'}_{m_2,m_4,m_2+m_4}\, C^{k_1'\ k_2'\ j}_{m_1+m_3,m_2+m_4,m}, \qquad (5.179)$$

where the entry in the generalized real orthogonal WCG matrix $C^{(\mathbf{j}')}$ of order $N(\mathbf{j}) = (2j_1+1)(2j_2+1)(2j_3+1)(2j_4+1)$ is 0, unless, as usual, $m = m_1 + m_2 + m_3 + m_4$. The elements of the transition probability matrix $S(U,V) = S(C_T^{(\mathbf{j}')\,tr}, I_{N(\mathbf{j})})$ are given by

$$S_{\mathbf{k}',j,m;\mathbf{m}}\left(C^{(\mathbf{j}')\,tr}, I_{N(\mathbf{j})}\right) = \left(\left(C_T^{(\mathbf{j}')}\right)_{\mathbf{k}',j,m;\mathbf{m}}\right)^2,$$
$$\mathbf{k}', j, m \in \mathbb{R}_T(\mathbf{j}'),\ \mathbf{m} \in \mathbb{C}_T(\mathbf{j}). \qquad (5.180)$$

The domains of definition $\mathbf{k} \in \mathbb{K}_T^{(j)}(\mathbf{j})$ and $\mathbf{k}' \in \mathbb{K}_T^{(j)}(\mathbf{j}')$ have the same cardinality $N_j(\mathbf{j})$ in specifying the rows of the matrices $C_T^{(\mathbf{j})}$ and $C_T^{(\mathbf{j}')}$ in (5.178) and (5.180), so that we have:

$$|\mathbb{R}_T(\mathbf{j})| = |\mathbb{R}_T(\mathbf{j}')| = |\mathbb{C}_T(\mathbf{j})| = N(\mathbf{j}). \qquad (5.181)$$

Thus, we have that the prepared binary coupled state $|T(\mathbf{j'\,k'})_{j\,m}\rangle$ is the measured state $|\mathbf{j\,m}\rangle$ with transition probability given by the matrix element (5.180) in row $\mathbf{k'},j,m$ and column $\mathbf{m}$ of the doubly stochastic matrix $S(C_T^{(\mathbf{j'})\,tr}, I_{N(\mathbf{j})})$.

We continue with the state vector expressions that reveal the role of the $9-j$ coefficients as transition probability amplitudes:

$$|T(\mathbf{j'\,k'})_{j\,m}\rangle = \sum_{(k_1,k_2)\in \mathbb{K}_T^{(j)}(\mathbf{j})} \left(R_{T,T}^{\mathbf{j;j'}}\right)_{\mathbf{k},j;\mathbf{k'},j} |T(\mathbf{j\,k})_{j\,m}\rangle, \quad (5.182)$$

$$\left(R_{T,T}^{\mathbf{j;j'}}\right)_{\mathbf{k},j;\mathbf{k'},j} = \langle T(\mathbf{j\,k})_{j\,m} | T(\mathbf{j'\,k'})_{j\,m}\rangle$$

$$= \left(C_T^{(\mathbf{j})}\,C_T^{(\mathbf{j'})\,tr}\right)_{\mathbf{k},j;\mathbf{k'},j} = \left\{\begin{array}{ccc|ccc} j_1 & j_3 & k_1 & j_1 & j_2 & k'_1 \\ j_2 & j_4 & k_2 & j_3 & j_4 & k'_2 \\ k_1 & k_2 & j & k'_1 & k'_2 & j \end{array}\right\}, \quad (5.183)$$

where this triangle coefficient is expressed in terms of usual notation for the Wigner $9-j$ coefficient by relation (5.159).

The domains of definition of the indices $\mathbf{k}=(k_1,k_2)$ and $\mathbf{k'}=(k'_1,k'_2)$ are given in (5.150) and (5.157); these two domains have the same cardinality given by the CG number $N_j(\mathbf{j})$. In relation (5.183), it is the same values of $\mathbf{j},j,m$ that occur, since $\mathbf{j'}$ is a permutation of $\mathbf{j}$. Thus, we select (fix) these parameters on both sides of relation (5.183). Then, the recoupling matrix

$$R_{T,T}^{\mathbf{j;j'}} = C_T^{(\mathbf{j})}\,C_T^{(\mathbf{j'})\,tr} \quad (5.184)$$

is a matrix of order $N_j(\mathbf{j})$, with (row; column) elements enumerated by $\mathbf{k},j;\mathbf{k'},j$, with $\mathbf{k}\in \mathbb{K}^{(j)}(\mathbf{j})$ and $\mathbf{k'}\in \mathbb{K}^{(j)}(\mathbf{j'})$.

The orthogonality of the recoupling matrix $R_{T,T}^{\mathbf{j;j'}}$ is stated by

$$R_{T,T}^{\mathbf{j;j'}\,tr}\,R_{T,T}^{\mathbf{j;j'}} = R_{T,T}^{\mathbf{j;j'}}\,R_{T,T}^{\mathbf{j;j'}\,tr} = I_{N_j(\mathbf{j})}. \quad (5.185)$$

This is just the matrix form of the orthogonality relations of the $9-j$ coefficients expressed in terms of triangle coefficients by

$$\sum_{k_1,k_2} \left\{\begin{array}{ccc|ccc} j_1 & j_3 & k_1 & j_1 & j_2 & k'_1 \\ j_2 & j_4 & k_2 & j_3 & j_4 & k'_2 \\ k_1 & k_2 & j & k'_1 & k'_2 & j \end{array}\right\}$$

$$\times \left\{\begin{array}{ccc|ccc} j_1 & j_3 & k_1 & j_1 & j_2 & k''_1 \\ j_2 & j_4 & k_2 & j_3 & j_4 & k''_2 \\ k_1 & k_2 & j & k''_1 & k''_2 & j \end{array}\right\} = \delta_{\mathbf{k'},\mathbf{k''}},$$

$$(5.186)$$

5.7. STATE VECTORS: UNCOUPLED AND COUPLED

$$\sum_{k'_1,k'_2} \left\{ \begin{array}{ccc|ccc} j_1 & j_3 & k_1 & j_1 & j_2 & k'_1 \\ j_2 & j_4 & k_2 & j_3 & j_4 & k'_2 \\ k_1 & k_2 & j & k'_1 & k'_2 & j \end{array} \right\}$$

$$\times \left\{ \begin{array}{ccc|ccc} j_1 & j_3 & k'''_1 & j_1 & j_2 & k'_1 \\ j_2 & j_4 & k'''_2 & j_3 & j_4 & k'_2 \\ k''_1 & k''_2 & j & k'_1 & k'_2 & j \end{array} \right\} = \delta_{\mathbf{k},\mathbf{k'''}}.$$

Again, similar to (5.172), the summation in the first expression is over all $(k_1, k_2) \in \mathbb{K}_T^{(j)}(\mathbf{j})$, while $(k'_1, k'_2), (k''_1, k''_2) \in \mathbb{K}_T^{(j')}(\mathbf{j})$; and the summation in the second expression is over all $(k'_1, k'_2) \in \mathbb{K}_T^{(j')}(\mathbf{j})$, while $(k_1, k_2), (k'''_1, k'''_2) \in \mathbb{K}_T^{(j)}(\mathbf{j})$. The triangle notation for $9-j$ coefficients (and for general n) make these distinctions in domains of definition particularly clear. It is, of course, again the property of having rows and columns of the recoupling matrix in (5.183) labeled by quantum numbers having distinct domains of definition that leads to such subtleties in the orthogonality relations.

The matrix of order $N_j(\mathbf{j})$ with elements given by the inner product $\langle T(\mathbf{j}\,\mathbf{k})_{jm} | T(\mathbf{j'}\,\mathbf{k'})_{jm} \rangle$ is the transition probability amplitude matrix for the prepared state $|T(\mathbf{j}\,\mathbf{k})_{jm}\rangle$ and the measured state $|T(\mathbf{j'}\,\mathbf{k'})_{jm}\rangle$. Thus, the transition probability amplitude matrix $W(U,V)$ has $U = C_T^{(\mathbf{j})\,tr}, V = C_T^{\mathbf{j'}\,tr}$; hence,

$$W\left(C_T^{(\mathbf{j})\,tr}, C_T^{(\mathbf{j'})\,tr}\right) = C_T^{(\mathbf{j})} C_T^{\mathbf{j'}\,tr} = R_{T,T}^{\mathbf{j};\mathbf{j'}}. \tag{5.187}$$

The transition probability matrix $S\left(C_T^{(\mathbf{j})\,tr}, C_T^{(\mathbf{j'})\,tr}\right)$ is then the matrix with element in row $\mathbf{k}, j$ and column $\mathbf{k'}, j$ given by

$$S_{\mathbf{k};j,\mathbf{k'},j}(C_T^{(\mathbf{j})\,tr}, C_T^{(\mathbf{j'})\,tr}) = \langle T(\mathbf{j}\,\mathbf{k})_{jm} | T(\mathbf{j'}\,\mathbf{k'})_{jm} \rangle$$

$$= \left(\left(R_{T,T}^{\mathbf{j};\mathbf{j'}}\right)_{\mathbf{k},j;\mathbf{k'},j}\right)^2 = \left(\left\{ \begin{array}{ccc|ccc} j_1 & j_3 & k_1 & j_1 & j_2 & k'_1 \\ j_2 & j_4 & k_2 & j_3 & j_4 & k'_2 \\ k_1 & k_2 & j & k'_1 & k'_2 & j \end{array} \right\}\right)^2. \tag{5.188}$$

Thus, the prepared coupled state

$$|T(\mathbf{j}\,\mathbf{k})_{jm}\rangle = \left| \begin{array}{cccc} j_1 & j_2 & j_3 & j_4 \\ k_1 & & & k_2 \\ & & jm & \end{array} \right\rangle \tag{5.189}$$

is the measured coupled state

$$|T(\mathbf{j'\,k'})_{jm}\rangle = \left| \begin{array}{c} \overset{j_1\quad j_3\quad j_2\quad j_4}{\diagdown\!\diagup\quad\diagdown\!\diagup} \\ k'_1 \qquad\qquad k'_2 \\ jm \end{array} \right\rangle \qquad (5.190)$$

with transition probability given by the square of the matrix elements of the recoupling matrix $\left(R_{T,T}^{\mathbf{j};\mathbf{j'}}\right)_{\mathbf{k},j;\,\mathbf{k'},j'}$, *or equivalently, the square of the (real) triangle coefficient of order six with left pattern determined by the corresponding labeled binary tree and similarly for the right pattern. This result is expressed in terms of the Wigner notation for the $9-j$ coefficient by relation (5.159). This probability is independent of m.*

It is also the case that every pair of irreducible labeled binary trees of order six must give a pair of coupled state vectors similar to that given by relations (5.182)-(5.190) with transformation coefficients that are the elements of the recoupling matrix and with the associated triangle coefficients of order six having the left and right triangle patterns as read-off the forks of the standard labeled binary trees. All such triangle coefficients correspond to a cubic graph on six points, there being two nonisomorphic cubic graphs. Depending on the structure of the cubic graph, some of the triangle coefficients give alternative expressions of the Wigner $9-j$ coefficient; some factorize into the product of two triangle coefficients of order four and give alternative expressions of the B-E identity. We discuss this *factoring phenomenon* in Sect. 5.8.5.

We next turn to a review of binary coupling schemes for general n and their probabilistic interpretations, including the techniques needed to calculate numerical values of the $3(n-1)-j$ coefficients.

5.8 General Binary Tree Couplings and Doubly Stochastic Matrices

5.8.1 Overview

In this section, we outline the general relationships between binary coupling of n angular momenta (as developed in [L]) and doubly stochastic matrices. As the examples in the last section show, there are two classes of doubly stochastic matrices that arise: (i) Those associated with the transformation of an uncoupled basis to a coupled basis; and (ii) those associated with pairs of such coupled basis sets. Each set of coupled basis vectors in (i) is a new orthonormal basis of the tensor product space $\mathcal{H}_{j_1} \otimes \mathcal{H}_{j_2} \otimes \cdots \otimes \mathcal{H}_{j_n}$ that is determined uniquely (up to arbitrary phase

5.8. BINARY TREES AND DOUBLY STOCHASTIC MATRICES

factors) by an associated complete set of mutually commuting Hermitian operators. This complete set always includes the subset of $n+2$ operators

$$\{\mathbf{J}^2(1), \mathbf{J}^2(2), \ldots, \mathbf{J}^2(n), \mathbf{J}^2, J_3\}, \tag{5.191}$$

together with a set of $n-2$ intermediate (squared) angular momenta

$$\{\mathbf{K}_T^2(1), \mathbf{K}_T^2(2), \ldots, \mathbf{K}_T^2(n-2)\}. \tag{5.192}$$

There is a distinct set of intermediate angular momenta for each binary bracketing of the sum of n angular momenta $\mathbf{J} = \mathbf{J}(1) + \mathbf{J}(2) + \cdots + \mathbf{J}(n)$ by $n-1$ parenthesis pairs, each such binary bracketing encoding exactly how pairs of angular momenta are to be coupled in accordance with the CG series rule for compounding two angular momenta. All $n!$ permutations of the angular momenta $\mathbf{J}(1), \ldots, \mathbf{J}(n)$ are also to be considered.

This structure can be described first of all by the abstract concept of shape (binary bracketing), which is a sequence of n points ∘ into which $n-1$ parenthesis pairs ( ) have been inserted by certain rules that lead to a total of a_n (Catalan number) distinct shapes, and second of all by the $n!$ ways of assigning the angular momentum quantum numbers $\mathbf{j} = (j_1, j_2, \ldots, j_n)$ to each shape. Thus, there are $n! a_n$ distinct labeled shapes to consider. The unlabeled shapes are, in turn, in one-to-one correspondence with the set of binary trees of order n; hence, the labeled shapes with the $n! a_n$ labeled binary trees, where a rule of assigning a set of labels to each such binary tree is specified and called the *standard rule*. But the unique (up to normalization conventions) orthonormal basis vectors corresponding to the diagonalization of each of the complete sets of operators (5.191)-(5.192) all span the tensor product space $\mathcal{H}_{j_1} \otimes \mathcal{H}_{j_2} \otimes \cdots \otimes \mathcal{H}_{j_n}$; hence, every pair of such orthonormal bases is related by a unitary transformation. Thus, not only does every labeled shape determine a coupled orthonormal set of vectors in terms of the same set of uncoupled orthonormal vectors of the tensor product space, but also each pair of sets is related by a linear transformation. Since the set (5.191) of mutually commuting Hermitian operators is always the same for each binary coupling scheme, it is the intermediate angular momenta set (5.192), which is one-to-one with the shape, that distinguishes the orthonormal set of binary coupled state vectors.

The linear transformation coefficients relating two such coupled bases are the triangle coefficients of order $2n-2$. Each triangle coefficient, in turn, defines a transition probability matrix of the Landé form $W(U,V) = U^\dagger V$, where U and V are the unitary transformations relating the respective binary coupling schemes to the uncoupled scheme — the basis of the tensor product space. Thus, with every binary coupling scheme there is associated a doubly stochastic matrix $S(U,V)$. Actually, because all relations between various bases can be chosen to be real, only real orthogonal matrices occur. Many of the doubly stochastic matrices that arise in the set of all such binary couplings are matrices with elements 0 and 1, but many are of great complexity — they include all the $3(n-1) - j$ coefficients.

The present review from [L] gives the summary above a uniform setting for the general problem and also the actual tools for carrying out the explicit calculations of all $3(n-1) - j$ coefficients, based on the notions of recoupling matrices, triangle coefficients, common forks and the reduction process, transformations of labeled shapes by elementary commutation and association operations, and cubic graphs. In essence, this is a study of how doubly stochastic matrices arise naturally from composite angular momentum systems, a viewpoint that complements the angular momentum properties themselves.

5.8.2 Uncoupled States

The uncoupled angular momentum basis defined by relations (1.13)-(1.16) serves as the standard basis in terms of which every binary coupled angular momentum state is defined. For convenience of reference, we summarize those relations:

Complete set of operators:

$$\mathbf{J}^2(1), J_3(1), \mathbf{J}^2(2), J_3(2), \ldots, \mathbf{J}^2(n), J_3(n). \tag{5.193}$$

Standard action and orthogonality:

$$\mathbf{J}^2(i)|\mathbf{j\,m}\rangle = j_i(j_i+1)|\mathbf{j\,m}\rangle,$$
$$J_3(i)|\mathbf{j\,m}\rangle = m_i|\mathbf{j\,m}\rangle,$$
$$J_+(i)|\mathbf{j\,m}\rangle = \sqrt{(j_i - m_i)(j_i + m_i + 1)}\,|\mathbf{j\,m}_{+1}(i)\rangle, \tag{5.194}$$
$$J_-(i)|\mathbf{j\,m}\rangle = \sqrt{(j_i + m_i)(j_i - m_i + 1)}\,|\mathbf{j\,m}_{-1}(i)\rangle,$$
$$\mathbf{m}_{\pm 1}(i) = (m_1, \ldots, m_i \pm 1, \ldots, m_n);$$
$$\langle \mathbf{j\,m}|\mathbf{j\,m'}\rangle = \delta_{\mathbf{m},\mathbf{m'}}, \text{ each pair } \mathbf{m}, \mathbf{m'} \in \mathbb{C}(\mathbf{j}). \tag{5.195}$$

Notations:

$$\mathbf{j} = (j_1, j_2, \ldots, j_n), \text{ each } j_i \in \{0, 1/2, 1, 3/2, \ldots\}, i = 1, 2, \ldots, n,$$
$$\mathbf{m} = (m_1, m_2, \ldots, m_n), \text{ each } m_i \in \{j_i, j_i - 1, \ldots, -j_i\},$$
$$i = 1, 2, \ldots, n,$$
$$\mathcal{H}_{\mathbf{j}} = \mathcal{H}_{j_1} \otimes \mathcal{H}_{j_2} \otimes \cdots \otimes \mathcal{H}_{j_n}, \tag{5.196}$$
$$|\mathbf{j\,m}\rangle = |j_1\,m_1\rangle \otimes |j_2\,m_2\rangle \otimes \cdots \otimes |j_n\,m_n\rangle,$$
$$\mathbb{C}(\mathbf{j}) = \{\mathbf{m}\,|\,m_i = j_i, j_i - 1, \ldots, -j_i; i = 1, 2, \ldots, n\}.$$

5.8.3 Generalized WCG Coefficients

The definition of all binary coupled states depends on the notion of a generalized WCG coefficient, which, in turn, is derived from a labeled binary tree. The i-th labeled fork in a labeled tree corresponding to the coupling of n angular momenta is of the following form and has an assigned WCG coefficient:

$$C\left(\begin{array}{cc} a_i\,\alpha_i & b_i\,\beta_i \\ & \\ & k_i\,q_i \end{array}\right) = C^{a_i\ b_i\ k_i}_{\alpha_i\ \beta_i\ q_i}, \quad q_i = \alpha_i + \beta_i. \tag{5.197}$$

The four types of forks (5.96) in this picture have labels as follows:

1. $(\diamond\ \diamond) = (\circ\ \circ) : (a_i\,\alpha_i)$ external; $(b_i\,\beta_i)$ external.
2. $(\diamond\ \diamond) = (\bullet\ \circ) : (a_i\,\alpha_i)$ internal; $(b_i\,\beta_i)$ external.
3. $(\diamond\ \diamond) = (\circ\ \bullet) : (a_i\,\alpha_i)$ external; $(b_i\,\beta_i)$ internal.
4. $(\diamond\ \diamond) = (\bullet\ \bullet) : (a_i\,\alpha_i)$ internal; $(b_i\,\beta_i)$ internal.

Thus, for each $T \in \mathbb{T}_n$, we have a standard labeled tree $T(\mathbf{j}\,\mathbf{m};\mathbf{k})_{j\,m}$ and an associated generalized WCG coefficient:

$$C_{T(\mathbf{j}\,\mathbf{m};\mathbf{k})_{j\,m}} = \prod_{i=1}^{n-1} C^{a_i\ b_i\ k_i}_{\alpha_i\ \beta_i\ \alpha_i+\beta_i}. \tag{5.198}$$

The product is over all $n - 1$ forks in the tree $T \in \mathbb{T}_n$, where the roots of the forks have the standard labels $k_1\,q_1, k_2\,q_2, \ldots, k_{n-2}\,q_{n-2}, k_{n-1}\,q_{n-1}$, with $k_{n-1}\,q_{n-1} = j\,m$ and $q_i = \alpha_i + \beta_i$.

Examples. Generalized WCG coefficients of labeled binary trees:

$$C\left(\begin{array}{c} j_1\,m_1\quad j_2\,m_2 \\ k_1\,q_1\quad\quad j_3\,m_3 \\ k_2\,q_2\quad\quad j_4\,m_4 \\ j\,m \end{array}\right) = C^{j_1\ j_2\ k_1}_{m_1\,m_2\,q_1} C^{k_1\ j_3\ k_2}_{q_1\,m_3\,q_2} C^{k_2\ j_4\ j}_{q_2\,m_4\,m},$$

$$C\left(\begin{array}{c} j_2\,m_2\quad j_3\,m_3 \\ k_1\,q_1\quad\quad j_4\,m_4 \\ j_1\,m_1\quad\quad k_2\,q_2 \\ j\,m \end{array}\right) = C^{j_2\ j_3\ k_1}_{m_2\,m_3\,q_1} C^{k_1\ j_4\ k_2}_{q_1\,m_4\,q_2} C^{j_1\ k_2\ j}_{m_1\,q_2\,m}, \tag{5.199}$$

$$C\left(\begin{array}{c} j_1\,m_1\quad j_2\,m_2\quad j_3\,m_3 \\ k_1\,q_1\quad\quad\quad j_4\,m_4 \\ \quad\quad k_2\,q_2 \\ j\,m \end{array}\right) = C^{j_1\ j_2\ k_1}_{m_1\,m_2\,q_1} C^{j_3\ j_4\ k_2}_{m_3\,m_4\,q_2} C^{k_1\ k_2\ j}_{q_1\,q_2\,m}.$$

The following properties of the projection quantum numbers q_i hold in these examples: Each q_i is a linear sum of the projection quantum numbers m_1, m_2, m_3, m_4 with $0-1$ coefficients such that $m_1+m_2+m_3+m_4 = m$. For example, in the second instance above, to obtain a nonzero result for specified (j_1, j_2, j_3, j_4), $j \in \{j_{\min}, j_{\min}+1, \ldots, j_{\max}\}$, $m \in \{j, j-1, \ldots, j\}$, the following relations must be satisfied:

(i) projection quantum sum rule: $q_1 = m_2+m_3$, $q_2 = m_2+m_3+m_4$, $m_1+m_2+m_3+m_4 = m$; and $q_1 \in \{k_1, k_1-1, \ldots, -k_1\}$ and $q_2 \in \{k_2, k_2-1, \ldots, -k_2\}$;

(ii) angular momentum addition rule: $k_1 \in \langle j_2, j_3 \rangle, k_2 \in \langle k_1, j_4 \rangle, k_2 \in \langle j_1, j \rangle$.

This already gives a set of linear relations that must necessarily be satisfied by the m_i if the generalized WCG coefficient is to be nonzero (even then anomalous zeros can occur). It is because of this complexity, that, for example, a generalized WCG coefficient is defined to be zero unless all the projection quantum sum rules on each WCG coefficient in the product are fulfilled; the details of implementing are left for each case at hand. The same is true for the (k_1, k_2) quantum labels, but we can specify their domains of definition precisely through the set $(k_1, k_2) \in \mathbb{K}_T^{(j)}(\mathbf{j})$, which is determined by multiple applications of the elementary angular momentum addition rule, as illustrate in (ii) above, and in the earlier examples above for $n = 2, 3, 4$.

The situation for general n is the following: The basis vectors $|\mathbf{j}\,\mathbf{m}\rangle$ for specified angular momentum quantum numbers $\mathbf{j}$ are enumerated by all values $\mathbf{m} \in \mathbb{C}(\mathbf{j})$, giving a number of orthonormal vectors equal to $\prod_{i=1}^{n}(2j_i+1)$. However, when we select $j \in \{j_{\min}, j_{\min}+1, \ldots, j_{\max}\}$, and then admit for the selected j all values $m \in \{j, j-1, \ldots, j\}$, the domain of definition of the generalized WCG coefficient (5.198) is constrained by the values that $\mathbf{m}$ can assume because each individual CG coefficient in the product that defines the generalized WCG must be defined; that is, the domain of definition depends on the binary tree $T \in \mathbb{T}_n$, as illustrated in examples (5.199). We introduce the notation $\mathbf{m} \in C_T(\mathbf{j})$ to designate these constraints. A more detailed description in terms of the binary tree itself is rather cumbersome. The situation for the domain of definition of the intermediate angular momenta $\mathbf{k}$ for prescribed $\mathbf{j}, j$ is also rather intricate, as indicated already in (5.126) for $n = 3$ and in (5.150) for $n = 4$. Careful attention to detail is required here because the set of such $\mathbf{k} \in \mathbb{K}_T^{(j)}(\mathbf{j})$ for each binary tree $T \in \mathbb{T}_n$ play a primary role in the definition of triangle patterns and their manifold properties in providing the pathway to a complete theory of the binary coupling of angular momenta.

5.8.4 Binary Tree Coupled State Vectors

We next adapt the α-coupled states introduced in relations (1.18)-(1.21) to the context of binary coupled states associated with a binary tree $T \in \mathbb{T}_n$. The coupled binary tree state vectors correspond to the labeled binary tree $T(\mathbf{j}\,\mathbf{m}; \mathbf{k})_{jm}$; they are the simultaneous eigenvectors (sometimes called ket-vectors) of the complete set of $2n$ mutually commuting Hermitian operators (5.192) with the following list of properties:

Definition of coupled states in terms of uncoupled states:

$$|T(\mathbf{j}\,\mathbf{k})_{jm}\rangle = \sum_{\mathbf{m} \in \mathbb{C}_T(\mathbf{j})} C_{T(\mathbf{j}\,\mathbf{m};\mathbf{k})_{jm}} |\mathbf{j}\,\mathbf{m}\rangle, \quad \text{each } \mathbf{k}, j, m \in \mathbb{R}_T(\mathbf{j}), \quad (5.200)$$

where the domains of definition of $\mathbf{k}, j, m \in \mathbb{R}_T(\mathbf{j})$ and $\mathbf{m} \in \mathbb{C}_T(\mathbf{j})$ are:

$$\mathbb{R}_T(\mathbf{j}) = \left\{ \mathbf{k}, j, m \,\middle|\, \begin{array}{l} j = j_{\min}, \ldots, j_{\max}; m = j, \ldots, -j; \\ \mathbf{k} \in \mathbb{K}_T^{(j)}(\mathbf{j}) \end{array} \right\}; \quad (5.201)$$

$$\mathbb{C}_T(\mathbf{j}) = \{\mathbf{m} \,|\, C_{T(\mathbf{j}\,\mathbf{m};\mathbf{k})_{jm}} \text{ is defined}\}.$$

Here $\mathbb{K}_T^{(j)}(\mathbf{j})$ is the domain of definition of the intermediate angular momentum quantum numbers $\mathbf{k} = (k_1, k_2, \ldots, k_{n-2})$ for specified $\mathbf{j}$ and j. Examples of $\mathbb{K}_T^{(j)}(\mathbf{j})$ are given by (5.106),(5.126), (5.150), where $\mathbb{K}_T^{(j)}(j_1, j_2)$ is empty for $n = 2$. The precise definition of $\mathbf{k} \in \mathbb{K}_T^{(j)}(\mathbf{j})$ for general n is given below, after a general triangle pattern has been defined (see (5.217)-(5.219)). We note, however, that the cardinality relation $|\mathbb{K}_T^{(j)}(\mathbf{j})| = N_j(\mathbf{j})$ (CG number) holds independently of $T \in \mathbb{T}_n$.

Eigenvector-eigenvalue relations:

$$\mathbf{J}^2(i)|T(\mathbf{j}\,\mathbf{k})_{jm}\rangle = j_i(j_i + 1)|T(\mathbf{j}\,\mathbf{k})_{jm}\rangle, \quad i = 1, 2, \ldots, n,$$
$$\mathbf{K}_T^2(i)|T(\mathbf{j}\,\mathbf{k})_{jm}\rangle = k_i(k_i + 1)|T(\mathbf{j}\,\mathbf{k})_{jm}\rangle, \quad i = 1, 2, \ldots, n-2, \quad (5.202)$$
$$\mathbf{J}^2|T(\mathbf{j}\,\mathbf{k})_{jm}\rangle = j(j+1)|T(\mathbf{j}\,\mathbf{k})_{jm}\rangle;$$

Standard action:

$$J_3|T(\mathbf{j}\,\mathbf{k})_{jm}\rangle = m|T(\mathbf{j}\,\mathbf{k})_{jm}\rangle,$$
$$J_+|T(\mathbf{j}\,\mathbf{k})_{jm}\rangle = \sqrt{(j-m)(j+m+1)}\,|T(\mathbf{j}\,\mathbf{k})_{j\,m+1}\rangle, \quad (5.203)$$
$$J_-|T(\mathbf{j}\,\mathbf{k})_{jm}\rangle = \sqrt{(j+m)(j-m+1)}\,|T(\mathbf{j}\,\mathbf{k})_{j\,m-1}\rangle.$$

For simplicity of notation, we keep the same $k_i(k_i+1)$ notation for the eigenvalues of the squared intermediate angular momentum operators $\mathbf{K}_T^2(i)$, although the k_i should be distinguished because the intermediate angular momenta $\mathbf{K}_T(i)$ are different for each shape of the tree $T \in \mathbb{T}_n$. This manifests itself in the different domains of definition for the $\mathbf{k}$ corresponding to trees of different shapes, namely, $\mathbf{k} \in \mathbb{K}_T^{(\mathbf{j})}(\mathbf{j})$.

It follows from the Hermitian property of each operator in the set of eigenvalue relations (5.202) and J_3 in (5.203) that the simultaneous eigenvectors $|T(\mathbf{j}\,\mathbf{k})_{j\,m}\rangle$ satisfy the following orthonormality relations:

$$\langle T(\mathbf{j}\,\mathbf{k})_{j\,m} | T(\mathbf{j}\,\mathbf{k}')_{j'\,m'}\rangle = \delta_{\mathbf{k},\mathbf{k}'}\delta_{j,j'}\delta_{m,m'}. \tag{5.204}$$

It is important to observe in this relation that it is the same binary tree T that occurs in the bra-ket vectors of the coupled states: $\mathbf{j}\,\mathbf{k}$ is the standard labeling of T, and $\mathbf{j}\,\mathbf{k}'$ is the same standard labeling with primes on the parts of $\mathbf{k} = (k_1, k_2, \ldots, k_{n-2})$. The orthonormality (5.204) of the coupled state vectors (5.200) and that of the uncoupled state vectors (5.195) gives the generalized WCG coefficients in terms of the inner product:

$$C_{T(\mathbf{j}\,\mathbf{m};\mathbf{k})_{j\,m}} = \langle \mathbf{j}\,\mathbf{m} | T(\mathbf{j}\,\mathbf{k})_{j\,m}\rangle = \langle T(\mathbf{j}\,\mathbf{k})_{j\,m} | \mathbf{j}\,\mathbf{m}\rangle, \tag{5.205}$$

since the generalized WCG coefficients are real. This relation is used to define the real orthogonal matrix $C_T^{(\mathbf{j})}$, with (row;column) elements

$$\mathbf{k},\,j,\,m \in \mathbb{R}_T(\mathbf{j});\ \mathbf{m} \in \mathbb{C}_T(\mathbf{j}) \tag{5.206}$$

by the inner product:

$$\left(C_T^{(\mathbf{j})}\right)_{\mathbf{k},j,m;\,\mathbf{m}} = \langle T(\mathbf{j}\,\mathbf{k})_{j\,m} | \mathbf{j}\,\mathbf{m}\rangle = C_{T(\mathbf{j}\,\mathbf{m};\mathbf{k})_{j\,m}}. \tag{5.207}$$

The row and column orthogonality of the real orthogonal matrix $C_T^{(\mathbf{j})}$ is expressed by

$$C_T^{(\mathbf{j})} C_T^{(\mathbf{j})\,tr} = I_{N_j(\mathbf{j})} \text{ and } C_T^{(\mathbf{j})\,tr} C_T^{(\mathbf{j})} = I_{N_j(\mathbf{j})}. \tag{5.208}$$

Since the rows and columns of the identity matrix are labeled, respectively, by the sets $\mathbb{R}_T(\mathbf{j})$ and $\mathbb{C}_T(\mathbf{j})$, the matrix elements of relations (5.208) give the following equivalent expressions of the orthogonality conditions:

$$\sum_{\mathbf{m}\in\mathbb{C}_T(\mathbf{j})} C_{T(\mathbf{j}\,\mathbf{m};\mathbf{k})_{j\,m}} C_{T(\mathbf{j}\,\mathbf{m};\mathbf{k}')_{j'\,m'}} = \delta_{\mathbf{k},\mathbf{k}'}\delta_{j,j'}\delta_{m,m'}, \tag{5.209}$$

$$\sum_{\mathbf{k},j,m\in\mathbb{R}_T(\mathbf{j})} C_{T(\mathbf{j}\,\mathbf{m};\mathbf{k})_{j\,m}} C_{T(\mathbf{j}\,\mathbf{m}';\mathbf{k})_{j\,m}} = \delta_{\mathbf{m},\mathbf{m}'}. \tag{5.210}$$

5.8. BINARY TREES AND DOUBLY STOCHASTIC MATRICES

The proof of these orthogonality can also be given directly from the definition (5.198) of the generalized WCG coefficients. For example, the proof of relation (5.209) matches together the same a_i, b_i pair in (5.198), uses the sum rule $m = m_1 + m_2 + \cdots + m_n$ for fixed m to express $\alpha_i + \beta_i = q_i$ in terms of other projection quantum numbers m_i, and then uses the orthogonality relation

$$\sum_{\alpha_i, \beta_i} C^{a_i\, b_i\, k_i}_{\alpha_i\, \beta_i\, q_i} C^{a_i\, b_i\, k'_i}_{\alpha_i\, \beta_i\, q_i} = \delta_{k_i, k'_i}. \tag{5.211}$$

The set of $\prod_{i=1}^{n}(2j_i + 1)$ orthonormal vectors $|T(\mathbf{j}\,\mathbf{k})_{j\,m}\rangle$ defined by (5.200) spans the tensor product space $\mathcal{H}_{j_1} \otimes \mathcal{H}_{j_2} \otimes \cdots \otimes \mathcal{H}_{j_n}$. But this set of orthonormal basis vectors can be partitioned into two important subsets that span sub vector spaces with properties of particular interest. These vector subspaces are defined in terms of their orthonormal basis sets as follows:

$$\mathbf{B}_{\mathbf{j}, \mathbf{k}, j} = \left\{ |T(\mathbf{j}\,\mathbf{k})_{j\,m}\rangle \,\Big|\, m \in \{j, j-1, \ldots, -j\} \right\}, \tag{5.212}$$

$$\mathbf{B}_{\mathbf{j}, j, m} = \left\{ |T(\mathbf{j}\,\mathbf{k})_{j\,m}\rangle \,\Big|\, \mathbf{k} \in \mathbb{K}_T^{(j)}(\mathbf{j}) \right\}. \tag{5.213}$$

The first basis $\mathbf{B}_{\mathbf{j}, \mathbf{k}, j}$ defines a vector subspace $\mathcal{H}_{\mathbf{j}, \mathbf{k}, j}$ of dimension $2j+1$ of the tensor product space $\mathcal{H}_{j_1} \otimes \cdots \otimes \mathcal{H}_{j_n}$, there being $N_j(\mathbf{j})$ such perpendicular spaces, each of which is the carrier space of the irreducible representation $D^j(U)$ of $SU(2)$ under $SU(2)$ frame rotations, since the total angular momentum $\mathbf{J}$ has the standard action on each of these subspaces. The second basis $\mathbf{B}_{\mathbf{j}, j, m}$ defines a vector subspace $\mathcal{H}_{\mathbf{j}, j, m}$ of dimension $N_j(\mathbf{j})$, there being such a perpendicular vector subspace of $\mathcal{H}_{j_1} \otimes \cdots \otimes \mathcal{H}_{j_n}$ for each $j = j_{\min}, j_{\min}+1, \ldots, j_{\max}$ and each $m = j, j-1, \ldots, -j$. Thus, the counting of all the vectors in either of the two subspace decompositions of $\mathcal{H}_{j_1} \otimes \cdots \otimes \mathcal{H}_{j_n}$ gives the formula between dimensions:

$$\prod_{i=1}^{n}(2j_i + 1) = \sum_{j=j_{\min}}^{j_{\max}} N_j(\mathbf{j})(2j+1). \tag{5.214}$$

The result of principal interest for the classification of coupled angular momentum states is the following, which extends the preceding results for $\mathbf{j} = (j_1, j_2, \ldots, j_n)$ to an arbitrary assignment of permutations of $\mathbf{j}$ to the shape of any binary tree $T \in \mathbb{T}_n$:

Let $|T(\mathbf{j}_\pi\,\mathbf{k})_{j\,m}\rangle$, each $T \in \mathbb{T}_n, \pi \in S_n$, denote any of the $n!a_n$ standard labeled binary coupling schemes. Then, each of these basis sets is an orthonormal basis of one-and-the-same vector space $\mathcal{H}_{\mathbf{j}, j, m}$ of order $N_j(\mathbf{j})$. Moreover, the inner product of coupled state vectors of each irreducible pair (no common forks) of coupled state vectors gives the irreducible triangle coefficients of order $2(n-1)$ that determine the real linear *transformation between the pair, as well as the domains of definition of the intermediate quantum numbers that enumerate the orthonormal basis vectors. The latter are uniquely determined by application of the elementary CG addition rule for two angular momenta to each of the $2(n-1)$ columns of each of the irreducible triangle coefficients.*

We elaborate further the concept of the *triangle pattern* corresponding to a given standard labeled binary tree $T(\mathbf{j}\,\mathbf{k})_j$, which, in turn, leads to the notion of a triangle coefficient. This has already been illustrated for $n = 3, 4$ in (5.125) and (5.149) from which the extension to the general case is obvious. Indeed, we have already set forth this structure in (5.133)-(5.135), which we repeat here in the present context, as used above. We read-off the standard labeled binary tree the $3 \times (n-1)$ matrix array $\Delta_T(\mathbf{j}\,\mathbf{k})_j$ defined in terms of labeled forks in the following manner:

$$\Delta_T(\mathbf{j}\,\mathbf{k})_j = (\Delta_1\,\Delta_1\,\cdots\,\Delta_{n-1}). \quad (5.215)$$

Each of the $n - 1$ columns Δ_i has three rows with entries read-off the $n - 1$ labeled forks of the standard labeled binary tree $T(\mathbf{j}\,\mathbf{k})_j$:

$$\Delta_i = \begin{pmatrix} f_{1i} \\ f_{2i} \\ k_i \end{pmatrix} \quad i = 1, 2, \ldots, n-1, \text{ with } k_{n-1} = j. \quad (5.216)$$

These are just the labels of the i-th fork, each fork being of one of the four types (5.197). Each column Δ_i is called a *triangle* because its entries satisfy the triangle rule for the addition of two angular momenta:

$$k_i \in \langle f_{1i}, f_{2i} \rangle = f_{1i} + f_{2i}, f_{1i} + f_{2i} - 1, \ldots, |f_{1i} - f_{2i}|, \quad (5.217)$$

which, of course, can be written in any of the alternative forms in (5.84). Recall that the standard labeling of the fork constituents of the standard labeled binary tree $T(\mathbf{j}\,\mathbf{k})_j$ by the sweeping rule: left-to-right across each level from top-to-bottom. Thus, the $3 \times (n-1)$ triangle pattern, also called a *fork matrix*, encodes exactly the same information as the labeled binary tree itself. This format is, however, often much more convenient.

We now state the domain of definition $\mathbf{k} \in \mathbb{K}_T^{(j)}(\mathbf{j})$:

$$\mathbb{K}_T^{(j)}(\mathbf{j}) = \left\{ (k_1, k_2, \ldots, k_{n-2}) \,\middle|\, \begin{array}{l} \text{each column } \Delta_i \text{ of the triangle} \\ \text{pattern } \Delta_T(\mathbf{j}\,\mathbf{k})_j \text{ is a triangle} \end{array} \right\}, \quad (5.218)$$

where in column $n - 1$, the triangle rule is to be written in one of the forms $f_{1,n-1} \in \langle f_{2,n-1i}, j \rangle$ or $f_{2,n-1} \in \langle f_{1,n-1}, j \rangle$, since j is specified (see the examples (5.131), (5.150),(5.157)).

We next consider pairs of binary trees in $\mathbb{T}_n$ and their associated sets of orthonormal basis vectors that span the tensor product space $\mathcal{H}_{j_1} \otimes \mathcal{H}_{j_2} \otimes \cdots \otimes \mathcal{H}_{j_n}$. Two such binary coupling schemes are:

5.8. BINARY TREES AND DOUBLY STOCHASTIC MATRICES

First coupling scheme for labeled binary tree $T \in \mathbb{T}_n$:

$$|T(\mathbf{j}\,\mathbf{k})_{jm}\rangle = \sum_{\mathbf{m} \in \mathbb{C}_T(\mathbf{j})} \left(C_T^{(j)}\right)_{\mathbf{k},\,\mathbf{j}\,m;\,\mathbf{m}} |\mathbf{j}\,m\rangle, \text{ each } \mathbf{k},\mathbf{j},\mathbf{m} \in \mathbb{R}_T(\mathbf{j}), \quad (5.219)$$

$$|\mathbf{j}\,m\rangle = \sum_{\mathbf{k},\mathbf{j},\mathbf{m} \in \mathbb{R}_T(\mathbf{j})} \left(C_T^{(j)\,tr}\right)_{\mathbf{m};\,\mathbf{k},\,\mathbf{j}\,m} |T(\mathbf{j}\,\mathbf{k})_{jm}\rangle, \text{ each } \mathbf{m} \in \mathbb{C}_T(\mathbf{j}). \quad (5.220)$$

Second coupling scheme for labeled binary tree $T' \in \mathbb{T}_n$:

$$|T'(\mathbf{j}_{\pi'}\,\mathbf{k}')_{jm}\rangle = \sum_{\mathbf{m} \in \mathbb{C}_{T'}(\mathbf{j})} \left(C_{T'}^{(\mathbf{j}_{\pi'})}\right)_{\mathbf{k}',\,\mathbf{j}\,m;\,\mathbf{m}} |\mathbf{j}\,m\rangle, \text{ each } \mathbf{k}',\mathbf{j},\mathbf{m} \in \mathbb{R}_{T'}(\mathbf{j}_{\pi'}),$$
$$(5.221)$$

$$|\mathbf{j}\,m\rangle = \sum_{\mathbf{k}',\mathbf{j},\mathbf{m} \in \mathbb{R}_{T'}(\mathbf{j}_{\pi'})} \left(C_{T'}^{(\mathbf{j}_{\pi'})\,tr}\right)_{\mathbf{m};\,\mathbf{k}',\,\mathbf{j}\,m} |T'(\mathbf{j}_{\pi'}\,\mathbf{k}')_{jm}\rangle, \text{ each } \mathbf{m} \in \mathbb{C}_{T'}(\mathbf{j}).$$
$$(5.222)$$

The sequences $\mathbf{j}_{\pi'} = (j_{\pi'_1}, j_{\pi'_2}, \ldots, j_{\pi'_n})$ and $\mathbf{m}_{\pi'} = (m_{\pi'_1}, m_{\pi'_2}, \ldots, m_{\pi'_n})$ are the same permutation $\pi' \in S_n$ of $\mathbf{j}$ and of $\mathbf{m}$; hence, the m_i and j_i are always matched in the labeled binary tree T' so that the summation is correctly given as over all projection quantum numbers $\mathbf{m} = (m_1, m_2, \ldots, m_n) \in \mathbb{C}_{T'}(\mathbf{j})$ for which the generalized WCG coefficients $C_{T'(\mathbf{j}_{\pi'}\,\mathbf{m}_{\pi'};\mathbf{k}')_{jm}}$ are defined. The forks in the respective labeled tree T, T' have the standard labels $\mathbf{k} = (k_1, k_2, \ldots, k_{n-2}) \in \mathbb{K}_T^{(j)}(\mathbf{j})$ and $\mathbf{k}' = (k'_1, k'_2, \ldots, k'_{n-2}) \in \mathbb{K}_{T'}^{(j)}(\mathbf{j}_{\pi'})$, which satisfy $|\mathbb{K}_T^{(j)}(\mathbf{j})| = |\mathbb{K}_{T'}^{(j)}(\mathbf{j}_{\pi'})| = N_j(\mathbf{j})$.

The state vectors (5.221) and (5.219) are related by

$$|T'(\mathbf{j}_{\pi'}\,\mathbf{k}')_{jm}\rangle = \sum_{\mathbf{k} \in \mathbb{K}_T^{(j)}(\mathbf{j})} \left(C_T^{(j)}\,C_{T'}^{(\mathbf{j}_{\pi'})\,tr}\right)_{\mathbf{k},\mathbf{j};\,\mathbf{k}',\mathbf{j}} |T(\mathbf{j}\,\mathbf{k})_{jm}\rangle,$$

$$= \sum_{\mathbf{k} \in \mathbb{K}_T^{(j)}(\mathbf{j})} \left(R_{T,T'}^{\mathbf{j};\mathbf{j}_\pi}\right)_{\mathbf{k},\mathbf{j};\,\mathbf{k}',\mathbf{j}} |T(\mathbf{j}\,\mathbf{k})_{jm}\rangle, \quad (5.223)$$

where we have defined the recoupling matrix by

$$R_{T,T'}^{\mathbf{j};\mathbf{j}_{\pi'}} = C_T^{(j)}\,C_{T'}^{(\mathbf{j}_{\pi'})\,tr}. \quad (5.224)$$

The transformation coefficients in (5.223) are independent of m, as shown earlier in (1.50)-(1.51).

In the way of comparison, relations (5.219)-(5.220) and (5.221)-(5.222) are an application of the relations between the states $|h\rangle, |k\rangle$, and $|k\rangle$

given by (5.14) and (5.22), respectively, in obtaining (5.223) versus (5.23), where we have the following correspondences:

$$|k\rangle \mapsto |\mathbf{j\,m}\rangle,\ |\mathbf{h}\rangle \mapsto |T(\mathbf{j\,k})_{jm}\rangle,\ |\mathbf{k}\rangle \mapsto |T'(\mathbf{j}_{\pi'}\,\mathbf{k'})_{jm}\rangle,$$
$$U \mapsto C_T^{(\mathbf{j})\,tr},\ V \mapsto C_{T'}^{(\mathbf{j}_{\pi'})\,tr},$$
$$W(U,V) \mapsto C_T^{(\mathbf{j})}\,C_{T'}^{(\mathbf{j}_{\pi'})\,tr}.$$
(5.225)

The recoupling matrix $R_{T,T'}^{\mathbf{j};\mathbf{j}_{\pi'}}$ is a real orthogonal matrix of order $N_j(\mathbf{j})$ in consequence of the orthonormality of the basis vectors on each side of relation (5.223) for specified $\mathbf{j}, j, m$:

$$R_{T,T'}^{\mathbf{j};\mathbf{j}_{\pi'}}\,R_{T,T'}^{\mathbf{j};\mathbf{j}_{\pi'}\,tr} = R_{T,T'}^{\mathbf{j};\mathbf{j}_{\pi'}\,tr}\,R_{T,T'}^{\mathbf{j};\mathbf{j}_{\pi'}} = I_{N_j(\mathbf{j})},$$
(5.226)

$$\sum_{\mathbf{k}' \in \mathbb{K}_T^{(j)}(\mathbf{j}_{\pi'})} \left(R_{T,T'}^{\mathbf{j};\mathbf{j}_{\pi'}}\right)_{\mathbf{k},j;\mathbf{k}',j} \left(R_{T,T'}^{\mathbf{j};\mathbf{j}_{\pi'}}\right)_{\mathbf{k}''',j;\mathbf{k}',j} = \delta_{\mathbf{k},\mathbf{k}'''},$$

$$\mathbf{k},\mathbf{k}''' \in \mathbb{K}_T^{(j)}(\mathbf{j}),$$
(5.227)

$$\sum_{\mathbf{k} \in \mathbb{K}_T^{(j)}(\mathbf{j})} \left(R_{T,T'}^{\mathbf{j};\mathbf{j}_{\pi'}}\right)_{\mathbf{k},j;\mathbf{k}',j} \left(R_{T,T'}^{\mathbf{j};\mathbf{j}_{\pi'}}\right)_{\mathbf{k},j;\mathbf{k}'',j} = \delta_{\mathbf{k}',\mathbf{k}''},$$

$$\mathbf{k}',\mathbf{k}'' \in \mathbb{K}_{T'}^{(j)}(\mathbf{j}_{\pi'}).$$

It is particularly significant to notice that it is the same total angular momentum quantum number j that occurs throughout these orthogonality relations. This signifies that it is $\mathbf{k} \in \mathbb{K}_T^{(j)}(\mathbf{j})$ and $\mathbf{k}' \in \mathbb{K}_{T'}^{(j)}(\mathbf{j}_{\pi'})$ alone that enumerate the rows and columns of the matrix elements of the recoupling matrix $R_{T,T'}^{\mathbf{j};\mathbf{j}_{\pi'}}$, which is of order $N_j(\mathbf{j})$.

Remark on notation. The notation for the elements of the recoupling matrix $\left(R_{T,T'}^{\mathbf{j};\mathbf{j}_{\pi'}}\right)_{\mathbf{k},j;\mathbf{k}'',j}$, in which j is fixed, obtains from carrying forward the notations in (5.219)-(5.220) and (5.221)-(5.222), and taking into account that the direct transformation in (5.223) is between binary coupled states diagonal in j because $\mathbf{J}^2$ and J_3 are diagonal on both eigenvectors. Thus, the matrix elements of the product of generalized WCG coefficient given by $C_T^{(\mathbf{j})}\,C_{T'}^{(\mathbf{j}_{\pi'})\,tr}$ are not only independent of m, they are also diagonal in j. Relation (5.223) can, of course, be applied to each $j = \{j_{\min}, j_{\min}+1, \ldots, j_{\max}\}$ and each corresponding $m = j, j-1, \ldots, j$ to obtain the block structure, as embodied in the direct sum relation (1.42). From this point of view, it would be more appropriate to intro-

5.8. BINARY TREES AND DOUBLY STOCHASTIC MATRICES

duce a notation such as

$$\left(R^{\mathbf{j};\mathbf{j}_{\pi'};j}_{T,T'}\right)_{\mathbf{k};\mathbf{k'}} = \left(R^{\mathbf{j};\mathbf{j}_{\pi'}}_{T,T'}\right)_{\mathbf{k},j;\mathbf{k'},j}. \tag{5.228}$$

But then the relation of the recoupling matrix to the product in (5.224) becomes obscure. We find it more natural to follow the notation in (5.223), keeping in mind that this notation accounts properly for the underlying block structure of the recoupling matrix $R^{\mathbf{j};\mathbf{j}_{\pi'}}_{T,T'}$ by keeping j the same in both the row and column indexing. The order of the matrix is, of course, $N_j(\mathbf{j})$ (CG number) for all $\pi' \in S_n$. □

In summary, we have all the following relations of the elements of the recoupling matrix to real inner products and triangle coefficients:

$$\left(R^{\mathbf{j};\mathbf{j}_{\pi'}}_{T,T'}\right)_{\mathbf{k},j;\mathbf{k'},j} = \left(C^{(\mathbf{j})}_T C^{(\mathbf{j}_{\pi'})\,tr}_{T'}\right)_{\mathbf{k},j;\mathbf{k'},j}$$
$$= \langle T(\mathbf{j}\,\mathbf{k})_{j\,m} | T(\mathbf{j}_{\pi'}\,\mathbf{k'})_{j\,m}\rangle = \langle T(\mathbf{j}_{\pi'}\,\mathbf{k'})_{j\,m} | T(\mathbf{j}\,\mathbf{k})_{j\,m}\rangle \tag{5.229}$$
$$= \left\{\Delta_T(\mathbf{j}\,\mathbf{k})_j \,\middle|\, \Delta_{T'}(\mathbf{j}_{\pi'}\,\mathbf{k'})_j\right\} = \left\{\Delta_{T'}(\mathbf{j}_{\pi'}\,\mathbf{k'})_j \,\middle|\, \Delta_T(\mathbf{j}\,\mathbf{k})_j\right\}.$$

It is in the relation of the elements of the recoupling matrix to triangle coefficients that the reduction and phase rules of the triangle coefficient given by (5.134) and (5.135) come into play. It is not required that the pair of coupled binary trees in relations (5.219) and (5.221) be irreducible; that is, have no common forks. The inner product and triangle coefficients can reduce to lower order triangle coefficients, even to a phase factor, because of common forks. It is only when the two coupling schemes are what we have called irreducible — no common forks — that we can obtain a new triangle coefficient, not already obtained from lower-order triangle coefficients. All such relations lend themselves to the probabilistic interpretation of the inner product of state vectors and the associated triangle coefficient, irreducible or not.

Binary tree couplings have two separate probabilistic interpretations, one associated with the coupling of binary states from the basis states $|\mathbf{j}\,\mathbf{m}\rangle$; the other associated with pairs of coupled binary states:

(i) The Landé transition probability amplitude matrix $W(U,V)$ and the transition probability matrix $S(U,V)$ for the prepared coupled state $|T(\mathbf{j}\,\mathbf{k})_{j\,m}\rangle$ and the measured uncouple state $|\mathbf{j}\,\mathbf{m}\rangle$ are the matrices of order $N(\mathbf{j}) = \prod_{i=1}^n (2j_i+1)$ with $U = C^{(\mathbf{j})\,tr}_T$ and $V = I_{N(\mathbf{j})}$ as follows:

$$W_{(U,V)} = U^\dagger V = C^{(\mathbf{j})}_T, \quad W_{\mathbf{k},j,m;\,\mathbf{m}}(U,V) = \left(C^{(\mathbf{j})}_T\right)_{\mathbf{k},j,m;\,\mathbf{m}},$$
$$S_{\mathbf{k},j,m;\,\mathbf{m}}(U,V) = \left(\left(C^{(\mathbf{j})}_T\right)_{\mathbf{k},j,m;\,\mathbf{m}}\right)^2. \tag{5.230}$$

(ii) The Landé transition probability amplitude matrix $W(U,V)$ and the transition probability matrix $S(U,V)$ for the prepared coupled state $|T(\mathbf{j\,k})_{jm}\rangle$ and the measured coupled state $|T'(\mathbf{j}_{\pi'}\,\mathbf{k}')_{jm}\rangle$ are the matrices of order $N_j(\mathbf{j})$ given by $U = C_T^{(\mathbf{j})\,tr}$ and $V = C_{T'}^{\mathbf{j}_{\pi'}\,tr}$ as follows:

$$W(U,V) = R_{T,T'}^{\mathbf{j};\mathbf{j}_{\pi'}} = C_T^{(\mathbf{j})}\,C_{T'}^{\mathbf{j}_{\pi'}\,tr},$$

$$W_{\mathbf{k},j,\mathbf{k}',j}(U,V) = \left(R_{T,T'}^{\mathbf{j};\mathbf{j}_{\pi'}}\right)_{\mathbf{k},j;\mathbf{k}',j} = \left\{\Delta_T(\mathbf{j\,k})_j \,\middle|\, \Delta_{T'}(\mathbf{j}_{\pi'}\,\mathbf{k}')_j\right\}, \quad (5.231)$$

$$S_{\mathbf{k},j;\mathbf{k}',j}(U,V) = \left(\left(R_{T,T'}^{\mathbf{j};\mathbf{j}_{\pi'}}\right)_{\mathbf{k},j;\mathbf{k}',j}\right)^2 = \left(\left\{\Delta_T(\mathbf{j\,k})_j \,\middle|\, \Delta_{T'}(\mathbf{j}_{\pi'}\,\mathbf{k}')_j\right\}\right)^2.$$

Of course, we could also choose $\mathbf{j}_\pi$ in place of $\mathbf{j}$ in (5.230) and (5.231).

The results (5.230)-(5.231) establish a principal result of this Chapter: The theory of doubly stochastic matrices in composite angular momentum systems and the binary coupling theory are one-to-one. *All relations between triangle coefficients of various orders carry over to doubly stochastic matrices and their probabilistic interpretation.*

The unifying structure underlying all relations between triangle coefficients is that of a recoupling matrix, as realized in the Landé form $W(U,V) = U^\dagger V$. As shown in (5.229) in the inner product relation, *the matrix elements of a recoupling matrix are transition probability amplitudes.*

The multiplication rule for transition probability amplitudes is expressed by

$$W(U,V)W(V,U') = W(U,U'), \text{ all } U,V,U' \in U(N), \quad (5.232)$$

for some appropriate unitary group $U(N)$, where the product is independent of the intermediate unitary matrix V. Each factor, in turn, is related to the transition probability by

$$\begin{aligned} S_{ik}(U,V) &= |W_{ik}(U,V)|^2, \\ S_{kj}(V,U') &= |W_{kj}(V,U')|^2, \\ S_{ij}(U,U') &= |W_{ij}(U,U')|^2, \end{aligned} \quad (5.233)$$

where $(i,k),(k,j),(i,j)$, for $i,k,j = 1,2,\ldots,N$ denote the (row, column) indexing of the unitary matrices being considered. These three relations are the fundamental rule for the interference of the respective amplitudes in (5.232). In matrix element form, relation (5.232) itself is given by

$$\sum_{k=1}^{N} W_{ik}(U,V)W_{kj}(V,U') = W_{ij}(U,U'). \quad (5.234)$$

5.8. BINARY TREES AND DOUBLY STOCHASTIC MATRICES

Relation (5.234) can also be expressed in terms of recoupling matrices and triangle coefficients, using (5.231), with general choices given, for example, by

$$U = C_T^{\mathbf{j}_\pi \, tr}, \; V = C_{T''}^{\mathbf{j}_\rho \, tr}, \; U' = C_{T'}^{\mathbf{j}_{\pi'} \, tr}. \tag{5.235}$$

Here T, T', T'' denote any three binary trees in the set $\mathbb{T}_n$ of binary trees of order n, and $\mathbf{j}_\pi, \mathbf{j}_\rho, \mathbf{j}_{\pi'}$ are arbitrary permutations of $\mathbf{j} = (j_1, j_2, \ldots, j_n)$.

We conclude: *Relation (5.232), which expresses the multiplication property for transition probability amplitudes and their interference, is one of the most basic rules in all of standard quantum theory.*

5.8.5 Racah Sum-Rule and Biedenharn-Elliott Identity as Transition Probability Amplitude Relations

It is useful to illustrate the application of the multiplication rule for transition probability amplitudes (5.232) in its simplest application, as we have done already in [6] and [L], in the derivation of the Racah sum-rule. The multiplication rule is realized as follows, where the third relation is the triangle coefficient expression of the identity:

$$W(U, V)W(V, U') = W(U, U'),$$

$$U = C_T^{((ab)c)\, tr}, \; V = C_T^{((bc)a)\, tr}, \; U' = C_T^{((ac)b)\, tr}, \tag{5.236}$$

$$\sum_{k''} \left\{ \begin{array}{cc|cc} a & k & b & k'' \\ b & c & c & a \\ k & j & k'' & j \end{array} \right\} \left\{ \begin{array}{cc|cc} b & k'' & a & k' \\ c & a & c & b \\ k'' & j & k' & j \end{array} \right\}$$

$$= \left\{ \begin{array}{cc|cc} a & k & a & k' \\ b & c & c & b \\ k & j & k' & j \end{array} \right\}.$$

The binary tree T is the one of shape $((\circ \circ) \circ)$, and we have used the shape notation to which have been assigned three permutations of $\mathbf{j} = (a, b, c)$ in the second line. We have also expressed the matrix elements of the first relation in terms of the triangle coefficients of order four, since this best expresses the identity in all of its varied forms under the identification with Racah coefficients.

We can now state the physical significance of the Racah sum-rule:

The Racah sum-rule (5.236) expresses, in its simplest form, the interference of transition probability amplitudes of the respective coupled angular momentum states, as given by the multiplication rule (5.232).

We also recall here the derivation of the Biedenharn-Elliott (B-E) identity and the derivation of the explicit form of the Wigner $9-j$ coefficient carried out in [L]. This is because of their foundational role in angular momentum coupling theory, but repeated now from the perspective of the application of the basic multiplication rule (5.232) for the interference of transition probability amplitudes. Also, these relations have historical significance in bringing out important properties of cubic graphs associated with irreducible pairs of labeled binary trees (no common forks) and their irreducible triangle coefficients. We do not rederive the B-E identity, but do rederive the expression for the Wigner $9-j$ coefficient to illustrate our present direct focus on the reduction and phase transformation rules (5.134)-(5.135). These rules are basic to our present algorithmic approach, which is also amply illustrated in Chapters 8 in the context of a physical problem — the Heisenberg magnetic ring.

The Biedenharn-Elliott identity is a consequence of the following multiplication of transition probabilistic amplitude matrices:

$$\begin{aligned} W(U_7, U_1) &= W(U_7, U_3)W(U_3, U_{10})W(U_{10}, U_4)W(U_4, U_1) \\ &= W(U_7, U_9)W(U_9, U_5)W(U_5, U_{11})W(U_{11}, U_2) \\ &\quad \times W(U_2, U_{12})W(U_{12}, U_{13})W(U_{13}, U_1). \end{aligned} \qquad (5.237)$$

Here we have defined the various real orthogonal matrices U_i in terms of the generalized WCG coefficients associated with the five binary trees $\mathbb{T}_4 = \{T_1, T_2, T_3, T_4, T_5\}$ of shapes

$$\mathrm{Sh}(T_1) = (((\circ\circ)\circ)\circ),\ \mathrm{Sh}(T_2) = (\circ((\circ\circ)\circ)),\ \mathrm{Sh}(T_3) = ((\circ\circ)(\circ\circ)),$$
$$\mathrm{Sh}(T_4) = ((\circ(\circ\circ))\circ),\ \mathrm{Sh}(T_5) = (\circ((\circ\circ)\circ)), \qquad (5.238)$$

(see (5.95)) by the following labeled shapes:

$$U_1 = C_{T_1}^{(((ab)c)d)\ tr}, \quad U_2 = C_{T_1}^{(((ab)d)c)\ tr}, \quad U_3 = C_{T_1}^{(((ac)b)d)\ tr},$$
$$U_4 = C_{T_1}^{(((ba)c)d)\ tr}, \quad U_5 = C_{T_2}^{(c(a(bd)))\ tr}, \quad U_6 = C_{T_3}^{((ab)(cd))\ tr},$$
$$U_7 = C_{T_3}^{((ac)(bd))\ tr}, \quad U_8 = C_{T_3}^{((ad)(bc))\ tr}, \quad U_9 = C_{T_3}^{((ca)(bd))\ tr},$$
$$U_{10} = C_{T_4}^{((b(ac))d)\ tr}, \quad U_{11} = C_{T_4}^{(c((ab)d))\ tr}, \quad U_{12} = C_{T_4}^{((d(ab))c)\ tr},$$
$$U_{13} = C_{T_5}^{(d((ab)c))\ tr}. \qquad (5.239)$$

The evaluation of the matrix elements of these relations, using relation (5.231) between matrix elements of recoupling matrices and triangle coefficients gives the B-E identity, as stated below in relation (5.251).

The derivation of the Wigner $9-j$ coefficient is a consequence of the following multiplication of transition probability amplitude matrices:

$$\begin{aligned} W(U_7, U_6) &= W(U_7, U_3)W(U_3, U_{10})W(U_{10}, U_4) \\ &\quad \times W(U_4, U_1)W(U_1, U_6), \end{aligned} \qquad (5.240)$$

5.8. BINARY TREES AND DOUBLY STOCHASTIC MATRICES

where the various unitary matrices (real orthogonal) are those defined in (5.239). The evaluation of the matrix elements of these relations, using relation (5.231) between matrix elements of recoupling matrices and triangle coefficients gives the explicit form of the Wigner $9-j$ coefficient in terms of triangle coefficients of order four, as presented in detail below in relations (5.260).

We first point out several significant features exhibited by the transition amplitude transformations that appear in (5.236)-(5.240) before presenting the B-E identity and Wigner $9-j$ relations: *The notions of a commutation C and an association A transformation within a general shape transformation and of a path are all present.* We first give the sequence of shape transformations associated with these three identities and then explain them, as taken directly from [L]:

1. Racah sum-rule:

    (a) first path:
    $$((ab)c) \xrightarrow{A} (a(bc)) \xrightarrow{C} ((bc)a) \xrightarrow{A} (b(ca)) \xrightarrow{C} (b(ac)) \xrightarrow{C} ((ac)b). \quad (5.241)$$

    (b) second path:
    $$((ab)c) \xrightarrow{C} (c(ab)) \xrightarrow{A} ((ca)b) \xrightarrow{C} ((ac)b). \quad (5.242)$$

2. Biedenharn-Elliott identity:

    (a) first path:
    $$\left((ac)(bd)\right) \xrightarrow{A} \left(((ac)b)d\right) \xrightarrow{C} \left((b(ac))d\right)$$
    $$\xrightarrow{A} \left(((ba)c)d\right) \xrightarrow{C} \left(((ab)c)d\right). \quad (5.243)$$

    (b) second path:
    $$\left((ac)(bd)\right) \xrightarrow{C} \left((ca)(bd)\right) \xrightarrow{A} \left(c(a(bd))\right) \xrightarrow{A} \left(c((ab)d)\right)$$
    $$\xrightarrow{C} \left(((ab)d)c\right) \xrightarrow{C} \left((d(ab))c\right) \xrightarrow{A} \left(d((ab)c)\right) \xrightarrow{C} \left(((ab)c)d\right). \quad (5.244)$$

3. Wigner $9-j$ coefficient:

    (a) first path:
    $$\left((ac)(bd)\right) \xrightarrow{A} \left(((ac)b)d\right) \xrightarrow{C} \left((b(ac))d\right)$$
    $$\xrightarrow{A} \left(((ba)c)d\right) \xrightarrow{C} \left(((ab)c)d\right) \xrightarrow{A} \left((ab)(cd)\right). \quad (5.245)$$

(b) second path:

$$\Big(((ab)c)d\Big) \xrightarrow{A} \Big((a(bc))d\Big) \xrightarrow{A} \Big(a((bc)d)\Big) \xrightarrow{C} \Big(a(d(bc))\Big)$$
$$\xrightarrow{A} \Big(a((db)c)\Big) \xrightarrow{C} \Big(((db)c)a\Big) \xrightarrow{C} \Big(((bd)c)a\Big). \tag{5.246}$$

These sequences of commutation mappings C and association mappings A from one binary bracketing (labeled shape) to another have the following definitions:

$$\text{commutation} \;:\; (xy) \xrightarrow{C} (yx);\; (x(yz)) \xrightarrow{C} ((yz)x), \tag{5.247}$$

$$\text{association} \;:\; (x(yz)) \xrightarrow{A} ((xy)z).$$

These are usual definitions in the sense that $(xy) \xrightarrow{C} (yx)$ means that the order of the contiguous elements x and y is to be reversed; and $((xy)z) \xrightarrow{A} (x(yz))$ means that the order of contiguous elements is to be maintained, but the parenthesis pair is to be shifted: This is exactly the meaning assigned to the mappings in each of relations (5.241)-(5.246), with the additional interpretation that x, y, z themselves can be any bracketing by parenthesis pairs of p, q, and r letters, respectively, in which the total of $p + q + r = n$ letters are all distinct. For example, $((ca)x) \xrightarrow{A} (c(ax))$ for $x = (bd)$ is an association: $\Big(((ca)(bd)\Big) \xrightarrow{A} \Big(c(a(bd))\Big)$. Thus, in so far as the action of a commutation or an association on a binary bracketing is concerned, any binary sub-bracketing of p objects by $p-1$ parenthesis pairs enclosed in a binary bracketing of $n > p$ objects by $n-1$ parenthesis pair may be considered as a single entity. For a single letter, we use the convention $(a) = a$.

We call the action of an commutation C or an association A on a general labeled shape (binary bracketing) an *elementary labeled shape transformation,* where we sometimes drop the term labeled. The simplest example is: $(ab) \xrightarrow{C} (ba)$. A sequence $w(A, C)$ of such elementary shape transformations is called a *shape transformation.* A shape transformation may be regarded as a *word* in the noncommuting letters A and C. Thus, a shape transformation maps a labeled shape of a binary tree $T \in \mathbb{T}_n$ to another labeled shape $T' \in \mathbb{T}_n$, which is written as

$$\text{Sh}_T(\mathbf{a}) \xrightarrow{w(A,C)} \text{Sh}_{T'}(\mathbf{b}), \tag{5.248}$$

where $\text{Sh}_T(\mathbf{a})$, $\mathbf{a} = (a_1, a_2, \ldots, a_n)$, denotes the shape of the initial binary bracketing and $\text{Sh}_{T'}(\mathbf{b})$, $\mathbf{b} = \mathbf{a}_\pi = (a_{\pi_1}, a_{\pi_2}, \ldots, a_{\pi_n})$, $\pi \in S_n$, the

5.8. BINARY TREES AND DOUBLY STOCHASTIC MATRICES

labeled shape of the final binary bracketing. But this relation is insufficient in itself to define the transformation to the final shape: It is also necessary to specify the subshape on which the elementary action of each letter in the word $w(A, C)$ is to be effected, as can be identified in the examples given above. For example, the shape transformation

$$\mathrm{Sh}_T\Big((ac)(bd)\Big) \stackrel{ACACA}{\longrightarrow} \mathrm{Sh}_T\Big((ab)(cd)\Big) \qquad (5.249)$$

is ambiguous until accompanied by the statement that the elementary shape transformations as read from right-to-left; that is, first A, then C, then A, then C, and finally A (in general, the word is not reversal symmetric) are to be applied sequentially to the initial shape $\mathrm{Sh}_T\Big((ac)(bd)\Big)$ in accord with the specified actions given by (5.245).

A second shape transformation from the initial to the final shape is given in the examples (5.241)-(5.246). This illustrates an important property of shape transformations: *There are several shape transformations that map a given initial shape to the same final shape.* We call the word that effects a given shape transformation a *path* — paths are not unique. We this aside for the moment, and consider it further in the results below.

All of the preceding results on shape transformations can, of course, be diagrammed directly by the standard labeled binary trees of order n that correspond to a given labeled shape, since shapes of binary bracketings and shapes of binary trees are one-to-one.

We next present the B-E identity as a transition probability amplitude transformation, following the form given for the Racah sum-rule, where we can simplify (5.243)-(5.244) considerably by combining the C–transformations with the A–transformations to obtain:

$$W(U, V) = W(U, V')W(V', V'')W(V'', V) = W(U, U')W(U', V),$$

$$U = C^{((ac)(bd))\,tr},\ V' = C^{(c(a(bd)))\,tr} = V'' = C^{(d((ab)c))\,tr}, \qquad (5.250)$$

$$V = C^{(((ab)c)d)\,tr},\ U' = C^{((b(ac))d)\,tr}.$$

Here we recognize that the binary tree T–type subscripts are not needed, since the shape of the binary tree is carried by the superscript shape notations (as also the case in (5.239)). The matrix elements of this form then yields rather quickly the B-E identity:

$$\left\{\begin{array}{ccc|ccc} a & b & k_1 & a & k'_1 & k'_2 \\ c & d & k_2 & b & c & d \\ k_1 & k_2 & j & k'_1 & k'_2 & j \end{array}\right\}$$

$$= \sum_k \left\{\begin{array}{cc|cc} a & k_1 & a & k \\ c & k_2 & k_2 & c \\ k_1 & j & k & j \end{array}\right\} \left\{\begin{array}{cc|cc} b & a & a & d \\ d & k_2 & b & k'_1 \\ k_2 & k & k'_1 & k \end{array}\right\} \left\{\begin{array}{cc|cc} d & k & k'_1 & k'_2 \\ k'_1 & c & c & d \\ k & j & k'_2 & j \end{array}\right\}$$

$$= \left\{ \begin{array}{cc|cc} b & k_1 & b & k_2' \\ d & k_2 & k_1 & d \\ k_2 & j & k_2' & j \end{array} \right\} \left\{ \begin{array}{cc|cc} a & b & a & k_1' \\ c & k_1 & b & c \\ k_1 & k_2' & k_1' & k_2' \end{array} \right\}. \tag{5.251}$$

This result agrees with (2.208) in [L] when phase transformations and interchange of left and right triangle patterns are made to bring the individual triangle coefficients of order four into the same form. The consolidation of the phase factors into the form (5.250) simplifies greatly the calculation compared to that given in [L].

The B-E identity (5.251) expresses the interference of the three transition probability amplitudes under the summation as a product of two transition probability amplitudes on the right-hand side. The full B-E identity, as expressed by the shape transformations (5.237), has a similar interpretation, but now in terms of the interference of elementary transition probability amplitudes corresponding to C and A shape transformations.

But why does this transition probability amplitude factor? The answer lies in the structure of the cubic graph on six points corresponding to the transition probability amplitude $W(U,V)$ given by the triangle coefficient on the left-hand side in (5.251):

$$W(U,V) \mapsto \left\{ \begin{array}{ccc|ccc} a & b & k_1 & a & k_1' & k_2' \\ c & d & k_2 & b & c & d \\ k_1 & k_2 & j & k_1' & k_2' & j \end{array} \right\}. \tag{5.252}$$

The cubic graph of this triangle coefficient is obtained from the six points defined by the triangles of the coefficient:

$$p_1 = (a, c, k_1), \; p_2 = (b, d, k_2), \; p_3 = (k_1, k_2, j),$$
$$p_4 = (a, b, k_1'), \; p_5 = (k_1', c, k_2'), \; p_6 = (k_2', d, j). \tag{5.253}$$

The rule for obtaining the cubic graph is: Draw a line between each pair of labeled points that share a common label. This gives the following cubic graph diagram:

$$W(U,V) \mapsto \begin{array}{c} \begin{array}{ccc} p_1 & p_3 & p_2 \end{array} \\ \bowtie\!\!\!\bowtie \\ \begin{array}{ccc} p_4 & p_5 & p_6 \end{array} \end{array} \tag{5.254}$$

This cubic graph has the property that if the three lines joining p_1 and p_3, p_2 and p_4, p_5 and p_6 are cut and the endpoints joined in the appropriate pairs, we obtain a factoring of the cubic graph (5.254) into a pair

5.8. BINARY TREES AND DOUBLY STOCHASTIC MATRICES

of cubic graphs on four points, as illustrated by the following diagram:

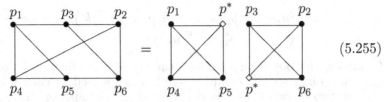

$$\tag{5.255}$$

The two ◇ points are "created" by joining the endpoints of the cut lines.

The phenomenon illustrated by the factoring of the cubic graph associated with the triangle coefficient of the B-E identity is the first example of the Yutsis factoring property (see Yutsis et al. [89], Yutsis and Bandzaitus [90], and [L, Sect. 3.4.1]):

Yutsis factoring theorem for cubic graphs (sufficiency version). *If a cubic graph is constituted of two subgraphs joined by two or three lines, and each of the subgraphs contains at least three points, then this cubic graph factors into two cubic graphs of lower order by the cut and join of lines. In the case of two lines labeled by k and k', there is a Kronecker delta contraction $\delta_{k,k'}$ that gives two disjoint cubic graphs that share the common line labeled k. In the case of three lines labeled by k, q, k', a new virtual point $p^* = (k\, q\, k')$ is created that gives two disjoint cubic graphs that share the common point p^*, provided the cut and join produces no loops (a line that closes on the same point) or double lines (two lines between the same point).*

It is known that there are two nonisomorphic cubic graphs on six points. In this case (not a general property), there is a irreducible triangle coefficient of order six corresponding to each of these two cubic graphs: One corresponds to the B-E identity; the other to the Wigner $9-j$ coefficient. (The definition of isomorphic cubic graphs is given below on p. 172.)

The question now also arises: "From the viewpoint of cubic graphs, why does the Racah sum-rule have the structure given by (5.236)?" There is only one cubic graph involved — the tetrahedron. The answer is obtained by mapping each triangle coefficient to a tetrahedron, as given by the diagram:

$$\tag{5.256}$$

The points in this diagram are given as follows:

$$p_1 = (a,b,k),\ p_2 = (k,c,j), p_1' = (a,c,k'), p_2' = (k',b,j),$$
$$p_1^* = (b,c,k''), p_2^* = (k'',a,j). \tag{5.257}$$

As above, the points are read off the triangle coefficients of order four, and the points with a common symbol are joined in each tetrahedron, and the two "created" points in the product are kept separated. But now the created points are saturated by being summed over. This diagram shows the tetrahedron as a "self-replicating" structure.

We next rederive the expression for the Wigner $9-j$ coefficient based on the first path (5.245) (it was also evaluated in [L])). We do this rederivation to emphasize again the basic role of the reduction and phase rules in (5.234)-(5.235). Once a path is chosen, these are the only rules needed to effect the evaluation. This method is used often in the application to the Heisenberg ring problem in Chapter 8.

Example. Here the Wigner $9-j$ coefficient is written in terms of recoupling matrices by using the first relation (5.245):

$$R^{((ac)(bd));((ab)(cd))} = R^{((ac)(bd));(((ac)b)d)} R^{(((ac)b)d);((b(ac))d)}$$
$$\times R^{((b(ac))d);(((ba)c)d)} R^{(((ba)c)d);(((ab)c)d)} R^{(((ab)c)d);((ab)(cd))}, \tag{5.258}$$

where again we drop the binary tree subscripts since the shape information is incorporated into the notation for the recoupling matrices themselves. We simply take matrix elements of this product of real orthogonal matrices to obtain the expression of the Wigner $9-j$ coefficients in terms of triangle coefficients of order six. The reduction rule and phase rule (5.134)-(5.135) are then applied directly to each matrix element factor to obtain the expression of the Wigner $9-j$ coefficient in terms of triangle coefficients of order four. This is fully illustrative of the general method.

We use directly relation (5.231) for the matrix elements of a recoupling matrix in terms of a triangle coefficient: We have the five relations corresponding to the matrix elements of each of the five factors in (5.258):

$$\left(R^{((ac)(bd));((ab)(cd))}\right)_{k_1,k_2,j;k_1',k_2',j} = \left\{\begin{array}{ccc|ccc} a & b & k_1 & a & c & k_1' \\ c & d & k_2 & b & d & k_2' \\ k_1 & k_2 & j & k_1' & k_2' & j \end{array}\right\};$$

$$\left(R^{((ac)(bd));(((ac)b)d)}\right)_{k_1,k_2,j;k_1'',k_2'',j} = \left\{\begin{array}{ccc|ccc} a & b & k_1 & a & k_1'' & k_2'' \\ c & d & k_2 & c & b & d \\ k_1 & k_2 & j & k_1'' & k_2'' & j \end{array}\right\}$$

5.8. BINARY TREES AND DOUBLY STOCHASTIC MATRICES

$$= \delta_{k_1,k_1''} \begin{Bmatrix} b & k_1 & k_1 & k_2'' \\ d & k_2 & b & d \\ k_2 & j & k_2'' & j \end{Bmatrix};$$

$$\left(R^{(((ac)b)d);((b(ac))d)}\right)_{k_1'',k_2'',j;k_1''',k_2''',j}$$

$$= \begin{Bmatrix} a & k_1'' & k_2'' & a & b & k_2''' \\ c & b & d & c & k_1''' & d \\ k_1'' & k_2'' & j & k_1''' & k_2''' & j \end{Bmatrix} = \delta_{k_1'',k_1'''}\delta_{k_2'',k_2'''}(-1)^{b+k_1''-k_2''};$$

$$\left(R^{((b(ac))d);(((ba)c)d)}\right)_{k_1''',k_2''',j;k_1'''',k_2'''',j} \qquad (5.259)$$

$$= \begin{Bmatrix} a & b & k_2''' \\ c & k_1''' & d \\ k_1''' & k_2''' & j \end{Bmatrix} \begin{Bmatrix} b & k_1'''' & k_2'''' \\ a & c & d \\ k_1'''' & k_2'''' & j \end{Bmatrix} = \delta_{k_2''',k_2''''} \begin{Bmatrix} a & b & b & k_1'''' \\ c & k_1''' & a & c \\ k_1''' & k_2''' & k_1'''' & k_2''' \end{Bmatrix};$$

$$\left(R^{(((ba)c)d);(((ab)c)d)}\right)_{k_1'''',k_2'''',j;k_1''''',k_2''''',j}$$

$$= \begin{Bmatrix} b & k_1'''' & k_2'''' & a & k_1''''' & k_2''''' \\ a & c & d & b & c & d \\ k_1'''' & k_2'''' & j & k_1''''' & k_2''''' & j \end{Bmatrix} = (-1)^{a+b-k_1''''}\delta_{k_1'''',k_1'''''}\delta_{k_2'''',k_2'''''};$$

$$\left(R^{(((ab)c)d);((ab)(cd))}\right)_{k_1''''',k_2''''',j;k_1',k_2',j}$$

$$= \begin{Bmatrix} a & k_1''''' & k_2''''' & a & c & k_1' \\ b & c & d & b & d & k_2' \\ k_1''''' & k_2''''' & j & k_1' & k_2' & j \end{Bmatrix} = \delta_{k_1''''',k_1'} \begin{Bmatrix} k_1' & k_2''''' & c & k_1' \\ c & d & d & k_2' \\ k_2''''' & j & k_2' & j \end{Bmatrix}.$$

It is always the case that an elementary C operation produces a phase factor and an elementary A operation produces a triangle coefficient of order four, possibly with a phase factor. The matrix elements (5.259) are to be summed over $k_1'', k_2''; k_1''', k_2'''; k_1'''', k_2''''; k_1''''', k_2'''''$, taking into account the Kronecker delta factors. The Kronecker delta factors require $k_1'' = k_1''' = k_1; k_2'' = k_2''' = k_2'''; k_1'''' = k_1''''' = k_1'$; hence, there is, a single summation index, which we choose to be $k = k_2''$. Thus, we obtain the following expression of the Wigner $9-j$ coefficient in terms of a double summation over three triangle coefficients of order four:

$$\left\{\begin{array}{ccc} a & c & b & d \\ k_1 & & & k_2 \\ & & j & \end{array}\middle| \begin{array}{ccc} a & b & c & d \\ k_1' & & & k_2' \\ & & j & \end{array}\right\}$$

$$= \sqrt{(2k_1+1)(2k_1'+1)(2k_2+1)(2k_2'+1)} \left\{\begin{array}{ccc} a & b & k_1' \\ c & d & k_2' \\ k_1 & k_2 & j \end{array}\right\}$$

$$= \left(R^{((ac)(bd));((ab)(cd))}\right)_{k_1,k_2,j;\,k_1',k_2',j} = \left\{\begin{array}{ccc} a & b & k_1 \\ c & d & k_2 \\ k_1 & k_2 & j \end{array}\middle| \begin{array}{ccc} a & c & k_1' \\ b & d & k_2' \\ k_1' & k_2' & j \end{array}\right\}$$

$$= \sum_{k} \left\{\begin{array}{ccc} b & k_1 \\ d & k_2 \\ k_2 & j \end{array}\middle| \begin{array}{ccc} b & k \\ k_1 & d \\ k & j \end{array}\right\} \left\{\begin{array}{ccc} a & b \\ c & k_1 \\ k_1 & k \end{array}\middle| \begin{array}{ccc} a & k_1' \\ b & c \\ k_1' & k \end{array}\right\}$$

$$\times \left\{\begin{array}{ccc} k_1' & k \\ c & d \\ k & j \end{array}\middle| \begin{array}{ccc} c & k_1' \\ d & k_2' \\ k_2' & j \end{array}\right\}. \tag{5.260}$$

The two phase factors in relations (5.259) are incorporated into the triangle coefficient of order four by use of (5.135); hence, no phase factors appear in the final relation (5.260). This relation for the Wigner $9-j$ coefficient agrees exactly with that given by relation (2.210) in [L], upon noting that the phase factor $(-1)^{a+k-k_1-k_1'} = (-1)^{a+b-k_1'}(-1)^{b+k_1-k} = (-1)^{a+b-k_1'}(-1)^{-b-k_1+k}$ in (2.210) can be incorporated into the triangle coefficients. In the present derivation, we have focused more strongly on the systematic rules based directly on the use of (5.134)-(5.135). □

The above calculation can be enhanced considerably by regrouping the factors in (5.256) to include the phase transformations in the recoupling matrices themselves to obtain (compare (5.250)):

$$R^{((ac)(bd));((ab)(cd))} = R^{((ac)(bd));((ac)b)d)} R^{((ac)b)d);((ba)c)d)}$$
$$\times R^{((ba)c)d);((ab)(cd))},$$

$$W(U,V) = W(U,V')W(V',V'')W(V'',V), \tag{5.261}$$

$$U = C^{((ac)(bd))\,tr}, V' = C^{((ac)b)d)\,tr},$$

$$V'' = C^{((ba)c)d)\,tr}, V = C^{((ab)(cd))\,tr}.$$

We observe that the number of recoupling matrices and transition probability amplitudes in relation (5.261) is equal to the number of elementary

5.8. BINARY TREES AND DOUBLY STOCHASTIC MATRICES

A−transformations, which is always the case for an irreducible triangle coefficient (see (5.259) and (5.265)). We call the number of such associations A the *length of the path* from the initial binary couple state vector to the final binary coupled state vector. These features hold for all pairs of irreducible coupled state vectors.

The notation for the recoupling matrices in all of relations (5.258)-(5.261) uses the shape of the pair of binary trees. It is often useful to diagram the labeled pair of binary trees and their common forks directly for each step corresponding to the reduction of a triangle coefficient. This gives a more vivid presentation of common forks and can be very useful in effecting error-free calculations.

It is also quite interesting that expressions for irreducible triangle coefficients can also be obtained in terms of generating functions (see [L]), which show that such triangle coefficients are always integers produced by complicated multi-summations times multiplying square-root factors, and all such square-root factors either get squared or can be canceled from the two sides of relations between irreducible triangle coefficients, leaving behind integer relationships. Such "cubic graph" integers, which are presented as summations over "tetrahedral integers" associated with the tetrahedron (triangle coefficient of order four) then sometimes factor into products of lower-order cubic graph integers, as implied by the Yutsis factoring theorem and illustrated by the B-E identity. In this sense, angular momentum theory is about the properties of such integers in which Diophantine theory eventually gets entangled — see the short summary in Ref. [45] and references therein.

The calculation of the Wigner $9 - j$ for the second path given in (5.246) parallels that given above for the first path. It is presented by the following pair of labeled binary trees with corresponding matrix elements of the recoupling matrix and irreducible triangle coefficients of order six:

$$\left\{ \begin{matrix} a & b \\ k_1 & c \\ k_2 & d \\ & j \end{matrix} \middle| \begin{matrix} b & d \\ k_1' & c \\ k_2' & a \\ & j \end{matrix} \right\} = \left(R^{(((ab)c)d);(((bd)c)a)} \right)_{k_1,k_2,j;\, k_1',k_2',j}$$

$$= \left\{ \begin{matrix} a & k_1 & k_2 \\ b & c & d \\ k_1 & k_2 & j \end{matrix} \middle| \begin{matrix} b & k_1' & k_2' \\ d & c & a \\ k_1' & k_2' & j \end{matrix} \right\}. \quad (5.262)$$

We give the full derivation of this version of the $9 - j$ coefficient because of its importance in the Heisenberg magnetic ring problem presented in Chapter 8. The shape transformation and associated recoupling matrix multiplications are the following:

$$\Big(((ab)c)d\Big) \xrightarrow{A} \Big((a(bc))d\Big) \xrightarrow{A} \Big(a((bc)d)\Big) \xrightarrow{C} \Big(a(d(bc))\Big)$$
$$\xrightarrow{A} \Big(a((db)c)\Big) \xrightarrow{C} \Big(((db)c)a\Big) \xrightarrow{C} \Big(((bd)c)a\Big); \quad (5.263)$$

$$R^{(((ab)c)d),(((bd)c)a)} = R^{(((ab)c)d);((a(bc))d)} R^{((a(bc))d);(a(d(bc)))}$$
$$\times R^{(a(d(bc)));(((bd)c)a)}.$$

The shortened product of recoupling matrices includes all the elementary transformations in the first line; hence, is a valid carrier of the same information. The matrix elements of the recoupling matrix product in (5.263) now give the following relation:

$$\left(R^{(((ab)c)d);(((bd)c)a)}\right)_{k_1,k_2,j;k'_1,k'_2,j}$$
$$= \sum_{k''_1,k''_2,k'''_1,k'''_2} \left(R^{(((ab)c)d);((a(bc))d)}\right)_{k_1,k_2,j;k''_1,k''_2,j}$$
$$\times \left(R^{((a(bc))d);(a(d(bc)))}\right)_{k''_1,k''_2,j;k'''_1,k'''_2,j}$$
$$\times \left(R^{(a(d(bc)));(((bd)c)a)}\right)_{k'''_1,k'''_2,j;k'_1,k'_2,j}. \quad (5.264)$$

The evaluation of the matrix elements of each factor now uses the reduction relations (5.134)-(5.135):

$$\left(R^{(((ab)c)d);((a(bc))d)}\right)_{k_1,k_2,j;k''_1,k''_2,j} = \begin{Bmatrix} a & k_1 & k_2 \\ b & c & d \\ k_1 & k_2 & j \end{Bmatrix} \begin{Bmatrix} b & a & k''_2 \\ c & k''_1 & d \\ k''_1 & k''_2 & j \end{Bmatrix}$$
$$= \delta_{k_2,k''_2} \begin{Bmatrix} a & k_1 & b & a \\ b & c & c & k''_1 \\ k_1 & k_2 & k''_1 & k_2 \end{Bmatrix};$$

$$\left(R^{((a(bc))d);(a(d(bc)))}\right)_{k''_1,k''_2,j;k'''_1,k'''_2,j} = \begin{Bmatrix} b & a & k''_2 \\ c & k''_1 & d \\ k''_1 & k''_2 & j \end{Bmatrix} \begin{Bmatrix} b & d & a \\ c & k'''_1 & k'''_2 \\ k'''_1 & k'''_2 & j \end{Bmatrix}$$
$$= \delta_{k''_1,k'''_1} \begin{Bmatrix} a & k''_2 & d & a \\ k''_1 & d & k''_1 & k'''_2 \\ k''_2 & j & k'''_2 & j \end{Bmatrix}; \quad (5.265)$$

5.8. BINARY TREES AND DOUBLY STOCHASTIC MATRICES 165

$$\left(R^{(a(d(bc)));(((bd)c)a)}\right)_{k_1''',k_2''',j;k_1',k_2',j} = \left\{\begin{array}{ccc|ccc} b & d & a & b & k_1' & k_2' \\ c & k_1''' & k_2''' & d & c & a \\ k_1''' & k_2''' & j & k_1' & k_2' & j \end{array}\right\}$$

$$= \delta_{k_2''',k_2'}(-1)^{a+k_2'-j}\left\{\begin{array}{cc|cc} b & d & b & k_1' \\ c & k_1''' & d & c \\ k_1''' & k_2' & k_1' & k_2' \end{array}\right\}.$$

We now assemble these results, account for the Kronecker delta factors, set $k = k_1''$ (summation parameter) and obtain the Wigner $9-j$ coefficient in its irreducible triangle coefficient form as follows:

$$\left\{\begin{array}{c}a \quad\circ\quad b \\ k_1 \bullet \quad \circ\, c \\ k_2 \bullet \quad \circ\, d \\ j \bullet \end{array}\middle| \begin{array}{c} b \quad\circ\quad d \\ k_1' \bullet \quad \circ\, c \\ k_2' \bullet \quad \circ\, a \\ j \bullet \end{array}\right\} = \left\{\begin{array}{ccc|ccc} a & k_1 & k_2 & b & k_1' & k_2' \\ b & c & d & d & c & a \\ k_1 & k_2 & j & k_1' & k_2' & j \end{array}\right\}$$

$$= \sum_k \left\{\begin{array}{cc|cc} a & k_1 & b & a \\ b & c & c & k \\ k_1 & k_2 & k & k_2 \end{array}\right\}\left\{\begin{array}{cc|cc} a & k_2 & d & k_2' \\ k & d & k & a \\ k_2 & j & k_2' & j \end{array}\right\}\left\{\begin{array}{cc|cc} b & d & b & k_1' \\ c & k & d & c \\ k & k_2' & k_1' & k_2' \end{array}\right\} \quad (5.266)$$

$$= (-1)^\kappa \sqrt{(2k_1+1)(2k_2+1)(2k_1'+1)(2k_2'+1)} \left\{\begin{array}{ccc} b & d & k_1' \\ a & j & k_2' \\ k_1 & k_2 & c \end{array}\right\}.$$

The phase factor in this relation is given by

$$(-1)^\kappa = (-1)^{a+b+k_2'-k_1'-k_2} = \omega(b,d,k_1')\omega(d,k_2,j)\omega(a,k_2',j), \quad (5.267)$$

where we use the property $\omega(a,b,c) = \omega(-a,-b,-c)$ of the primitive phase factor (5.136) to obtain the integer $\kappa = a + b + k_2' - k_1' - k_2$. The relation between the triangle coefficient of order six on the left and the $9-j$ coefficient (Wigner's notation) on the right was stated, without proof, by (2.219) in [L]. Such a relation, up to phase factor, must be valid because the six triangles in the Wigner $9-j$ coefficient are the same as in the triangle coefficient; hence, they define the same cubic graph. The rule for determining the phase factor is given in the next section.

5.8.6 Symmetries of the $6-j$ and $9-j$ Coefficients

The only symmetry we use in transforming triangle coefficients of order $2n - 2$ and in the calculation of relations related to them is the primitive phase transformations defined by (5.136) — these phases are intrinsic to the definition of a triangle coefficient and its symbol. Only these primitive symmetries, which leave invariant the triangle relations of a triple of quantum labels (a, b, k) satisfying the addition of angular momentum rule, have a role in our presentation.

It is a well-known fact that the $6-j$ coefficient possesses 144 transformations of its six labels that give back the same coefficient, possibly multiplied by a phase. The symmetries are best presented as the 4!3! row and column exchanges of the 4×3 Bargmann array (see (3.314) in Ref. [6]). All these symmetries can, of course, be transferred to the notation of the irreducible triangle coefficient of order four. But since we make no use of this, we forgo this exercise.

It is also well-known that the Wigner $9-j$ coefficient in (5.260) or (5.266) possesses 72 symmetries that give back the same coefficient, possibly multiplied by a phase. These are the six row permutations, six column permutations, and transposition of the 3×3 array. It is quite easy to verify that the transposition symmetry is the interchange of the left and right patterns of the triangle coefficient. The symmetries of the Wigner $9-j$ are all realized by the basic expression of the $9-j$ coefficient in terms of the $6-j$ coefficient as given by relation (3.319) in Ref. [6]. This relation is given in a nice matrix-element notation as follows:

$$\left\{\begin{array}{ccc} j_{11} & j_{12} & j_{13} \\ j_{21} & j_{22} & j_{23} \\ j_{31} & j_{32} & j_{33} \end{array}\right\} = \sum_k (-1)^{2k}(2k+1) \left\{\begin{array}{ccc} j_{11} & j_{21} & j_{31} \\ j_{32} & j_{33} & k \end{array}\right\}$$

$$\times \left\{\begin{array}{ccc} j_{12} & j_{22} & j_{32} \\ j_{21} & k & j_{23} \end{array}\right\} \left\{\begin{array}{ccc} j_{13} & j_{23} & j_{33} \\ k & j_{11} & j_{12} \end{array}\right\}. \quad (5.268)$$

The $6-j$ coefficient in this result is defined in terms of the Racah coefficient by (5.142). The only symmetries of the $6-j$ coefficient that we require here are the permutational symmetries of the angular momentum labels that result from the six permutations of the columns and the interchange of any pair of elements in the top row with the corresponding pair in the lower row (three operations), all of which leave the value of the $6-j$ coefficient unchanged.

It must be possible to bring a given relation that expresses an irreducible triangle coefficient of order six in terms of a summation over three irreducible triangle coefficients of order four (including a possible phase factor) into the same form as the right-hand side of (5.268), provided the triangles of the given triangle coefficient of order six agree with those of the $9-j$ symbol on the left-hand side of (5.268). Relations (5.139), (5.142), and primitive phase factor transformations are used to bring the right-hand side of the given relation to the exact form of the right-hand side of (5.268), with possible overall multiplication of the summation by a product of primitive phase factors. This then identifies the j_{ik} in terms of the parameters of the triangle coefficient and gives the phase factor relation between the triangle coefficient of order

5.8. BINARY TREES AND DOUBLY STOCHASTIC MATRICES

six and Wigner's $9-j$ coefficient. This is how the phase factor in (5.266) was determined.

The calculation of the Racah sum-rule, B-E identity, and the Wigner $9-j$ coefficient detailed above is fully illustrative of the general method for the calculation of relations between triangle coefficients of order four, and of all $3(n-1)-j$ coefficients. From the structure of these model calculations, we can now describe the general procedure.

5.8.7 General Binary Tree Shape Transformations

Shape transformations map one labeled shape to another labeled shape of a binary tree by application of the basic involution operations of commutation C and association A of symbols. As involutions, such shape transformations are invertible. These shape transformations were introduced in [L], but this is such an important calculation tool that we review the methodology again, now focusing strongly on the role of the reduction and phase rules (5.134)-(5.135).

We now continue with the development of relations (5.219)-(5.224), beginning with some slightly modified notations suitable for the general case:

$$\mathbf{j}_1 = \mathbf{j}_{\pi^{(1)}},\ \mathbf{k}_1 = (k_1^{(1)}, k_2^{(1)}, \ldots, k_{n-2}^{(1)}),$$
$$\mathbf{j}_2 = \mathbf{j}_{\pi^{(2)}},\ \mathbf{k}_2 = (k_1^{(2)}, k_2^{(2)}, \ldots, k_{n-2}^{(2)}), \quad (5.269)$$
$$\mathbf{j}_3 = \mathbf{j}_{\pi^{(3)}},\ \mathbf{k}_3 = (k_1^{(3)}, k_2^{(3)}, \ldots, k_{n-2}^{(3)});$$

$$\mathbf{B}_1 = \{|T_1(\mathbf{j}_1\,\mathbf{k}_1)_{jm}\rangle\,|\,\mathbf{k}_1, j, m \in \mathbb{R}_{T_1}(\mathbf{j}_1)\},$$
$$\mathbf{B}_2 = \{|T_2(\mathbf{j}_2\,\mathbf{k}_2)_{jm}\rangle\,|\,\mathbf{k}_2, j, m \in \mathbb{R}_{T_2}(\mathbf{j}_2)\}, \quad (5.270)$$
$$\mathbf{B}_3 = \{|T_2(\mathbf{j}_3\,\mathbf{k}_3)_{jm}\rangle\,|\,\mathbf{k}_3, j, m \in \mathbb{R}_{T_3}(\mathbf{j}_3)\},$$

where $\pi^{(1)}, \pi^{(2)}, \pi^{(3)}$ are arbitrary permutations in S_n.

Using now the above notations, we have from relation (5.248) the following summary result:

Each shape transformation

$$Sh_{T_1}(\mathbf{j}_1) \stackrel{w(A,C)}{\longrightarrow} Sh_{T_2}(\mathbf{j}_2) \quad (5.271)$$

determines a unique orthonormal transformation of the final coupled basis vectors $\mathbf{B}_2$ *in terms of the initial coupled basis vectors* $\mathbf{B}_1$:

$$|T_2(\mathbf{j}_2\,\mathbf{k}_2)_{jm}\rangle = \sum_{\mathbf{k}_1}\left\{\Delta_{T_1}(\mathbf{j}_1\,\mathbf{k}_1)_j\Big|\Delta_{T_2}(\mathbf{j}_2\,\mathbf{k}_2)_j\right\}|T_1(\mathbf{j}_1\,\mathbf{k}_1)_{jm}\rangle. \quad (5.272)$$

The triangle coefficients in this relation, which are defined by the inner product,

$$\left\langle T_1(\mathbf{j}_1\,\mathbf{k}_1)_{jm} \middle| T_2(\mathbf{j}_2\,\mathbf{k}_2)_{jm} \right\rangle = \left(R^{\mathbf{j}_1;\mathbf{j}_2}_{T_1,T_2} \right)_{\mathbf{k}_1,j;\mathbf{k}_2,j}$$
$$= \left\{ \Delta_{T_1}(\mathbf{j}_1\,\mathbf{k}_1)_j \middle| \Delta_{T_2}(\mathbf{j}_2\,\mathbf{k}_2)_j \right\}, \quad (5.273)$$

are uniquely determined by the word $w(A,C)$ that effects the shape transformation.

As emphasized earlier, the sequential action on the initial shape $\mathrm{Sh}_{T_1}(\mathbf{j}_1)$ of each elementary shape operation A and C that occurs in the word $w(A,C)$ must be uniquely specified in (5.271).

We next show how to calculate the triangle coefficient in (5.273) by using the multiplicative properties of the recoupling matrix $R^{\mathbf{j}_1;\mathbf{j}_2}_{T_1,T_2}$, the reduction property (5.134) and the phase property (5.135).

The simple multiplication property of recoupling matrices and their accompanying expression in terms of transition probability matrices, as expressed by

$$R^{\mathbf{j}_1;\mathbf{j}_2}_{T_1,T_2} R^{\mathbf{j}_2;\mathbf{j}_3}_{T_2,T_3} = R^{\mathbf{j}_1;\mathbf{j}_3}_{T_1,T_3},$$
$$(5.274)$$
$$W(U_1,U_2)W(U_2,U_3) = W(U_1,U_3),$$

are the source of all relations among triangle coefficients of order $2(n-1)$, including the determination of all $3(n-1)-j$ coefficients. This relationship is, or course, true for arbitrary coupled states with orthonormal basis sets defined in relations (5.269)-(5.270) — pairs of coupled state vectors need not be irreducible; that is, the shapes of pairs of binary trees can contain common forks.

There is a great deal of informational content encoded in the product rules (5.274): They are valid for an arbitrary number of coupled binary trees; that is, we can iterate (5.274) to obtain a product, with an accompanying product of transition probability amplitude matrices, of the following forms:

$$R^{\mathbf{j}_1;\mathbf{j}_h}_{T_1,T_h} = R^{\mathbf{j}_1;\mathbf{j}_2}_{T_1,T_2} R^{\mathbf{j}_2;\mathbf{j}_3}_{T_2,T_3} \cdots R^{\mathbf{j}_{h-1};\mathbf{j}_h}_{T_{h-1},T_h}, \; h \geq 3,$$
$$(5.275)$$
$$W(U_1,U_h) = W(U_1,U_2)W(U_2,U_3)\cdots W(U_{h-1},U_h),$$

where $\mathbf{j}_r = \mathbf{j}_{\pi^{(r)}},\, \pi^{(r)} \in S_n$, in which the permutations for $r = 1, 2, \ldots, h$ are completely arbitrary. However, should we choose $\pi^{(r+1)} = \pi^{(r)}$ and

5.8. BINARY TREES AND DOUBLY STOCHASTIC MATRICES

$T_{r+1} = T_r$ for some r, the corresponding recoupling matrix would be the unit matrix: $R^{\mathbf{j}_r;\mathbf{j}_r}_{T_r,T_r} = I_{N_j(\mathbf{j})}$; this situation is to be avoided.

The matrix elements of each recoupling matrix factor in (5.275) is given by the triangle coefficient relation:

$$\left(R^{\mathbf{j};\mathbf{j}_{r+1}}_{T_r,T_{r+1}}\right)_{\mathbf{k}_r,j;\mathbf{k}_{r+1},j} = \left\{\Delta_{T_r}(\mathbf{j}_r\,\mathbf{k}_r)_j \,\middle|\, \Delta_{T_{r+1}}(\mathbf{j}_{r+1}\,\mathbf{k}_{r+1})_j\right\}. \tag{5.276}$$

In this relation, the intermediate angular momenta are the standard intermediate angular momentum labels $\mathbf{k}_r = (k_1^{(r)}, k_2^{(r)}, \ldots, k_{n-2}^{(r)})$ and $\mathbf{k}_{r+1} = (k_1^{(r+1)}, k_2^{(r+1)}, \ldots, k_{n-2}^{(r+1)})$ of the binary trees of labeled shapes $\mathrm{Sh}_{T_r}(\mathbf{j}_r)$ and $\mathrm{Sh}_{T_{r+1}}(\mathbf{j}_{r+1})$.

We next show that in the general relation (5.275) there always exists an h such that the following relation holds:

$$\mathrm{Sh}_{T_r}(\mathbf{j}_r) \xrightarrow{w_r} \mathrm{Sh}_{T_{r+1}}(\mathbf{j}_{r+1}), \; r = 1, 2, \ldots, h-1, \tag{5.277}$$

where w_r is either an elementary commutation C–transformation or an elementary association A–transformation. Thus, we have the following sequence of shape transformations in one-to-one correspondence with relations (5.277):

$$\mathrm{Sh}_{T_1}(\mathbf{j}_1) \xrightarrow{w_1} \mathrm{Sh}_{T_2}(\mathbf{j}_2) \xrightarrow{w_2} \mathrm{Sh}_{T_3}(\mathbf{j}_3) \xrightarrow{w_3} \cdots \xrightarrow{w_{h-1}} \mathrm{Sh}_{T_h}(\mathbf{j}_h). \tag{5.278}$$

In this relation, each $w_r = C$ effects a phase transformation of the corresponding triangle coefficient in consequence of (5.135), and each $w_s = A$ effects a transformation of the corresponding triangle coefficient to an irreducible triangle coefficient of order four, possibly multiplied by a phase factor. Thus, we obtain the shape transformation from the initial shape to the final shape as expressed by

$$\mathrm{Sh}_{T_1}(\mathbf{j}_1) \xrightarrow{w_{h-1}\cdots w_2 w_1} \mathrm{Sh}_{T_h}(\mathbf{j}_h). \tag{5.279}$$

The convention for the application of the shape transformation $w = w_{h-1}\cdots w_2 w_1$ is: first w_1 to $\mathrm{Sh}_{T_1}(\mathbf{j}_1)$, then w_2 to $\mathrm{Sh}_{T_2}(\mathbf{j}_2), \ldots$. A proof of relation (5.278) is given in [L, pp. 153-154] based on a well-known decomposition theorem of an arbitrary binary tree into a sum over products of pairs of binary subtrees (see Comtet [22, p. 55]). But, in fact, a relation of the form (5.278) must hold, since, if it is not true, the factoring can be continued until the elementary form is achieved.

In summary, we have the following result:

There exists a shape transformation by elementary A and C transformations from each initial shape $\mathrm{Sh}_{T_1}(\mathbf{j_1})$ to each final shape $\mathrm{Sh}_{T_h}(\mathbf{j_h})$, as given by relation (5.279), such that the recoupling matrix for the corresponding state vectors $|T_1(\mathbf{j_1}\,\mathbf{k_1})_{jm}\rangle$ and $|T_h(\mathbf{j_h}\,\mathbf{k_h})_{jm}\rangle$ is given by a product of $h-1$ recoupling matrices with a corresponding product of transition probability amplitudes, as expressed by the two relations:

$$R_{T_1,T_h}^{\mathbf{j_1};\mathbf{j_h}} = R_{T_1,T_2}^{\mathbf{j_1};\mathbf{j_2}}\, R_{T_2,T_3}^{\mathbf{j_2};\mathbf{j_3}} \cdots R_{T_{h-1},T_h}^{\mathbf{j_{h-1}};\mathbf{j_h}},\ h \geq 3, \tag{5.280}$$

$$W(U_1, U_h) = W(U_1, U_2) W(U_2, U_3) \cdots W(U_{h-1}, U_h).$$

In contrast to the general relation (5.275), all transformations in this relation are effected by elementary A and C transformations; hence, the significance of h is that of denoting the total number of elementary transformations in A and C. Also, there is no requirement that the labeled binary trees for the initial state and the final state be irreducible (also the case in (5.275), although this will most often be the case in our applications.

Once the recoupling matrix on the left-hand side in (5.280) has been factored into a product of factors related by elementary shape transformations as in the right-hand side, the various recoupling matrix factors can be recombined so as to contain a number of factors equal to the number of occurrences of the A−transformations, as illustrated by the Racah sum-rule (5.236), the B-E identity (5.250); and the Wigner $9-j$ coefficient in (5.261) and (5.263). Thus, it is sufficient to partition an arbitrary word in A and C into subwords as indicated by the parenthesis pairs in $(\cdots A \cdots)(A \cdots) \cdots (A \cdots)$, where each parenthesis pair contains exactly one A, and $\cdots$ denotes any number of C−transformations, including none, as appropriate to the given word. The full product (5.280) of recoupling matrices can now be written in the shortened form containing only A−transformations:

$$R_{T_1,T_s}^{\mathbf{j_1};\mathbf{j_s}} = R_{T_1,T_2}^{\mathbf{j_1};\mathbf{j_2}}\, R_{T_2,T_3}^{\mathbf{j_2};\mathbf{j_3}} \cdots R_{T_{s-1},T_s}^{\mathbf{j_{s-1}};\mathbf{j_s}},\ 3 \leq s < h, \tag{5.281}$$

$$W(U_1, U_s) = W(U_1, U_2) W(U_2, U_3) \cdots W(U_{s-1}, U_s).$$

Here we have conveniently kept the same $\mathbf{j}_i$−notation as in (5.280) for denoting the recoupling matrix factors, but modified their definitions to be in accord with the word factoring $(\ldots A \ldots)(A \cdots) \cdots (A \cdots)$: Thus, $\mathbf{j}_1$ corresponds to the initial shape $Sh_{T_1}(\mathbf{j}_1)$ in (5.280), while $\mathbf{j}_2, \mathbf{j}_3, \ldots, \mathbf{j}_{s-1}$ correspond, respectively, to the product shapes containing all the various C operations in accord with the parenthesis pairs containing single A operations, and the last $\mathbf{j}_s$ corresponds to the final shape $Sh_{T_h}(\mathbf{j}_h)$. The positive integer $s-1$ is now the length of the path, since it counts the number of A−transformations.

It is also the case that, since C−transformations always effect transformations of the form (5.135), it is always possible to write the matrix element expression of the recoupling matrix relation in a form that contains no phase factors as shown explicitly by the Racah sum rule (5.236), the B-E identity (5.251), and the two Wigner $9-j$ coefficient identities (5.260) and (5.266).

Relation (5.281) is the principal result for the theory of binary coupling of n angular momenta:

It is the basic result that allows for the calculation of every irreducible triangle coefficient of order $2(n-1)$ in terms of irreducible triangle coefficients of order four.

No input whatsoever from the theory of cubic graphs is required in implementing the explicit calculation of an irreducible triangle coefficient in (5.281). The role of cubic graphs concerns the additional structure of the multiple summation that results when matrix elements of the recoupling matrix are taken. We are assured from the Yutsis factoring theorem that this complicated summation has one of two possible forms, as controlled by the corresponding cubic graph.

Either the irreducible triangle coefficient obtained from (5.281) factors into a product of lower-order triangle coefficients or it is the expression of a $3(n-1)-j$ coefficient. It factors if and only if the corresponding cubic graph is one that can be presented as the joining of two subgraphs by two or by three lines (two cases), and such that by the cut-and-join operation on the joining lines, there results two cubic graphs of angular momentum type of lower order; otherwise, it defines a $3(n-1)-j$ coefficient. The probability transition amplitude matrix $W(U_1, U_s)$ shares these two features.

We recall again briefly from [L] some of the definitions and terminology for cubic graphs, as applied to angular momentum theory. We have already described and illustrated earlier how a cubic graph is obtained from an irreducible triangle coefficient of order $2(n-1)$ by the rule of taking the $2(n-1)$ triangles (columns) of the triangle coefficient as points and joining all points that share a common label. This mapping always yields a graph with $2(n-1)$ points, $3(n-1)$ lines, with each point being of degree 3. This is a cubic graph $C^*_{2(n-1)}$ on $2(n-1)$ points. We refer to [L, p. 198] for the discussion of such graphs that arise in the binary coupling theory of angular momenta — called *cubic graphs of angular momentum type*. Such cubic graphs are, by definition, just the set that can arise from irreducible triangle coefficients.

Thus, while cubic graphs are not part of the calculational technique based on (5.281) and the reduction and phase rules (5.134) and (5.135), they are pivotal for understanding the structure of irreducible triangle coefficients. This then applies also to the corresponding Landé transition probability amplitudes which, in turn, yield the transition probabilities for coupled angular momentum states corresponding to complete sets of mutually commuting Hermitian operators.

The properties of cubic graphs are also used to classify the $3(n-1)-j$ coefficients that arise into "types," as we next describe.

Let $\mathbb{C}_{2n-2}$ denote the set all graphs with $2(n-1)$ points, $3(n-1)$ lines, with each point of degree 3. The set $\mathbb{C}_{2n-2}$ can be partitioned into equivalence classes by the following rule: Two cubic graphs in the set $\mathbb{C}_{2n-2}$ are called *equivalent* if there exists a labeling of each of their $2(n-1)$ points by the integers $1, 2, \ldots 2n-2$ that preserves adjacency of points, where *adjacent points* are, by definition, any pair of points joined by the same line. All equivalent graphs in the set $\mathbb{C}_{2n-2}$ are said to be *isomorphic*; graphs belonging to different equivalence classes, *non isomorphic*; that is,

Two cubic graphs in $\mathbb{C}_{2n-2}$ are isomorphic if and only if there exists a labeling of their respective points by a permutation of $(1, 2, \ldots, 2n-2)$ such that adjacency of points is preserved; otherwise, they are nonisomorphic.

It is the property of isomorphism versus nonisomorphism of cubic graphs that is used to define the "type" of a $3(n-1)-j$ coefficient: Two $3(n-1)-j$ coefficients are of the same type if and only the triangle coefficient of order $2(n-1)$ that defines them corresponds to isomorphic cubic graphs; of different types if they correspond to nonisomorphic cubic graphs. Not all cubic graphs are of angular momentum type $\mathbb{C}^*_{2n-2}$.

We point out that a counting formula for the number of nonisomorphic cubic graphs on $2n-2$ points is unknown, as is the number of nonisomorphic cubic graphs on $2n-2$ points of angular momentum type; that is, that arise in the construction described above based on the properties of recoupling matrices and their reduction (see Yutsis et al. [89], Yutsis and Bandzaitus [90], Ref. [6], and [L] for explicit numbers and further discussion). Interestingly, a formula is known for the number of so-called <u>labeled</u> cubic graphs (see Read [64], Chen and Louck [20], and [L, p. 226]).

5.8.8 Summary

We summarize the procedure for computing all irreducible triangle coefficients of order $2(n-1)$ and the corresponding Landé transition probability amplitude matrices:

1. Construct the coupled state vectors: Select a pair of binary trees $T_1, T_2 \in \mathbb{T}_n$ corresponding to the addition of n angular momenta as defined by two distinct binary bracketing schemes, where the binary trees T_1 and T_2 can be the same or different, but the labeled shapes are distinct and contain no common forks — they are an irreducible

5.8. BINARY TREES AND DOUBLY STOCHASTIC MATRICES 173

pair. Construct the coupled state vectors for each binary tree in terms of the uncoupled state vectors — these state vectors are then the simultaneous eigenvectors of the two distinct complete sets of commuting Hermitian operators given by relations (5.77)-(5.78) for $T = T_1$ and $T = T_2$. (Example: Wigner $9-j$ coefficient, relations (5.143)-(5.157).)

2. Determine the triangle coefficients: Express the T_2-coupled state vectors as linear combination of the T_1- coupled state vectors, which is always possible, since they span the same vector space. The inner product of these state vectors then defines the irreducible triangle coefficients of order $2n-2$. Express the triangle coefficients in terms of the matrix elements of the recoupling matrix associated with the pair of T_1- and T_2-coupled state vectors. (Example: Wigner $9-j$ coefficient, relations (5.158)-(5.159), (5.182)-(5.183).)

3. Find a path — one always exists: Determine a sequence of elementary shape transformations leading from the shape of the left triangle pattern to the shape of the right triangle pattern, and write the recoupling matrix and the transition amplitude matrix as a product the corresponding elementary recoupling and the transition amplitude matrices. Compress these products to those matrices representing just the $A-$transformations. (Examples: Biedenharn-Elliott relations (5.243)-(5.244), (5.250)-(5.251); Wigner $9-j$ coefficient relations (5.245)-(5.246), (5.261), (5.263).)

4. Calculate the matrix elements of the product in Item 3: This gives the irreducible triangle coefficient of order $2n-2$ as a summation over products of irreducible triangle coefficients of order four, the number of which is equal to the length of the path, which is always the number of $A-$transformations in the path; these are fully known objects in terms of Racah coefficients. (Examples: Biedenharn-Elliott identity (5.251); Wigner $9-j$ coefficient expressions (5.260) and (5.266).)

5. Construct the cubic graph. Take the column triangles of the irreducible triangle coefficient as the vertices of a cubic graph and construct the cubic graph C^*_{2n-2}. Determine if the cubic graph C^*_{2n-2} is of the Yutsis type; that is, if it can be factored into two cubic graphs of lower order by the cut-and-join procedure on two joining lines or on three joining lines. If so, then the irreducible triangle coefficients and transition probability amplitudes coefficients factor into products of lower-order products of these quantities; otherwise, it determines a new $3(n-1)-j$ coefficient.

6. Determine the type of the new $3(n-1)-j$ coefficient in Item 5 by determining to which isomorphic equivalence class it belongs.

7. Give the probabilistic interpretation: In relation (5.281), the prepared state $|T_1(\mathbf{j_1\,k_1})_{jm}\rangle$ is the measured state $|T_s(\mathbf{j_s\,k_s})_{jm}\rangle$ with transition probability given by

$$S_{\mathbf{k_1},j;\mathbf{k_s},j}(U_1,U_s) = \left|\left(R^{\mathbf{j_1 \cdot j_s}}_{T_1,T_s}\right)_{\mathbf{k_1},j;\mathbf{k_s},j}\right|^2 \qquad (5.282)$$

$$= \left|\left\{\Delta_{T_1}(\mathbf{j_1\,k_1})_j \,\middle|\, \wedge_{T_s}(\mathbf{j_s\,k_s})_j\right\}\right|^2,$$

where the transition probability amplitude matrix itself is given by the elements of the recoupling matrix $W(U_1, U_s) = R^{\mathbf{j_1 \cdot j_s}}_{T_1,T_s}$. The matrix $W(U_1, U_s)$ is, of course, a doubly stochastic matrix of the Landé form.

8. Observe: *The transition from the prepared state to the measured state is brought about by a shape transformation; this property should have a significant role in the geometry of quantum mechanics, since time itself does not appear (see Bengtsson and Życkowski [4] and Chruśinński [21]).*

Remarks and unsolved problems: We have found no systematic procedure for determining a path — there are many. From the viewpoint of numerical calculations, it is desirable to realize the product form by the least number of A–transformations. With each path, minimal or not, there is associated a product of transition probability amplitudes. These, of course, are all discrete objects. Different paths imply different summation forms for one and the same coefficient — sort of a generalized hypergeometric series behavior. The understanding of these phenomena requires further study. Moreover, it is the case that a counting formula for number of cubic graphs of angular momentum type, as well as for unlabeled general cubic graphs, is unknown. Finally, we have not found a physical interpretation of the expansion of a doubly stochastic matrix in terms of permutations matrices, as we discuss briefly in the concluding section.

5.8.9 Expansion of Doubly Stochastic Matrices into Permutation Matrices

To conclude this Chapter, we still must come back to the expansion of doubly stochastic matrices in terms of permutation matrices, in search of the role of such permutation matrices in the probabilistic interpretation of doubly stochastic matrices.

All possible expansions of a given doubly stochastic matrix A as a sum of permutation matrices are obtained by finding all possible real

5.8. BINARY TREES AND DOUBLY STOCHASTIC MATRICES

coefficients $a_\pi, \pi \in S_n$, that solve the system of n^2 equations given by

$$\sum_{\pi \in S_n} a_\pi \delta_{i,\pi_j} = a_{ij}, \quad \sum_{\pi \in S_n} a_\pi = 1. \tag{5.283}$$

The Birkhoff expansions are obtained as a subset of all solutions of these relations by imposing the extra conditions $a_\pi > 0$. If we write the Birkhoff solutions of (5.283) in the form

$$A = \sum_{\pi \in S_n} a_\pi P_\pi, \quad \sum_{\pi \in S_n} a_\pi = 1, \text{ each } a_\pi \geq 0, \tag{5.284}$$

then it would appear that each coefficient a_π should have a probabilistic interpretation, but we have not found such. Accordingly, we present just a few examples to illustrate some of the problems.

We present four examples of the expansion theorem (5.283) for doubly stochastic matrices, as well as the corresponding unique expansion into the basis set for $\pi \in \Sigma_n(e, p)$ (see Sect. 1.5 and Chapters 2 and 3). These examples illustrate several features that must be taken into account in assigning any probabilistic interpretation to the expansion coefficients a_π.

Example 1. The spin-1/2 doubly stochastic matrix $S(C^{tr}, I_4)$ in (5.70):

$$\begin{pmatrix} 1 & 0 & 0 & 0 \\ 0 & 1/2 & 1/2 & 0 \\ 0 & 0 & 0 & 1 \\ 0 & 1/2 & 1/2 & 0 \end{pmatrix} = \tfrac{1}{2}P_{1,2,4,3} + \tfrac{1}{2}P_{1,4,2,3}$$

$$= -\tfrac{1}{2}P_{1,2,3,4} + P_{1,2,4,3} + \tfrac{1}{2}P_{1,3,2,4} - \tfrac{1}{2}P_{2,3,4,1} + \tfrac{1}{2}P_{2,4,3,1}. \tag{5.285}$$

The second expansion gives the **unique** coordinates in the basis set $\Sigma_4(e,p)$ (see (3.18)-(3.21)).

Example 2. Arbitrarily many expansions:

$$\tfrac{1}{8}\begin{pmatrix} 8 & 0 & 0 & 0 \\ 0 & 3 & 3 & 2 \\ 0 & 2 & 2 & 4 \\ 0 & 3 & 3 & 2 \end{pmatrix} = \tfrac{x}{8}P_{1,2,3,4} + \tfrac{(3-x)}{8}P_{1,2,4,3} + \tfrac{(2-x)}{8}P_{1,3,2,4}$$

$$+ \tfrac{x}{8}P_{1,3,4,2} + \tfrac{(x+1)}{8}P_{1,4,2,3} + \tfrac{(2-x)}{8}P_{1,4,3,2} \tag{5.286}$$

$$= -\tfrac{1}{8}P_{1,2,3,4} + \tfrac{1}{2}P_{1,2,4,3} + \tfrac{3}{8}P_{1,3,2,4} + \tfrac{1}{4}P_{1,3,4,2} - \tfrac{3}{8}P_{2,3,4,1} + \tfrac{3}{8}P_{2,4,3,1},$$

where the parameter $x \in \mathbb{R}$ is arbitrary, and for $-1 < x < 2$, all the a_π are positive. The second expansion containing the two negative coefficients gives the **unique** coordinates in the basis set $\Sigma_4(e, p)$.

CHAPTER 5. DOUBLY STOCHASTIC MATRICES IN AMT

Example 3. Birkhoff expansion and $\Sigma_4(e,p)$ basis expansion agree:

$$\frac{1}{2}\begin{pmatrix} 1 & 0 & 1 \\ 1 & 1 & 0 \\ 0 & 1 & 1 \end{pmatrix} = \frac{1}{2}P_{1,2,3} + \frac{1}{2}P_{2,3,1}. \tag{5.287}$$

We consider yet another example for $n = 4$ to show that the properties exhibited by (5.285), which is essentially an $n = 3$ result, carry forward to $n = 4$. For this example, we choose $\mathbf{a} = \mathbf{b}$ and $a_3 = b_3 = 0$ in the doubly stochastic matrix given by (5.74). This gives the following example:

Example 4. Two parameter Birkhoff expansion ($\mathbf{a} = \mathbf{b}$ and $a_3 = b_3 = 0$ in (5.74)):

$$\frac{1}{4}\begin{pmatrix} 1 & 1 & 1 & 1 \\ 2 & 0 & 0 & 2 \\ 1 & 1 & 1 & 1 \\ 0 & 2 & 2 & 0 \end{pmatrix}$$

$$= xP_{1,3,4,2} + \left(\tfrac{1}{4}-x\right)P_{1,4,3,2} + \left(\tfrac{1}{4}-x\right)P_{2,3,4,1} + xP_{2,4,3,1}$$

$$+ yP_{2,1,4,3} + \left(\tfrac{1}{4}-y\right)P_{2,4,1,3} + \left(\tfrac{1}{4}-y\right)P_{3,1,4,2} + yP_{3,4,1,2}$$

$$= -\tfrac{1}{2}P_{1,2,3,4} + \tfrac{1}{4}P_{2,1,3,4} + \tfrac{1}{4}P_{2,3,1,4} - \tfrac{1}{2}P_{2,3,4,1}$$

$$+ \tfrac{1}{2}P_{1,3,4,2} + \tfrac{1}{4}P_{3,2,4,1} + \tfrac{1}{4}P_{1,2,4,3} + \tfrac{1}{2}P_{2,4,3,1}. \tag{5.288}$$

The first identity is valid for all $x, y \in \mathbb{R}$, as verified directly; the second is an application of (3.18)-(3.21) for the coordinates in terms of the basis $\pi \in \Sigma_4(e,p)$. The two parameter solution in x and y has all positive coefficients in the expansion for $0 < x < 1/4$ and $0 < y < 1/4$.

The examples above raise the following issues: (i) Since a Birkhoff expansion can have an uncountable number of realizations in terms of permutation matrices, it seems unlikely that a probabilistic meaning can be assigned to the positive coefficient a_π multiplying a given permutation matrix $P_\pi, \pi \in S_n$, that occurs in the expansion (5.284). (ii) The coordinates $\alpha_{(i,j)}$ in the basis $\mathbb{P}_{\Sigma_n(e,p)}$ are unique, and because $\alpha_{(i,j)} = a_{ij}$, for $(i,j) \neq (1,1)$ and $(i,j) \neq (1,n)$, the probabilistic meaning of all but the two coordinates $\alpha_{(1,1)}$ and $\alpha_{(1,n)}$ is known. But this leaves open the meaning of the two coordinates $\alpha_{(1,1)}$ and $\alpha_{(1,n)}$, which can be negative, zero, or positive. We have not found satisfactory answers to these issues.

Chapter 6

Magic Squares

6.1 Review

We recall that a magic square A of order n is a matrix having arbitrary line-sum $r \in \mathbb{N}$, in which all n^2 elements are nonnegative integers $a_{ij} \in \mathbb{N}$. Equivalently stated, we have that:

Each magic square is a doubly stochastic matrix A/r, in which the entries are the rational numbers a_{ij}/r, each $a_{ij} \in \mathbb{N}$. Magic squares are a special class of discrete doubly stochastic matrices.

We introduce the notations:

$$\mathbb{M}_n(r) = \{\text{set of magic squares of order } n \text{ and line-sum } r\},$$
$$H_n(r) = |\mathbb{M}_n(r)|, \tag{6.1}$$

where the notation for the cardinality is that of Stanley [75].

Relations (3.5)-(3.6) give the unique coordinates of any magic square $A \in \mathbb{M}_n(r)$ in the basis $\mathbb{P}_{\Sigma_n(e,p)}$ of permutation matrices. The converse problem is (see Chapter 3, especially, (3.1)-(3.4) for notations):

Determine all $b_n = (n-1)^2 + 1$ coordinates $\alpha_{(1,1)}, \alpha_{(1,n)}, a_{ij} \in \mathbb{N}, (i,j) \in \mathbb{I}_n^{(12)}$ that solve the Diophantine equation

$$\alpha_{(1,1)} + \alpha_{(1,n)} + \sum_{(i,j) \in \mathbb{I}_n^{(12)}} a_{ij} = r, \tag{6.2}$$

where the down-diagonals $Diag_{1,1} = (a_{11}, a_{22}, \ldots, a_{nn})$ and $Diag_{1,n} = (a_{21}, a_{32}, \ldots, a_{n\,n-1}, a_{n\,1})$ are related to the coordinates $\alpha_{(1,1)}$ and $\alpha_{(1,n)}$

by the following relations:

$$a_{kk} = \alpha_{(1,1)} + \sum_{(i,j)\in \mathbb{I}^{(n)}_{(k,k)}} a_{ij} \geq 0, \; k=1,2,\ldots,n,$$

$$a_{k\,k-1} = \alpha_{(1,n)} + \sum_{(i,j)\in \mathbb{I}^{(n)}_{(k,k-1)}} a_{ij} \geq 0, \; k=2,3,\ldots,n, \quad (6.3)$$

$$a_{1n} = \alpha_{(1,n)} + \sum_{(i,j)\in \mathbb{I}^{(n)}_{(1,n)}} a_{ij} \geq 0.$$

We introduce the following amendments of notations to place the problem of solving the Diophantine equations (6.2)-(6.3) into a standard setting. We replace $b_n - 2 = n(n-2)$ by an arbitrary nonnegative integer k, the $n(n-2)$ elements $a_{ij}, (i,j) \in \mathbb{I}_n^{(12)}$ by the parts of the sequence $x = (x_1, x_2, \ldots, x_k)$ (identified in any convenient way), move the coordinate parts $\alpha_{(1,1)}$ and $\alpha_{(1,n)}$ to the right-hand side, and replace $r - \alpha_{(1,1)} - \alpha_{(1,n)}$ by an arbitrary nonnegative integer m. This casts (6.2)-(6.3) into the following form:

Determine all nonnegative integral solutions $x = (x_1, x_2, \ldots, x_k)$ of

$$x_1 + x_2 + \cdots + x_k = m; \; 0 \leq x_i \leq r, \; i = 1, 2, \ldots, k; \; 0 \leq m \leq kr, \quad (6.4)$$

where the nonnegative integers k, r, m are given.

This then is the Diophantine equation with the stated restrictions that we are led to consider. We denote the set of all solutions of (6.4) by $\mathbb{C}_{m,k}(r)$ and the cardinality of the set by $c_{m,k}(r) = |\mathbb{C}_{m,k}(r)|$. A sequence x satisfying (6.4) is said to be a composition of m into k nonnegative parts, each part $\leq r$, or, briefly, a *restricted composition*. The goal then is to give a constructive procedure for generating all such sequences and a counting formula for their number. We summarize in Appendix A some of the classical results relating to problems of this sort. In particular, we give the generating function for the numbers $c_{m,k}(r)$, as well as their explicit form.

Let us be quick to note, however, that a counting formula for these restricted compositions does not yet give the number $H_n(r)$ of magic squares. This is because we have $m = r - \alpha_{(1,1)} - \alpha_{(1,n)}$, and the domain of definition of the coordinate parts $\alpha_{(1,1)}$ and $\alpha_{(1,n)}$ that determine the allowed values of m are required; in addition, there is the degeneracy associated with the sum $\alpha_{(1,1)} + \alpha_{(1,n)}$; that is, various values of the pair $(\alpha_{(1,1)}, \alpha_{(1,n)})$ can give the same sum. Thus, there is a multiplicity factor $\text{Mult}_{m,k}(r)$ associated with each restricted composition in the set

6.1. REVIEW

$\mathbb{C}_{m,k}(r)$. Since $k = b_n - 2 = n(n-2)$, the number of magic squares of order n and line-sum r is the sum of these multiplicity factors:

$$H_n(r) = \sum_{m=0}^{(2n-3)r} \text{Mult}_{m,b_n-2}(r), \qquad (6.5)$$

where a multiplicity factor $\text{Mult}_{m,b_n-2}(r)$ is defined to be 0 should a given integer m in the closed interval $[0, (2n-3)r]$ not occur. We have used the inequalities $-(n-2)r \leq \alpha_{(1,1)} \leq r$, $-(n-2)r \leq \alpha_{(1,n)} \leq r$, $-2(n-2)r \leq \alpha_{(1,1)} + \alpha_{(1,n)} \leq r$ to obtain the upper limit in the summation in (6.5), these bounds themselves being obtained from the definitions of the pair $(\alpha_{(1,1)}, \alpha_{(1,n)})$ of coordinate parts, the property $0 \leq a_{ij} \leq r$, and the fact that the sum of all coordinates is the line-sum r. It is not out intention to effect the further refinements needed to determine the unknown multiplicity factors in relation (6.5), since Stanley [75, Chapter 4] has addressed in a general way how Diophantine equations of the type posed by (6.4) can be solved by the use of rational polynomial generating functions. We have presented the above results to show the relation between Diophantine equations and the coordinates assigned to each magic square in the basis set $\mathbb{P}_{\Sigma_n(e,p)}$ of permutation matrices. A formula for $H_n(r)$ gives, of course, the number of such coordinates sets. We return to Stanley's discussion of magic squares and the generating function for $H_n(r)$ below in Sect. 6.3, after giving examples of the above relations, and presenting the well-known Regge example of magic squares and angular momentum theory.

Examples. The simplest nontrivial example is the set of magic squares of order 3 and line-sum $r = 1$. This set contains the six permutation matrices $P_{1,2,3}, P_{2,1,3}, P_{2,3,1}, P_{1,3,2}, P_{3,2,1}, P_{3,1,2}$, which have, respectively, the coordinates in the basis $\Sigma_3(e,p)$ given by $(1,0,0,0,0), (0,1,0,0,0)$, $(0,0,1,0,0), (0,0,0,1,0), (0,0,0,0,1), (-1,1,-1,1,1)$.

The next simplest example is the set of magic squares $\mathbb{M}_3(2)$ of order 3 and line-sum 2. This set contains twenty-one magic squares which by direct enumeration gives the following set of coordinates as given in terms of the coordinate notation in (3.5)-(3.6):

five permutations of $(2,0,0,0,0)$;

ten permutations of $(1,1,0,0,0)$; $\qquad (6.6)$

$(0,1,-1,1,1), \; (-1,1,0,1,1), \; (-1,1,-1,1,2),$

$(-1,1,-1,2,1), \; (-1,2,-1,1,1), \; (-2,2,-2,2,2).$

These coordinates, in turn, give the solutions of relations (6.4), where $(x_1, x_2, x_3) = (a_{12}, a_{23}, a_{31})$:

$(0,0,0)[4]$; $(1,0,0)[2]$; $(0,1,0)[2]$; $(0,0,1)[1]$;

three permutations of $(2,0,0)[1]$ and of $(1,1,0)[1]$; (6.7)

$(1,1,1)[2]$; three permutations of $(1,1,2)[1]$; $((2,2,2)[1]$.

The value of m in these relations is the sum $x_1 + x_2 + x_3$, and the multiplicity factor $\text{Mult}_{m,3}(2)$ is indicated in the bracket [ ] following each sequence. Thus, the number of solutions of (6.4) for $k = 3, r = 2$, and $m = 0, 1, 2, 3, 4, 6$, respectively, is $1, 3, 6, 3, 1, 1$, which gives a total of 15. The value $m = 5$ does not occur, although the sequence $(2,2,1)$ and its permutations satisfies relation (6.4). It is therefore assigned multiplicity 0 in relation (6.5). Thus, the sum of the multiplicity numbers in (6.5) gives $H_3(2) = 4 + 2 + 2 + 1 + 3 + 3 + 2 + 3 + 1 = 21$. □

As remarked above, we will follow Stanley [75] in developing general formulas for magic squares. (See also Carter and Louck [17] for many results on the expansion of magic squares in terms of the basis $\mathbb{P}_{\Sigma_n(e,p)}$.) Before turning to some relevant results of Stanley, we next present, in the interest of physics, an observation of Regge [65], and a generalization of his result, extended to magic squares of order n.

6.2 Magic Squares and Addition of Angular Momenta

Physicists are familiar with magic squares for the case of $n = 3$, as discovered by Regge [65], for enumerating the coupled angular momentum states for two angular momenta (see also [6] and [L]). This perspective of the problem is presented in considerable detail in [L], as we next review briefly. Each magic square A of order 3 can be expressed in terms of the angular momentum quantum numbers (j_1, m_1) and (j_2, m_2) of two independent systems and the total angular momentum quantum numbers (j, m) by

$$A = \begin{pmatrix} a_{11} & a_{12} & a_{13} \\ a_{21} & a_{22} & a_{23} \\ a_{31} & a_{32} & a_{33} \end{pmatrix} = \begin{pmatrix} j_1 + m_1 & j_2 + m_2 & j - m \\ j_1 - m_1 & j_2 - m_2 & j + m \\ j_2 - j_1 + j & j_1 - j_2 + j & j_1 + j_2 - j \end{pmatrix}.$$
(6.8)

The angular momentum quantum numbers are given in terms of the elements of A by

$$j_1 = (a_{11} + a_{21})/2, \ j_2 = (a_{12} + a_{22})/2, \ j = (a_{13} + a_{23})/2;$$
(6.9)

$$m_1 = (a_{11} - a_{21})/2, \ m_2 = (a_{12} - a_{22})/2, \ m = (a_{23} - a_{13})/2.$$

6.2. MAGIC SQUARES AND ADDITION OF ANGULAR MOMENTA

The angular momentum quantum numbers satisfy the standard domain of definition rules:

$$\text{for each } j_1 \in \{0, 1/2, 1, 3/2, ...\}; \ m_1 = j_1, j_1 - 1, \ldots, -j_1;$$
$$\text{for each } j_2 \in \{0, 1/2, 1, 3/2, ...\}; \ m_2 = j_2, j_2 - 1, \ldots, -j_2; \quad (6.10)$$
$$\text{for each } j \in \langle j_1, j_2 \rangle = \{j_1 + j_2, j_1 + j_2 - 1, \ldots, |j_1 - j_2|\},$$
$$m = m_1 + m_2 = j, j - 1, \ldots, -j.$$

Thus, to enumerate all magic squares of line-sum $J = j_1 + j_2 + j$, all values of (j_1, j_2, j) are selected that satisfy these domain rules and also sum to a given positive integer J, which can be chosen arbitrary; and, for each such selection, the projection quantum numbers (m_1, m_2, m) assume all values consistent with these domain rules.

All magic squares of order 3 are enumerated in the manner of (6.8)-(6.10) and have coordinates in the basis $\mathbb{P}_{\Sigma_3(e,p)}$ (see (3.11)-(3.13)) given by

$$x_A = (j_1 - j - m_2, j_2 + m_2, j_1 - j_2 - m, j + m, j_2 - j_1 + j), \quad (6.11)$$

which sum to the line-sum $r = J = j_1 + j_2 + j$.

The number of magic squares that have the fixed line-sum J is obtained as follows. Define the sets Δ_J and $M(j_1, j_2, j)$ by

$$\Delta_J = \{(j_1, j_2, j) \,|\, j \in \langle j_1, j_1 \rangle; j_1 + j_2 + j = J\},$$
$$(6.12)$$
$$M(j_1, j_2, j) = \{(m_1, m_2) \,|\, -j_1 \le m_1 \le j_1;$$
$$-j_2 \le m_2 \le j_2; -j \le m_1 + m_2 \le j\}.$$

Thus, Δ_J is just the set of triplets (j_1, j_2, j), with $j \in \langle j_1, j_2 \rangle = \{j_1 + j_2, j_1 + j_2 - 1, \ldots, |j_1 - j_1|\}$ of angular momentum quantum numbers that give the specified line-sum J, and $M(j_1, j_2, j)$ is just the set of triplets of projection quantum numbers (m_1, m_2, m) that give a magic square for each (j_1, j_2, j). It follows that the total number of magic squares of line-sum J is given by

$$\sum_{(j_1, j_2, j) \in \Delta_J} |M(j_1, j_2, j)| = \binom{J+5}{5} - \binom{J+2}{5}. \quad (6.13)$$

We note that the significance of the factor $|M(j_1, j_2, j)|$ is that it is the multiplicity factor associated with each set of angular momenta $(j_1, j_2, j) \in \Delta_J$. Also, while we have proved the summation formula on the left, which uniquely determines the number of magic square, we have not proved that the sum is that shown. This result is known from completely different considerations as given in Stanley [75, p. 92].

For the Regge magic squares, all solution matrices (6.8) of the Diophantine equation (6.4) are enumerated in two steps: First, all angular momentum labels (j_1, j_2, j) are listed that belong to the set Δ_J; second, for each such member of this set, the collection of sequences corresponding to the value of the projection quantum numbers in the multiplicity set $M(m_1, m_2, m)$ are listed; and, finally, the sum of the multiplicity numbers is formed. This gives a total of $\binom{J+5}{5} - \binom{J+2}{2}$ such sequences that solve the Diophantine equation. We find the following to be a quite remarkable result:

The set of values that the angular momentum of a single quantum system can possess and the compounding of two such angular momenta to the total angular momentum that a composite system with two such constituents can possess, including the half-integral values, is implied by the enumeration of all magic squares of order 3 with no reference whatsoever to quantum theory, since the enumeration can be given in terms of the a_{ij} and then the quantum labels fully determined.

It is clear structurally why the Regge mapping (6.8) for compounding two angular momenta to the magic square problem for $n = 3$ works. The nonnegative requirement on the entries enforces exactly the angular momentum conditions on the quantum numbers; in particular, the Clebsch-Gordan rule is just the condition that the entries in row 3 be nonnegative. Moreover, the one-to-one mapping is from the 9 dependent parameters $(a_{ij})_{1 \leq i,j \leq n}$ having fixed line-sum to a set of $b_3 = 5$ independent parameters (j_1, j_2, j, m_1, m_2), whose domain of definition enumerates exactly all magic squares of line-sum $J = j_1 + j_2 + j$. These observations do not, of course, explain the deeper question as to why the rules of quantum angular momentum should have such a one-to-one correspondence with magic squares of order 3.

A generalization of the Regge matrix (6.8) can be given: We write each magic square A of line-sum r as follows:

$$A = (Q + M)/(n-1), \quad Q = \begin{pmatrix} q_1 & q_2 & \cdots & q_n \\ q_1 & q_2 & \cdots & q_n \\ \vdots & \vdots & \cdots & \vdots \\ q_1 & q_2 & \cdots & q_n \\ s_1 & s_2 & \cdots & s_n \end{pmatrix},$$

(6.14)

$$M = \begin{pmatrix} m_{11} & m_{12} & \cdots & m_{1n} \\ m_{21} & m_{22} & \cdots & m_{2n} \\ \vdots & \vdots & \cdots & \vdots \\ m_{n-1\,1} & m_{n-1\,2} & \cdots & m_{n-1\,n} \\ 0 & 0 & \cdots & 0 \end{pmatrix}.$$

6.2. MAGIC SQUARES AND ADDITION OF ANGULAR MOMENTA

The matrices M and Q of order n have the following definitions:

1. The matrix M has line-sum equal to 0; that is, the sum of the elements m_{ij} in each row and column sums to 0, the last row, of course, has all elements equal to 0, as shown.

2. The matrix Q of order n has line-sum $(n-1)r$, with row 1 through row $n-1$ all equal to the same sequence given by $q = (q_1\, q_2\, \cdots\, q_n)$; and row n is given by $s = (s_1\, s_2\, \cdots\, s_n)$, where

$$s_j = (n-1)(r-q_j),\ r-q_j \geq 0,\ j = 1, 2, \ldots, n. \qquad (6.15)$$

But now, because M has line-sum 0, we find that the elements of Q are uniquely determined in terms of those of $A = (a_{ij})$ by

$$q_j = \sum_{i=1}^{n-1} a_{ij},\ j = 1, 2, \ldots, n, \qquad (6.16)$$

while those of M are uniquely determined in terms of those of A and the q_j by

$$m_{ij} = (n-1)a_{ij} - q_j,\ i = 1, 2, \ldots, n-1;\ j = 1, 2, \ldots, n, \qquad (6.17)$$

where it is very important in this relation to enforce the integral conditions:

$$(q_j + m_{ij})/(n-1) \in \mathbb{N},\ i = 1, 2, \ldots, n-1; j = 1, 2, \ldots, n. \qquad (6.18)$$

We next observe that the matrix Q is a magic square that is fully defined on its own by its special form and the requirement that it has line-sum $(n-1)r$. This is a well-defined class of magic squares with no reference whatsoever to A or M. We denote the set of all magic squares of the form Q by Dom $\mathbb{Q}_r^{(n)}$, and call each matrix $Q \in$ Dom $\mathbb{Q}_r^{(n)}$ a *dominant magic square*.

But now we have that $q_1 + q_2 + \cdots + q_n = (n-1)r$ implies $s_1 + s_2 + \cdots + s_n = (n-1)r$, so that we have immediately that number of dominant magic squares is given by the number of nonnegative solutions of $q_1 + q_2 + \cdots + q_n = (n-1)r$, each $q_i \leq r$. This is the number of nonnegative solutions to $x_1 + x_2 + \cdots + x_n = r$, where $x_j = r - q_j$; hence, $0 \leq x_j \leq r$, for each $j = 1, 2, \ldots, n$. This counting number is obtained by application of the well-known (see Appendix A) binomial coefficient that counts the number of sequences $x = (x_1, x_2, \ldots, x_n)$, each x_i a nonnegative integer, that solve the relation $x_1 + x_2 + \cdots + x_n = r$, where r is a prescribed nonnegative integer. Such sequences are called compositions of k into r nonnegative parts ; the set of all such sequences is denoted $\mathbb{C}_{n,r}$; and the cardinality of the set is given by

$$|\text{Dom}\,\mathbb{Q}_r^{(n)}| = |\mathbb{C}_{n,r}| = \binom{n+r-1}{r}. \tag{6.19}$$

Each matrix $Q \in \text{Dom}\,\mathbb{Q}_r^{(n)}$ has, of course, the coordinates in the basis $\mathbb{P}_{\Sigma_n(e,p)}$ as given by (3.5)-(3.6).

The matrix M has the following description. Let $q = (q_1, q_2, \ldots, q_n)$ be a sequence such that $(r - q_1, r - q_2, \ldots, r - q_n) \in \mathbb{C}_{n,r}$, is a composition of r into n nonnegative parts. Then, for each such sequence $q \in \text{Dom}\,\mathbb{Q}_r^{(n)}$, the matrix M is a matrix of order n with row $n = (0,0,\ldots,0)$ and line-sum $= 0$, such that all it other elements are negative, zero, and positive integers that satisfy $(q_j + m_{ij}/(n-1) \in \mathbb{N}$ and $-q_j \leq m_{ij} \leq (n-1)r - q_j$, $i = 1, 2, \ldots, n-1; j = 1, 2, \ldots, n$. We call a matrix M satisfying these conditions, which are imposed by the dominant magic square Q, an *adjunct matrix* to Q. The set of adjunct matrices to Q is denoted $\text{Adj}\,_Q$; it is defined by

The set Adj_Q *of matrices adjunct to* $Q \in Dom\mathbb{Q}_r^{(n)}$ *is the set of all matrices* M *with row* $n = (0,0,\ldots,0)$ *and line-sum* $= 0$, *such that all its other elements* m_{ij}, $i = 1, 2, \ldots, n-1; j = 1, 2, \ldots, n$, *are negative, zero, and positive integers that satisfy* $(q_j + m_{ij})/(n-1) \in \mathbb{N}$, *the set of nonnegative integers.*

Summary. Every magic square matrix A of line-sum r has a unique decomposition $A = (Q+M)/(n-1)$ into a unique dominant magic square $Q \in \text{Dom}\,\mathbb{Q}_r^{(n)}$ and a unique adjunct matrix $M \in \text{Adj}\,_q$. Conversely, each dominant matrix $Q \in \text{Dom}\,\mathbb{Q}_r^{(n)}$ and each adjunct matrix $M \in \text{Adj}\,_Q$ determines a unique magic square matrix $A = (Q+M)/(n-1)$ of line-sum r. The counting formula for the number $H_n(r) = |\mathbb{A}_r^{(n)}|$ of magic squares of order n and fixed line-sum r is:

$$H_n(r) = \sum_{Q \in \text{Dom}\,\mathbb{Q}_r^{(n)}} |\text{Adj}\,_Q|. \tag{6.20}$$

Relations (6.14)-(6.20) developed above apply, of course, to every magic square of order n. We call relation (6.14), with the definitions of the dominant matrix $Q \in \text{Dom}\,\mathbb{Q}_r^{(n)}$ and its adjunct matrix $M \in \text{Adj}\,_Q$, the *Regge form of a magic square*. The main result shown above is the following:

6.2. MAGIC SQUARES AND ADDITION OF ANGULAR MOMENTA

Every magic square can be written in the Regge form

$$A = (Q + M)/(n - 1), \qquad (6.21)$$

where the dominant magic square matrix Q and the adjunct matrix M are elements of the set of matrix pairs defined by

$$\text{Dom}\, Q_r^{(n)} \times \text{Adj}_Q = \left\{(Q, M) \,\middle|\, Q \in \text{Dom}\, Q_r^{(n)};\ M \in \text{Adj}_Q \right\}. \qquad (6.22)$$

□

The elements $(q_1, q_2, \ldots, q_n)$ that define each $Q \in \text{Dom}\, Q_r^{(n)}$ are angular momentum j_i–like quantum numbers, and the elements m_{ij} that define each $M \in \text{Adj}_Q$ are projection m_i–like quantum numbers. But there is, of course, no reason to think such labels characterize any real physical system as they do for the Regge $n = 3$ case.

Just as in relation (6.13), we have given no method of effecting the summation in (6.20) to obtain an explicit formula for the number $H_n(r)$, although we have given an explicit rule for calculating each $|\text{Adj}_Q|$ that could be implemented on a computer. An example of the hand calculation for $n = 3, r = 2$ follows:

Example. We list here the dominant magic square matrices and their adjunct matrices for the next to simplest case: $n = 3, r = 2$. The dominant magic squares are six in number, as given by

$$\text{Dom}\, Q_2^{(3)} = \begin{pmatrix} q_1 & q_2 & q_3 \\ q_1 & q_2 & q_3 \\ -q_1 + q_2 + q_3 & q_1 - q_2 + q_3 & q_1 + q_2 - q_3 \end{pmatrix}, \qquad (6.23)$$

$(q_2, q_2, q_3) \in \{(0, 2, 2), (2, 0, 2), (2, 2, 0), (1, 1, 2), (1, 2, 1), (2, 1, 1)\}.$

For each composition (q_1, q_2, q_3), the adjunct set is obtained by direct enumeration of all M of the form:

$$M = \begin{pmatrix} m_{11} & m_{12} & -(m_{11} + m_{12}) \\ -m_{11} & -m_{12} & (m_{11} + m_{12}) \\ 0 & 0 & 0 \end{pmatrix}, \qquad (6.24)$$

where the conditions on the elements of this matrix are:

$$\frac{1}{2}\Big(q_1 \pm m_{11},\ q_2 \pm m_{12},\ 2q_3 \pm (m_{11} + m_{12})\Big) \in \mathbb{N}^3. \qquad (6.25)$$

The set of M matrices satisfying these conditions for each of the dominant matrices $\text{Dom}\, Q_2^{(3)}$ has the following pairs (m_{11}, m_{12}) of values:

$$(0,2,2) : (m_{11}, m_{12}) \in \{(0,-2), (0,0), (0,2)\},$$
$$(2,0,2) : (m_{11}, m_{12}) \in \{(-2,0), (0,0), (2,0)\},$$
$$(2,2,0) : (m_{11}, m_{12}) \in \{(-2,2), (0,0), (2,-2)\}, \quad (6.26)$$
$$(1,1,2) : (m_{11}, m_{12}) \in \{(-1,-1), (-1,1), (1,-1), (1,1)\},$$
$$(1,2,1) : (m_{11}, m_{12}) \in \{(-1,0), (-1,2), (1,-2), (1,0)\},$$
$$(2,1,1) : (m_{11}, m_{12}) \in \{(-2,1), (0,-1), (0,1), (2,-1)\}.$$

The twenty-one magic squares of order 3 and line-sum 2 are now obtained from the formula $A = (Q + M)/2$. This number, of course, agrees with that given by (6.13) for $J = r = 2$, since the decomposition $A = (Q + M)/2$ is essentially just the original Regge matrix for $n = 3$.

The twenty-one sets of coordinates for the magic squares in the basis set of permutation matrices $\mathbb{P}_{\Sigma_3(e,p)}$ are the following, a unique set for each $A = (Q + M)/2$:

$$\begin{aligned} x_A &= (x_Q + x_M)/2, \\ x_Q &= (q_1 - q_3, q_2, q_1 - q_2, q_3, -q_1 + q_2 + q_3), \quad (6.27) \\ x_M &= (-m_{12}, m_{12}, -m_{11} - m_{12}, m_{11} + m_{12}, 0). \end{aligned}$$

Thus, the coordinate x_Q takes on the six values corresponding to (q_1, q_2, q_3) as given for the six dominant matrices Q in (6.23), while, for each such x_Q, the coordinate x_M takes on the values corresponding to the pair (m_{11}, m_{12}) as given for the adjunct matrices M by the appropriate line from (6.26). □

6.3 Rational Generating Function of $H_n(r)$

The methods of Stanley [75] deal with the multiplicity problem inherent in relation (6.20) in a completely different manner, as we next outline. The key relation is the following (proved by Stanley [75, p. 209]):

$$\sum_{r \geq 0} \binom{r + d - k}{d} t^r = \frac{t^k}{(1-t)^{d+1}}. \quad (6.28)$$

We present the following synthesis of relations found and proved by Stanley (his Sect. 4.6) in his comprehensive development for counting the number of solutions of linear Diophantine equations by rational generating functions. Relation (6.28) is the key property for deducing relation (6.31) below from the two relations in Item 1:

6.3. GENERATING FUNCTION

1. We state Stanley's result for magic squares in two steps:
   (a) Let $P_n(t)$ be a polynomial of degree $b_n - n$ in t that satisfies the reciprocity relation $x^{b_n-n}P_n(\frac{1}{t}) = P_n(t)$, and has positive integral coefficient of t^k given by $a_{nk}, k = 0, 1, \ldots, b_n - n$:

   $$P_n(t) = \sum_{k=0}^{b_n-n} a_{nk} t^k. \tag{6.29}$$

   In consequence of the reciprocity property, the coefficients have the symmetry $a_{n,b_n-n-k} = a_{n,k}$, $k = 0, 1, \ldots, b_n - k$.

   (b) There exists a polynomial $P_n(t)$ such that the rational generating function for the number $H_n(r)$ of magic squares of order n and line-sum r is given by

   $$\frac{P_n(t)}{(1-t)^{b_n}} = \sum_{r \geq 0} H_n(r) t^r. \tag{6.30}$$

2. The two relations (6.29)-(6.30) are now joined by application of (6.28), which gives the number $H_n(r)$ of magic squares of order n and line-sum r in terms of the coefficients a_{nk} as

   $$H_n(r) = \sum_{k=0}^{b_n-n} a_{nk} \binom{b_n - k + r - 1}{b_n - 1}. \tag{6.31}$$

   We use the notation $b_n = n(n-2) + 2$ in formulas (6.29)-(6.31) to accentuate the meaning of the integer b_n as the number of linearly independent permutation matrices of order n in a basis set.

3. The coefficients a_{nk} are related to the values of the numbers $H_n(r)$ through the following triangular relations, as follows from (6.31):

$$\begin{aligned}
a_{n0} &= H_n(0) = 1, \\
a_{n1} &= H_n(1) - a_{n0}\binom{b_n}{1}, \\
a_{n2} &= H_n(2) - a_{n1}\binom{b_n}{1} - a_{n0}\binom{b_n+1}{2}, \\
&\vdots \\
a_{kn} &= H_n(k) - \sum_{s=1}^{k} a_{n\,k-s}\binom{b_n+s-1}{s}.
\end{aligned} \tag{6.32}$$

These relations could be combined into a single formula for a_{nk} by substituting a_{n1} in the second relation, the resulting a_{n2} in the third relation, etc., but their triangular relationship is more easily seen in this form.

Relation (6.31) shows that the number $H_n(r)$ is a polynomial in r of degree $b_n - 1 = (n-1)^2$. We make this explicit by defining the polynomial $H_n(x)$ in x by

$$H_n(x) = \sum_{k=0}^{b_n-n} a_{nk} \binom{x+b_n-k-1}{b_n-1}, n \geq 1. \qquad (6.33)$$

Because of the reciprocity property $x^{b_n-x} H_n(x) = H_n(1/x)$, this polynomial is completely determined by its values

$$H_n(0) = 1, H_n(1) = n!, H_n(2), \ldots, H_n(k_n), k_n = \lceil (b_n - n + 1)/2 \rceil - 1, \qquad (6.34)$$

where $\lceil y \rceil$ denotes the least integer $\geq y$, for real y. Also, $H_n(x)$ has the zeros given by

$$x \in \{-1, -2, \ldots, -n+1\}, n \geq 2 \text{ (empty for } n=1). \qquad (6.35)$$

These are just the zeros for which each of the individual binomial functions under the summation in (6.33) has value 0; hence, these zeros do not depend on the coefficients a_{nk}. As usual, the binomial function is the polynomial of degree m given by $\binom{z}{m} = z(z-1) \cdots (z-m+1)/m!$, $m \in \mathbb{N}$, $z \in \mathbb{R}$.

The zeros of $H_n(x)$ emerge automatically from the form (6.33); however, we still require $k_n + 1$ values of $H_n(x)$ for $x = 0, 1, \ldots, k_n$ to determine the coefficients a_{nk} as related by the triangular relations (6.32). Thus, the polynomials $H_n(x)$ are determined for all values of $x \in \mathbb{R}$ once the coefficients

$$a_{nk}, k = 0, 1, \ldots, k_n \qquad (6.36)$$

are determined. This result, applied at level $n-1$, has the following trivial consequence that nonetheless is important in deriving below an algorithm for generating all polynomials $H_n(x)$:

If the value $H_{n-1}(k)$ is known for $k = 0, 1, \ldots, k_{n-1}$, then the value $H_{n-1}(k)$ is known for $k = k_{n-1}+1, k_{n-1}+2, \ldots, k_n$.

We know generally that $H_n(0) = 1$ and $H_n(1) = n!$; hence, $a_{n0} = 1$ and $a_{n1} = n! - b_n$.

It is worth nothing that the polynomial property $H_n(x)$ traces it origin all the way back to the binomial coefficient in (6.28), since the

6.3. GENERATING FUNCTION

coefficients a_{nk} in (6.33) are independent of r. The multiplicity problem can be rephrased in the following manner:

The nonnegative integer coefficient a_{nk} in the polynomial $H_n(x)$ presented by (6.33) is the number of occurrences of the basis binomial polynomial $\binom{x+b_n-k-1}{b_n-1}$.

Thus, the entire problem is now refocused on the deeper meaning and determination of these coefficients.

The following examples of the polynomials (6.29) are given by Stanley [75, p. 234]:

Examples.

$$\begin{aligned}
P_1(t) &= P_2(t) = 1, \\
P_3(t) &= 1 + t + t^2, \\
P_4(t) &= 1 + 14t + 87t^2 + 148t^3 + 87t^4 + 14t^5 + t^6, \\
P_5(t) &= 1 + 103t + 4306t^2 + 63110t^3 \\
&\quad + 388615t^4 + 1115068t^5 + 1575669t^6 \\
&\quad + 1115068t^7 + 388615t^8 + 63110t^9 \\
&\quad + 4306t^{10} + 103t^{11} + t^{12}.
\end{aligned} \qquad (6.37)$$

The a_{nk} coefficients in these five polynomials are fully determined by (6.31)-(6.32) and the following values of $H_n(x)$:

$$\begin{aligned}
&H_1(0) = H_2(0) = 1; \ H_3(0) = 1, H_3(1) = 6; \\
&H_4(0) = 1, H_4(1) = 24, H_4(2) = 282, H_4(3) = 2008; \\
&H_5(0) = 1, H_5(1) = 120, H_5(2) = 6210, H_5(3) = 153040, \\
&H_5(4) = 2224955, H_5(5) = 22069251, H_5(6) = 164176640.
\end{aligned} \qquad (6.38)$$

These numbers were obtained from Stanley's values of the a_{nk} in the $P_n(x), n = 3, 4, 5$, from the reversible mapping (6.32); hence, these numbers are implied by Stanley's. But given these numbers, we can reverse the calculation to obtain back the a_{nk} in (6.37). □

Stanley's methods allow the calculation of the positive integers a_{nk} that define the polynomials $P_n(t)$, but we pursue a (seemingly) different method here that makes contact with still other very interesting aspects of combinatorics. We state the rule in full in language familiar from the combinatorial theory of symmetric functions (see Stanley [75] and Macdonald [50]):

The number of magic squares of order n and line-sum r is given by the following sum of squares of Kostka numbers:

$$H_n(r) = \sum_{\lambda \vdash nr} \left(K_\lambda^{(n)}(r)\right)^2, \qquad (6.39)$$

where $K_\lambda^{(n)}(r)$ denotes the Kostka number corresponding to the partition λ of nr into n parts having weight (or content) $r^{[n]} = (r, r, \ldots, r)$ (r repeated n times). These Kostka numbers are generated from the recurrence relation as follows:

$$K_\lambda^{(n)}(r) = \sum_{\mu \prec \lambda} K_\mu^{(n-1)}(r), \qquad (6.40)$$

where the starting point for the iteration is $K_{\mu_1}^{(1)}(r) = \delta_{\mu_1, r}$.

We need to explain the details of the notations, conventions, and the origin of relations (6.39)-(6.40). We use the convention of including zero as a part of a partition (see Appendix A). Thus, we have that the partition λ of nr into n parts (denoted $\lambda \vdash nr$) is defined to be the sequence of nonnegative integers $(\lambda_1, \lambda_2, \ldots, \lambda_n)$ such that $\lambda_1 \geq \lambda_2 \geq \cdots \geq \lambda_n \geq 0$ and $\lambda_1 + \lambda_2 + \cdots + \lambda_n = nr$. The notation $r^{[n]}$ denotes the sequence of length n in which r is repeated n times. We have here abbreviated the general notation $K(\lambda, \alpha)$ for a Kostka number of weight α to that used in (6.39), since $\alpha = r^{[n]}$ (see (4.17)).

Relation (6.40) is just the well-known Robinson-Schensted-Knuth [34] relation (see (4.17)), specialized to the partitions and weights shown. The general RSK relation from which (6.39) is derived is discussed at great length in [L], both from the point of view of Gelfand-Tsetlin (GT) patterns and Young-Weyl semistandard tableaux of shape λ filled-in with $1, 2, \ldots, n$ repeated, as required to obtain a semistandard tableau.

Relation (6.40) is just the well-known formula for enumerating all GT patterns having two rows of the form

$$\begin{pmatrix} \lambda_1 & \lambda_2 & \cdots & \lambda_n \\ \mu_1 & \mu_2 & \cdots & \mu_{n-1} \end{pmatrix}, \qquad (6.41)$$

where, for the problem at hand, and each given partition $\lambda \vdash nr$, the partition $\mu \vdash (n-1)r$ is to have the property that its parts fall between the successive parts of the partition λ; that is,

$$\mu_1 \in [\lambda_2, \lambda_1], \mu_2 \in [\lambda_3, \lambda_2], \ldots, \mu_{n-1} \in [\lambda_n, \lambda_{n-1}], \qquad (6.42)$$

where $[a, b]$ denotes the closed interval of points $x \in \mathbb{R}$ such that $a \leq x \leq b$, for $a, b \in \mathbb{R}$; in the application to the two-rowed GT pattern (6.41), the

6.3. GENERATING FUNCTION

intervals always have integral endpoints, and the included points in the interval are always to be integral. We denote the so-called betweenness conditions on the pattern (6.41) by $\mu \prec \lambda$:

Relation (6.40) is just the obvious statement that if the number of GT patterns $K_\mu^{(n-1)}(r)$ corresponding to all partitions $\mu \vdash (n-1)r$ and weight $r^{[n-1]}$ are known, then the number of GT patterns $K_\lambda^{(n)}(r)$ corresponding to partition $\lambda \vdash nr$ and weight $r^{[n]}$ is obtained by effecting the summation over all two-rowed GT patterns such that $\mu \prec \lambda$. Relation (6.40) is just the Robinson-Schensted-Knuth formula, applied in this context.

This result can also be stated in terms of standard tableaux and Pieri's rule (see Stanley [75] and [L]).

We can now state a numerical algorithm whereby the formula for the polynomial $H_n(x)$ can be generated.

Numerical algorithm for obtaining $H_n(x)$. The polynomial $H_n(x)$ is generated from the polynomial $H_{n-1}(x)$ by implementing the following finite numerical procedure, where we assume that we have already obtained the polynomial $H_{n-1}(x)$ in full:

1. (a) Enumerate the set of all $\lambda \vdash nr$ having n parts (see Appendix A), and select a specific λ.
   (b) Enumerate all the partitions $\mu \prec \lambda$ that satisfy the betweenness relations (6.42) (all two-rowed GT patterns (6.41) that satisfy betweenness);
   (c) Calculate the Kostka numbers $K_\lambda^{(n)}(r), r = 0, 1, \ldots, k_n$, from relation (6.40), using the Kostka numbers $K_\mu^{(n-1)}(r), r = 0, 1, \ldots, k_{n-1}$ already tabulated in the calculation of the polynomial $H_{n-1}(x)$, together with those for $r = k_{n-1}, k_{n-1}+1, \ldots, k_n$, as obtained by direct enumeration of the two-rowed GT patterns (6.41) for the given partition λ.
   (d) Carry out this calculation for each of the partitions λ of nr into n parts enumerated in Item (1a).
   (e) Carry out the sum of squares of the Kostka numbers calculated in Item (1c), thus obtaining the set of numbers $H_n(r), r = 0, 1, \ldots, k_n$.
   (f) Calculate the numerical coefficients $a_{nk}, k = 0, 1, \ldots, k_n$, from the triangular relations (6.32), using the special $H_n(r)$ obtained in Item (1e). These coefficients then give the polynomial $H_n(x)$ in (6.33).

2. The full calculation (thus, providing a check of the results for $n = 3, 4, 5$ given by (6.37)-(6.38)) can be started with the initial data

for $n = 2$ given by

$$K^{(2)}_{(r+s,r-s)}(r) = 1, \ r \geq 1, \text{ for each partition}$$
$$(\lambda_1, \lambda_2) = (r+s, r-s), \ s = 0, 1, \ldots, r; \quad (6.43)$$
$$H_2(x) = x + 1.$$

Examples. It is convenient for the implementation of the numerical algorithm to have in one place the explicitly known polynomials $H_n(x), n = 1, 2, 3, 4, 5$:

$$H_1(1) = 1, \quad H_2(x) = x + 1,$$

$$H_3(x) = \binom{x+4}{4} + \binom{x+3}{4} + \binom{x+2}{4},$$

$$H_4(x) = \binom{x+9}{9} + 14\binom{x+8}{9} + 87\binom{x+7}{9} + 148\binom{x+6}{9}$$
$$+ 87\binom{x+5}{9} + 14\binom{x+4}{9} + \binom{x+3}{9}, \quad (6.44)$$

$$H_5(x) = \binom{x+16}{16} + 103\binom{x+15}{16} + 4306\binom{x+14}{16}$$
$$+ 63110\binom{x+13}{16} + 388615\binom{x+12}{16} + 1115068\binom{x+11}{16}$$
$$+ 1575669\binom{x+10}{16} + 1115068\binom{x+9}{16} + 388615\binom{x+8}{16}$$
$$+ 63110\binom{x+7}{16} + 4306\binom{x+6}{16} + 103\binom{x+5}{19} + \binom{x+4}{16}.$$

These relations for $H_n(x)$ are uniquely determined by the values:

$$H_1(0) = 1; \ H_2(0) = 1; \ H_3(0), H_3(1); \ H_4(0), H_4(1), H_4(2), H_4(3);$$
$$(6.45)$$
$$H_5(0), H_5(1), H_5(2), H_5(3), H_5(4), H_5(5), H_5(6).$$

We know, in general, that $H_n(0) = a_{n0}, H_n(1) = n! = a_{n1} + b_n$. The relation for $H_3(x)$ is in the original form obtained by MacMahon [52] (compare with (6.13) with J replaced by x). □

6.3. GENERATING FUNCTION

Example. We illustrate, step by step, the calculation of $H_4(2) = 282$ by the numerical algorithm stated above:

(a). The set of partitions $\{\lambda\}$ of 8 into 4 parts is:

$$(8,0,0,0), (7,1,0,0), (6,2,0,0), (6,1,1,0), (5,3,0,0),$$
$$(5,2,1,0), (5,1,1,1), (4,4,0,0), (4,3,1,0), (4,2,2,0), \qquad (6.46)$$
$$(4,2,1,1), (3,3,2,0), (3,3,1,1), (3,2,2,1), (2,2,2,2).$$

(b)-(d). We select one of these partitions and write out, or at least count, all two-rowed GT patterns

$$\begin{pmatrix} \lambda_1 & \lambda_2 & \lambda_3 & \lambda_4 \\ & \mu_1 & \mu_2 & \mu_3 & \end{pmatrix}, \qquad (6.47)$$

where $\mu = (\mu_1, \mu_2, \mu_3)$ is a partition of $(n-1)r = (3)(2) = 6$, such that the betweenness conditions (6.42) are satisfied. This gives the following results, where we indicate the number of patterns, which is the Kostka number $K_\lambda^{(4)}(2)$, for each λ in the square bracket following the partition:

$$(8,0,0,0)[1], (7,1,0,0)[3], (6,2,0,0)[6], (6,1,1,0)[3],$$
$$(5,3,0,0)[6], (5,2,1,0)[8], (5,1,1,1)[1], (4,4,0,0)[3],$$
$$(4,3,1,0)[7], (4,2,2,0)[6], (4,2,1,1)[3], (3,3,2,0)[3],$$
$$(3,3,1,1)[2], (3,2,2,1)[3], (2,2,2,2)[1]. \qquad (6.48)$$

(e) The sum of squares of the Kostka numbers in square brackets is effected to obtain:

$$\begin{aligned} H_4(2) &= 1^2 + 3^2 + 6^2 + 3^2 + 6^2 + 8^2 + 1^2 + 3^2 \\ &+ 7^2 + 6^2 + 3^2 + 3^2 + 2^2 + 3^2 + 1^2 = 282. \ \square \end{aligned} \qquad (6.49)$$

The polynomial $H_n(x)$ giving the number of magic squares of order n and line-sum r is thus known for every n for which the numerical algorithm outlined above (or some other method) can be reasonably carried out to obtain the integers $a_{nk}, k = 0, 1, \ldots, k_n$. The formulas for $n = 4, 5$ were obtained by Stanley, but not made explicit in the forms (6.44). The main point of the algorithm is trivial, but powerful: Because we are dealing with a polynomial, a finite number of numerical calculations leads to the general polynomial answer.

The results obtained in this section fit well within our goal of extending the results obtained in [L] to applications. It is quite nice to see that the Regge form of magic squares, with its classical origin in angular momentum theory, motivates a new perspective of magic squares. Magic squares, of course, occur throughout the application of symmetry techniques in quantum theory, often as objects for enumerating terms in complex, but elegant formulas, such as special cases of the matrix Schur functions. To date, such a nice exposition of the properties of the "generalized magic squares," the matrices $\mathbb{M}^p_{n\times n}(\alpha,\alpha')$ for which each line-sum is specified by a weight of a GT pattern or a semistandard Young tableau — the magical objects that enumerate homogeneous polynomials that have specified homogeneity properties in their n^2 indeterminate variables; give the terms in a general matrix Schur function; and enter into the general Robinson-Schensted-Knuth formula (4.17) — has not been achieved. The interconnections with seemingly disparate topics such as the theory of partitions, of GT patterns, and of Young tableaux is also very satisfying.

Chapter 7

Alternating Sign Matrices

7.1 Introduction

We recall:

Definition. An alternating sign matrix of order n has line-sum 1 with elements $a_{ij} \in \mathcal{A} = \{-1, 0, 1\}$, such that when a given row is read from left-to-right across the columns or a given column is read from top-to-bottom down the rows and all zeros in that row (column) are ignored, the row (column) contains either a single 1, or is an alternating series of odd length of the exact form $1, -1, 1, -1, \ldots, 1, -1, 1$.

We denote the set of alternating sign matrices of order n by $\mathrm{A}S_n$. This set includes all $n!$ permutation matrices $P_\pi, \pi \in S_n$. The first simple examples are the following:

Examples. For $n = 1, 2, 3$, the alternating sign matrices are:

$$n = 1: \;(1); \quad n = 2: \; \begin{pmatrix} 1 & 0 \\ 0 & 1 \end{pmatrix}, \begin{pmatrix} 0 & 1 \\ 1 & 0 \end{pmatrix};$$

(7.1)

$$n = 3: \; \text{all six } P_\pi, \pi \in S_3, \text{ and } \begin{pmatrix} 0 & 1 & 0 \\ 1 & -1 & 1 \\ 0 & 1 & 0 \end{pmatrix}.$$

Thus, the number of alternating sign matrices for $n = 1, 2, 3$ is $1, 2, 7$. □

Our continuing interest in alternating sign matrices is founded in the fascinating role that special Gelfand-Tsetlin patterns play in their enumeration. These patterns abound in many symmetry related problems in physical theory, including angular momentum theory of composite systems, as expounded on in many papers with Lawrence C. Biedenharn (see Ref. [6]). This is, perhaps, because of their one-to-one relationship with Young semistandard tableaux, the method usually preferred by combinatorialists, and the pervasive occurrence of the latter across many boundaries in mathematics. Nonetheless, as emphasized in [L], the GT patterns stand on their own, and because of our own commitment to this approach, we choose to follow it to its natural conclusions. We find now that it is the skew-symmetric GT patterns that enter more deeply into the analysis; we will show that these patterns provide an understanding of the zeros of the Zeilberger polynomials $Z_n(x)$ defined by

$$Z_n(x) = a_n \binom{x+n-2}{n-1}\binom{-x+2n-1}{n-1}, \ n \geq 1. \tag{7.2}$$

Here the multiplicative number a_n is given by

$$a_n = d_n \bigg/ \binom{3n-2}{n-1}, \tag{7.3}$$

where d_n is the Andrews-Zeilberger number defined by

$$d_n = \prod_{j=0}^{n-1} \frac{(3j+1)!}{(n+j)!}, \ n \geq 1. \tag{7.4}$$

The first paper by Zeilberger [91] was on the determination of the number d_n, also found earlier by Andrews [2] in a different context. The second paper by Zeilberger [92] led to the formulation (7.2) in [L]. Thus, relations (7.2)-(7.4) are complete, based on Zeilberger's work. We pursue the problem further from the point of view of the GT pattern approach with the goal of using this method to establish the full proof of (7.2)-(7.4) without benefit of Zeilberger's results for reasons mentioned above; this also places the problem under the full purview of Young tableaux, which are one-to-one with the GT patterns used here. The interesting history and documentation of the proof of the counting formula for d_n given by relation (7.4) can be found in a number of interesting articles: Bressoud and Propp [12], Bressoud [13], Propp [61].

It is to be observed, then, that, unlike the determination of the magic square polynomials $H_n(x)$ defined by (6.33), the polynomial part of relation (7.2) is simple — it is the multiplying coefficient a_n in its dependence on the Andrews-Zeilberger number d_n that is complicated. Nonetheless, it is much in the same spirit as for the counting formula for the magic square polynomials (6.33) that we approach the counting problem for alternating sign matrices — the establishment of zeros. (We have not seen how to adapt the methods of Stanley to this proof.)

7.2. STANDARD GELFAND-TSETLIN PATTERNS

The method given here establishes the following results:

Theorem on Zeilberger polynomials. *There exists a set of polynomials $Z_n(x), n \in \mathbb{N}$ that satisfy the following three properties:*

1. *Values of Andrews-Zeilberger numbers in terms of $Z_n(x)$:*

$$d_n = \sum_{i=1}^{n} Z_n(i) \text{ and } Z_n(n) = d_{n-1}. \tag{7.5}$$

2. *Complete set of zeros:*

$Z_n(x) = 0$, if and only if
$x \in \{-n+2, -n+3, \ldots, -1, 0; n+1, n+2, \ldots, 2n-1\}$. (7.6)

These three properties imply that $Z_n(x)$ is exactly the Zeilberger polynomials with the properties (7.2)-(7.4).

Proof. The proof is given as follows: Use the binomial coefficient identity (proved in [L]) give by

$$\sum_{i=1}^{n} Z_n(i) = a_n \sum_{i=1}^{n} \binom{i+n-2}{n-1}\binom{-i+2n-1}{n-1} = a_n \binom{3n-2}{n-1}. \tag{7.7}$$

In consequence of the first and second relations in (7.5), we have the identities

$$d_n = a_n \binom{3n-2}{n-1} \text{ and } d_{n-1} = a_n \binom{2n-2}{n-1}. \tag{7.8}$$

Eliminate a_n between these relations to obtain

$$d_n = d_{n-1} \binom{3n-2}{n-1} \bigg/ \binom{2n-2}{n-1}, \, n \geq 2, \, d_1 = 1. \tag{7.9}$$

Iterate this relation to obtain (7.4). □

The result just proved on Zeilberger polynomials places the burden of proof of (7.2)-(7.4) on establishing the validity of Items 1 and 2. The remainder of this Chapter reviews the background from [L] and gives the new results needed to prove Items 1-2.

7.2 Standard Gelfand-Tsetlin Patterns

We repeat from Sect. 4.3 the definition of the general Gelfand-Tsetlin (GT) patterns; it is these patterns that are specialized in the next section to establish the bijection between strict Gelfand-Tsetlin (GT) patterns and alternating sign matrices (AS).(Many more details can be found in [L], as well as the bijection between such general patterns and Young-Weyl semistandard tableaux):

1. Gelfand-Tsetlin (GT) patterns:

    (a) λ is a partition with n parts: $\lambda = (\lambda_1, \lambda_2, \ldots, \lambda_n)$, each $\lambda_i \in \mathbb{N}$, with $\lambda_1 \geq \lambda_2 \geq \cdots \geq \lambda_n \geq 0$.

    (b) $\binom{\lambda}{m}$ is a *stacked array of partitions*:

$$\binom{\lambda}{m} = \begin{pmatrix} \lambda_1 & \lambda_2 & \cdots & \lambda_j & \cdots & & \lambda_{n-1} & \lambda_n \\ m_{1,n-1} & m_{2,n-1} & \cdots & m_{j,n-1} & \cdots & & m_{n-1,n-1} & \\ & & & \vdots & & & & \\ & & m_{1,i} & m_{2,i} & \cdots & m_{i,i} & & \\ & & & \vdots & & & & \\ & & & m_{1,2} & m_{2,2} & & & \\ & & & & m_{1,1} & & & \end{pmatrix} \tag{7.10}$$

Row i is a partition denoted $\mathbf{m}_i = (m_{1,i}, m_{2,i}, \ldots, m_{i,i})$ with i parts for each $i = 1, 2, \ldots, n$, and $\mathbf{m}_n = \lambda$. Adjacent rows $(\mathbf{m}_{i-1}, \mathbf{m}_i), i = 2, 3, \ldots, n$, are required to satisfy the "betweenness conditions:"

$$m_{j,i-1} \in [m_{j,i}, m_{j+1,i}], \ 2 \leq i \leq n; \ 1 \leq j \leq i-1.$$
$$[a, b], a \leq b, \text{ denotes the closed interval of} \tag{7.11}$$
$$\text{nonnegative integers } a, a+1, \ldots, b.$$

2. Lexical GT patterns (patterns they satisfy betweenness rule (7.11)):

$$\mathbb{G}_\lambda = \left\{ \binom{\lambda}{m} \;\middle|\; m \text{ is a lexical pattern} \right\}. \tag{7.12}$$

The single symbol m in $\binom{\lambda}{m}$ denotes the entire $(n-1)$–rowed pattern on the right-hand side of (7.10).

3. Weight α of a GT pattern (sometimes called *content*):

$$\alpha = (\alpha_1, \alpha_2, \ldots, \alpha_n), \ \alpha_i = (m_{1,i} + m_{2,i} + \cdots + m_{i,i})$$
$$- (m_{1,i-1} + m_{2,i-1} + \cdots + m_{i-1,i-1}), \tag{7.13}$$
$$i = 1, 2, \ldots, n; \ \alpha_1 = m_{1,1}.$$

A weight is a sequence of nonnegative integers; its multiplicity is the Kostka number $K(\lambda, \alpha)$.

7.2. STANDARD GELFAND-TSETLIN PATTERNS

4. Cardinality recurrence relation:

$$|\mathbb{G}_\lambda| = \sum_{\mu_1 \in [\lambda_1,\lambda_2], \mu_2 \in [\lambda_2,\lambda_3], \ldots, \mu_{n-1} \in [\lambda_{n-1},\lambda_n]} |\mathbb{G}_\mu|, \qquad (7.14)$$

where $\lambda = \mathbf{m}_n$ and $\mu = \mathbf{m}_{n-1}$. The starting point of the iteration is $\mathbf{m}_1 = 1$, which gives $|\mathbb{G}_{(\lambda_1,\lambda_2)}| = \lambda_1 - \lambda_2 + 1$, $|\mathbb{G}_{(\lambda_1,\lambda_2,\lambda_3)}| = (\lambda_1-\lambda_2+1)(\lambda_1-\lambda_3+2)(\lambda_2-\lambda_3+1)/2, \ldots$. In principal, the iteration can be carried upward to deduce the Weyl dimension formula given in relations (4.15).

5. Cardinality invariance operations:

   (i). Translation invariance: The number of patterns in $\mathbb{G}_\lambda$ is invariant to the shift of all entries in the pattern by the same integer $h \in \mathbb{Z}$, which implies

   $$|\mathbb{G}_{\lambda+h}| = |\mathbb{G}_\lambda|, \qquad (7.15)$$

   where $\lambda + h$ denotes $(\lambda_1 + h, \lambda_2 + h, \ldots, \lambda_n + h)$.

   (ii). Sign-reversal-shift conjugation: The number of patterns in $\mathbb{G}_\lambda$ is invariant to the operation of writing each row of the pattern backwards with minus signs and translating by $h = \lambda_1 + \lambda_n$. This operation effects the transformation of shape λ to shape λ^*, where $\lambda_i^* = -\lambda_{n-i+1} + \lambda_1 + \lambda_n, i = 1, 2, \ldots, n$. Thus, we have the identity

   $$|\mathbb{G}_\lambda| = |\mathbb{G}_{\lambda^*}|. \qquad (7.16)$$

   This operation is called sign-reversal-shift conjugation; it is an *involution:* repetition of the operation restores the original pattern. It is not necessary that these operations have physical content — the point is that they leave the cardinality $\mathbb{G}_\lambda$ invariant.

7.2.1 A-Matrix Arrays

We next describe a mapping of the set $\mathbb{G}_\lambda$ of GT patterns onto a set of $n \times \lambda_1$ matrix arrays, which we call A–*arrays*. Thus, given a GT pattern $\binom{\lambda}{m} \in \mathbb{G}_\lambda, \lambda_n > 0$, we define the corresponding A–array by the following rule:

$$\binom{\lambda}{m} \mapsto A\binom{\lambda}{m} = \begin{pmatrix} A_n \\ \vdots \\ A_2 \\ A_1 \end{pmatrix}, \qquad (7.17)$$

where the $n \times \lambda_1$ matrix $A\binom{\lambda}{m}$ has n rows A_i defined by

$$
\begin{aligned}
A_1 &= e_{m_{1,1}}, \\
A_2 &= e_{m_{1,2}} + e_{m_{2,2}} - e_{m_{1,1}} \\
A_3 &= e_{m_{1,3}} + e_{m_{2,3}} + e_{m_{3,3}} - e_{m_{1,2}} - e_{m_{2,2}}, \\
&\vdots \qquad\qquad \vdots \\
A_n &= e_{m_{1,n}} + e_{m_{2,n}} + \cdots + e_{m_{n,n}} \\
&\quad - e_{m_{1,n-2}} - e_{m_{2,n-1}} - \cdots - e_{m_{n-1,n-1}},
\end{aligned} \tag{7.18}
$$

where each e_k denotes the *unit row vector (matrix)* of length λ_1 with 1 in part k and 0 elsewhere. We choose the length of each e_k to be exactly λ_1, and also always take $\lambda_n \geq 1$. This is because then each $m_{i,j}$ in relations (7.18) satisfies $\lambda_1 \geq m_{i,j} \geq 1$, hence, each unit row matrix e_k is defined. Choosing the length of e_k greater than λ_1 adjoins columns of zeros to the right-end of the array (7.17) and contributes nothing to the structure of such A–matrices — columns of zeros can still occur to the left. (The option of using row vectors, in contrast to column vectors, enhances the display of a GT pattern and its associated A–matrix.) We also use the following notations for row n interchangeably:

$$\lambda = (\lambda_1 \lambda_2, \ldots, \lambda_n) = (m_{1,n}, m_{2,n}, \ldots, m_{n,n}), \tag{7.19}$$

as already indicated in the compact notation (7.11). To summarize: We always take the A–array to be a matrix of size $n \times \lambda_1$ and $\lambda_n = m_{n,n} \geq 1$ (all positive parts); hence, the unit row vectors that occur in (7.18) are always in the set $\{e_1, e_2, \ldots, e_{\lambda_1}\}$. We denote the set of A–array matrices corresponding to all $\binom{\lambda}{m} \in \mathbb{G}_\lambda$ by the notation $\mathbb{A}_\lambda$.

An important property of the mapping defined by (7.17)-(7.19) is:

Bijection property. The mapping

$$\binom{\lambda}{m} \mapsto A\binom{\lambda}{m} = \begin{pmatrix} A_n \\ \vdots \\ A_2 \\ A_1 \end{pmatrix}, \tag{7.20}$$

where the A_i are defined by (7.18), is a bijection.

Proof. The system of n relations (7.18) is invertible, as shown as follows: We start with row A_1, which uniquely determines $m_{1,1}$; then row A_2, which uniquely determines $(m_{1,2}, m_{2,2})$, since $m_{1,2} \geq m_{1,1} \geq m_{2,2}$; then row A_3, which uniquely determines $(m_{1,3}, m_{2,3}, m_{3,3})$, since $m_{1,3} \geq m_{1,2} \geq m_{2,3} \geq m_{2,2} \geq m_{3,3}$; etc. The general proof can also be given by

7.2. STANDARD GELFAND-TSETLIN PATTERNS

showing that the relations

$$e_{m'_{1,i}} + e_{m'_{2,i}} + \cdots + e_{m'_{i,i}}$$
$$= e_{m_{1,i}} + e_{m_{2,i}} + \cdots + e_{m_{i,i}}, \text{ each } i = 1, 2, \ldots, n, \quad (7.21)$$

imply, for $\binom{\lambda}{m'} \in \mathbb{G}_\lambda$ and $\binom{\lambda}{m} \in \mathbb{G}_\lambda$, that the identity $\binom{\lambda}{m'} = \binom{\lambda}{m}$ holds. We next show this. The most general form of the right-hand side of relation (7.21) for given i is obtained from the following system of relations among the $m_{1,i}, m_{2,i}, \ldots, m_{i,i}$:

k_1 equal terms: $\quad m_{1,i} = m_{2,i} = \cdots = m_{k_1,i}$,

k_2 equal terms: $\quad m_{k_1+1,i} = m_{k_1+2,i} = \cdots = m_{k_1+k_2,i}$,

$$\vdots \quad (7.22)$$

k_r equal terms: $\quad m_{k_1+k_2+\cdots+k_{r-1}+1,i} = m_{k_1+k_2+\cdots+k_{r-1}+2,i}$
$$= \cdots = m_{k_1+k_2+\cdots+k_r,i},$$

where the sequence of positive integers $(k_1, k_2, \ldots, k_r)$, each $k_i \geq 1$, and r that appear in (7.22) must satisfy the following relations:

$$k_1 + k_2 + \cdots + k_r = i, \; r = 1, 2, \ldots, i. \quad (7.23)$$

But we can also conclude from (7.22)-(7.23) that the form of the right-hand side of (7.21) is

$$e_{m_{1,i}} + e_{m_{2,i}} + \cdots + e_{m_{i,i}}$$
$$= (\cdots k_1 \cdots) + (\cdots k_2 \cdots) + \cdots + (\cdots k_r \cdots), \quad (7.24)$$

where $\cdots$ represent sequences of $0's$ (possibly none); that is, the only nonzero parts in these sequences are the positive integers $k_1, k_2, \ldots, k_r$, and the positions of these nonzero parts are all distinct. But from each sequence having this structure, the unit row matrices in the sum $e_{m'_{1,i}} + e_{m'_{2,i}} + \cdots + e_{m'_{i,i}}$, with $m'_{1,i} \geq m'_{2,i} \geq \cdots \geq m'_{i,i}$ can be read off uniquely; that is, $\mathbf{m}'_i = (m'_{1,i}, m'_{2,i}, \ldots, m'_{i,i}) = \mathbf{m}_i$. Since this result holds for each $i = 1, 2, \ldots, n$, we conclude that $\binom{\lambda}{m'} = \binom{\lambda}{m}$. $\square$

It follows from the above proof that the cardinality of the respective sets $\mathbb{G}_\lambda$ and $\mathbb{A}_\lambda$ are equal:

$$|\mathbb{G}_\lambda| = |\mathbb{A}_\lambda| = \text{Dim}\lambda, \quad (7.25)$$

where Dimλ is the Weyl dimension formula (4.15).

We have given the mapping (7.20) to illustrate its generality within the framework of GT patterns — it is not particular to the application to *strict GT patterns* that we consider next.

7.2.2 Strict Gelfand-Tsetlin Patterns

A *strict* GT pattern is a GT pattern such that λ is a strict partition, that is, $\lambda_1 > \lambda_2 > \cdots > \lambda_n > 0$, and such that each of the $n-1$ rows, $m_{n-1}, m_{n-2}, \ldots, m_2$ is also a strict partition; that is, each of the GT patterns at every row-level $2, 3, \ldots, n$ is a strict GT pattern. We denote the set of all strict Gelfand-Tsetlin patterns of shape λ by $\text{Str}\,\mathbb{G}_\lambda$, and an element of this set by $\text{Str}\binom{\lambda}{m}$. The mapping given by (7.17)-(7.18) is still applicable, since it applies to all patterns $\binom{\lambda}{m} \in \mathbb{G}_\lambda$. We now write this mapping as

$$\text{Str}\binom{\lambda}{m} \mapsto \text{Str}A\binom{\lambda}{m}. \tag{7.26}$$

This mapping is now to be applied to every strict GT pattern to yield a strict $A\binom{\lambda}{m}$ pattern:

$$\binom{\lambda}{m} \in \text{Str}\,\mathbb{G}_\lambda \text{ and } A\binom{\lambda}{m} \in \text{Str}\,\mathbb{A}_\lambda. \tag{7.27}$$

A noted above, this mapping is exactly the same as that given by (7.17)-(7.18), since this mapping covers every possible lexical GT pattern — now each row of the pattern is required to be a strict partition. Thus, the mapping is still one-to-one; that is, the cardinality condition

$$|\text{Str}\,\mathbb{G}_\lambda| = |\text{Str}\,\mathbb{A}_\lambda| \tag{7.28}$$

still holds, but we lose the equality with the Weyl dimension formula, since strict patterns are a subset of the full set. Indeed, the cardinality of these sets is unknown — such a result would be a generalization of the Andrews-Zeilberger number d_n (see (7.29) below).

7.3 Strict Gelfand-Tsetlin Patterns for $\lambda = (n\,n-1\,\cdots\,2\,1)$

The relation between the set of strict A-arrays corresponding to the set of strict GT patterns with $\lambda = (n\,n-1\,\cdots\,2\,1)$, the set of alternating sign matrices of order n, and the Andrews-Zeilberger number d_n is

$$\text{Str}\,\mathcal{A}_{n\,n-1\,\cdots\,1} = \mathcal{A}\,\mathcal{S}_n, \quad d_n = |\text{Str}\,\mathbb{G}_{n\,n-1\,\cdots\,1}|. \tag{7.29}$$

The proof is given directly from (7.18), now applied to the set $\text{Str}\,\mathbb{G}_{n\,n-1\,\cdots\,1}$ by showing that each A-array corresponding to a member of this set is an alternating sign matrix, and conversely, and then using (7.28). Results equivalent to (7.29) are due to Robbins *et al.* [66] (see also Mills *et al.* [56]), but the proof of formula (7.4) for d_n was not achieved

7.3. SPECIAL STRICT GT PATTERNS

For $\lambda_1 = n$, all matrices introduced in Sects. 7.2.1-7.2.2 are of order n. We have presented the material in the previous sections on GT patterns with the goal of deriving the Andrews-Zeilberger number d_n by actually effecting the counting of $|\text{Str}\,\mathbb{G}_{n\,n-1\cdots 1}|$ along the lines given in the Introduction, Sect. 7.1; namely, through the properties of the zeros of the Zeilberger polynomial $Z_n(x)$. We continue this effort after giving two examples of the mapping (7.17)-(7.18), now applied to strict GT patterns for $\lambda = (n \cdots 2\,1)$.

Examples. Application of the mapping (7.17)-(7.18) to two strict GT patterns with partitions $\lambda = (4\,3\,2\,1)$ and $\lambda = (5\,4\,3\,2\,1)$ are quite easy to effect to obtain the following results:

$$\begin{pmatrix} 4 & 3 & 2 & 1 \\ & 4 & 3 & 1 \\ & & 4 & 2 \\ & & & 3 \end{pmatrix} \mapsto \begin{pmatrix} 0 & 1 & 0 & 0 \\ 1 & -1 & 1 & 0 \\ 0 & 1 & -1 & 1 \\ 0 & 0 & 1 & 0 \end{pmatrix}, \qquad (7.30)$$

$$\begin{pmatrix} 5 & 4 & 3 & 2 & 1 \\ & 5 & 4 & 2 & 1 \\ & & 4 & 3 & 1 \\ & & & 3 & 2 \\ & & & & 2 \end{pmatrix} \mapsto \begin{pmatrix} 0 & 0 & 1 & 0 & 0 \\ 0 & 1 & -1 & 0 & 1 \\ 1 & -1 & 0 & 1 & 0 \\ 0 & 0 & 1 & 0 & 0 \\ 0 & 1 & 0 & 0 & 0 \end{pmatrix}. \qquad (7.31)$$

$\square$

An important feature of all GT patterns of shape $\lambda = \mathbf{m}_n$ is that each pattern naturally splits into the determination of all patterns of shape $\mu = \mathbf{m}_{n-1}$, such that $\mu \prec \lambda$, as given explicitly by (7.14). In particular, the number of GT patterns in the set $\text{Str}\,\mathbb{G}_{n\,n-1\cdots 1}$ is the sum of the number of patterns in each of the n subsets $\text{Str}\,\mathbb{G}_{\mu^{(i)}}$, where $\mu^{(i)} \prec (n\,n-1\cdots 1), i = 1, 2, \ldots, n$, is any of the strict partitions with $n-1$ nonzero parts that satisfies the betweenness conditions for the two-rowed GT pattern:

$$\begin{pmatrix} n & n-1 & \cdots & 1 \\ & \mu^{(i)} & & \end{pmatrix}. \qquad (7.32)$$

There are n such strict partitions $\mu^{(i)}$ that satisfy the betweenness conditions; they can be enumerated by the notation

$$\mu^{(i)} = (n\,n-1\,\cdots\,\widehat{i}\,\cdots\,2\,1), \; i = 1, 2, \ldots, n, \qquad (7.33)$$

where the symbol $\widehat{i}$ designates that integer i is missing from the sequence $(n\,n-1\cdots 1)$. For example, for $n = 4$, we have $\mu^{(1)} = (4\,3\,2)$, $\mu^{(2)} = (4\,3\,1)$, $\mu^{(3)} = (4\,2\,1)$, $\mu^{(4)} = (3\,2\,1)$. The general result (7.32) is discussed in greater detail in [L], and illustrated by examples. We denote the set of strict GT patterns corresponding to the partition $\mu^{(i)}$ by $\operatorname{Str} \mathbb{G}_{\mu^{(i)}}$, and the cardinality of this set by $Z_n(i)$, which is then given by

$$Z_n(i) = |\operatorname{Str} \mathbb{G}_{\mu^{(i)}}|. \tag{7.34}$$

It is important in our present approach that relation (7.34) **defines** $Z_n(i)$ without reference to any other quantity; this positive integer is uniquely determined by counting the number of lexical strict GT patterns of shape $\mu^{(i)}$. Since the rules for a strict partition and for betweenness are precisely set forth, this counting can always be effected. Indeed, we know from Zeilberger's result that this counting must give the value of the Zeilberger polynomial $Z_n(x)$, evaluated at $x = i$. We avoid, of course, using this result, because it is our goal to prove it. But we already have the proof for $i = 1$ and $i = n$ of the following relations:

$$\mu^{(1)} = (n \; n-1 \; \cdots \; 2) \text{ and } \mu^{(n)} = (n-1 \; n-2 \; \cdots \; 2\; 1);$$
$$d_{n-1} = Z_n(1) = Z_n(n) = |\operatorname{Str} \mathbb{G}_{\mu^{(1)}}| = |\operatorname{Str} \mathbb{G}_{\mu^{(n)}}|. \tag{7.35}$$

In writing out the relation for $\mu^{(1)}$, we have anticipated the shift-invariance of the number of GT patterns proved below (it is an obvious property). Of course, the explicit form of d_{n-1} remains undetermined in relations (7.35). We also have immediately the following identity, since d_n is the summation over all GT patterns in the sets $\mathbb{G}_{\mu^{(i)}}, i = 1, 2, \ldots, n$:

$$d_n = |\operatorname{Str} \mathbb{G}_{n\, n-1 \cdots 1}| = \sum_{i=1}^{n} Z_n(i). \tag{7.36}$$

We have thus proved, using only elementary properties of strict GT patterns for $\lambda = (n \cdots 2\, 1)$ that both relations in (7.5) are valid. This leaves the proof of the set of zeros property (7.6) to which we next turn.

7.3.1 Symmetries

We begin with the observation that the cardinality invariance of general GT patterns given in (7.15)-(7.16) applies also to the set $\operatorname{StrG}_{(n\ldots 2\, 1)}$ of strict partitions. We have the invariance under arbitrary shifts $h \in \mathbb{Z}$:

shift invariance: $|\operatorname{StrG}_{n\ldots 2\, 1)}| = |\operatorname{StrG}_{n+h\cdots 2+h\, 1+h}|. \tag{7.37}$

The situation for sign-reversal-shift conjugation must be considered in more detail. The mapping operation gives:

sign-reversal-shift operation:

$$\begin{pmatrix} n \; n-1 \; \cdots \; 1 \\ \mu^{(i)} \end{pmatrix} \mapsto \begin{pmatrix} n \; n-1 \; \cdots \; 1 \\ \mu^{(-i+n+1)} \end{pmatrix}, i = 1, 2, \ldots, n. \tag{7.38}$$

7.3. SPECIAL STRICT GT PATTERNS

This relation is direct consequence of the rule for applying the sign-reversal-shift operation, which, for the case at hand, is: Write each row in this two-rowed array in reverse order with minus signs and add $n+1$. Thus, in consequence of definition (7.34), we have the relation:

$$Z_n(-i+n+1) = Z_n(i), \; i = 1, 2, \ldots, n. \tag{7.39}$$

Relation (7.36) is invariant under the substitution $i \mapsto -i+n+1$.

Summary. The positive integers $Z_n(i), i = 1, 2, \ldots, n$, defined by $Z_n(i) = |\mathrm{Str}\mathbb{G}_{\mu^{(i)}}|$ are unique; they are determined by direct counting of the number of strict patterns in the set of strict GT patterns $\mathrm{Str}\mathbb{G}_{\mu^{(i)}}$. In addition, these numbers satisfy the sign-reversal-shift symmetry $Z_n(-i+n+1) = Z_n(i)$. This is an important property for giving the proof of the zeros in Item 2 in (7.6).

First, we give some examples:

Examples. The following results were calculated and presented in [L]. The strict GT patterns in the set $\mathbb{G}_{\mu^{(i)}}$, each $i = 1, 2, \ldots, n$, can clearly be enumerated by directly writing out all patterns such that each row is a strict partition satisfying the betweenness relations: *Each number $Z_n(i)$ defined by the cardinality of $\mathbb{G}_{\mu^{(i)}}$ is the unique positive integer determined by this direct enumeration.* The cases for $n = 1, 2, 3, 4, 5$ were calculated by this method and verified directly to be given by the values $Z_n(i)$ of the five polynomials listed below. They serve as models for generalization.

The polynomials $Z_n(x)$ for $n = 1, 2, 3, 4, 5$ are:

$$\begin{aligned}
Z_1(x) &= 1, \\
Z_2(x) &= \frac{1}{2}\binom{x}{1}\binom{-x+3}{1}, \\
Z_3(x) &= \frac{1}{3}\binom{x+1}{2}\binom{-x+5}{2}, \\
Z_4(x) &= \frac{7}{20}\binom{x+2}{3}\binom{-x+7}{3}, \\
Z_5(x) &= \frac{3}{5}\binom{x+3}{3}\binom{-x+9}{3}.
\end{aligned} \tag{7.40}$$

These polynomials have the following zeros:

$$n=2: x=0,3; \; n=3: x=-1,0,4,5; \tag{7.41}$$
$$n=4: x=-2,-1,0,5,6,7; \; n=5: x=-3,-2,-1,0,6,7,8,9. \; \square$$

These example polynomials satisfy the sign-reversal-shift conjugation rule $Z_n(x) = Z_n(-x+n+1)$, and give examples of the full set of zeros of $Z_n(x)$:

$$\{-n+2, -n+3, \ldots, -1, 0; n+1, n+2, \ldots, 2n-1\}. \qquad (7.42)$$

That these zeros split into the two sets separated by a semicolon is a consequence of the sign-reversal-shift conjugation symmetry. We observe also that the values between the two sets of zeros are at $i = 1, 2, \ldots, n$, which are just the points in relation (7.36) at which the $Z_n(i)$ are evaluated: The $2(n-1)$ zeros in (7.42) occur sequentially, and symmetrically, at the first possible integral points at each side of the central values for which $Z_n(i)$ is positive. □

The remaining issue is, of course, whether or not it is possible to find polynomials $P(x)$ of some given degree dependent on n that have the following properties for all $n \geq 2$:

$$P(-x+n+1) = P(x) \text{ and } P(i) = Z_n(i),\ i = 1, 2, \ldots, n, \qquad (7.43)$$

where the $Z_n(i)$ are **given** positive integers that satisfy $Z_n(i) = Z_n(-i+n+1)$ for all $i = 1, 2, \ldots, n$. This is similar to the Lagrange interpolation problem, which, of course, has a well-known solution. We turn next to this issue.

7.4 Sign-Reversal-Shift Invariant Polynomials

Sign-reversal-shift symmetry plays a major role in establishing the property of Zeilberger polynomials stated in Item 2, relation (7.6). We show in this section that a polynomial with the symmetry $P(x) = P(-x+n+1)$ always exists that "fits" a certain prescribed set of values, and derive the most general polynomial $P(x)$ with these properties.

We begin by introducing the family of polynomials $Z_n^{(d)}(x)$ of degree $2(d-1)$ defined by

$$Z_n^{(d)}(x) = \binom{x+d-2}{d-1}\binom{-x+d+n-1}{d-1}, d \geq 1,\ Z_n^{(1)}(x) = 1. \qquad (7.44)$$

This polynomial is sign-reversal-shift invariant for each $d = 1, 2, \ldots$:

$$Z_n^{(d)}(x) = Z_n^{(d)}(-x+n+1). \qquad (7.45)$$

It also has the set of $2(d-1)$ zeros given by

$$-d+2, -d+3, \ldots, -1, 0; n+1, n+2, \ldots, n+d-1. \qquad (7.46)$$

7.4. SIGN-REVERSAL-SHIFT INVARIANT POLYNOMIALS

The family of polynomials $Z_n^{(d)}(x)$, $d = 1, 2, \ldots$ are basis polynomials for all polynomials that have sign-reversal-shift invariance, as we next prove:

Basis theorem. *Every real polynomial that is sign-reversal-shift invariant has even degree $2(k-1)$ and the following form:*

$$P_k(x) = \sum_{d=1}^{k} a_d Z_n^{(d)}(x), \qquad (7.47)$$

where $a_1, a_2, \ldots, a_k$ are arbitrary real coefficients with $a_k \neq 0$.

Proof. Let $Q_m(x)$ be an arbitrary polynomial of degree $m-1$ given by

$$Q_m(x) = \sum_{r=1}^{m} b_r x^{r-1}, \; b_m \neq 0, \qquad (7.48)$$

that satisfies $Q_m(x) = Q_m(-x + n + 1)$. Then, equating powers of x on both sides of this relation gives the following system of relations:

$$\sum_{s=0}^{m-r} b_{r+s} \binom{r+s-1}{s}(n+1)^s = (-1)^{r-1} b_r, \; r = 1, 2, \ldots, m. \qquad (7.49)$$

These relations have exactly the structure needed to prove the form (7.47). To see this, we rewrite them in the following form corresponding to r even and r odd, and, for convenience, also set $y = n + 1$:

$$2b_r + \sum_{s=1}^{m-r} b_{r+s} \binom{r+s-1}{s} y^s = 0, \; r \text{ even}. \qquad (7.50)$$

$$\sum_{s=1}^{m-r} b_{r+s} \binom{r+s-1}{s} y^s = 0, \; r \text{ odd}. \qquad (7.51)$$

The summation term in each of these relations is empty for $r = m$. Thus, we find first for $r = m = $ even that $b_m = 0$, so that it is required that m be odd in (7.48):

In all that follows, m is odd: the degree $m-1$ of the polynomial $Q_m(x)$ in (7.48) must be even.

We now proceed as follows: set $r = m - 1$ in (7.50) and $r = m - 2$ in (7.51), with the result that each relation gives $b_{m-1} = -\frac{1}{2} b_m \binom{m-1}{1} y$; set $r = m - 3$ in (7.50) and $r = m - 4$ in (7.51), with the result that each relation gives $b_{m-3} = -\frac{1}{2}b_{m-2}\binom{m-3}{1}y + \frac{1}{4}b_m\binom{m-1}{3}y^3$, after using the first relation to eliminate b_{m-1}; etc. Thus, the system of relations is triangular in structure, and can be solved completely by a continuation of this iterative procedure. This leads to the unique expression of each $b_{m-1}, b_{m-3}, \ldots, b_2$ coefficient in terms of the $b_m, b_{m-2}, \ldots, b_1$ coefficients. It is must be the case that the even relation (7.50) for $r = m - k$ and the odd relation (7.51) for $r = m - k - 1$ with k odd give exactly the same result. This must hold because otherwise a linear relation between the b_{odd}, which are $(m+1)/2$ in number, will obtain, and this is inconsistent with the fact that there exists a sign-reversal-shift symmetric polynomial of even degree $m - 1$ in which there are $(m + 1)/2$ unconstrained parameters. This is the polynomial on the right-hand side of (7.47) for $2(k - 1) = m - 1$ with coefficients $a_1, a_2, \ldots, a_k$, where $k = (m + 1)/2$. Since relations (7.49) are necessary and sufficient conditions that an arbitrary polynomial $Q_m(x)$ of even degree $m-1$ satisfy sign-reversal-shift symmetry, we conclude that every polynomial that has sign-reversal-shift symmetry is of even degree and has the form (7.47). □

The basis theorem for sign-reversal-shift symmetric polynomials expressed by (7.47) places very strong constraints on a polynomial with this symmetry. We can now consider the analog of the Lagrange interpolation formula for sign-reversal-shift symmetric polynomials. For the statement of this result, we use the standard notations for an arbitrary positive integer n:

$$\lceil \tfrac{n}{2} \rceil = n/2,\ n\text{ even};\quad \lceil \tfrac{n}{2} \rceil = (n+1)/2,\ n\text{ odd};$$
$$\lfloor \tfrac{n}{2} \rfloor = n/2,\ k\text{ even};\quad \lfloor \tfrac{n}{2} \rfloor = (n-1)/2,\ n\text{ odd}. \qquad (7.52)$$

Let $z_1, z_2, \cdots, z_n$ be a set of real numbers that satisfy $z_i = z_{-i+n+1}$; that is, $z_1 = z_n, z_2 = z_{n-1}, \ldots, z_n = z_1$, but are otherwise unspecified. Consider now the system of equations obtained from (7.47) by the requirement $P_k(i) = z_i, i = 1, 2, \ldots, n$. Because the equations for i and for $-i + n + 1$ are identical, there obtains the following system of $\lceil \tfrac{n}{2} \rceil$ relations in k unknowns:

$$\sum_{d=1}^{k} a_d \binom{i+d-2}{d-1}\binom{-i+d+n-1}{d-1} = z_i,\ i = 1, 2, \ldots, \lceil \tfrac{n}{2} \rceil. \qquad (7.53)$$

We thus arrive at the interpolation rule:

Interpolation rule: *For arbitrarily given $z_1, z_2, \ldots, z_n$ satisfying $z_{-i+n+1} = z_i$, find the set of all solutions $a_1, a_2, \ldots, a_k$ of the $\lceil \tfrac{n}{2} \rceil$ relations (7.53), thus obtaining the set of all sign-reversal-shift polynomials $P_k(x)$ defined by (7.47) that take on the values $P_k(i) = z_i, i = 1, 2, \ldots, \lceil \tfrac{n}{2} \rceil$.*

It is not our intent to classify all sign-reversal-shift polynomials that can occur by the interpolation rule. For our purposes, it is sufficient to

7.4. SIGN-REVERSAL-SHIFT INVARIANT POLYNOMIALS

note that for $k = n$ there are many solutions to the set of $\lceil \frac{n}{2} \rceil$ equations in n unknowns. Indeed, by showing that the matrix of binomial coefficients of order $\lceil \frac{n}{2} \rceil$ obtained by selecting any $\lceil \frac{n}{2} \rceil$ distinct columns from the n columns that define the linear transformation in $a_1, a_2, \ldots, a_n$ in the left-hand side is nonsingular, the system of equations has infinitely many solutions for the coefficients $a_1, a_2, \ldots, a_n$ for $n \geq 3$; these solutions are described as follows: Select any subset containing $\lceil \frac{n}{2} \rceil$ distinct coefficients from the full set of n coefficients $a_1, a_2, \ldots, a_n$; and move the remaining $\lfloor \frac{n}{2} \rfloor$ coefficients to the right-hand side of (7.53). Then, the resulting set of $\lceil \frac{n}{2} \rceil$ equations in $\lceil \frac{n}{2} \rceil$ unknowns that remains on the left-hand side can be solved in terms of the $1 + \lfloor \frac{n}{2} \rfloor$ quantities now on the right-hand side.

Such a general interpolation formula would appear not to be very useful. But it does establish the following result:

The sign-reversal-shift invariant polynomial $P_n(x)$ of degree $2(n-1)$ is uniquely defined by its values at the following points:

$$P_n(0), P_n(-1), P_n(-2), \ldots, P_n(-n+1). \tag{7.54}$$

Proof. We have that

$$P_n(-x) = \sum_{d=1}^{n} (-1)^{d-1} \binom{x}{d-1} \binom{x+d+n-1}{d-1}, \tag{7.55}$$

where we have used the property $\binom{-z}{a} = (-1)^a \binom{z+a-1}{a}$ of the binomial function to transform the first factor in the definition of $Z_n^{(d)}(x)$ in (7.44). Evaluating (7.55) at $x = i$ for $i = 0, -1, \ldots, -n+2$, gives

$$P_n(-i) = \sum_{d=1}^{i+1} (-1)^{d-1} \binom{i}{d-1} \binom{i+d+n-1}{d-1}. \tag{7.56}$$

This is a triangular system of relations, which gives: $P_n(0) = a_1$, $P_n(-1) = a_1 - a_2 \binom{1}{1} \binom{n+2}{1}$, $P_n(-2) = a_1 - a_2 \binom{2}{1} \binom{n+3}{1} + a_3 \binom{2}{2} \binom{n+4}{2}, \ldots$; hence, the coefficients $a_1, a_2, \ldots, a_n$ are uniquely determined, respectively, in terms of $P_n(0)$; then $P_n(0), P_n(-1)$; then $P_n(0), P_n(-1), P_n(-2)$; $\ldots$; then $P_n(0), P_n(-1), \ldots, P_n(-n+1)$. □

The above results are all a consequence of the basis theorem for sign-reversal-shift polynomials, and the implied consequences of the interpolation rule, as applied to $k = n$. It is useful to summarize these results:

CHAPTER 7. ALTERNATING SIGN MATRICES

Summary. *Given any set of data points*

$$z_1, z_2, \ldots, z_n \text{ that satisfy } z_i = z_{-i+n+1} \text{ for } i = 1, 2, \ldots, n, \quad (7.57)$$

there always exists a sign-reversal-shift polynomial of degree $2(n-1)$ defined by

$$P_n(x) = \sum_{d=1}^{n} a_d \binom{x+d-2}{d-1} \binom{-x+d+n-1}{d-1} \quad (7.58)$$

that takes on the values (7.57); that is,

$$P_n(i) = \sum_{n=1}^{n} a_d \binom{i+d-2}{d-1} \binom{-i+d+n-1}{d-1} = z_i. \quad (7.59)$$

The polynomial is uniquely determined by its values at the n points

$$P_n(0), P_n(-1), \ldots, P_n(-i), \ldots, P_n(-n+1). \quad (7.60)$$

We can now apply this result to the case at hand, where we choose $z_i = Z_n(i) = Z_n(-i+n-1), i = 1, 2, \ldots, n$, which are the positive numbers given uniquely by the strict GT pattern counting formula $Z_n(i) = |\mathrm{StrG}_{\mu^{(i)}}|$:

Existence of a sign-reversal-shift polynomial. *Given the set of data points $Z_n(i), i = 1, 2, \ldots, n$, there exists a sign-reversal-shift polynomial $P_n(x)$ of degree $2(n-1)$ that gives these values, namely,*

$$P_n(x) = \sum_{d=1}^{n} a_d \binom{x+d-2}{d-1} \binom{-x+d+n-1}{d-1},$$

$$\sum_{d=1}^{n} a_d \binom{i+d-2}{d-1} \binom{-i+d+n-1}{d-1} = Z_n(i). \quad (7.61)$$

The Andrews-Zeilberger number d_n is given by

$$d_n = \sum_{d=1}^{n} a_d S_{d,n}, \quad S_{d,n} = \sum_{i=1}^{n} \binom{i+d-2}{d-1} \binom{-i+d+n-1}{d-1}, \quad (7.62)$$

where the values of the coefficients $a_1, a_2, \ldots, a_n$ are uniquely determined by a set of triangular relations in terms of the set of n values of $P_n(x)$ given by

$$P_n(0), P_n(-1), \ldots, P_n(-n+1). \quad (7.63)$$

Thus, the existence of a sign-reversal-shift polynomial of degree $2(n-1)$ that gives d_n has been proved; it only remains to prove that the explicit zeros in question are those asserted in the assumption:

Assumption of zeros. *Assume $P_n(-i) = 0$, for $i = 0, 1, \ldots, n-2$ in relation (7.61). Then, it follows that $a_1 = a_2 = \cdots = a_{n-1} = 0$, and the polynomial $P_n(x)$ becomes the Zeilberger polynomial*

$$Z_n(x) = a_n \binom{x+n-2}{n-1}\binom{-x+2n-2}{n-1}, \qquad (7.64)$$

$$a_n = d_n \Big/ \binom{3n-2}{n-1}.$$

The coefficient a_n is determined by the procedure (7.7)-(7.9).

Remark. That the formula for obtaining d_n is polynomial in structure is a consequence of the basis theorem for sign-reversal-shift invariant polynomials, and the proof that such a polynomial can always reproduce the data provided by the counting formula $Z_n(i) = |\text{Str}\mathbb{G}_{\mu^{(i)}}|$. The polynomial structure is no longer an issue. But there are many sign-reversal-shift invariant polynomials that can reproduce this data. The uniqueness of the polynomial can only be achieved by further specification of its properties, such as its value at other points, which is unknown, but, of course, always a definite number. The choice of the points (7.60) is a convenient one, based on the special examples (7.40) and the resulting triangular nature of the equations to be solved; the value 0 at these points remains to be proved. The burden of our approach is now shifted to proving that the zeros are those stated in the **assumption of zeros**.

7.5 The Requirement of Zeros

The proof that the sign-reversal-shift polynomial that replicates the uniquely calculated values $Z_n(i) = |\text{Str}\mathbb{G}_{\mu^{(i)}}|$ must possess exactly the set of zeros (7.57) is equivalent to obtaining the **exact** form of $Z_n(i)$ as a polynomial in the indexing number i, but we have been unable to obtain a proof of this property using our approach. This result, in our view, is a major contribution of Zeilberger's computer-based proof.

We resolve the problem of zeros by placing it in a broader context. Such a possibility exists because standard GT triangular patterns have a natural extension to what are known as skew GT patterns, which, in fact, are the exact analogs of skew Young tableaux, with which they are in one-to-one correspondence. In this section, we reformulate the counting problem for alternating sign matrices in terms of a certain class of skew GT patterns. We will show that this requires exactly that each value in (7.60) be zero: *The zeros problem is resolved by embedding the set of strict GT patterns in a larger set of semistrict skew GT patterns.*

A skew GT pattern is an extension of the triangular GT pattern, as follows: We introduce k partitions, each with n parts, and denoted by

$$\mathbf{m}_i = (m_{1,i}\ m_{2,i}\ \ldots\ m_{n,i}),\ i = 1, 2, \ldots, k, \tag{7.65}$$

where we use the same $m_{j,i}$ notation for the parts as in the GT pattern (7.10), now extended to n parts for each partition. These k partitions are now stacked as shown on the right in the following display, where the $k \times n$ size of the array is suppressed in the notation (see [L, Sect. 11.3.3]):

$$\begin{pmatrix} \lambda/\mu \\ m \end{pmatrix} = \begin{pmatrix} \mathbf{m}_k \\ \mathbf{m}_{k-1} \\ \cdot \\ \cdot \\ \cdot \\ \mathbf{m}_1 \end{pmatrix}. \tag{7.66}$$

Each row is shifted one unit to the right from the row above so that the parts in row $\mathbf{m}_{i-1}$ fall between those in row $\mathbf{m}_i$ for $i = 2, 3, \ldots, k$, except for part n, where we impose $m_{n,i} \geq m_{n,i-1} \geq 0$ for $i = 2, 3, \ldots, k$. Thus, the left margin is staggered, just as in the triangular array $\binom{\lambda}{m}$ shown in (7.10), but now all patterns extend n parts to the right. As with ordinary GT patterns, this staggered display is intended to suggest the betweenness relations. We denote these betweenness relations by

$$\mathbf{m}_k \sqsupseteq \mathbf{m}_{k-1} \sqsupseteq \cdots \sqsupseteq \mathbf{m}_1. \tag{7.67}$$

The betweenness conditions encode the closed interval conditions:

$$m_{j,i-1} \in [m_{j,i}, m_{j+1,i}],\ 2 \leq i \leq k;\ 1 \leq j \leq n-1,$$
$$\text{with } m_{n,k} \geq m_{n,k-1} \geq m_{n,2} \geq m_{n,1} \geq 0. \tag{7.68}$$

These are just the betweenness conditions (7.11), now extended to include n entries in each row. The notation $\binom{\lambda/\mu}{m}$ on the left in (7.66) denotes that the partition $\lambda = \mathbf{m}_k$ and the partition $\mu = \mathbf{m}_1$ are now to be any pair of **specified** partitions such that the betweenness conditions between row k (the top row) and row 1 (the bottom row) are not empty, that is, relations (7.68) have solutions. Thus, the notation λ/μ specfies these conditions, which are $\lambda_1 \geq \mu_1, \lambda_2 \geq \mu_2, \ldots, \lambda_n \geq \mu_n \geq 0$, which we write as $\lambda \geq \mu$. Then, $\mathbf{m}_k = \lambda$ is entered at the top of the array (row k), and $\mathbf{m}_1 = \mu$ at the bottom (row 1). The single symbol m in $\binom{\lambda/\mu}{m}$ then enumerates all k rows; there is no row corresponding to λ/μ since this symbol just specifies the conditions on λ and μ. Then, for such specified λ and μ, all patterns are allowed that satisfy the betweenness conditions. Such a pattern is called a lexical skew GT pattern. The set of all such patterns is denoted by $\mathbb{G}_{\lambda/\mu}$.

7.5. THE REQUIREMENT OF ZEROS

Example. A single example for $k = 3$ and $n = 4$ makes clear the general case:

$$\binom{\lambda/\mu}{m} = \begin{bmatrix} \lambda_1 & \lambda_2 & \lambda_3 & \lambda_4 \\ & m_{1,2} & m_{2,2} & m_{3,2} & m_{4,2} \\ & & \mu_1 & \mu_2 & \mu_3 & \mu_4 \end{bmatrix}, \quad (7.69)$$

where $\lambda = (\lambda_1, \lambda_2, \lambda_3, \lambda_4) = (m_{1,3}, m_{2,3}, m_{3,3}, m_{4,3})$, $\mu = (\mu_1, \mu_2, \mu_3, \mu_4) = (m_{1,1}, m_{2,1}, m_{3,1}, m_{4,1})$. Then, for prescribed partitions λ and μ such that $\lambda_1 \geq \mu_1, \lambda_2 \geq \mu_2, \lambda_3 \geq \mu_3, \lambda_4 \geq \mu_4 \geq 0$, all possible patterns are enumerated such that the following betweenness conditions on row 2 are fulfilled:

$$m_{1,2} \in [\lambda_1, \lambda_2], \ m_{2,2} \in [\lambda_2, \lambda_3], \ m_{2,3} \in [\lambda_3, \lambda_4];$$
$$m_{2,2} \in [\mu_1, \mu_2], \ m_{3,2} \in [\mu_2, \mu_3], \ m_{4,2} \in [\mu_3, \mu_4]. \quad \square \quad (7.70)$$

We next give the symmetries of the skew GT patterns that correspond to Item 5 in (7.15)-(7.16):

1. Cardinality invariance operations:

   (a) Translation invariance: The number of patterns in $\mathbb{G}_{\lambda/\mu}$ is invariant to the shift of all entries in the pattern by the same integer $h \in \mathbb{Z}$, which implies

   $$|\mathbb{G}_{(\lambda+h)/(\mu+h)}| = |\mathbb{G}_{\lambda/\mu}|, \quad (7.71)$$

   where $\lambda + h = (\lambda_1 + h, \lambda_2 + h, \ldots, \lambda_n + h)$ and $\mu + h = (\mu_1 + h, \mu_2 + h, \ldots, \mu_n + h)$.

   (b) Sign-reversal-shift conjugation: The number of patterns in $\mathbb{G}_{\lambda/\mu}$ is invariant to the operation of writing each row of the pattern backwards with minus signs and translating by $h = \lambda_1 + \lambda_n$, where, in addition, we must reverse the staggering so that row $k-1$ is shifted one unit to the left of row k, row $k-2$ one unit to the left of row $k-1$, ..., row 1 one unit to the left of row 2. Thus, the right margin is staggered, just as in the triangular array $\binom{\lambda}{m}$ shown in (7.10), but now all patterns extend n units to the left. We denote this new skew GT pattern obtained by the operation of sign-reversal-shift conjugation by

   $$\binom{\lambda^*/\mu^*}{m^*} = \begin{pmatrix} m_k^* \\ m_{k-1}^* \\ \cdot \\ \cdot \\ \cdot \\ m_1^* \end{pmatrix}. \quad (7.72)$$

Row i in this $k \times n$ array is given by
$$\mathbf{m}_i^* = (m_{1,i}^* \, m_{2,i}^* \, \cdots \, m_{n,i}^*),$$
$$m_{j,i}^* = -m_{n-i+1} + \lambda_1 + \lambda_n, \quad j = 1, 2, \ldots, n. \quad (7.73)$$

We call this pattern the *sign-reversal-shift conjugate pattern*, or simply the conjugate to the pattern (7.66). The set of all such patterns for prescribed λ^*/μ^* is denoted $\mathbb{G}_{\lambda^*/\mu^*}$. It is again the case that we have equality of cardinalities:
$$|\mathbb{G}_{\lambda/\mu}| = |\mathbb{G}_{\lambda^*/\mu^*}|. \quad (7.74)$$

Skew GT patterns of the sort $\binom{\lambda/\mu}{m}$ introduced above, their variations and relation to skew semistandard tableaux are discussed in greater detail in [L], although the operation of sign-reversal-shift conjugation is not introduced there.

The idea now is to choose λ and μ in the skew GT patterns described above in such a way that the set $\mathbb{G}_{\lambda/\mu}$, when appropriately restricted, gives exactly the set of strict GT patterns Str $\mathbb{G}_{n\,n-1\cdots 1}$. (We have given the above general setting because it important to know the background against which this special class makes its appearance.)

We choose λ and μ as follows:
$$\lambda = (3n-2\ 3n-3\ \cdots\ 2\ 1),$$
$$\mu = (3n-2\ 3n-3\ \cdots\ 2n\ n-1\ \cdots\ 1\ 1^{[n]}). \quad (7.75)$$

Thus, λ is a strict partition with no jumps between the strictly decreasing parts, and μ has this same strict property for its first $n-1$ parts, but then is followed by the strict partition $(n-1\ n-2\ \cdots\ 2\ 1)$ having $n-1$ parts, and then by n 1's, denoted by $1^{[n]} = (1\,1\,\cdots\,1)$. Thus, the length of λ, as well as μ, is $3n-2$. These two partitions constitute row $n+1$ and row 1 of a set of GT patterns $\mathbb{G}_{\lambda/\mu}$ of size $(n+1) \times (3n-2)$.

Examples. $n = 2, 3$:
$$\begin{pmatrix} 4 & 3 & 2 & 1 \\ & 4 & \bullet & 1 & 1 \\ & & 4 & 1 & 1 & 1 \end{pmatrix} \text{ and } \begin{pmatrix} 7 & 6 & 5 & 4 & 3 & 2 & 1 \\ & 7 & 6 & \bullet & \bullet & 2 & 1 & 1 \\ & & 7 & 6 & \bullet & 2 & 1 & 1 & 1 \\ & & & 7 & 6 & 2 & 1 & 1 & 1 & 1 \end{pmatrix}. \quad (7.76)$$

The entries in the two embedded subpatterns
$$\begin{pmatrix} 3 & 2 \\ & \bullet \end{pmatrix} \text{ and } \begin{pmatrix} 5 & 4 & 3 \\ & \bullet & \bullet \\ & & \bullet \end{pmatrix} \quad (7.77)$$

7.5. THE REQUIREMENT OF ZEROS

give lexical patterns for the assignment of the • points in one-to-one correspondence with the equivalent shifted patterns:

$$\begin{pmatrix} 2 & 1 \\ & \bullet \end{pmatrix} \text{ and } \begin{pmatrix} 3 & 2 & 1 \\ & \bullet & \bullet \\ & & \bullet \end{pmatrix}. \tag{7.78}$$

Thus, the count of patterns is $|\mathbb{G}_{2\,1}|$ and $|\mathbb{G}_{3\,2\,1}|$. □

The property exhibited by the example above extends to the general case for n for the choices of λ and μ given by (7.75), which, of course, was the reason for their selection, as we next show.

The choice of λ and μ given by (7.75), together with the betweenness conditions, force row i of the skew pattern to be

$$\mathbf{m}_i = (3n-2\ 3n-3\ \cdots\ 2n\ \bullet\ \bullet\ \cdots\ \bullet\ n-1\cdots 1\ 1^{[n-i+1]}),$$
$$i = 2, \ldots, n. \tag{7.79}$$

This partition contains the strict partition $(3n-2\ 3n-3\ \cdots\ 2n)$ containing $n-1$ parts, followed by exactly $i-1$ unfilled points ($\bullet\ \bullet\ \cdots\ \bullet$), followed by the partition $(n-1\ n-2\ \cdots\ 1\ 1^{[n-i+1]}$ containing $2n-i$ fixed parts, thus giving a length of $3n-2$. Thus, the central pattern corresponding to the n parts from row $n+1$ given by $(2n-1\ 2n-2\ \cdots\ n)$ has the form given by the triangular pattern

$$\begin{pmatrix} 2n-1 & 2n-2 & \cdots & n+1 & n \\ & \bullet & & & \bullet \\ & & \vdots & & \\ & & \bullet & & \end{pmatrix} \equiv \begin{pmatrix} n & n-1 & \cdots & 2 & 1 \\ & \bullet & & & \bullet \\ & & \vdots & & \\ & & \bullet & & \end{pmatrix}. \tag{7.80}$$

The • points in each row of these two triangular patterns are the only free places for an entry in the full lexical skew pattern. In terms of the general standard notations $m_{j,i}$ in (7.66) (illustrated in (7.69)), this is the ordinary GT pattern filled-in with entries as follows:

$$\begin{pmatrix} \lambda \\ m \end{pmatrix} = \begin{pmatrix} 2n-1 & 2n-2 & \cdots & n-1 & n \\ m_{n,n} & m_{n+1,n} & \cdots & m_{2n-3,n} & m_{2n-2,n} \\ & & \vdots & & \\ & m_{n,i} & \cdots & m_{n+i-2,i} & \\ & & \vdots & & \\ & & m_{n,3} & m_{n+1,3} & \\ & & m_{n,2} & & \end{pmatrix}. \tag{7.81}$$

We thus arrive at the skew GT pattern of size $(n+1) \times (3n-2)$ in which the labels in the full top row $n+1$ are a strict partition with no jumps that contains the central strict partition $(2n-1 \cdots n)$ as the shape of the pattern (7.81), in which all lexical entries can occur; that is, there are no strict conditions yet imposed on the rows of the pattern (7.81). The full pattern (7.81) of shape $(2n-1 \cdots n)$ is exactly embedded as the triangular array of $\bullet$ points in the skew GT pattern as follows:

$$\begin{pmatrix} 3n-2 & 3n-3 & \cdots & 2n & 2n-1 & \cdots & n & n-1 & n-2 & \cdots & 1 & & & & \\ & 3n-2 & 3n-3 & \cdots & 2n & \bullet & \cdots & & \bullet & n-1 & n-2 & \cdots & 1 & 1 & \\ & & \vdots & & & & \vdots & & & & & \vdots & & & \\ & & & 3n-2 & 3n-3 & \cdots & 2n & \bullet & n-1 & n-2 & \cdots & 1 & 1 & \cdots & 1 & 1 \\ & & & & 3n-2 & 3n-3 & \cdots & 2n & n-1 & n-2 & \cdots & 1 & 1 & & \cdots & & 1 & 1 \end{pmatrix}$$
(7.82)

We next define a *semistrict partition* to be a partition of the form $(\nu, 1^{[k]})$, $k \geq 2$, where ν is a strict partition. We can now impose the constraints on the full GT pattern (7.82) that each of its rows be a semistrict partition. Since each row i in (7.82) is a partition of the form

$$3n-2 \ 3n-3 \ \cdots \ 2n \ m_{n,i} \ m_{n+1,i} \ \cdots \ m_{n+i-2,i} \ n-1 \cdots \ 1 \ 1^{[n-i+1]},$$
(7.83)

the semistrict condition on the rows of (7.82) requires that the pattern (7.81), which gives the entries of the $\bullet$ points in (7.82), be strict:

$$m_{n,i} > m_{n+1,i} > \cdots > m_{n+i-2,i}, i = 2, 3, \ldots, n.$$
(7.84)

Thus, each row i of (7.82) given by (7.83) is semistrict, since it is strict from part 1 to part $2n+i-3$, followed by $1^{[n-i+1]}$; that is, the full skew pattern (7.82) is semistrict.

We denote the set of all semistrict skew patterns (7.82) by $\text{SStr}\,\mathbb{G}_n$, where λ and μ are fixed at the strict partition $\lambda = (3n-2\ 3n-3\ \cdots\ 1)$, and semistrict partition $\mu = (3n-2\ 3n-3\ \cdots\ 2n\ n-1\ n-2\ \cdots\ 1\ 1^{[n]})$, for all $n \geq 2$. Thus, for each semistrict skew GT pattern in the set $\text{SStr}\,\mathbb{G}_n$, the triangular GT patterns defined by (7.81) are all strict GT patterns in the set $\text{Str}\,\mathbb{G}_{2n-1\,2n-2\cdots n}$, which, under the shift by $h = -n+1$, is exactly the set of patterns $\text{Str}\,\mathbb{G}_{n\,n-1\cdots 1}$. Thus, we have succeeded in embedding the semistrict GT patterns for partition $(2n-1 \cdots n)$ within the special class $\text{SStr}\,\mathbb{G}_n$ of semistrict skew GT patterns. Since the GT patterns in the strict sets $\text{Str}\,\mathbb{G}_{2n-1\,\ldots\,n}$ and $\text{Str}\,\mathbb{G}_{n\,\ldots\,1}$ have the same count, we have the identity with the Andrews-Zeilberger number given by

$$|\text{SStr}\,\mathbb{G}_n| = d_n.$$
(7.85)

But we can now go further with the analysis and consider the effect of sign-reversal-shift symmetry. The set of all $(n+1) \times (3n-2)$ skew patterns for which λ is the strict partition $\lambda = (3n-2\ 3n-3\ \cdots\ 1)$ can only have a row n of the form (see (7.32)-(7.33)):

$$\nu^{(k)} = (3n-2\ 3n-3\ \cdots\ \widehat{k}\ \cdots\ 1), \ k = 1, 2, \ldots, 3n-2.$$
(7.86)

7.5. THE REQUIREMENT OF ZEROS

These strict partitions enters into the above enumeration of patterns in the following manner. Consider the set of skew patterns of size $(n+1) \times (3n-2)$, where the partition μ is no longer specified, but is allowed to assume all values that give lexical skew patterns for the prescribed λ. We denote this set of skew GT patterns by $\text{Skew}\,\mathbb{G}_\lambda$. The counting of the number of patterns in the set $\text{Skew}\,\mathbb{G}_\lambda$ is the well-defined problem of simply enumerating the patterns. The important property is that the number of patterns in $\text{Skew}\,\mathbb{G}_\lambda$ must have the property of being invariant under $k \mapsto -k + 3n - 1$, which leaves the partition λ unchanged. Under the mapping $k \mapsto -k + 3n - 1$, the two-rowed pattern containing row $n+1$ and n is transformed to

$$\begin{pmatrix} 3n-2 & 3n-3 & \cdots & 1 \\ & \nu^{(k)} & & \end{pmatrix} \mapsto \begin{pmatrix} 3n-2 & 3n-3 & \cdots & 1 \\ & \nu^{(-k+3n-1)} & & \end{pmatrix}, \quad (7.87)$$

where, for $k = 1, 2, \ldots, 3n - 2$, the patterns on the right are just the enumeration of the strict partitions $\nu^{(1)}, \nu^{(2)}, \ldots, \nu^{(3n-2)}$ on the left in reverse order. In particular, the mapping (7.87), restricted to $k = n, n+1, \ldots, 2n-1$, gives just the possible strict partitions that can occur in the two-rowed pattern corresponding to the mapping

$$\begin{pmatrix} 2n-1 & 2n-2 & \cdots & n \\ & \rho^{(k)} & & \end{pmatrix} \mapsto \begin{pmatrix} 2n-1 & 2n-2 & \cdots & 1 \\ & \rho^{(-k+3n-1)} & & \end{pmatrix},$$

$$\rho^{(k)} = (2n-1 \cdots \widehat{k} \cdots n), \, k = n, n+1, \ldots, 2n-1. \quad (7.88)$$

Again, for $k = n, n+1, \ldots, 2n-1$, the partitions $\rho^{(n)}, \rho^{(n+1)}, \ldots, \rho^{(2n-1)}$ are repeated in reverse order in the pattern to the right. Finally, if the mapping (7.88) is shifted down by $-n + 1$, we obtain:

$$\begin{pmatrix} n & n-1 & \cdots & 1 \\ & \mu^{(k)} & & \end{pmatrix} \mapsto \begin{pmatrix} n & n-1 & \cdots & 1 \\ & \mu^{(-k+n+1)} & & \end{pmatrix},$$

$$\mu^{(k)} = (n \cdots \widehat{k} \cdots 1), \, k = 1, 2, \ldots, n, \quad (7.89)$$

which for $k = 1, 2, \ldots, n$ gives the strict partitions $\mu^{(1)}, \ldots, \mu^{(2)}, \ldots, \mu^{(n)}$ in reverse order in the pattern to the right. This last set of patterns, of course, are just those of interest for the counting problem of alternating sign matrices:

It is convenient to summarize, for ease of reference, the definitions of the skew GT patterns given above, where here λ is taken to be the specified strict partition $\lambda = (3n - 2 \cdots 2\, 1)$:

$$\text{Skew}\,\mathbb{G}_\lambda = \left\{ \begin{array}{l} \text{all skew GT patterns of size } (n+1) \times \\ (3n-2) \text{ in which } \mu \sqsubset \lambda \text{ runs over all} \\ \text{partitions that give lexical patterns} \end{array} \right\}. \quad (7.90)$$

$$\text{SStr}\,\mathbb{G}_n = \left\{\begin{array}{l} \text{SStr}\,\mathbb{G}_n \subset \mathbb{G}_{\lambda/\mu},\ \lambda = (3n-2\ \cdots\ 1), \\ \mu = (3n-2\ \cdots\ 2n\ n-1\ \cdots\ 1\ 1^{[n]}), \\ \mathbf{m}_i = (3n-2\ \cdots\ 2n\ \kappa_i\ n-1\ \cdots\ 1\ 1^{[n]}), \\ \kappa_i = (m_{n,i}\,m_{n+1,i}\,\cdots\,m_{n+i-2,i}), \\ 2n > m_{n,i} > m_{n+1,i} > \cdots > m_{n+i-2,i} > n-1, \\ i = 2, 3, \ldots, n \end{array}\right\} \tag{7.91}$$

We next appeal to the basis theorem (7.47) on sign-reversal-shift polynomials: There must exist a sign-reversal-shift polynomial $Q(x)$ with the property

$$Q(x) = Q(-x + 3n - 1). \tag{7.92}$$

This polynomial, when evaluated at the points $x = k = 1, 2, \ldots 3n-2$, must give a set of unique positive numbers q_k satisfying the relations:

$$Q(k) = q_k,\ \text{with}\ q_{-k+3n-1} = q_k,\ \text{each}\ k = 1, 2, \ldots, 3n-2, \tag{7.93}$$

$$\sum_{k=1}^{3n-2} q_k = |\text{Skew}\,\mathbb{G}_\lambda|. \tag{7.94}$$

The situation is similar for the subset of semistrict patterns $\text{SStr}\,\mathbb{G}_n \subset \mathbb{G}_{\lambda/\mu} \subset \text{Skew}\,\mathbb{G}_\lambda$, for which the subset of values of k is given by $k = n, n+1, \ldots, 2n-1$, and which give the strict partitions $\rho^{(k)}$ in (7.88) for row n, whose collection of all semistrict subpatterns for $k = n, n+1, \ldots, 2n-1$ enumerate all patterns in $\text{SStr}\,\mathbb{G}_n$, and gives $|\text{SStr}\,\mathbb{G}_n| = d_n$. Indeed, it follows from the equivalence of these patterns to those in (7.89) that

$$Q(j) = Z_n(j - n + 1),\ j = n, n+1, \ldots, 2n - 1. \tag{7.95}$$

But, since $Q(x)$ is a polynomial, it has values for all x, in particular, for $x = 1, 2, \ldots, n-1; 2n, 2n+1, \ldots, 3n-2$. But the values at these points must all be 0, because there are no patterns in $\text{SStr}\,\mathbb{G}_n$ corresponding to these values. Thus, $Q(x)$ must have the following $2(n-1)$ zeros:

$$Q(1) = Q(2) = \cdots = Q(n-1) = \\ Q(2n) = Q(2n+1) = \cdots = Q(3n-2) = 0. \tag{7.96}$$

The sign-reversal-shift polynomial $Q(x)$ must have the $2(n-1)$ zeros at $k = 1, 2, \ldots, n-1; 2n, 2n+1, \ldots, 3n-2$, because there are no patterns in the set $\text{SStr}\,\mathbb{G}_n$ for these values of k, even though k is allowed to assume all values $k \in \{1, 2, \ldots, 3n-2\}$ in the unrestricted set $\text{Skew}\,\mathbb{G}_\lambda$.

But there is a unique polynomial with the sign-reversal-shift invariance $x \mapsto -x + 3n - 1$ property (up to a multiplicative constant) that fits the above required zero values; namely,

$$Q(x) = a \binom{x-1}{n-1} \binom{-x+3n-2}{n-1}. \tag{7.97}$$

7.6. THE INCIDENCE MATRIX FORMULATION

We now define the polynomial $P_n(x)$ of degree $2(n-1)$ by

$$P_n(x) = (a_n/a)Q(x+n-1) = a_n \binom{x+n-2}{n-1}\binom{-x+2n-2}{n-1}, \quad (7.98)$$

which then has the zeros given by

$$P_n(-n+2) = P_n(-n+3) = \cdots = P_n(0) = 0,$$
$$P_n(n+1) = P_n(n+2) = \cdots = P_n(2n-1) = 0. \quad (7.99)$$

We conclude:

(i) There exists a sign-reversal-shift polynomial that can fit the criteria given in (7.5)-(7.6) required of the Zeilberger polynomials. (ii) This polynomial is the polynomial $P_n(x)$, which is unique up to a multiplicative constant, given by (7.98) that reproduces the data associated with the counting problem of the set of semistrict skew GT patterns $SStr\mathbb{G}_n \subset Skew\mathbb{G}_\lambda$, where the multiplicative constant is uniquely determined by the procedure of relations (7.7)-(7.9). This gives the final result for the Zeilberger polynomial:

$$Z_n(x) = d_n \binom{x+n-2}{n-1}\binom{-x+2n-2}{n-1} \bigg/ \binom{3n-2}{n-1}. \quad (7.100)$$

The counting of alternating sign matrices based on zeros of the counting polynomial has now been solved by GT pattern methods.

7.6 The Incidence Matrix Formulation

There is still another way of counting the number of alternating sign matrices bases on an incidence matrix that has a natural formulation originating from the A-matrix mapping given in Sect. 7.2.1. We present this because the incidence matrix is a $(0-1)$−matrix, and such matrices are of general interest in combinatorics (see Brualdi and Ryser [14]). The problem also has a simple Diophantine equation formulation.

The bijective mapping to $(0-1)$−matrices is clear from the bijective map (7.26) and the proof of (7.20):

$$\operatorname{Str}X\binom{\lambda}{m} = \begin{pmatrix} X_n \\ \vdots \\ X_2 \\ X_1 \end{pmatrix} = \begin{pmatrix} A_n - A_{n-1} \\ \vdots \\ A_2 - A_1 \\ A_1 \end{pmatrix}, X_i = e_{m_{1,i}} + e_{m_{2,i}} + \cdots + e_{m_{i,i}},$$

$$(7.101)$$

is a $(0-1)$−matrix in consequence of the strict conditions

$$\mathbf{m}_i = (m_{1,i}, m_{2,i}, \cdots, m_{i,i}), \; m_{1,i} > m_{2,i} > \cdots > m_{i,i} \quad (7.102)$$

on each of its rows $i = 1, 2, \ldots, n$. We recall that each e_j is a unit row matrix with n parts with 1 in part j and 0 elsewhere. Thus, if use the notation $m_{i,j} \in \binom{\lambda}{m}$ to denote that the nonnegative integer $m_{i,j}$ is part j of the element in row i of the GT pattern $\binom{\lambda}{m}$, then we have that the elements of the matrix $\mathrm{Str}X\binom{\lambda}{m}$ are given by

$$x_{ij} = \begin{cases} 1, & \text{if } j \in \mathbf{m}_{n-i+1}, \\ 0, & \text{otherwise.} \end{cases} \quad (7.103)$$

We call $\mathrm{Str}X\binom{\lambda}{m}$ the *incidence matrix* of the GT pattern $\binom{\lambda}{m}$. Its row and column elements x_{ij} are indexed by $i = 1, 2, \ldots, n; j = 1, 2, \ldots, \lambda_1$.

We denote the set of all $n \times \lambda_1$ incident matrices defined above by $\mathrm{Str}\mathbb{X}_\lambda$; it is bijective with the set $\mathrm{Str}\,\mathbb{G}_\lambda$ of strict GT patterns, as we next show:

The 1's in the matrix have the following distribution in this $n \times \lambda_1$ $(0-1)-$ matrix:

row n and column $m_{1,1}$;

row $n-1$ and columns $m_{1,2}, m_{2,2}$;

$$\vdots \qquad \vdots \qquad (7.104)$$

row 2 and columns $m_{1,n-1}, m_{2,n-1}, \ldots, m_{n-1,n-1}$;

row 1 and columns $\lambda_1, \lambda_1, \ldots, \lambda_n$.

It follows that the incidence matrix $X \in \mathrm{Str}\,\mathbb{X}_\lambda$ has row and column linesums r_X and c_X in row 1 (top), row 2, $\ldots$, row n (bottom) and column 1 (left), column 2, $\ldots$, column λ_1 (right) given by

$$r_X = (n,, n-1, \ldots, 1), \quad c_X = (c_1, c_2, \ldots, c_{\lambda_1}),$$
$$c_j = \text{total number of } j's \text{ that appear as entries} \quad (7.105)$$
in the strict GT pattern $\mathrm{Str}\binom{\lambda}{m}$.

The inverse problem as to when a $(0-1)-$matrix is the adjacency matrix of a strict GT pattern can be stated as follows: Let $\lambda = (\lambda_1, \lambda_2, \ldots, \lambda_n)$ be an arbitrary strict partition. Let X be an $n \times \lambda_1$ $(0-1)-$matrix with n 1's in row 1; $n-1$ 1's in row 2; $\ldots$; one 1 in row n (ordered $1, 2, \ldots, n$ from top-to-bottom in the usual way). Then, $X \in \mathbb{X}_\lambda$, if and only if the row index i of x_{ij} and the $n-i+1$ column indices j where the 1's occur, as read from right-to-left, hence, $j_1 > j_2 > \cdots > j_{n-i+1}$, give rows defined by $\mathbf{m}_i = (j_1, j_2, \ldots, j_{n-i+1})$, $i = 1, 2, \ldots, n$, $\mathbf{m}_n = \lambda$, such that the following conditions are satisfied:

Rows $\mathbf{m}_i$, $i = 1, 2, \ldots, n$, constitute the n rows

of a **lexical** strict GT pattern $\mathrm{Str}\binom{\lambda}{m}$. $\quad (7.106)$

7.6. THE INCIDENCE MATRIX FORMULATION

The set of incident matrices $X \in \text{Str}\mathbb{X}_\lambda$ can also be found by determining all possible solutions over the integers $\{0, 1\}$ of the following linear system of Diophantine equations:

$$\sum_{j=1}^{\lambda_1} x_{ij} = n - i + 1, \quad i = 1, 2, \ldots, n. \tag{7.107}$$

But all solutions that do not satisfy the criteria (7.106) for a lexical strict GT pattern by the inverse read-out procedure must be rejected. Thus, the simplicity of the Diophantine relations (7.107) is compounded by the complexity of the constraining conditions (7.106) for a solution of the desired problem. This behavior is, perhaps, to be expected, considering that the complexity of the original problem (7.2)-(7.4) is in the multiplying factor. Nonetheless, the procedure of solving (7.107) is fully defined. We denote the set of all solutions of the constrained Diophantine equations, (7.107) subject to conditons (7.106), over $\{0, 1\}$ by $\mathbb{D}_\lambda$.

Examples. $n = 2, \lambda = (2\,1); n = 3, \lambda = (3\,2\,1)$.

(i). $n = 2, \lambda = (2\,1)$: The solutions of the two Diophantine equations in four unknowns over $\{0, 1\}$ and their maps to GT patterns are:

$$\begin{pmatrix} 1 & 1 \\ 1 & 0 \end{pmatrix} \mapsto \begin{pmatrix} 2 & & 1 \\ & 1 & \end{pmatrix}, \quad \begin{pmatrix} 1 & 1 \\ 0 & 1 \end{pmatrix} \mapsto \begin{pmatrix} 2 & & 1 \\ & 2 & \end{pmatrix}. \tag{7.108}$$

There is no redundancy.

(ii). $n = 3, \lambda = (3\,2\,1)$: The solutions of the three Diophantine equations in nine unknowns over $\{0, 1\}$ and their maps to GT patterns are:

$$\begin{pmatrix} 1 & 1 & 1 \\ 1 & 1 & 0 \\ 1 & 0 & 0 \end{pmatrix} \mapsto \begin{pmatrix} 3 & & 2 & & 1 \\ & 2 & & 1 & \\ & & 1 & & \end{pmatrix}, \quad \begin{pmatrix} 1 & 1 & 1 \\ 1 & 1 & 0 \\ 0 & 1 & 0 \end{pmatrix} \mapsto \begin{pmatrix} 3 & & 2 & & 1 \\ & 2 & & 1 & \\ & & 2 & & \end{pmatrix},$$

$$\begin{pmatrix} 1 & 1 & 1 \\ 1 & 1 & 0 \\ 0 & 0 & 1 \end{pmatrix} \mapsto \begin{pmatrix} 3 & & 2 & & 1 \\ & 2 & & 1 & \\ & & 3 & & \end{pmatrix};$$

$$\begin{pmatrix} 1 & 1 & 1 \\ 1 & 0 & 1 \\ 1 & 0 & 0 \end{pmatrix} \mapsto \begin{pmatrix} 3 & & 2 & & 1 \\ & 3 & & 1 & \\ & & 1 & & \end{pmatrix}, \quad \begin{pmatrix} 1 & 1 & 1 \\ 1 & 0 & 1 \\ 0 & 1 & 0 \end{pmatrix} \mapsto \begin{pmatrix} 3 & & 2 & & 1 \\ & 3 & & 1 & \\ & & 2 & & \end{pmatrix},$$

$$\begin{pmatrix} 1 & 1 & 1 \\ 1 & 0 & 1 \\ 0 & 0 & 1 \end{pmatrix} \mapsto \begin{pmatrix} 3 & & 2 & & 1 \\ & 3 & & 1 & \\ & & 3 & & \end{pmatrix};$$

$$\begin{pmatrix} 1 & 1 & 1 \\ 0 & 1 & 1 \\ 1 & 0 & 0 \end{pmatrix} \mapsto \begin{pmatrix} 3 & & 2 & & 1 \\ & 3 & & 2 & \\ & & 1 & & \end{pmatrix}, \quad \begin{pmatrix} 1 & 1 & 1 \\ 0 & 1 & 1 \\ 0 & 1 & 0 \end{pmatrix} \mapsto \begin{pmatrix} 3 & & 2 & & 1 \\ & 3 & & 2 & \\ & & 2 & & \end{pmatrix},$$

$$\begin{pmatrix} 1 & 1 & 1 \\ 0 & 1 & 1 \\ 0 & 0 & 1 \end{pmatrix} \mapsto \begin{pmatrix} 3 & & 2 & & 1 \\ & 3 & & 2 & \\ & & 3 & & \end{pmatrix}. \qquad \square \tag{7.109}$$

Two patterns in this set, the third and seventh, are nonlexical and are rejected, leaving behind $d_3 = 7$ acceptable solutions.

Given the incident matrix $X\binom{\lambda}{m} \in \text{Str} \, \mathbb{X} \lambda$, the matrix $A\binom{\lambda}{m} \in \text{Str}\mathbb{G}_\lambda$ in the original mapping (7.20) of strict GT patterns is *uniquely* given by

$$A\binom{\lambda}{m} = \begin{pmatrix} X_n - X_{n-1} \\ \vdots \\ X_2 - X_1 \\ X_1 \end{pmatrix}. \tag{7.110}$$

We have thus established the following equalities of cardinalities:

$$|\text{Str} \, \mathbb{G}_\lambda| = |\text{Str} \, \mathbb{A}_\lambda| = |\text{Str} \, \mathbb{X}_\lambda| = |\mathbb{D}_\lambda|. \tag{7.111}$$

The counting formula for these cardinalities is unknown for the case of a general strict λ. The specialization to $\lambda = (n \; n-1 \; \cdots \; 1)$ gives, of course, the result

$$\begin{aligned} |\text{Str} \, \mathbb{G}_{n \; n-1 \; \cdots \; 1}| &= |\text{Str} \, \mathbb{A}_{n \; n-1 \; \cdots \; 1}| \\ &= |\text{Str} \, \mathbb{X}_{n \; n-1 \; \cdots \; 1}| = |\mathbb{D}_{n \; n-1 \; \cdots \; 1}| = d_n. \end{aligned} \tag{7.112}$$

Example. We give one additional example of the relation of the mapping from strict GT patterns for $\text{Str}\mathbb{G}_{n \; n-1 \; \cdots \; 1}$ to the corresponding incident matrix to the corresponding alternating sign matrix to illustrate the simplicity of the rules: First (7.101), and then the inverse rule (7.106) as a check, followed by the difference rule (7.110):

$$\begin{pmatrix} 5 & 4 & 3 & 2 & 1 \\ & 5 & 4 & 2 & 1 \\ & & 4 & 3 & 1 \\ & & & 3 & 2 \\ & & & & 2 \end{pmatrix} \mapsto \begin{pmatrix} 1 & 1 & 1 & 1 & 1 \\ 1 & 1 & 0 & 1 & 1 \\ 1 & 0 & 1 & 1 & 0 \\ 0 & 1 & 1 & 0 & 0 \\ 0 & 1 & 0 & 0 & 0 \end{pmatrix} \mapsto \begin{pmatrix} 0 & 0 & 1 & 0 & 0 \\ 0 & 1 & -1 & 0 & 1 \\ 1 & -1 & 0 & 1 & 0 \\ 0 & 0 & 1 & 0 & 0 \\ 0 & 1 & 0 & 0 & 0 \end{pmatrix}. \tag{7.113}$$

Remarks. We have included here the results on $(0-1)$-matrices to indicate yet another viewpoint of the counting role of the Andrews-Zeilberger numbers d_n — one that affords perhaps still another way for their derivation from these fundamental combinatorial objects. We point out again that the work of Andrews [2], Kuperberg [38], Mills *et al.* [56], Robbins *et al.* [66], and, perhaps, others give still other perspectives. The viewpoints presented here on GT patterns also serve that role.

Chapter 8

The Heisenberg Magnetic Ring

8.1 Introduction

Few problems in quantum physics have received more attention than the one relating to ferromagnetism addressed by Bethe in his paper in 1931 (Bethe [5]). The model Hamiltonian H for a class of such problem, called the Heisenberg ring, is an Hermitian operator that relates to the components of n angular momenta in a very special way. As required of such Hamiltonians, it is an invariant under the action of the diagonal subgroup of the direct product of n copies of the unitary group $SU(2)$, as given explicitly below. This action, in turn, is a consequence of the requirement that all Hamiltonians of composite physical systems in physical 3-space, $\mathbb{R}^3$, must be invariant under $SU(2)$ frame rotations, as described in detail in [L]. We take the Hamiltonian for the Heisenberg ring to be (see Lulek [48])

$$H = \sum_{k=1}^{n} \mathbf{J}(k) \cdot \mathbf{J}(k+1), \qquad (8.1)$$

where, by definition, the cyclic condition $\mathbf{J}(n+1) = \mathbf{J}(1)$ is imposed, and $\mathbf{J}(k) = (J_1(k), J_2(k), J_3(k))$ is the quantum angular momentum of the k-th system in Cartesian space $\mathbb{R}^3$, where the systems are arranged at equally spaced angular points on a circular ring, which is the reason for identifying $\mathbf{J}(n+1) = \mathbf{J}(1)$. The dot product is defined by

$$\mathbf{J}(k) \cdot \mathbf{J}(k+1) = \sum_{i=1}^{3} J_i(k) J_i(k+1). \qquad (8.2)$$

It is customary to diagonalize H on the tensor product vector space $\mathcal{H}_{j_1} \otimes \mathcal{H}_{j_1} \otimes \cdots \otimes \mathcal{H}_{j_n}$ with orthonormal basis set $|\mathbf{j}\,\mathbf{m}\rangle$ on which the individual angular momenta $\mathbf{J}(k)$, $k = 1, 2, \ldots, n$, have the standard action (see Chapters 1 and 5); this basis is fully defined (up to normalization phase conventions) as the simultaneous eigenvectors of the complete set of $2n$ commuting Hermitian angular momentum operators $\mathbf{J}^2(k), J_3(k), k = 1, 2, \ldots, n$. But the problem can equally well be formulated from the viewpoint that we are dealing with a composite system in which the total angular momentum

$$\mathbf{J} = \mathbf{J}(1) + \mathbf{J}(2) + \cdots + \mathbf{J}(n) \qquad (8.3)$$

is conserved; that is, we view the system as composite at the outset.

The total angular momentum $\mathbf{J}$ commutes with the Hamiltonian H, as it must, because H is an invariant under the $SU(2)$–rotational symmetry generated by $\mathbf{J}$, which is equivalent to the viewpoint of SU(2) frame rotations:

$$[H, \mathbf{J}] = 0^{op} \quad (0 \text{ operator}). \qquad (8.4)$$

Moreover, H also commutes with the square of each of the constituent angular momenta:

$$[H, \mathbf{J}^2(k)] = 0^{op},\ k = 1, 2, \ldots, n. \qquad (8.5)$$

Thus, it is natural to consider the diagonalization of H in the basis corresponding to a complete set of mutually commuting Hermitian angular momentum operators that contains the $n+2$ operators $\mathbf{J}^2(k), \mathbf{J}^2, J_3, k = 1, 2, \ldots n,$; that is, the basis corresponding to the binary coupled states developed at length in [L] and reviewed here in Chapters 1 and 5: *The full apparatus of binary coupling theory comes into play.*

It is also the case that the Hamiltonian H is invariant under the action of the cyclic subgroup $C_n \subset S_n$ of the symmetric group S_n defined by the set of n cyclic permutations:

$$C_n = \{g_i = (i, i+1, \ldots, n, 1, 2, \ldots, i-1)|\, i = 1, 2, \ldots, n\}, \qquad (8.6)$$

where $g_1 = (1, 2, \ldots, n) = e =$ identity, and we use the one-line notation for a permutation $\pi = (\pi_1, \pi_2, \ldots, \pi_n)$ of the integers $1, 2, \ldots, n$; that is, for the substitution $1 \mapsto \pi_1, 2 \mapsto \pi_2, \ldots, n \mapsto \pi_n$. The group C_n is an abelian group whose elements are the n cyclic permutations of the identity permutation.

The problem thus posed is to determine the eigenvector-eigenvalue states of the Hamiltonian H on binary coupled angular momentum states such that these eigenvectors also transform irreducibly under the action of the cyclic group.

A very detailed procedure is required for putting in place the steps necessary for effecting the calculations outlined above. It is useful to

8.1. INTRODUCTION

recall briefly the angular momentum structures involved, and to lay out the course of action.

1. Complete set of mutually commuting Hermitian operators. A set of mutually commuting Hermitian operators $H_1, H_2, \ldots, H_n$ is complete on the Hilbert space $\mathcal{H}$, if and only if the set of simultaneous normalized eigenvectors is an orthonormal basis of $\mathcal{H}$.

2. Coupling schemes. There are many binary coupling schemes containing the set $\mathbf{J}^2(k), \mathbf{J}^2, J_3, k = 1, 2, \ldots n$, of $n+2$ mutually commuting Hermitian operators as a subset, each such set being completed by $n-2$ Hermitian operators that are the squares of the $n-2$ intermediate angular momenta $\mathbf{K}_T^2(i), i = 1, 2, \ldots, n-2$, associated with a given binary tree $T \in \mathbb{T}_n$, as discussed in detail in Chapter 5. The main point is: H can be diagonalized on any binary coupling scheme that one chooses; the completeness of the set of mutually commuting Hermitian operators defining the coupling scheme means that each sequence of angular momentum quantum numbers $(j_1, j_2, \ldots, j_n)$, the total angular momentum quantum number j, together with the $n-2$ quantum numbers $(k_1, k_2, \ldots, k_{n-2})$ of the $n-2$ intermediate (squared) angular momenta determine a unique Hermitian matrix representation of H of order $N_j(\mathbf{j})$ for each j.

3. Exact solutions $n = 2, 3, 4$. The spectrum of H can be given exactly for these cases. We identify in detail the eigenvector-eigenvalue relations, the associated binary coupled states, and the associated WCG, $6-j$ and $9-j$ coefficients.

4. Structure of all n even solutions ($n \geq 6$). For n even, the Heisenberg ring Hamiltonian H can be written as the sum of two parts, $H = H_1 + H_2$, where H_1 and H_2 are diagonal, respectively, on two distinct binary coupling schemes, called scheme I and scheme II. It is shown how this structure can be used to diagonalize H itself.

5. Structure of all n odd solutions ($n \geq 5$). For n odd, the Heisenberg ring Hamiltonian H has a structure different than that for n even; H is now written as the sum of three parts, $H = H_1 + H_2 + H_3$, where H_1, H_2, H_3 are diagonal, respectively, on three distinct binary coupling schemes, called scheme I_o, scheme II_o, and scheme III_o. It is shown how this structure can be used to diagonalize H itself.

6. Implementation of Items 4 and 5. This requires a great deal of development on the shapes and paths associated with the irreducible triangle coefficients of order $2n-2$ corresponding to the various coupling schemes, which, in turn, determine the $3(n-1)-j$ coefficients that relate the coupling schemes in question, it being these coefficients that enter into the diagonalization of H itself. The methods used are just those as summarized in Chapter 5, Sect. 5.8.8.

7. Implementation of the action of the cyclic group. The final step entails the determination of the simultaneous eigenvectors of H that takes into account the invariance of H under the action of the cyclic group C_n.

The purpose of the present chapter is to put in place all of the steps above, thus demonstrating the composite system approach to the solution of the Heisenberg ring problem by the use of the binary theory of addition of angular momenta. As seen from the outline above, this is a sizeable task, especially with respect to the details required. From our review of the published literature, it appears that there are some structural aspects of the Heisenberg problem still to be learned from this viewpoint. This paper is mathematical in its approach, and no attempt has been made to relate it to the abundance of physically oriented approaches. This is a nontrivial task, requiring, perhaps, a lengthy review by experts in the field; it will not be attempted here.

The Bethe Ansatz method for addressing the Heisenberg problem (and generalizations), beginning with the historical work of Hulthén [31], is the subject of hundreds of papers. Forty some papers appear in the proceedings of biennial conferences in Poland entitled *Symmetry and Structural Properties of Condensed Matter,* beginning in 1990. Many of these originate with W. J. Caspers [18], T. Lulek [48], and their students and associates. Because most of these papers do not relate directly to the methods of the binary coupling theory of angular momenta developed here, we reference the general proceedings of the conferences themselves [77], giving only those selected author references that suit our needs. It is, however, in the same spirit, essential to bring attention to the combinatorial approach to the Bethe Ansatz based on Young tableaux as described in Kirillov and Reshetikhin [33], and references therein.

8.2 Matrix Elements of H in the Uncoupled and Coupled Bases

We repeat here from Chapter 5, for the convenience of the reader, and for the purpose of putting the Heisenberg magnetic ring Hamiltonian in the full context of the binary coupling theory of n angular momenta based on the structure of binary trees and their shapes, many of the relevant results. This also helps give the present chapter some semblance of independence from the previous Chapters.

The set of $2n$ mutually commuting Hermitian operators

$$\mathbf{J}^2(1), J_3(1), \mathbf{J}^2(2), J_3(2), \ldots, \mathbf{J}^2(n), J_3(n) \tag{8.7}$$

is a complete set of operators in the tensor product space $\mathcal{H}_\mathbf{j}$ in that

8.2. MATRIX ELEMENTS OF H

the set of simultaneous eigenvectors $|\mathbf{j\,m}\rangle, \mathbf{m} \in \mathbb{C}(\mathbf{j})$ is an orthonormal basis; that is, there is no degeneracy left over. The action of the angular momentum operators $\mathbf{J}(i), i = 1, 2, \ldots, n$, is the standard action:

$$\mathbf{J}^2(i)|\mathbf{j\,m}\rangle = j_i(j_i+1)|\mathbf{j\,m}\rangle,$$
$$J_3(i)|\mathbf{j\,m}\rangle = m_i|\mathbf{j\,m}\rangle,$$
$$J_\pm(i)|\mathbf{j\,m}\rangle = \sqrt{(j_i \mp m_i)(j_i \pm m_i + 1)}\,|\mathbf{j\,m}_{\pm 1}(i)\rangle, \qquad (8.8)$$
$$\mathbf{m}_{\pm 1}(i) = (m_1, \ldots, m_i \pm 1, \ldots, m_n).$$

The orthonormality of the basis functions is expressed by

$$\langle \mathbf{j\,m} | \mathbf{j\,m}' \rangle = \delta_{\mathbf{m},\mathbf{m}'}, \text{ each pair } \mathbf{m}, \mathbf{m}' \in \mathbb{C}(\mathbf{j}). \qquad (8.9)$$

This basis $\mathbf{B_j}$ of the tensor product space $\mathcal{H}_\mathbf{j} = \mathcal{H}_{j_1} \otimes \mathcal{H}_{j_2} \otimes \cdots \otimes \mathcal{H}_{j_n}$ of dimension $N(\mathbf{j}) = \prod_{i=1}^n (2j_i + 1)$ is the coupled basis.

The Hamiltonian H commutes with the n squared angular momentum operators $\mathbf{J}^2(k), k = 1, 2, \ldots, n$. Hence, it must be possible to simultaneously diagonal the $2n$ Hermitian operators (8.7) and H. Moreover, the matrix elements of H on the orthonormal basis $\mathbf{B_j}$ are given by

$$\langle \mathbf{j\,m}'|H|\mathbf{j\,m}\rangle = \sum_{k=1}^{n}\sum_{i=1}^{3} \langle \mathbf{j\,m}' | J_i(k)J_i(k+1)|\mathbf{j\,m}\rangle. \qquad (8.10)$$

These matrix elements are fully determined from the standard action (8.8) of the various angular momentum operators. Thus, the problem in the uncoupled basis is to diagonalize the Hermitian matrix of order $\prod_{i=1}^n (2j_i + 1)$ with elements given by (8.10). This is the standard formulation of the problem addressed by Bethe, and many others, over the years, following his lead and methods.

But it is clear that we can also formulate the Heisenberg problem in terms of the subspaces of $\mathcal{H}_\mathbf{j}$ such that the action of the total angular momentum $\mathbf{J}$ is irreducible and standard, maintaining still the diagonal form of the square $\mathbf{J}^2(k), k = 1, 2, \ldots, n$ of the n constituent angular momenta. Before considering the role of binary coupling of angular momenta, it is useful to view the problem in this more general setting.

The $SU(2)$-rotational symmetry of the Hamiltonian H implies that there exist simultaneous eigenvectors $|\mathbf{j}; j\,m\rangle, m = j, j-1, \ldots, -j$ such that the following relations are satisfied:

$$\mathbf{J}^2|\mathbf{j}; j\,m\rangle = j(j+1)|\mathbf{j}; j\,m\rangle, \quad J_3|\mathbf{j}; j\,m\rangle = m|\mathbf{j}; j\,m\rangle,$$
$$J_\pm|\mathbf{j}, j\,m\rangle = \sqrt{(j \mp m)(j \pm m + 1)}|\mathbf{j}; j\,m \pm 1\rangle,$$
(8.11)
$$\mathbf{J}^2(i)|\mathbf{j}; j\,m\rangle = j_i(j_i + 1)|\mathbf{j}; j\,m\rangle, \quad i = 1, 2, \ldots, n;$$
$$H|\mathbf{j}; j\,m\rangle = \lambda_j(\mathbf{j})|\mathbf{j}; j\,m\rangle.$$

Since H is an $SU(2)$-rotational invariant, its eigenvalues are independent of m. The notation for these simultaneous eigenvectors is, of course, incomplete; they should carry an extra symbol, as in (1.20)-(1.21), to designate there are many such sets of simultaneous orthonormal eigenvectors for given angular momenta $\mathbf{j} = (j_1, j_2, \ldots, j_n)$ and for given $j = j_{\min}, j_{\min} + 1, \ldots, j_{\max}$ that satisfy these relations. We suppress these labels, but they must always be kept in mind. The problem, then, is to determine all eigenvalues $\lambda_j(\mathbf{j})$ and eigenvectors $|\mathbf{j}; j\,m\rangle$ that satisfy these relations. The existence of such a set of eigenvectors is, of course, assured because we are dealing with a set of mutually commuting Hermitian operators: There must exist a unitary transformation of the basis $\mathbf{B_j}$ that gives all such eigenvectors satisfying relations (8.11), where the eigenvalue $\lambda_j(\mathbf{j})$ is real. The special feature of relations (8.11) is that the total angular momentum has the standard action on the eigenvectors, and, most significantly: *the Heisenberg ring magnet is treated as a composite whole*. This signifies that the tensor product space $\mathcal{H}_\mathbf{j}$ has been decomposed into a direct sum of spaces on which the action of the total angular momentum is irreducible and standard. Independently of how this is effected, the following direct sum relation must hold:

$$\mathcal{H}_\mathbf{j} = \sum_{j=j_{\min}}^{j_{\max}} \oplus N_j(\mathbf{j}) \mathcal{H}_{\mathbf{j};j}, \qquad (8.12)$$

where $\mathcal{H}_{\mathbf{j};j}$ is the vector space with orthonormal basis $\{|\mathbf{j}, j\,m\rangle\,|\,m = j, j-1, \ldots, -j\}$ of dimension $2j+1$ on which the total angular momentum has the standard irreducible action (8.11), and $N_j(\mathbf{j})$ is the Clebsch-Gordan (CG) number giving the number of times each such irreducible space occurs. Relation (8.12) is accompanied by

$$\prod_{i=1}^n (2j_i + 1) = \sum_j N_j(\mathbf{j})(2j+1), \qquad (8.13)$$

which expresses the equality of dimensions of the vector spaces in (8.12). We refer to any set of orthonormal basis vectors satisfying relations (8.11) as a *total angular momentum basis*.

8.2. MATRIX ELEMENTS OF H

It is important to show how the square of the total angular momentum operator $\mathbf{J}^2$ relates to the $\mathbf{J}^2(k)$, $k = 1, 2, \ldots, n$, and to H. Thus, the square of $\mathbf{J}^2 = (\mathbf{J}(1) + \cdots + \mathbf{J}(n)) \cdot (\mathbf{J}(1) + \cdots + \mathbf{J}(n))$ gives the following relation:

$$\mathbf{J}^2 - (\mathbf{J}^2(1) + \mathbf{J}^2(2) + \cdots + \mathbf{J}^2(n)) = 2H + 2L, \quad (8.14)$$

where the operator L is defined as the sum of dot products given by

$$L = \sum_{(k,l) \in \mathbb{L}} \mathbf{J}(k) \cdot \mathbf{J}(l). \quad (8.15)$$

The indexing set $\mathbb{L}$ is given by

$$\begin{aligned}\mathbb{L} &= \{(i,j) \mid 1 \leq i < j \leq n\} - \{(1,2), (2,3), \ldots, (n-1,n), (1,n)\} \\ &= \{(1,r), (2, r+1), \ldots, (n-r+1, n) \mid 3 \leq r \leq n-1\}, \quad (8.16)\end{aligned}$$

where $\mathbb{L} = \emptyset$ for $n = 2, 3$. Thus, it is the case that the diagonalization of the set of operators (8.11) also diagonalizes L. While we make no use of this fact, it seems worth pointing out. We also note that relations between dot products of angular momenta and added pairs of angular momenta $\mathbf{J}(k\,l) = \mathbf{J}(k) + \mathbf{J}(l), k \neq l$, can be transformed to various combinations by using the identity

$$\mathbf{J}^2(k\,l) = \mathbf{J}^2(k) + \mathbf{J}^2(l) + 2\mathbf{J}(k) \cdot \mathbf{J}(l), k \neq l. \quad (8.17)$$

However, coupled angular momenta that share a common index do not commute, since

$$[\mathbf{J}(k\,l), \mathbf{J}(k\,h)] = i\mathbf{J}(k) \cdot (\mathbf{J}(l) \times \mathbf{J}(h)), \; l \neq h. \quad (8.18)$$

The problem of diagonalizing H (hence, also K) can be reduced to that of diagonalizing H on the space $\mathcal{H}_{\mathbf{j};j}$ of dimension $2j+1$, for each $j = j_{\min}, j_{\min}+1, \ldots, j_{\max}$, but to effect this requires the introduction of the additional quantum labels $\boldsymbol{\alpha}$ needed to complete the mutually commuting Hermitian operator set $\mathbf{J}^2(k), \mathbf{J}^2, J_3$ in the total angular momentum basis $|\mathbf{j}; jm\rangle$ in (8.11). This role is, of course, filled by the binary coupled states and their associated intermediate angular momenta, as we next recall from relations (5.200)-(5.218):

The general orthonormal coupled states for each $T \in \mathbb{T}_n$ are:

$$|T(\mathbf{j}\,\mathbf{k})_{j\,m}\rangle = \sum_{\mathbf{m} \in \mathbb{C}_T(\mathbf{j})} C_{T(\mathbf{j}\,\mathbf{m};\mathbf{k})_{j\,m}} |\mathbf{j}\,\mathbf{m}\rangle, \text{ each } \mathbf{k}, j, m \in \mathbb{R}_T(\mathbf{j}); \quad (8.19)$$

$$\langle T(\mathbf{j}\,\mathbf{k})_{j\,m} | T(\mathbf{j}\,\mathbf{k}')_{j'\,m'}\rangle = \delta_{\mathbf{k},\mathbf{k}'} \delta_{j,j'} \delta_{m,m'}. \quad (8.20)$$

The eigenvector-eigenvalue relations and standard action are given by the relations:

$$\mathbf{J}^2(i)|T(\mathbf{j\,k})_{j\,m}\rangle = j_i(j_i+1)|T(\mathbf{j\,k})_{j\,m}\rangle,\ i=1,2,\ldots,n,$$
$$\mathbf{K}_T^2(i)|T(\mathbf{j\,k})_{j\,m}\rangle = k_i(k_i+1)|T(\mathbf{j\,k})_{j\,m}\rangle,\ i=1,2,\ldots,n-2, \quad (8.21)$$
$$\mathbf{J}^2|T(\mathbf{j\,k})_{j\,m}\rangle = j(j+1)|T(\mathbf{j\,k})_{j\,m}\rangle;$$

$$J_3|T(\mathbf{j\,k})_{j\,m}\rangle = m|T(\mathbf{j\,k})_{j\,m}\rangle,$$
$$J_+|T(\mathbf{j\,k})_{j\,m}\rangle = \sqrt{(j-m)(j+m+1)}\,|T(\mathbf{j\,k})_{j\,m+1}\rangle, \quad (8.22)$$
$$J_-|T(\mathbf{j\,k})_{j\,m}\rangle = \sqrt{(j+m)(j-m+1)}\,|T(\mathbf{j\,k})_{j\,m-1}\rangle.$$

For given $(\mathbf{j},j)$, the domain of definition of the intermediate quantum labels $\mathbf{k}=(k_1,k_2,\ldots,k_{n-2})$ is:

$$\mathbf{k}\in\mathbb{K}_T^{(j)}(\mathbf{j}) \quad (8.23)$$
$$=\left\{(k_1,k_2,\ldots,k_{n-1})\,\Big|\,\begin{array}{l}\text{each column of the triangle coefficient}\\ \Delta_T(\mathbf{j\,k})_j \text{ is a triangle}\end{array}\right\}.$$

The total angular momentum state vector $|\mathbf{j};j\,m\rangle$ in (8.11) on which H is diagonal can at most be a linear combination of the coupled state vectors $|T(\mathbf{j\,k})_{j\,m}\rangle$ for all $T\in\mathbb{T}_n$ and $\mathbf{k}\in\mathbb{K}_T^{(j)}(\mathbf{j})$; that is, there must exist complex numbers $A_T(\mathbf{j\,k})$ such that

$$|\mathbf{j};j\,m\rangle = \sum_{T\in\mathbb{T}_n}\sum_{\mathbf{k}\in\mathbb{K}_T^{(j)}(\mathbf{j})} A_T(\mathbf{j\,k})\,|T(\mathbf{j\,k})_{j\,m}\rangle. \quad (8.24)$$

Before addressing the general problem thus posed, we give the results for $n=2,3,4$, in which cases *the exact solution for the spectrum of H can be obtained using only elementary angular momentum theory.*

8.3 Exact Solution of the Heisenberg Ring Magnet for $n=2,3,4$

The above relations in a total angular momentum description for $n=2,3,4$ are summarized as follows:

$$n=2: H = \mathbf{J}^2 - \mathbf{J}^2(1) - \mathbf{J}^2(2)$$
$$= \left(\mathbf{J}(1)+\mathbf{J}(2)\right)\cdot\left(\mathbf{J}(1)+\mathbf{J}(2)\right) - \mathbf{J}^2(1) - \mathbf{J}^2(2)$$
$$= \mathbf{J}(1)\cdot\mathbf{J}(2) + \mathbf{J}(2)\cdot\mathbf{J}(1) = 2\mathbf{J}(1)\cdot\mathbf{J}(2). \quad (8.25)$$

8.3. EXACT SOLUTION

$$n = 3 : 2H = \mathbf{J}^2 - \mathbf{J}^2(1) - \mathbf{J}^2(2) - \mathbf{J}^2(3)$$
$$= \Big(\mathbf{J}(1) + \mathbf{J}(2) + \mathbf{J}(3)\Big) \cdot \Big(\mathbf{J}(1) + \mathbf{J}(2) + \mathbf{J}(3)\Big)$$
$$- \mathbf{J}^2(1) - \mathbf{J}^2(2) - \mathbf{J}^2(3) \quad (8.26)$$
$$= 2\Big(\mathbf{J}(1) \cdot \mathbf{J}(2) + \mathbf{J}(2) \cdot \mathbf{J}(3) + \mathbf{J}(3) \cdot \mathbf{J}(1)\Big).$$

$$n = 4 : 2H = \mathbf{J}^2 - \mathbf{J}^2(1) - \mathbf{J}^2(2) - \mathbf{J}^2(3) - \mathbf{J}^2(4) = 2\mathbf{K}'(1) \cdot \mathbf{K}'(2)$$
$$= 2\Big(\mathbf{J}(1) \cdot \mathbf{J}(2) + \mathbf{J}(2) \cdot \mathbf{J}(3) + \mathbf{J}(3) \cdot \mathbf{J}(4) + \mathbf{J}(4) \cdot \mathbf{J}(1)\Big)$$
$$= \mathbf{J}^2 - \mathbf{K}'^2(1) - \mathbf{K}'^2(2); \quad (8.27)$$
$$\mathbf{K}'(1) = \mathbf{J}(1) + \mathbf{J}(3), \quad \mathbf{K}'(2) = \mathbf{J}(2) + \mathbf{J}(4).$$

We next give the exact diagonalization of H for each of the cases above, where the background on the binary coupling scheme needed to effect the diagonal form has been given in Sect. 5.7:

1. Diagonalization of H for $n = 2$: coupled basis. The eigenvectors are:

$$\left| \begin{array}{cc} j_1 & j_2 \\ & \\ & jm \end{array} \right\rangle = |(j_1 j_2) jm\rangle$$
$$= \sum_{m_1, m_2} C^{j_1 \; j_2 \; j}_{m_1 m_2 m} |j_1 m_1\rangle \otimes |j_2 m_2\rangle. \quad (8.28)$$

The eigenvalues of $H = \mathbf{J}(1) \cdot \mathbf{J}(2) + \mathbf{J}(2) \cdot \mathbf{J}(1)$ are:

$$H|(j_1 j_2) jm\rangle$$
$$= (j(j+1) - j_1(j_1+1) - j_2(j_2+1))|(j_1 j_2) jm\rangle, \quad (8.29)$$

where, for given angular momenta $j_1, j_2 \in \{0, 1/2, 1, 3/2, \ldots\}$, the values of j are given by $j \in \{|j_1 - j_2|, |j_1 - j_2| + 1, \ldots, j_1 + j_2\}$.

2. Diagonalization of H for $n = 3$: coupled basis. The eigenvectors are:

$$\left| \begin{array}{cc} j_1 & j_2 \\ k & j_3 \\ jm & \end{array} \right\rangle = |T((j_1 j_2 j_3); k_1)_{jm}\rangle$$

$$= \sum_{m_1, m_2, m_3} C_{T(\mathbf{j}\,\mathbf{m}; k_1)_{jm}} |j_1 m_1\rangle \otimes |j_2 m_2\rangle \otimes |j_3 m_3\rangle, \quad (8.30)$$

The eigenvalues of $H = \mathbf{J}(1) \cdot \mathbf{J}(2) + \mathbf{J}(2) \cdot \mathbf{J}(3) + \mathbf{J}(3) \cdot \mathbf{J}(1)$ are:

$$H \left| \begin{array}{c} j_1 \quad j_2 \\ k \quad j_3 \\ j\,m \end{array} \right\rangle \qquad (8.31)$$

$$= \tfrac{1}{2}\bigl(j(j+1) - j_1(j_1+1) - j_2(j_2+1) - j_3(j_3+1)\bigr) \left| \begin{array}{c} j_1 \quad j_2 \\ k \quad j_3 \\ j\,m \end{array} \right\rangle.$$

The transformation coefficients $C_{T(\mathbf{j}\,m;k)_{j\,m}}$ are the real orthogonal generalized WCG coefficients defined by relation (5.124).

There are two ways to enumerate the orthonormal vectors defined by (8.31). Let $j_1, j_2 \in \{0, 1/2, 1, \ldots\}$: (i) Let k assume all values $k \in \langle j_1, j_2 \rangle = \{j_1 + j_2, j_1 + j_2 - 1, \ldots, |j_1 - j_2|\}$; then, j all values in $\langle k, j_3 \rangle$; then, m all values in $\{j, j-1, \ldots, -j\}$. The total number of orthonormal vectors enumerated by (8.30) is then $N(j_1, j_2, j_3) = (2j_1+1)(2j_2+1)(2j_3+1)$. (ii) Preselect $j \in \{j_{\min}, j_{\min}+1, \ldots, j_{\max}\}$ and $m \in \{j, j-1, \ldots, -j\}$. Then, k can be any value such that $k \in \langle j_1, j_2 \rangle$ and $k \in \langle j_3, j \rangle$; that is, $k \in \mathbb{K}_T^{(j)}(\mathbf{j})$, where

$$\mathbb{K}_T^{(j)}(\mathbf{j}) = \{k \mid k \in \langle j_1, j_2 \rangle \cap \langle j_3, j \rangle\}. \qquad (8.32)$$

The cardinality of this set is then the CG number $N_j(j_1, j_2, j_3)$, which is the total number of orthonormal vectors enumerated by (8.31) under these constraints.

All other irreducible binary coupling schemes for $n = 3$ can also be used to diagonalize H; these give, of course, the same set of eigenvalues. The Racah coefficients relate all other irreducible binary coupling schemes to the one given by (8.30). For example, we have the relations between coupled state vectors and triangle coefficients as follows:

$$\left| \begin{array}{c} j_2 \quad j_3 \\ k' \quad j_1 \\ j\,m \end{array} \right\rangle = \sum_{k \in \mathbb{K}_T^{(j)}(\mathbf{j})} \left\{ \begin{array}{cc|cc} j_1 & k & j_2 & k' \\ j_2 & j_3 & j_3 & j_1 \\ k & j & k' & j \end{array} \right\} \left| \begin{array}{c} j_1 \quad j_2 \\ k \quad j_3 \\ j\,m \end{array} \right\rangle,$$

$$(8.33)$$

$$\left\{ \begin{array}{cc|cc} j_1 & k & j_2 & k' \\ j_2 & j_3 & j_3 & j_1 \\ k & j & k' & j \end{array} \right\} = (-1)^{j_1+k'-j}\sqrt{(2k+1)(2k'+1)}\,W(j_1 j_2 j j_3; k k').$$

The action of H on the vector (8.33) is the same as that in (8.31).

8.3. EXACT SOLUTION

3. Diagonalization of H for $n = 4$: coupled basis:

(i). Eigenvectors:

$$\left| \begin{array}{cccc} j_1 & j_3 & j_2 & j_4 \\ k'_1 & & k'_2 \\ & jm & \end{array} \right\rangle = |T(\mathbf{j}'\, \mathbf{k}')_{jm}\rangle$$

$$= \sum_{m_1,m_2,m_3,m_3} C_{T(\mathbf{j}'\,\mathbf{m}';\mathbf{k}')_{jm}} |j_1\, m_1\rangle \otimes |j_2\, m_2\rangle \otimes |j_3\, m_3\rangle \otimes |j_4\, m_4\rangle,$$
(8.34)

where $\mathbf{j}' = (j_1, j_3, j_2, j_4)$, $\mathbf{m}' = (m_1, m_3, m_2, m_4)$, $\mathbf{k}' = (k'_1, k'_2)$. This basis state vector is the same as the binary coupled vector given by (5.151)-(5.157), where the coupling scheme and the complete set of mutually commuting Hermitian operators are given by (5.151) and (5.154). These relations, together with the eigenvalues of H, are summarized here again because of their importance for the present problem:

(ii). Coupling scheme:

$$\mathbf{J} = \mathbf{K}'(1) + \mathbf{K}'(2), \quad \mathbf{K}'(1) = \mathbf{J}(1) + \mathbf{J}(3), \quad \mathbf{K}'(2) = \mathbf{J}(2) + \mathbf{J}(4), \quad (8.35)$$

$$\mathbf{J}^2, J_3, \mathbf{J}^2(1), \mathbf{J}^2(2), \mathbf{J}^2(3), \mathbf{J}^2(4), \mathbf{K}'^2(1), \mathbf{K}'^2(2). \quad (8.36)$$

(iii). Simultaneous eigenvector-eigenvalues of angular momentum operators and the Hamiltonian H:

$$\begin{aligned} \mathbf{J}^2 |T(\mathbf{j}'\,\mathbf{k}')_{jm}\rangle &= j(j+1)|T(\mathbf{j}'\,\mathbf{k}')_{jm}\rangle, \\ J_3 |T(\mathbf{j}'\,\mathbf{k}')_{jm}\rangle &= m|T(\mathbf{j}'\,\mathbf{k}')_{jm}\rangle, \quad (8.37) \\ \mathbf{J}^2(i)|T(\mathbf{j}'\,\mathbf{k}')_{jm}\rangle &= j_i(j_i+1)|T(\mathbf{j}'\,\mathbf{k}')_{jm}\rangle, \quad i=1,2,3,4, \\ \mathbf{K}'^2(i)|T(\mathbf{j}'\,\mathbf{k}')_{jm}\rangle &= k'_i(k'_i+1)|T(\mathbf{j}'\,\mathbf{k}')_{jm}\rangle, \quad i=1,2. \end{aligned}$$

$$\begin{aligned} H|T(\mathbf{j}'\,\mathbf{k}')_{jm}\rangle &= (\mathbf{K}'(1)\cdot\mathbf{K}'(2))|T(\mathbf{j}'\,\mathbf{k}')_{jm}\rangle \\ &= \frac{1}{2}(\mathbf{J}^2 - \mathbf{K}'^2(1) - \mathbf{K}'^2(2))|T(\mathbf{j}'\,\mathbf{k}')_{jm}\rangle \quad (8.38) \\ &= \frac{1}{2}\Big(j(j+1) - k'_1(k'_1+1) - k'_2(k'_2+1)\Big)|T(\mathbf{j}'\,\mathbf{k}')_{jm}\rangle. \end{aligned}$$

Again, there are two ways to enumerate the orthonormal vectors defined by (8.34), in which $j_1, j_2, j_3, j_4 \in \{0, 1/2, 1, \ldots\}$ are selected: (i) k'_1 and k'_2 assume all values $k'_1 \in \langle j_1, j_3 \rangle, k'_2 \in \langle j_2, j_4 \rangle$, and then $j\, m$ all values in $j \in \langle k'_1, k'_2 \rangle$, $m \subset \{j, j-1, \ldots, -j\}$. The total number of orthonormal vectors enumerated by (8.34) is then $N(j_1, j_2, j_3, j_4) = (2j_1+1)(2j_2+1)(2j_3+1)(2j_4+1)$; (ii) $j\, m$ are preselected $j \in \{j_{\min}, j_{\min}+1, \ldots, j_{\max}\}$, $m \in \{j, j-1, \ldots, -j\}$; hence, k'_1, k'_2 can be any values such that $k'_1, k'_2 \in \mathbb{K}_T^{(j)}(\mathbf{j}')$; that is,

$$\mathbb{K}_T^{(j)}(\mathbf{j}') = \left\{ k'_1, k'_2 \,\middle|\, \begin{array}{l} k'_1 \in \langle j_1, j_3 \rangle,\ k'_2 \in \langle j_2, j_4 \rangle \\ k'_1 \in \langle k'_2, j \rangle,\ k'_2 \in \langle k'_1, j \rangle \end{array} \right\}. \tag{8.39}$$

The cardinality of the set $\mathbb{K}_T^{(j)}(\mathbf{j}')$ is then the CG number $N_j(j_1, j_2, j_3, j_4)$, which is the total number of orthonormal vectors enumerated by (8.34) under these constraints.

For $n = 2, 3, 4$, the binary coupling theory of angular momenta solves completely the problem of diagonalizing and giving the full spectrum of H. The spectrum does not depend on the projection quantum number m, a result that is true for general n because H is an $SU(2)$ rotational invariant.

It is impossible for $n \geq 5$ to write H as a linear combination of the complete set of mutually commuting Hermitian operators (5.191)-(5.192) for any single binary coupling scheme. (The proof of this result is quite detailed; it is placed in Appendix B.) Binary coupling schemes can still be used for $n \geq 5$ to diagonalize H, but now several coupling schemes are required. For $n = 2f$ ($f \geq 2$), the Hamiltonian can be written as a sum $H = H_1 + H_2$ of two noncommuting parts, H_1 and H_2, each containing f parts, and these two parts can be separately diagonalized on two distinct binary coupling schemes. For $n = 2f + 1$ ($f \geq 2$), the Hamiltonian can be written as a sum $H = H_1 + H_2 + \mathbf{J}(n) \cdot \mathbf{J}(1)$ of three noncommuting parts, H_1 being exactly the same H_1 as in the $n = 2f$ case, and H_2 for $n = 2f$ and $n = 2f + 1$ each containing f terms, but differing by a single term; these three parts can be diagonalized separately on three different binary coupling schemes. Despite the noncommutativity of the parts H_1, H_2, and H_3, the diagonalization property on distinct coupling schemes makes it possible to give a reasonable formulation of the problem of diagonalizing H itself. The two-part decomposition property of H for n even is, of course, simpler to develop; hence, we carry it out first. To put this into effect, we give the class of binary trees corresponding to $n = 4, 6, \ldots$, as well as the corresponding cubic graphs for pairs of such binary trees. The situation for n odd is similar for the three-part Hamiltonian H. But the development is even longer than that for even n, and has a quite different structure. We keep the developments for the even and odd cases separated for the most part to avoid confusion.

8.4 The Heisenberg Ring Hamiltonian: Even n

We define the Hamiltonians H_1 and H_2 for even $n = 2f, f \geq 2$, as follows:

$$H_1 = \sum_{\text{all odd } i \leq n-1} \mathbf{J}(i) \cdot \mathbf{J}(i+1), \quad H_2 = \sum_{\text{all even } i \leq n} \mathbf{J}(i) \cdot \mathbf{J}(i+1);$$

$$H = H_1 + H_2. \qquad (8.40)$$

The Hamiltonians H_1 and H_2 each contain f dot product terms, each of which is a Hermitian operator on the space $\mathcal{H}_j$. The Hamiltonians H_1 and H_2 have the following distinctive features: they do not commute; each of the dot products constituting the separate sums mutually commute because they share no common angular momenta terms; and they can be diagonalized on separate binary coupling schemes. In this section, we demonstrate fully this n–even result. We first need the required collection of binary trees and associated coupling schemes.

There are only two types of forks in each of the coupling schemes, denoted I and II, that diagonalize H_1 and H_2 in (8.40). (In general, there are four types of forks as given by (5.96).) Now we require just two:

$$f^\circ = \vee, \quad f^\bullet = \vee, \qquad (8.41)$$

where there are $f = n/2$ forks of type f° and $f - 1$ of type $f^\bullet$, giving a total number of forks equal to $2f - 1$.

We state at the outset the family of unlabeled binary trees that enter into the separated diagonalization of the Hamiltonians H_1 and H_2:

$$T_2 = f^\circ = \vee, \quad T_4 = \vee\vee$$

$$T_6 = \quad , \quad T_8 =$$

$$\vdots \qquad \vdots$$

$$T_{2f} = \qquad\qquad\qquad (8.42)$$

We define binary coupling schemes I and II as follows in terms of the standard labeling of the binary trees in (8.42):

1. Binary coupling scheme I:

    (i) External angular momenta: The pairwise addition of angular momenta is given by $\mathbf{J}(2i-1) + \mathbf{J}(2i), i = 1, 2, \ldots, n/2$, as assigned left-to-right respectively, to the (∘ ∘) endpoints of the $n/2$ forks of type f°.

    (ii) Intermediate angular momenta:
    The $2f-1$ roots of the binary trees in (8.42) are labeled by the standard rule by $\mathbf{k} = (k_1, k_2, \ldots, k_{2f-2})$ and j. Since there are two • points at each level $1, 2, \ldots, f-1$, the pairs of labels (k_{2i-1}, k_{2i}) occur at level $f-i$ for $i = 1, 2, \ldots, f-1$; and j at level 0. The addition of angular momenta associated with this standard labeling is

    $$\mathbf{K}(1) = \mathbf{J}(1) + \mathbf{J}(2); \; \mathbf{K}(2i-1) = \mathbf{J}(2i+1) + \mathbf{J}(2i+2),$$
    $$i = 1, 2, \ldots, f-1,$$
    $$\mathbf{K}(2i+1) = \mathbf{K}(2i-1) + \mathbf{K}(2i), i = 1, 2, ,\ldots, f-2, \quad (8.43)$$
    $$\mathbf{J} = \mathbf{K}(2f-3) + \mathbf{K}(2f-2).$$

2. Binary coupling scheme II:

    (i) External angular momenta: The pairwise addition of angular momenta is given by $\mathbf{J}(2i) + \mathbf{J}(2i+1), i = 1, 2, \ldots, n/2$, where $\mathbf{J}(n+1) = \mathbf{J}(1)$, as assigned left-to-right respectively, to the (∘ ∘) endpoints of the $n/2$ forks of type f°.

    (ii) Intermediate angular momenta:
    The $2f-1$ roots of the forks in (8.42) are labeled by the standard rule by $\mathbf{k}' = (k'_1, k'_2, \ldots, k'_{2f-2})$ and j. Since there are two • points at each level $1, 2, \ldots, f-1$, the pairs of labels (k'_{2i-1}, k'_{2i}) occur at level $f-i$ for $i = 1, 2, \ldots, f-1$; and j at level 0. The addition of angular momenta associated with this standard labeling is

    $$\mathbf{K}'(1) = \mathbf{J}(2) + \mathbf{J}(3); \; \mathbf{K}'(2i) = \mathbf{J}(2i+2) + \mathbf{J}(2i+3),$$
    $$i = 1, 2, \ldots, f-1,$$
    $$\mathbf{K}(2i+1) = \mathbf{K}(2i-1) + \mathbf{K}(2i), i = 1, 2, ,\ldots, f-2, \quad (8.44)$$
    $$\mathbf{J} = \mathbf{K}(2f-3) + \mathbf{K}(2f-2).$$

The labeling of the two coupling schemes is fully explicit. Throughout this chapter, we use the following notations for the sequences of external

8.4. HEISENBERG HAMILTONIAN: EVEN n

and internal (intermediate) angular momenta in schemes I and II, where $n = 2f$:

$$\begin{aligned} \mathbf{j}_1 &= (j_1, j_2, \ldots, j_n), \; \mathbf{m}_1 = (m_1, m_2, \ldots, m_n), \\ \mathbf{k}_1 &= \mathbf{k} = (k_1, k_2, \ldots, k_{n-2}); \\ \mathbf{j}_2 &= (j_2, j_3, \ldots, j_n, j_1), \; \mathbf{m}_2 = (m_2, m_3, \ldots, m_n, m_1), \\ \mathbf{k}_2 &= \mathbf{k}' = (k'_1, k'_2, \ldots, k'_{n-2}). \end{aligned} \quad (8.45)$$

We next write out the coupled state vectors corresponding to schemes I and II, using the above notations:

scheme I :

$$|T(\mathbf{j}_1 \, \mathbf{k}_1)_j \, m\rangle = \sum_{\mathbf{m} \in \mathbb{C}_T(\mathbf{j})} C_{T(\mathbf{j}_1 \, \mathbf{m}_1; \, \mathbf{k}_1)_j \, m} \, |\mathbf{j} \, \mathbf{m}\rangle. \quad (8.46)$$

scheme II :

$$|T(\mathbf{j}_2 \, \mathbf{k}_2)_j \, m\rangle = \sum_{\mathbf{m} \in \mathbb{C}_T(\mathbf{j})} C_{T(\mathbf{j}_2 \, \mathbf{m}_2; \, \mathbf{k}_2)_j \, m} \, |\mathbf{j} \, \mathbf{m}\rangle. \quad (8.47)$$

The T in the notation $\mathbb{C}_T(\mathbf{j})$ for the domain of the projection quantum numbers denotes the general tree $T = T_{2f}$ in diagrams (8.42), and the domain of definition $\mathbb{C}_T(\mathbf{j})$ is defined in Sect. 5.8.4 (see (5.201)).

Using these notations, we obtain the complete set of mutually commuting Hermitian operators and eigenvector-eigenvalue relations for scheme I, which diagonalizes H_1, and for scheme II, which diagonalizes H_2:

scheme I:

$$\mathbf{J}^2 | T(\mathbf{j}_1 \, \mathbf{k}_1)_j \, m\rangle = j(j+1) | T(\mathbf{j}_1 \, \mathbf{k}_1)_j \, m\rangle;$$
$$J_3 | T(\mathbf{j}_1 \, \mathbf{k}_1)_j \, m\rangle = m | T(\mathbf{j}_1 \, \mathbf{k}_1)_j \, m\rangle;$$
$$\mathbf{J}^2(i) | T(\mathbf{j}_1 \, \mathbf{k}_1)_j \, m\rangle = j_i(j_i+1) | T(\mathbf{j}_1 \, \mathbf{k}_1)_j \, m\rangle, \, i = 1, 2, \ldots, n;$$
$$\mathbf{K}_T^2(i) | T(\mathbf{j}_1 \, \mathbf{k}_1)_j \, m\rangle = k_i(k_i+1) | T(\mathbf{j}_1 \, \mathbf{k}_1)_j \, m\rangle, \, i = 1, 2, \ldots, n-2;$$
$$(8.48)$$

$$2H_1 = -\sum_{i=1}^{n} \mathbf{J}^2(i) + \mathbf{K}^2(1) + \sum_{i=1}^{f-1} \mathbf{K}^2(2i),$$

$$H_1 | T(\mathbf{j}_1 \, \mathbf{k}_1)_j \, m\rangle = \lambda(\mathbf{j}_1 \, \mathbf{k}_1) | T(\mathbf{j}_1 \, \mathbf{k}_1)_j \, m\rangle,$$

$$2\lambda(\mathbf{j}_1 \, \mathbf{k}_1) = -\sum_{i=1}^{n} j_i(j_i+1) + k_1(k_1+1) + \sum_{i=1}^{f-1} k_{2i}(k_{2i}+1).$$

scheme II:

$$\mathbf{J}^2|T(\mathbf{j}_2\,\mathbf{k}_2)_{jm}\rangle = j(j+1)|T(\mathbf{j}_2\,\mathbf{k}_2)_{jm}\rangle;$$
$$J_3|T(\mathbf{j}_2\,\mathbf{k}_2)_{jm}\rangle = m|T(\mathbf{j}_2\,\mathbf{k}_2)_{jm}\rangle;$$
$$\mathbf{J}^2(i)|T(\mathbf{j}_2\,\mathbf{k}_2)_{jm}\rangle = j_i(j_i+1)|T(\mathbf{j}_2\,\mathbf{k}_2)_{jm}\rangle, i=1,2,\ldots,n;$$
$$\mathbf{K'}_T^2(i)|T(\mathbf{j}_2\,\mathbf{k}_2)_{jm}\rangle = k'_i(k'_i+1)|T(\mathbf{j}_2\,\mathbf{k}_2)_{jm}\rangle, i=1,2,\ldots,n-2;$$

(8.49)

$$2H_2 = -\sum_{i=1}^{n}\mathbf{J}^2(i) + \mathbf{K'}^2(1) + \sum_{i=1}^{f-1}\mathbf{K'}^2(2i),$$

$$H_2|T(\mathbf{j}_2\,\mathbf{k}_2)_{jm}\rangle = \lambda(\mathbf{j}_2\,\mathbf{k}_2)|T(\mathbf{j}_2\,\mathbf{k}_2)_{jm}\rangle,$$

$$2\lambda(\mathbf{j}_2\,\mathbf{k}_2) = -\sum_{i=1}^{n}j_i(j_i+1) + k'_1(k'_1+1) + \sum_{i=1}^{f-1}k'_{2i}(k'_{2i}+1).$$

We note that the eigenvalues of both H_1 and H_2 on their respective diagonalizing coupled bases (8.46)-(8.47) depend in the same way, first on $(j_1, j_2, \ldots, j_n)$, and then on their respective $(k_1, k_2, k_4, \ldots, k_{2f-2})$ and $(k'_1, k'_2, k'_4, \ldots, k'_{2f-2})$. The domains of definition of the latter intermediate quantum numbers are, of course, different. Thus, quite remarkably, the coupled bases, scheme I and scheme II, provide quite comprehensive expressions for the eigenvalues of H_1 and H_2. These two Hamiltonians do not, however, commute, and we consider next this aspect of the problem.

It is useful to put the problem posed by the diagonalization of $H = H_1 + H_2$, where H_1 and H_2 are Hermitian, noncommuting operators, in a broader context to see the underlying structures. Thus, consider two orthonormal bases $\mathbf{B}_1 = \{|u_1\rangle, |u_2\rangle, \ldots, |u_d\rangle\}$ and $\mathbf{B}_2 = \{|v_1\rangle, |v_2\rangle, \ldots, |v_d\rangle\}$ of one and the same finite-dimensional space $\mathcal{H}_d$ of dimension d, such that H_1 is diagonal on the basis $\mathbf{B}_1$ and H_2 is diagonal on the basis $\mathbf{B}_2$. Moreover, let the two bases be related by a real orthogonal matrix R with element in row i and column j given by R_{ij} :

$$|u_j\rangle = \sum_{i=1}^{d} R_{ij}|v_i\rangle. \qquad (8.50)$$

Let D_1 and D_2 denote the diagonal matrices of order d representing H_1 and H_2 on the respective bases $\mathbf{B}_1$ and $\mathbf{B}_2$, as given by

$$(D_1)_{ij} = \delta_{i,j}\langle u_i|H_1|u_i\rangle,\ (D_2)_{ij} = \delta_{i,j}\langle v_i|H_2|v_i\rangle,\ i,j=1,2\ldots,d. \quad (8.51)$$

It is an easy exercise to prove that the **matrix representation** $H_{\mathbf{B}_1}$ of

8.4. HEISENBERG HAMILTONIAN: EVEN n

H on the basis $\mathbf{B}_1$ is given by

$$H_{\mathbf{B}_1} = D_1 + R^{tr} D_2 R. \tag{8.52}$$

This result could equally well have been formulated in terms of the matrix representation $H_{\mathbf{B}_2}$ of H on the basis $\mathbf{B}_2$.

The eigenvector-eigenvalue structure presented by relations (8.50)-(8.51) is just that of (8.46)-(8.49) under the identifications of vector spaces and matrices given by

$$|u_{\mathbf{k}_1}\rangle = |T(\mathbf{j}_1 \, \mathbf{k}_1)_{jm}\rangle, \; |v_{\mathbf{k}_2}\rangle = |T(\mathbf{j}_2 \, \mathbf{k}_2)_{jm}\rangle; \tag{8.53}$$

$$R_{\mathbf{k}_2, \mathbf{k}_1} = \langle v_{\mathbf{k}_2} | u_{\mathbf{k}_1} \rangle = \langle T(\mathbf{j}_2 \, \mathbf{k}_2)_{jm} | T(\mathbf{j}_1 \, \mathbf{k}_1)_{jm} \rangle.$$

These identifications of the elements of R are valid because in relations (8.50)-(8.51) the quantum numbers $\mathbf{j}_1, \mathbf{j}_2, j, m$ may be taken as fixed, while the domain of definition of internal quantum numbers $\mathbf{k}_1$ and $\mathbf{k}_2$ specify the orthonormal vectors in the respective spaces.

The exact relation that replaces (8.50) for the Heisenberg ring problem is that between coupling scheme I and coupling scheme II given by

$$|T(\mathbf{j}_2 \, \mathbf{k}_2)_{jm}\rangle = \sum_{\mathbf{k}_1 \in \mathbb{K}_T^{(j)}(\mathbf{j}_1)} \left(R_{T,T}^{\mathbf{j}_1; \mathbf{j}_2} \right)_{\mathbf{k}_1, j; \mathbf{k}_2, j} |T(\mathbf{j}_1 \, \mathbf{k}_1)_{jm}\rangle. \tag{8.54}$$

The eigenvector-eigenvalue relations are, of course, those given by (8.48)-(8.49), which replace (8.51). Finally, the relation that replaces (8.52) is described in detail by the following collection of relations in Items 1-3:

$$H_{\mathbf{B}_1} = D^{\mathbf{j}_1} + R_{T,T}^{\mathbf{j}_2; \mathbf{j}_1 \, tr} D^{\mathbf{j}_2} R_{T,T}^{\mathbf{j}_2; \mathbf{j}_1}, \tag{8.55}$$

where we have the following definitions:

1. The matrix representation of H on the basis

$$\mathbf{B}_1 = \left\{ |T(\mathbf{j}_1 \, \mathbf{k}_1)_{jm}\rangle \, | \, \mathbf{k}_1 \in \mathbb{K}_T^{(j)}(\mathbf{j}_1) \right\} \tag{8.56}$$

is denoted by $H_{\mathbf{B}_1}$ and defined by

$$(H_{\mathbf{B}_1})_{\mathbf{k}'_1, j; \mathbf{k}_1, j} = \langle T(\mathbf{j}_1 \, \mathbf{k}'_1)_{jm} | H | T(\mathbf{j}_1 \, \mathbf{k}_1)_{jm} \rangle,$$

$$\mathbf{k}'_1, \mathbf{k}_1 \in \mathbb{K}_T^{(j)}(\mathbf{j}_1). \tag{8.57}$$

2. The diagonal matrix $D^{\mathbf{j}_1}$ is the matrix representation of H_1 on the basis $\mathbf{B}_1$; hence, it has matrix elements given by

$$\left(D^{\mathbf{j}_1} \right)_{\mathbf{k}'_1, j; \mathbf{k}_1, j} = \langle T(\mathbf{j}_1 \, \mathbf{k}'_1)_{jm} | H_1 | T(\mathbf{j}_1 \, \mathbf{k}_1)_{jm} \rangle$$

$$= \delta_{\mathbf{k}'_1, \mathbf{k}_1} \lambda(\mathbf{j}_1 \, \mathbf{k}_1), \; \mathbf{k}'_1, \mathbf{k}_1 \in \mathbb{K}_T^{(j)}(\mathbf{j}_1). \tag{8.58}$$

3. The diagonal matrix D^{j_2} is the matrix representation of H_2 on the basis $\mathbf{B}_2$ defined by

$$\mathbf{B}_2 = \left\{ |T(\mathbf{j_2} \, \mathbf{k_2})_{jm}\rangle \mid \mathbf{k_2} \subset \mathbb{K}_T^{(j)}(\mathbf{j}_2) \right\}; \qquad (8.59)$$

hence, it has matrix elements given by

$$\left(D^{j_2} \right)_{\mathbf{k}_2', j; \mathbf{k}_2, j} = \langle T(\mathbf{j}_2 \, \mathbf{k}_2')_{jm} | H_2 | T(\mathbf{j}_2 \, \mathbf{k}_2)_{jm} \rangle$$

$$= \delta_{\mathbf{k}_2', \mathbf{k}_2} \lambda(\mathbf{j}_2 \, \mathbf{k}_2), \quad \mathbf{k}_2', \mathbf{k}_2 \in \mathbb{K}_T^{(j)}(\mathbf{j}_2). \quad (8.60)$$

The recoupling matrix relating different binary tree coupling schemes is the basic quantity that allows us to determined fully the matrix representation of the Heisenberg Hamiltonian. We recall that the definitions of the indexing sets $\mathbb{K}_T^{(j)}(\mathbf{j}_1)$ and $\mathbb{K}_T^{(j)}(\mathbf{j}_2)$ have been given generally in terms of irreducible triangle coefficients of order $2(n-1)$ in (5.218), with special examples in relations (5.90), (5.126), and (5.157). The triangle coefficients are expressions of the matrix elements of the corresponding recoupling matrix, as given by (5.229). It is always the case that $|\mathbb{K}_T^{(j)}(\mathbf{j}_1)| = |\mathbb{K}_T^{(j)}(\mathbf{j}_2)| = N_j(\mathbf{j})$. The main point proved above is:

All matrices in (8.54)-(8.60) are of order $N_j(\mathbf{j})$, the CG number. The problem of diagonalizing H has been split into one of diagonalizing the original matrix in (8.10) of order $N(\mathbf{j}) = \prod_{i=1}^{n}(2j_i + 1)$ to that of diagonalizing a collection of matrices of dimension $N_j(\mathbf{j})$, there being such a matrix for each $j = j_{\min}, j_{\min} + 1, \ldots, j_{\max}$. The problem of the original diagonalization in the uncoupled basis has been transferred to the determination of the $3(n-1)-j$ coefficients that enter into the coupled bases of scheme I and scheme II and the symmetric matrix $H_{\mathbf{B}_1}$ defined by (8.55) and its subsequent diagonalization by a real orthogonal matrix:

$$R^T H_{\mathbf{B}_1} R = D. \qquad (8.61)$$

There is, of course, such a matrix relation for each $N_j(\mathbf{j})$, the order of the matrix $H_{\mathbf{B}_1}$.

8.4.1 Summary of Properties of Recoupling Matrices

It is appropriate to summarize here the results thus far obtained for the recoupling matrix $R_{T,T}^{\mathbf{j}_1;\mathbf{j}_2}$ and its matrix elements, drawing, as necessary, from the review in Chapter 5. This summary sets forth in Items (i)-(v) the basic objects entering the problem, and serves as an outline of the steps still necessary to obtain the explicit matrix elements of the recoupling matrix $\mathcal{H}_{\mathbf{B}_1}$ in relation (8.55):

8.4. HEISENBERG HAMILTONIAN: EVEN n

(i). Relation to generalized WCG coefficients:

$$R_{T,T}^{j_2;j_1} = C_T^{j_2} C_T^{j_1\,tr}, \qquad (8.62)$$

in consequence of which we have the transposition and multiplication relations:

$$R_{T,T}^{j_2;j_1\,tr} = R_{T,T}^{j_1;j_2} \quad \text{and} \quad R_{T,T}^{j_2;j_1} R_{T,T}^{j_1;j_3} = R_{T,T}^{j_2;j_3}, \qquad (8.63)$$

(ii). Orthogonality relations: These come in all the versions $R_{T,T}^{j_1;j_2\,tr} R_{T,T}^{j_1;j_2} = R_{T,T}^{j_1;j_2} R_{T,T}^{j_1;j_2\,tr} = R_{T,T}^{j_2;j_1\,tr} R_{T,T}^{j_2;j_1} = R_{T,T}^{j_2;j_1} R_{T,T}^{j_2;j_1\,tr} = I_{N_j(\mathbf{j})}$. One example of the matrix element expression is

$$\sum_{\mathbf{k}_2 \in \mathbb{K}_T^{(j)}(\mathbf{j}_2)} \left(R_{T,T}^{j_2;j_1\,tr}\right)_{\mathbf{k}_1'',\mathbf{j};\mathbf{k}_2,\mathbf{j}} \left(R_{T,T}^{j_2;j_1}\right)_{\mathbf{k}_2,\mathbf{j};\mathbf{k}_1,\mathbf{j}} = \delta_{\mathbf{k}_1'',\mathbf{k}_1}, \quad \mathbf{k}_1'', \mathbf{k}_1 \in \mathbb{K}_T^{(j)}(\mathbf{j}_1).$$

$$(8.64)$$

The point is: Care must be taken in evaluating matrix elements because they are enumerated by different indexing sets, $\mathbb{K}_T^{(j)}(\mathbf{j}_1)$ and $\mathbb{K}_T^{(j)}(\mathbf{j}_2)$, which, however, have the same cardinality, the CG number $N_j(\mathbf{j})$.

(iii). Relation to inner products and triangle coefficients:

$$\left(R_{T,T}^{j_1;j_2}\right)_{\mathbf{k}_1,\mathbf{j};\mathbf{k}_2,\mathbf{j}} = \langle T(\mathbf{j}_1\,\mathbf{k}_1)_{jm} \mid T(\mathbf{j}_2\,\mathbf{k}_2)_{jm} \rangle$$

$$= \left\{ \Delta_T(\mathbf{j}_1\mathbf{k}_1)_j \mid \Delta_T(\mathbf{j}_2\mathbf{k}_2)_j \right\} \qquad (8.65)$$

$$= \left(R_{T,T}^{j_2;j_1}\right)_{\mathbf{k}_2,\mathbf{j};\mathbf{k}_1,\mathbf{j}} = \langle T(\mathbf{j}_2\,\mathbf{k}_2)_{jm} \mid T(\mathbf{j}_1\,\mathbf{k}_1)_{jm} \rangle$$

$$= \left\{ \Delta_T(\mathbf{j}_2\mathbf{k}_2)_j \mid \Delta_T(\mathbf{j}_1\mathbf{k}_1)_j \right\}.$$

(iv). Explicit triangle coefficients: The left and right triangle arrays associated with the triangle coefficients above are the explicit arrays:

$$\Delta_T(\mathbf{j}_1\mathbf{k}_1)_j = \begin{pmatrix} j_1 & j_3 & k_1 & j_5 & k_3 & \cdots & k_{2f-5} & j_{2f-1} & k_{2f-3} \\ j_2 & j_4 & k_2 & j_6 & k_4 & \cdots & k_{2f-4} & j_{2f} & k_{2f-2} \\ k_1 & k_2 & k_3 & k_4 & k_5 & \cdots & k_{2f-3} & k_{2f-2} & j \end{pmatrix}; \qquad (8.66)$$

$$\Delta_T(\mathbf{j}_2\mathbf{k}_2)_j = \begin{pmatrix} j_2 & j_4 & k_1' & j_6 & k_3' & \cdots & k_{2f-5}' & j_{2f} & k_{2f-3}' \\ j_3 & j_5 & k_2' & j_7 & k_4' & \cdots & k_{2f-4}' & j_{j_1} & k_{2f-2}' \\ k_1' & k_2' & k_3' & k_4' & k_5' & \cdots & k_{2f-3}' & k_{2f-2}' & j \end{pmatrix}. \qquad (8.67)$$

(v). Degeneracy of eigenvalues: The eigenvalues of H_1 and H_2 are degenerate because the eigenvalues of H_1 given by $\lambda(\mathbf{j_1\,k_1})$ in relations (8.48) and of H_2 given by $\lambda(\mathbf{j_2\,k_2})$ in relations (8.49) depend (for fixed j_i)only on the intermediate quantum numbers associated with the coupling of the given angular momenta $\mathbf{J}_i, i = 1, 2, \ldots, n$, and not on the other intermediate k_i quantum labels. Thus, the degeneracy of eigenvalues of H_1 is enumerated by the values that $k_3, k_5, \ldots, k_{2f-3}$ can assume in the left triangle coefficient (8.66), when all other labels are held fixed, and, similarly, for H_2 and the right triangle coefficient (8.67). This property can be incorporated into relations (8.52) and (8.61) to determine more precisely the form of R, the real orthogonal matrix that diagonalizes the symmetric matrix $H_{\mathbf{B}_1}$. This degeneracy is a consequence of the property that while the angular momentum states are complete, the number of $SU(2)$-invariant operators that commutes with the complete set of angular momentum operators is not — there are other Hamiltonian-like $SU(2)$-invariant operators that not only commute with the Heisenberg Hamiltonian H, but themselves constitute a mutually commuting set. This is an important result, but we make no use of it, and do not develop it further (see, however, Sect. 8.7 for the effect of the cyclic group symmetry). Because the matrix $H_{\mathbf{B}_1}$ is real symmetric, it is diagonalizable by a real orthogonal matrix R : *There are no complex matrices or numbers needed for the determination of the eigenvalue spectrum of H.*

8.4.2 Maximal Angular Momentum Eigenvalues

The CG number $C_{j_{\max}}(\mathbf{j}) = 1$ for $j = j_{\max} = j_1 + j_2 + \cdots + j_n$. For $j = j_{\max}$, the space $\mathcal{H}_{\mathbf{j};j}$ is one-dimensional, and the orthogonal recoupling matrix in (8.55) and R in (8.61) are both unity; that is, all the matrices are one-dimensional. This is a consequence of the fact that all intermediate quantum numbers in the sequences $\mathbf{k}_1$ and $\mathbf{k}_2$ are forced to their maximum values. In particular, we have the following values of the intermediate quantum numbers that enter into the eigenvalues of H_1 and H_2 in (8.48) and (8.49) for all $i = 1, 2, \ldots, n/2$:

$$k_i = j_{2i-1,2i} = j_{2i-1} + j_{2i},\ k'_i = j_{2i,2i+1} = j_{2i} + j_{2i+1},\ (j_{n+1} = j_1). \quad (8.68)$$

We now obtain the following result $H_{\mathbf{B}_1}(\max)$ for the maximum eigenvalue of $H_{\mathbf{B}_1}$ in relation (8.55) and its corresponding eigenvector:

$$H_{\mathbf{B}_1}(\max) = \lambda_{\max} = D^{\mathbf{j}_1}(\max) + D^{\mathbf{j}_2}(\max) = -\sum_{i=1}^{n} j_i(j_i + 1)$$

$$+\tfrac{1}{2}\sum_{i=1}^{n/2} j_{2i-1,2i}(j_{2i-1,2i} + 1) + \tfrac{1}{2}\sum_{i=1}^{n/2} j_{2i,2i+1}(j_{2i,2i+1} + 1); \quad (8.69)$$

$$|\mathbf{j}\,j\rangle = |j_1\,j_1\rangle \otimes |j_2\,j_2\rangle \otimes \cdots \otimes |j_n\,j_n\rangle.$$

8.4. HEISENBERG HAMILTONIAN: EVEN n

This expression for the eigenvalue must agree with relation (8.38) for $n = 4, j = j_{\max} = j_1 + j_2 + j_3 + j_4$. Thus, the following identity must hold:

$$j(j+1) - (j_1+j_3)(j_1+j_3+1) - (j_2+j_4)(j_2+j_4+1)$$
$$= -2\sum_{i=1}^{4} j_i(j_i+1) + (j_1+j_2)(j_1+j_2+1) + (j_3+j_4)(j_3+j_4+1)$$
$$+ (j_2+j_3)(j_2+j_3+1) + (j_4+j_1)(j_4+j_1+1). \qquad (8.70)$$

This relation is verified to be correct; this is reassuring, considering the rather intricate derivation of relation (8.55). (It is also the case for $n = 4$ that we must be able to effect the full diagonalization (8.61) to obtain the eigenvalues given by (8.38). This is carried out in Sect. 8.4.5 as a special case of the general theory.)

In general, states of lowest angular momentum j are degenerate — the CG number need not be unity, so that the calculation of other eigenvalues requires the diagonalization of a matrix $H_{\mathbf{B}_1}$ of order ≥ 2.

8.4.3 Shapes and Paths for Coupling Schemes I and II

We continue in this section with the description of the binary trees that enter into the problem of diagonalizing the Heisenberg Hamiltonian $H = H_1 + H_2$ for n even, as described generally in Chapter 5. In particular, we have summarized in Sect. 5.8.8 the full set of steps required. We have completed Items 1-2, and continue now with Item 3, the enumeration of shapes and determination of paths.

The shapes of the respective binary trees in diagrams (8.42) are

$$(\circ\,\circ), \quad \big((\circ\,\circ)(\circ\,\circ)\big), \quad \Big(\big((\circ\,\circ)(\circ\,\circ)\big)(\circ\,\circ)\Big),$$
$$\vdots \qquad \vdots \qquad (8.71)$$
$$\Big(\underbrace{\cdots\big(\big((\circ\,\circ)(\circ\,\circ)\big)(\circ\,\circ)\big)(\circ\,\circ)\big)\cdots(\circ\,\circ)}_{f-1}\Big),$$

where the general shape of T_{2f} has $2f-1$ parenthesis pairs. The standard labeling of a pair of binary trees from diagrams (8.42) of shape (8.71) corresponding to binary coupling schemes I and II gives the following triangle coefficient of order $2(2f-1)$, which we repeat from the left and

244   CHAPTER 8. THE HEISENBERG MAGNETIC RING

right patterns given by relations (8.66)-(8.67):

$$\left\{\Delta_T(\mathbf{j}_1\,\mathbf{k}_1)_j \middle| \Delta_T(\mathbf{j}_2\,\mathbf{k}_2)_j\right\} = \left\{\begin{array}{ccccccccc} j_1 & j_3 & k_1 & j_5 & k_3 & \cdots & j_{2f-1} & k_{2f-3} \\ j_2 & j_4 & k_2 & j_6 & k_4 & \cdots & j_{2f} & k_{2f-2} \\ k_1 & k_2 & k_3 & k_4 & k_5 & \cdots & k_{2f-2} & j \\ j_2 & j_4 & k'_1 & j_6 & k'_3 & \cdots & j_{2f} & k'_{2f-3} \\ j_3 & j_5 & k'_2 & j_7 & k'_4 & \cdots & j_1 & k'_{2f-2} \\ k'_1 & k'_2 & k'_3 & k'_4 & k'_5 & \cdots & k'_{2f-2} & j \end{array}\right\}. \qquad (8.72)$$

As remarked and illustrated in many places in this volume, we take the irreducible triangle coefficient (8.72) as the symbol for the corresponding $3(2f-1)-j$ coefficient, and sometimes use the pair of standard labeled binary trees from (8.42) themselves. The shapes (binary bracketings) (8.71) and the standard labels in the triangle coefficient (8.72) give us the setting against which to calculate the coefficients themselves, the next step being to identify a path in terms of elementary $A-$ and $C-$operations that effect the transformation between schemes I and II. However, because we have the triangle coefficient before us, this is a convenient place to interrupt and give the extraordinary cubic graph associated with this irreducible triangle coefficient; we continue in the next section with shape transformations.

The cubic graph for schemes I and II

The standard rule for obtaining the cubic graph from the irreducible triangle coefficient (8.72) has two instructions:

(i). Read-off left-to-right across the left pattern and the right pattern the triples of angular momenta triangles given as follows:

$p_1 = (j_1, j_2, k_1),$

$p_3 = (k_1, k_2, k_3), p_5 = (k_3, k_4, k_5), \ldots, p_{2f-3} = (k_{2f-5}, k_{2f-4}, k_{2f-3}),$

$$p_{2f-1} = (k_{2f-3}, k_{2f-2}, j); \qquad (8.73)$$

$p_2 = (j_3, j_4, k_2), p_4 = (j_5, j_6, k_4), \ldots, p_{2f-2} = (j_{2f-1}, j_{2f}, k_{2f-2});$

$p'_1 = (j_2, j_3, k'_1),$

$p'_3 = (k'_1, k'_2, k'_3), p'_5 = (k'_3, k'_4, k'_5), \ldots, p'_{2f-3} = (k'_{2f-5}, k'_{2f-4}, k'_{2f-3}),$

$$p'_{2f-1} = (k'_{2f-3}, k'_{2f-2}, j); \qquad (8.74)$$

$p'_2 = (j_4, j_5, k'_2), p'_4 = (j_6, j_7, k'_4), \ldots, p'_{2f-2} = (j_{2f}, j_1, k'_{2f-2}).$

These triangles define the $2(2f-1)$ points of the cubic graph.

8.4. HEISENBERG HAMILTONIAN: EVEN n

(ii). Draw a line between each pair of points that contains a common angular momentum label.

The resulting configuration of points and lines can be presented in the plane by the following general diagram:

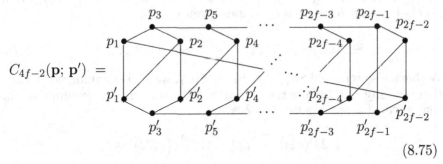

(8.75)

This graph of $2(2f-1)$ points and $3(3f-1)$ lines with 3 lines incident on each point are the properties that define a cubic graph $C_{2(2f-1)}$, when all labels are removed.

The rules (i) and (ii) still hold for pairs of irreducible triangle coefficients in which the columns have labels (triangles) corresponding to all four types of labeled forks (5.96), but now Cayley's [19] trivalent trees are involved, and the diagrams for cubic graphs are much more complicated (see [L]).

8.4.4 Determination of the Shape Transformations

We now continue down the path toward calculating the $3(2f-1)-j$ coefficients for which we now have the symbols, both in terms of the triangle coefficients of order $2(2f-1)$ and the associated cubic graph of order $2(2f-1)$. As emphasized repeatedly, these triangle coefficients are already uniquely defined in terms of the sums over products of the known WCG coefficients, there being such a product for each of the labeled binary trees that define binary coupling schemes I and II. The achievement of binary coupling theory is the expression of all such $3(2f-1)-j$ coefficients as summations over the basic triangle coefficients of order four (Racah coefficients up to phase and dimension factors). This step requires the factorization of the recoupling matrix into a product of elementary $A-$ and $C-$shape transformations, followed by the application of the reduction rule (5.134) and the phase rule (5.135).

Shape transformations map one shape to another shape by application of commutation C and association A of symbols: these are involution operations that act on the letters and parenthesis pairs in a labeled shape.

Since $C^2 = A^2 = \mathbb{I}$ = identity operation, such mappings are invertible. Such transformation are introduced and explained in [L] and reviewed in Chapter 5. We rephrase here from those general results those needed for the special shapes given by coupling schemes I and II. In describing the shape transformations in question, it is convenient to economize the notation by writing, for example, $((j_1 j_2)(j_3 j_4)) = ((12)(34))$; that is, each j_i is replaced by i. Thus, we seek a transformation $w_{2f}(A, C)$ that effects the transformation of shapes given by

$$\text{Sh}_{2f}(2, 3, \ldots, 2f, 1) \xrightarrow{w_{2f}(A,C)} \text{Sh}_{2f}(1, 2, \ldots, 2f). \qquad (8.76)$$

We have also replaced $\text{Sh}_{T_{2f}}$ by simply Sh_{2f}, which is unambiguous, since there is only one binary tree of order $2f$ involved. An example of the shape transformations for $f = 2$ is:

$$((23)(41)) \xrightarrow{C} ((23)(14)) \xrightarrow{A} \left(((23)1)4\right)$$

$$\xrightarrow{C} \left((1(23))4\right) \xrightarrow{A} \left(((12)3)4\right) \xrightarrow{A} ((12)(34)). \qquad (8.77)$$

We write this transformation as

$$((23)(41)) \xrightarrow{w_4} ((12)(34)), \; w_4 = w_4(A, C) = AACAC. \qquad (8.78)$$

It is very important here and below to observe that we write the word $AACAC$ in (8.78) in the reverse order of their application shown in (8.77); it is the intent that C is to be applied first to $((23)(41))$, followed by A, followed by C, followed by A, followed by A, as read right-to-left, which is the usual rule for the application of sequences of operators to the objects on which they act.

It is always necessary to give the subshapes on which each of the elementary A and C shape transformations act. Thus, in the example (8.77), these actions are the following:

$$(xy) \xrightarrow{C} (yx), \; x = 4, y = 1;$$

$$((x(yz)) \xrightarrow{A} ((xy)z)), \; x = (23), y = 1, z = 4;$$

$$(xy) \xrightarrow{C} (yx), \; x = (23), y = 1; \qquad (8.79)$$

$$(x(yz)) \xrightarrow{A} ((xy)z), \; x = 1, y = 2, z = 3;$$

$$((xy)z) \xrightarrow{A} (x(yz)), \; x = (12), y = 3, z = 4.$$

Without this detailed information, it is impossible to transcribe, in complicated cases, the meaning of the general transformation (8.76).

8.4. HEISENBERG HAMILTONIAN: EVEN n

It is useful to illustrate shape transformations for the binary trees in (8.42) of unlabeled shapes given by (8.71) for $n = 2f = 6, 8$; this also shows the recurring nature of these transformations:

Examples. The labeled shape transformations for $f = 3, 4$:

1. The transformation $w_6 = w_6(A, C)$:

$$\Big(\big((23)(45)\big)(61)\Big) \xrightarrow{A} \Big((23)\big((45)(61)\big)\Big) \xrightarrow{w} \Big((23)\big((41)(56)\big)\Big)$$
$$\xrightarrow{A} \Big(\big((23)(41)\big)(56)\Big) \xrightarrow{w_4} \Big(\big((12)(34)\big)(56)\Big). \tag{8.80}$$

Here w is the shape transformation given by

$$((45)(61)) \xrightarrow{w} ((41)(56)). \tag{8.81}$$

Thus, we have that

$$\Big(\big((23)(45)\big)(61)\Big) \xrightarrow{w_6} \Big(\big((12)(34)\big)(56)\Big), \tag{8.82}$$

$$w_6 = w_4 \, A w A.$$

The shape transformation w occurs repeatedly, and it is identified generally in (8.87) below (see also (8.85)). Of course, in the shape transformation (8.80), the subshapes on which A and C act must be identified, as in (8.79). But this is quite easy to do by inspection. For example, the first and last A transformations in (8.80) are, respectively, given by

$$((xy)z) \xrightarrow{A} (x(yz)), \quad x = (23), y = (45), z = (61); \tag{8.83}$$

$$(x(yz)) \xrightarrow{A} ((xy)z), \quad x = (23), y = (41), z = (56).$$

2. The transformation $w_8 = w_8(A, C)$:

$$\Big(\big((23)(45)\big)(67)\big)(81)\Big) \xrightarrow{A} \Big(\big((23)(45)\big)\big)((67)(81))\Big)$$
$$\xrightarrow{w} \Big(\big(((23)(45))\big)((61)(78))\big)\Big) \xrightarrow{A} \Big(\big((23)(45)\big)(61)\big)(78)\Big)$$
$$\xrightarrow{w_6} \Big(\big((12)(34)\big)(56)\big)(78)\Big). \tag{8.84}$$

Here w is the shape transformation given by

$$((67)(81)) \xrightarrow{w} ((61)(78)), \tag{8.85}$$

which has the same structure as that given in (8.81). Thus, we have that

$$\Big(\big((23)(45)\big)(67)\big)(81)\Big) \xrightarrow{w_8} \Big(\big((12)(34)\big)(56)\big)(78)\Big), \tag{8.86}$$

$$w_8 = w_6\, AwA = w_4\, AwA\, AwA.$$

Again, the subshapes on which A and C act must be identified. But this is quite easy to do by inspection from (8.84).

The shape transformation w is defined on an arbitrary labeled shape $((ab)(cd))$ by

$$((ab)(cd)) \xrightarrow{C} ((cd)(ab)) \xrightarrow{A} (c(d(ab))) \xrightarrow{A} (c((da)b))$$
$$\xrightarrow{C} (((da)b)c) \xrightarrow{A} ((da)(bc)) \xrightarrow{C} ((ad)(bc)); \tag{8.87}$$

$$((ab)(cd)) \xrightarrow{w} ((ad)(bc)),$$
$$w = CACAAC. \qquad \Box$$

The general result is now easily inferred:

$$\mathrm{Sh}_{2f}(2,3,\ldots,2f-1,2f,1) \xrightarrow{w_{2f}} \mathrm{Sh}_{2f}(1,2,\ldots,2f-1,2f), \tag{8.88}$$

$$w_{2f} = w_4\,\underbrace{AwA\,AwA\,\cdots\,AwA}_{f-2} = w_4\,(AwA)^{f-2}.$$

Proof. For each $f \geq 4$, we have the following shape transformations:

$$\mathrm{Sh}_{2f}(2,3,\ldots,2f-1,2f,1) = \Big(\big(X_{2f-4}(ab)\big)(cd)\Big)$$
$$\xrightarrow{A} \Big(\big(X_{2f-4}\big((ab)(cd)\big)\big)\Big) \xrightarrow{w} \Big(\big(X_{2f-4}\big((ad)(bc)\big)\big)\Big) \tag{8.89}$$
$$\xrightarrow{A} \Big(\big(X_{2f-4}(ad)\big)(bc)\Big) \xrightarrow{w_{2f-2}} \mathrm{Sh}_{2f}(1,2,\ldots,2f-1,2f),$$

where $a = 2f-2, b = 2f-1, c = 2f, d = 1$, and

$$X_{2f-4} = \underbrace{\Big(\cdots\big(\big((23)(45)\big)(67)\big)\cdots\Big)(2f-4\,2f-3)}_{f-3}. \qquad \Box \tag{8.90}$$

8.4. HEISENBERG HAMILTONIAN: EVEN n

It follows from (8.87)-(8.88) that the length of the shape transformation w_{2f} (number of $A's$) is given by

$$|w_{2f}| = 5f - 7, f \geq 2. \qquad (8.91)$$

Before turning to the calculation of the $2(2f-1)-j$ coefficients associated with the shape transformation (8.88), we return to the relation between the exact solution of the Heisenberg Hamiltonian eigenvector-eigenvalues for $n=4$ and the binary coupling schemes I and II.

8.4.5 The Transformation Method for $n = 4$

The basis vectors for the coupling schemes I and II discussed in the present section are not those for the coupling scheme given by (8.34) for $n = 4$ that gives the exact eigenvalue spectrum. It is useful to show how the transformations between these three coupling schemes are related, not only to illustrate the various constructions of basis vectors and transformation coefficients, but also because we are led to a new relation between $9 - j$ coefficients from which also follows a new relation between Racah coefficients (see (8.102) and (8.104) below).

There are three distinct sets of coupled binary tree bases involved: Two in coupling schemes I and II that occur in (8.48)-(8.49) for $n = 4$, which expresses the matrix $H_{\mathbf{B}_1}$ representing H in the basis $\mathbf{B}_1$ in terms of the diagonal matrix D^{j_1} representing H_1 in the basis $\mathbf{B}_1$ and diagonal matrix D^{j_2} representing H_2 in the basis $\mathbf{B}_2$; and the basis $\mathbf{B}'$ in which H is diagonal:

$$\mathbf{B}_1 = \left\{ \left| \begin{array}{c} j_1 \ j_2 \ j_3 \ j_4 \\ j_{12} \diagdown \diagup j_{34} \\ jm \end{array} \right\rangle \, \middle| \, \begin{array}{c} j_{12} \in \langle j_1, j_2 \rangle, j_{34} \in \langle j_3, j_4 \rangle, \\ j_{12} \in \langle j_{34}, j \rangle, j_{34} \in \langle j_{12}, j \rangle \end{array} \right\},$$

(8.92)

$$\mathbf{B}_2 = \left\{ \left| \begin{array}{c} j_2 \ j_3 \ j_4 \ j_1 \\ j_{23} \diagdown \diagup j_{41} \\ jm \end{array} \right\rangle \, \middle| \, \begin{array}{c} j_{23} \in \langle j_2, j_3 \rangle, j_{41} \in \langle j_4, j_1 \rangle, \\ j_{23} \in \langle j_{41}, j \rangle, j_{41} \in \langle j_{23}, j \rangle \end{array} \right\};$$

$$\mathbf{B}' = \left\{ \left| \begin{array}{c} j_1 \ j_3 \ j_2 \ j_4 \\ j_{13} \diagdown \diagup j_{24} \\ jm \end{array} \right\rangle \, \middle| \, \begin{array}{c} j_{13} \in \langle j_1, j_3 \rangle, j_{24} \in \langle j_2, j_4 \rangle, \\ j_{13} \in \langle j_{24}, j \rangle, j_{24} \in \langle j_{13}, j \rangle \end{array} \right\}. \qquad (8.93)$$

It is convenient here to use the indicated notations for intermediate quantum numbers. In the enumeration of the basis vectors the quantum numbers j_1, j_2, j_3, j_4 and $j \in \{j_{\min}, j_{\min}+1, \ldots, j_{\max}\}$ are all taken as specified, as is $m \in \{j, j-1, \ldots, -j\}$. The intermediate quantum numbers take the values shown, and enumerate a number of orthonormal vectors in each basis set equal to the CG number $N_j(j_1, j_2, j_3, j_4)$; that is, the cardinality of the basis sets $\mathbf{B_1}, \mathbf{B_2}, \mathbf{B'}$ is the same CG number. The binary trees in these three sets of basis vectors all have the shape $\mathrm{Sh}_T = ((\circ\circ)(\circ\circ))$ and correspond to the following coupling schemes:

$$\begin{aligned}\text{scheme I}: \mathbf{j}_1 &= ((j_1 j_2)(j_3 j_4)), \\ \text{scheme II}: \mathbf{j}_2 &= ((j_2 j_3)(j_4 j_1)), \\ \text{diagonal}: \mathbf{j'} &= ((j_1 j_3)(j_2 j_4)),\end{aligned} \qquad (8.94)$$

where diagonal refers to the diagonal coupling scheme given in great detail in relations (8.34)-(8.39). (We reserve $\mathbf{j}_3 = ((j_3 j_4)(j_1 j_2))$ for subsequent use as a cyclic permutation of $\mathbf{j}_1$.)

The determination of the matrix elements of

$$H_{\mathbf{B}_1} = D^{\mathbf{j}_1} + R^{\mathbf{j}_2;\mathbf{j}_1 \, tr} D^{\mathbf{j}_2} R^{\mathbf{j}_2;\mathbf{j}_1}, \quad R^{\mathbf{j}_2;\mathbf{j}_1 \, tr} = R^{\mathbf{j}_1;\mathbf{j}_2} \qquad (8.95)$$

requires those of the recoupling matrix $R^{\mathbf{j}_1;\mathbf{j}_2}$. (since T is always the same binary tree shown in (8.92)-(8.93), we drop it from the notation in (8.95).) The matrix elements are given by the irreducible triangle coefficient of order six defined by

$$\left(R^{\mathbf{j}_1;\mathbf{j}_2}\right)_{j_{12},j_{34},j;j_{23},j_{41},j} = \left\{\begin{array}{ccc|ccc} j_1 & j_3 & j_{12} & j_2 & j_4 & j_{23} \\ j_2 & j_4 & j_{34} & j_3 & j_1 & j_{41} \\ j_{12} & j_{34} & j & j_{23} & j_{41} & j \end{array}\right\}. \qquad (8.96)$$

We give the explicit expression for this form of the $9-j$ coefficient below in relation (8.110).

The diagonal matrix $D^{\mathbf{j'}}$ with elements given by (8.38) and representing H on the basis $\mathbf{B'}$ is obtained from the matrix $H_{\mathbf{B}_1}$ by an orthogonal similarity transformation R, as shown by (8.61): $R^{tr} H_{\mathbf{B}_1} R = D^{\mathbf{j'}}$. But we can now show that R can be chosen to be $R = R^{\mathbf{j}_1;\mathbf{j'}}$:

$$R^{\mathbf{j}_1;\mathbf{j'} \, tr} H_{\mathbf{B}_1} R^{\mathbf{j}_1;\mathbf{j'}} = D^{\mathbf{j'}}. \qquad (8.97)$$

Proof. The relation between basis vectors of $\mathbf{B'}$ and $\mathbf{B}_1$ is:

$$\left|\begin{array}{c} j_1 \; j_3 \; j_2 \; j_4 \\ j_{13} \quad j_{24} \\ jm \end{array}\right\rangle = \sum_{j_{12},j_{34}} \left(R^{\mathbf{j}_1;\mathbf{j'}}\right)_{j_{12},j_{34},j;j_{13},j_{24},j} \left|\begin{array}{c} j_1 \; j_2 \; j_3 \; j_4 \\ j_{12} \quad j_{34} \\ jm \end{array}\right\rangle. \qquad (8.98)$$

8.4. HEISENBERG HAMILTONIAN: EVEN n

The matrix elements of the recoupling matrix $R^{j_1;j'}$ are given in terms of the triangle coefficient of the transformation by

$$\left(R^{j_1;j'}\right)_{j_{12},j_{34},j;j_{13},j_{24},j} = \left\{ \begin{array}{ccc|ccc} j_1 & j_3 & j_{12} & j_1 & j_2 & j_{13} \\ j_2 & j_4 & j_{34} & j_3 & j_4 & j_{24} \\ j_{12} & j_{34} & j & j_{13} & j_{24} & j \end{array} \right\}. \quad (8.99)$$

Thus, the diagonal matrix $D^{j'}$ is given by $D^{j'} = R^{j_1;j'\,tr} H_{\mathbf{B}_1} R^{j_1;j'}$, and we find that the similarity transformation by the real orthogonal matrix $R = R^{j_1;j'}$ diagonalizes the matrix $H_{\mathbf{B}_1}$ in (8.97). □

Relation (8.97) can be written in a symmetric form by substituting $H_{\mathbf{B}_1}$ from (8.55) into (8.97) and using relations $R^{j_1;j'\,tr} = R^{j';j_1}$ and $R^{j_2;j_1} R^{j_1;j'} = R^{j_2;j'}$ from (8.63) to obtain:

$$R^{j';j_1} D^{j_1} R^{j_1;j'} + R^{j';j_2} D^{j_2} R^{j_2;j'} = D^{j'}. \quad (8.100)$$

It is quite interesting that this relation implies a new relation between $9-j$ coefficients, and its specialization to $j = 0$ implies a new relation between Racah coefficients. Since the $9-j$ and $6-j$ coefficients are among the most important coefficients in angular momentum theory, it is worthwhile to give these new identities.

The diagonal matrices appearing in (8.100) are given, respectively, by (8.48), (8.49), (8.38). If we write diagonal elements of a matrix in the notation $D_{i,j} = \delta_{i,j} D_i$, then we have the following diagonal elements:

$$\left(D^{j_1}\right)_{j_{12},j_{34},j} = \tfrac{1}{2}\left(-\sum_{i=1}^{4} j_i(j_i+1) + j_{12}(j_{12}+1) + j_{34}(j_{34}+1)\right),$$

$$\left(D^{j_2}\right)_{j_{23},j_{41},j} = \tfrac{1}{2}\left(-\sum_{i=1}^{4} j_i(j_i+1) + j_{23}(j_{23}+1) + j_{41}(j_{41}+1)\right),$$

$$\left(D^{j'}\right)_{j_{13},j_{24},j} = \tfrac{1}{2}\left(j(j+1) - j_{13}(j_{13}+1) - j_{24}(j_{24}+1)\right). \quad (8.101)$$

We next take matrix elements of (8.100), exercising caution to provide the proper summation indices over intermediate quantum numbers. This gives the following identity among $9-j$ coefficients, where we optimally leave the relation in terms of triangle coefficients of order six:

$$\sum_{j_{12},j_{34}} \Big(j_{12}(j_{12}+1) + j_{34}(j_{34}+1)\Big)$$

$$\times \left\{ \begin{array}{ccc|ccc} j_1 & j_3 & j_{12} & j_1 & j_2 & j_{13} \\ j_2 & j_4 & j_{34} & j_3 & j_4 & j_{24} \\ j_{12} & j_{34} & j & j_{13} & j_{24} & j \end{array} \right\} \left\{ \begin{array}{ccc|ccc} j_1 & j_3 & j_{12} & j_1 & j_2 & j'_{13} \\ j_2 & j_4 & j_{34} & j_3 & j_4 & j'_{24} \\ j_{12} & j_{34} & j & j'_{13} & j'_{24} & j \end{array} \right\}$$

$$+ \sum_{j_{23},j_{41}} \Big(j_{23}(j_{23}+1) + j_{41}(j_{41}+1)\Big)$$

$$\times \left\{ \begin{array}{ccc|ccc} j_2 & j_4 & j_{23} & j_1 & j_2 & j_{13} \\ j_3 & j_1 & j_{41} & j_3 & j_4 & j_{24} \\ j_{23} & j_{41} & j & j_{13} & j_{24} & j \end{array} \right\} \left\{ \begin{array}{ccc|ccc} j_2 & j_4 & j_{23} & j_1 & j_2 & j'_{13} \\ j_3 & j_1 & j_{41} & j_3 & j_4 & j'_{24} \\ j_{23} & j_{41} & j & j'_{13} & j'_{24} & j \end{array} \right\}$$

$$= \delta_{j_{13},j'_{13}} \delta_{j_{24},j'_{24}}$$

$$\times \Big(2\sum_{i=1}^{4} j_i(j_i+1) + j(j+1) - j_{13}(j_{13}+1) - j_{24}(j_{24}+1)\Big). \qquad (8.102)$$

In deriving this relation, we have used the orthogonality of the triangle coefficients to move the terms $\sum_i j_i(j_i+1)$ in the diagonal matrices on the left-hand side to the right-hand side. We have also use the symmetry of the triangle coefficients under interchange of left and right patterns to obtain this final form.

The transition from $9-j$ coefficients to $6-j$ (Racah) coefficients is made by setting $j=0$. This relation is the following, when expressed in terms of triangle coefficients:

$$\left\{ \begin{array}{ccc|ccc} a & c & k_1 & a & b & k'_1 \\ b & d & k_2 & c & d & k'_2 \\ k_1 & k_2 & 0 & k'_1 & k'_2 & 0 \end{array} \right\} = \delta_{k_1,k_2} \delta_{k'_1,k'_2} \left\{ \begin{array}{cc|cc} a & k_1 & a & k'_1 \\ b & c & c & b \\ k_1 & d & k'_1 & d \end{array} \right\}. \qquad (8.103)$$

Choosing $j=0$ in the left-hand side of the $9-j$ coefficient (8.103) imposes additional conditions on the domain of definition of the assigned external angular momenta a,b,c,d; namely, that a,b,c,d must be such that all the triangle rules for the triples $(a,b,k_1), (c,d,k_1), (a,c,k'_1), (b,d,k'_1)$ are satisfied. If the selected angular momenta a,b,c,d do not satisfy these triangle conditions, setting $j=0$ is a contradiction to the domain of definition of j for specified $a,b,c,d \in \{0,1/2,1,\ldots\}$. Caution must be exercised in setting $j=0$ in relations between $9-j$ coefficients to ensure that no such contradictions arise.

The following relation between triangle coefficients of order 4 can be derived from (8.102) by setting $j=0$: both sides are 0, unless $j_{13} = j_{24}$

8.4. HEISENBERG HAMILTONIAN: EVEN n

and $j'_{13} = j'_{24}$, in which case the relation reduces to the following:

$$\sum_k k(k+1) \left\{ \begin{array}{cc|cc} a & k & a & k'' \\ b & c & c & b \\ k & d & k'' & d \end{array} \right\} \left\{ \begin{array}{cc|cc} a & k & a & k''' \\ b & c & c & b \\ k & d & k''' & d \end{array} \right\} \quad (8.104)$$

$$+ \sum_{k'} k'(k'+1) \left\{ \begin{array}{cc|cc} b & k' & a & k'' \\ c & d & c & b \\ k' & a & k'' & d \end{array} \right\} \left\{ \begin{array}{cc|cc} b & k' & a & k''' \\ c & d & c & b \\ k' & a & k''' & d \end{array} \right\}$$

$$= \delta_{k'',k'''} \Big(a(a+1) + b(b+1) + c(c+1) + d(d+1) - k''(k''+1) \Big).$$

This result is a consequence of setting $j = 0, j_1 = a, j_2 = b, j_3 = c, j_4 = d, j_{12} = k, j_{23} = k', j_{13} = k'', j'_{13} = k'''$ in (8.102) and using property (8.103).

The validity of (8.104) can be checked for the maximal case $d = a + b + c$, where all quantum numbers (including the summations) are forced to their maximum values, $k = a+b, k' = b+c, k'' = k''' = a+c$; and each triangle coefficient equals 1. Thus, the left-hand sum is just the two terms $(a+b)(a+b+1) + (b+c)(b+c+1)$, which agrees with the right-hand side. Relation (8.104) can, of course, be expressed in terms of conventional Racah coefficients or $6-j$ coefficients.

A similar result where $6-j$ coefficients play a role in diagonalizing a special symmetric matrix, which itself is a special symmetric $6-j$ coefficient, has been given by Rose and Yang [69]. The Racah sum rule is a consequence of the first multiplication rule in the relations

$$R^{(ab)c;(bc)a} R^{(bc)a;(ac)b} = R^{(ab)c;(ac)b};$$

(8.105)

$$R^{(bc)(da);(ab)(cd)} R^{(ab)(cd);(ac)(bd)} = R^{(bc)(da);(ac)(bd)}$$

for recoupling matrices; it is the important result for the derivation of the Rose and Yang identity, just as the second relation is for the derivation of (8.102) from which follows (8.104).

8.4.6 The General $3(2f-1)-j$ Coefficients

The $9-j$ coefficient factor w_4

The Wigner $9-j$ coefficient for the path $w_4 = AACAC$ given by (8.106)-(8.107) below occurs as a factor in the path w_{2f} given by (8.88) for the general $3(2f-1)-j$ coefficient. We first give the expression for this coefficient in terms of triangle coefficients of order four before turning to the general case, summarizing again the properties of w_4:

$$((23)(41)) \xrightarrow{w_4} ((12)(34)),$$
$$w_4 = w_4(A,C) = AACAC;$$
(8.106)

$$((23)(41)) \xrightarrow{C} ((23)(14)) \xrightarrow{A} \left(((23)1)4\right)$$
$$\xrightarrow{C} \left((1(23))4\right) \xrightarrow{A} \left(((12)3)4\right) \xrightarrow{A} ((12)(34)).$$

$$R^{((23)(41));((12)(34))} = R^{((23)(41));((23)(14))} R^{((23)(14));(((23)1)4)}$$
$$\times R^{(((23)1)4);((1(23))4)} R^{((1(23))4);(((12)3)4)} R^{(((12)3)4);((12)(34))}, \quad (8.107)$$

$$R^{((23)(41));((12)(34))} = R^{((23)(41));((23)1)4)} R^{((23)1)4;((12)3)4)}$$
$$\times R^{((12)3)4);((12)(34))},$$
$$W(U,V) = W(U,V')W(V',V'')W(V'',V), \quad (8.108)$$
$$U = C^{((23)(41))\ tr}, \quad V' = C^{((23)1)4)\ tr},$$
$$V'' = C^{((12)3)4)\ tr}, \quad V = C^{((12)(34))\ tr}.$$

It is not necessary to carry out the calculation resulting from the matrix elements of relation (8.107), since the result can be obtained from relation (5.260) by setting $a = j_2, b = j_3, c = j_1, d = j_4$, which gives the following relation:

$$\sqrt{(2k_1+1)(2k_2+1)(2k'_1+1)(2k'_2+1)} \begin{Bmatrix} j_2 & j_3 & k'_1 \\ j_1 & j_4 & k'_2 \\ k_1 & k_2 & j \end{Bmatrix} = \begin{Bmatrix} j_2 & j_3 & k_1 \\ j_1 & j_4 & k_2 \\ k_1 & k_2 & j \end{Bmatrix} \begin{Bmatrix} j_2 & j_1 & k'_1 \\ j_3 & j_4 & k'_2 \\ k'_1 & k'_2 & j \end{Bmatrix}$$
$$= \sum_r \begin{Bmatrix} j_3 & k_1 & j_3 & r \\ j_4 & k_2 & k_1 & j_4 \\ k_2 & j & r & j \end{Bmatrix} \begin{Bmatrix} j_2 & j_3 & j_2 & k'_1 \\ j_1 & k_1 & j_3 & j_1 \\ k_1 & r & k'_1 & r \end{Bmatrix} \begin{Bmatrix} k'_1 & r & j_1 & k'_1 \\ j_1 & j_4 & j_4 & k'_2 \\ r & j & k'_2 & j \end{Bmatrix}. \quad (8.109)$$

By obvious primitive phase transformations of this relation in both the left- and right-hand sides, interchange of left and right patterns in a triangle coefficient, and the reordering of the factors under the summation, it can be brought to the form (8.110) below. This is the natural form that occurs when the matrix elements of relation (8.107)-(8.108) are evaluated directly. We have done the latter calculation as well, thus obtaining two independent verifications of the relation as follows:

8.4. HEISENBERG HAMILTONIAN: EVEN n

$$\left(R^{((23)(41));((12)(34))}\right)_{k'_1,k'_2,j;k_1,k_2,j}$$

$$= \begin{Bmatrix} j_2 & j_4 & k'_1 \\ j_3 & j_1 & k'_2 \\ k'_1 & k'_2 & j \end{Bmatrix} \begin{Bmatrix} j_1 & j_3 & k_1 \\ j_2 & j_4 & k_2 \\ k_1 & k_2 & j \end{Bmatrix} = \begin{Bmatrix} j_1 & j_3 & k_1 \\ j_2 & j_4 & k_2 \\ k_1 & k_2 & j \end{Bmatrix} \begin{Bmatrix} j_2 & j_4 & k'_1 \\ j_3 & j_1 & k'_2 \\ k'_1 & k'_2 & j \end{Bmatrix}$$

$$= \sum_r \begin{Bmatrix} j_4 & k'_1 & k'_1 & r \\ j_1 & k'_2 & j_1 & j_4 \\ k'_2 & j & r & j \end{Bmatrix} \begin{Bmatrix} j_2 & k'_1 & j_1 & j_3 \\ j_3 & j_1 & j_2 & k_1 \\ k'_1 & r & k_1 & r \end{Bmatrix} \begin{Bmatrix} j_3 & r & j_3 & k_1 \\ k_1 & j_4 & j_4 & k_2 \\ r & j & k_2 & j \end{Bmatrix}. \quad (8.110)$$

Of course, we also have from (8.109) the relation between triangle coefficients of order six and the Wigner $9-j$ coefficients:

$$\begin{Bmatrix} j_1 & j_3 & k_1 \\ j_2 & j_4 & k_2 \\ k_1 & k_2 & j \end{Bmatrix} \begin{Bmatrix} j_2 & j_4 & k'_1 \\ j_3 & j_1 & k'_2 \\ k'_1 & k'_2 & j \end{Bmatrix} = (-1)^{j_1+j_2-k_1}(-1)^{j_1+j_4-k'_2}$$

$$\times \sqrt{(2k_1+1)(2k_2+1)(2k'_1+1)(2k'_2+1)} \begin{Bmatrix} j_2 & j_3 & k'_1 \\ j_1 & j_4 & k'_2 \\ k_1 & k_2 & j \end{Bmatrix}. \quad (8.111)$$

8.4.7 The General $3(2f-1) - j$ Coefficients Continued

We next give the formula for the general $3(2f-1)-j$ coefficient, based on the shape transformation (8.88). But now, as for w_4 above, we calculate first the $9-j$ coefficient for the repeated subpath $w = CACAAC$. The relevant properties of w are:

$$((ab)(cd)) \overset{C}{\to} ((cd)(ab)) \overset{A}{\to} (c(d(ab))) \overset{A}{\to} (c((da)b))$$

$$\overset{C}{\to} (((da)b)c) \overset{A}{\to} ((da)(bc)) \overset{C}{\to} ((ad)(bc));$$

$$((ab)(cd)) \overset{w}{\to} ((ad)(bc)), \quad w = CACAAC.$$

$$R^{((ab)(cd));((ad)(bc))} \quad (8.112)$$

$$= R^{((ab)(cd));(c(d(ab)))} R^{(c(d(ab)));(c((da)b))} R^{(c((da)b));((ad)(bc))};$$

$$W(U,V) = W(U,V')W(V',V'')W(V'',V),$$

$$U = C^{((ab)(cd))\,tr}, \quad V' = C^{(c(d(ab)))\,tr},$$

$$V'' = C^{(c((da)b))\,tr}, \quad V = C^{((ad)(bc))\,tr}.$$

We have used the rule (5.281) in writing the product of recoupling matrices in the indicated form that incorporates the C−transformations.

We obtain the following result for these triangle coefficients:

$$\left(R^{((ab)(cd));((ad)(bc))}\right)_{k_1,k_2,j;\,k_1',k_2',j}$$

$$= \begin{Bmatrix} a & c & k_1 \\ b & d & k_2 \\ k_1 & k_2 & j \end{Bmatrix} \begin{vmatrix} a & b & k_1' \\ d & c & k_2' \\ k_1' & k_2' & j \end{vmatrix} = \begin{Bmatrix} a & b & k_1' \\ d & c & k_2' \\ k_1' & k_2' & j \end{Bmatrix} \begin{vmatrix} a & c & k_1 \\ b & d & k_2 \\ k_1 & k_2 & j \end{vmatrix}$$

$$= (-1)^{c+d-k_2}\sqrt{(2k_1+1)(2k_2+1)(2k_1'+1)(2k_2'+1)} \begin{Bmatrix} a & d & k_1' \\ b & c & k_2' \\ k_1 & k_2 & j \end{Bmatrix}$$

$$= \sum_r \begin{Bmatrix} c & k_1 & d \\ d & k_2 & k_1 \\ k_2 & j & r \end{Bmatrix} \begin{Bmatrix} a & d & a \\ b & k_1 & d \\ k_1 & r & k_1' \end{Bmatrix} \begin{Bmatrix} k_1' & c & b \\ b & r & c \\ r & j & k_2' \end{Bmatrix} \begin{Bmatrix} k_1' & c & b \\ r & c & k_2' \\ j & k_2' & j \end{Bmatrix}. \quad (8.113)$$

This result is obtained from (5.260) by interchanging c with d and (k_1,k_2) with (k_1',k_2'), making the appropriate phase factor modifications by the rule (5.135), and using the invariance under interchange of left and right triangle patterns. It was also independently derived directly from the matrix elements of the product of recoupling matrices given in (8.112).

The next step is to implement the shape transformation and accompanying recoupling matrix product given by

$$w_{2f} = w_{2f-2}\,A w A,$$

$$S_{2f}(2) \xrightarrow{A} S_{2f}(3) \xrightarrow{w} S_{2f}(4) \xrightarrow{A} S_{2f}(5) \xrightarrow{w_{2f-2}} S_{2f}(1), \quad (8.114)$$

where these abbreviated shapes are defined by (see (8.76)):

$$\mathrm{Sh}_{2f}(2) = \mathrm{Sh}_{2f}(2,3,\ldots,2f-1,2f,1) = \left(\left(X_{2f-4}(ab)\right)(cd)\right),$$

$$\mathrm{Sh}_{2f}(3) = \left(\left(X_{2f-4}\big((ab)(cd)\big)\right)\right),\ \mathrm{Sh}_{2f}(4) = \left(\left(X_{2f-4}\big((ad)(bc)\big)\right)\right),$$

$$\mathrm{Sh}_{2f}(5) = \left(\left(X_{2f-4}(ad)\right)(bc)\right),\ \mathrm{Sh}_{2f}(1) = \mathrm{Sh}_{2f}(1,2,\ldots,2f-1,2f).$$

$$(8.115)$$

We have also written $a = 2f-2,\ b = 2f-1,\ c = 2f,\ d = 1$, and defined the shape X_{2f-4} by

$$X_{2f-4} = \underbrace{\Big(\cdots\Big(\big((23)(45)\big)(67)\Big)\cdots\Big)}_{f-3}(2f-4\,2f-3). \quad (8.116)$$

8.4. HEISENBERG HAMILTONIAN: EVEN n

Thus, corresponding to the shape transformation (8.114), we have the following product of recoupling matrices:

$$R^{\mathrm{Sh}_{2f}(2);\,\mathrm{Sh}_{2f}(1)} = R^{\mathrm{Sh}_{2f}(2);\,\mathrm{Sh}_{2f}(3)} R^{\mathrm{Sh}_{2f}(3);\,\mathrm{Sh}_{2f}(4)}$$
$$\times R^{\mathrm{Sh}_{2f}(4);\,\mathrm{Sh}_{2f}(5)} R^{\mathrm{Sh}_{2f}(5);\,\mathrm{Sh}_{2f}(1)}. \quad (8.117)$$

We next take matrix elements $(\mathbf{k}',\,j;\,\mathbf{k},\,j) = (\mathbf{k}_2,\,j;\,\mathbf{k}_1,\,j)$ of (8.117):

$$\left(R^{\mathrm{Sh}_{2f}(2);\,\mathrm{Sh}_{2f}(1)}\right)_{\mathbf{k}',j;\mathbf{k},j} \quad (8.118)$$

$$= \left\{\Delta_{T_{2f}}(\mathbf{j}_2\,\mathbf{k}_2)_j\,\big|\,\Delta_{T_{2f}}(\mathbf{j}_1\,\mathbf{k}_1)_j\right\} = \left\{\Delta_{T_{2f}}(\mathbf{j}_1\,\mathbf{k}_1)_j\,\big|\,\Delta_{T_{2f}}(\mathbf{j}_2\,\mathbf{k}_2)_j\right\}$$

$$= \sum_{\mathbf{k}'',\mathbf{k}''',\mathbf{k}''''} \left(R^{\mathrm{Sh}_{2f}(2);\,\mathrm{Sh}_{2f}(3)}\right)_{\mathbf{k}',j;\mathbf{k}'',j} \left(R^{\mathrm{Sh}_{2f}(3);\,\mathrm{Sh}_{2f}(4)}\right)_{\mathbf{k}'',j;\mathbf{k}''',j}$$

$$\times \left(R^{\mathrm{Sh}_{2f}(4);\,\mathrm{Sh}_{2f}(5)}\right)_{\mathbf{k}''',j;\mathbf{k}'''',j} \left(R^{\mathrm{Sh}_{2f}(5);\,\mathrm{Sh}_{2f}(1)}\right)_{\mathbf{k}'''',j;\mathbf{k},j}.$$

The explicit forms of the irreducible triangle coefficients on the left-hand side of (8.118) are given by (see (8.66)-(8.67)):

$$\left\{\Delta_{T_{2f}}(\mathbf{j}_1\,\mathbf{k}_1)_j\,\big|\,\Delta_{T_{2f}}(\mathbf{j}_2\,\mathbf{k}_2)_j\right\} = \left\{\begin{array}{ccccccccc} j_1 & j_3 & k_1 & j_5 & k_3 & \cdots & j_{2f-1} & k_{2f-3} \\ j_2 & j_4 & k_2 & j_6 & k_4 & \cdots & j_{2f} & k_{2f-2} \\ k_1 & k_2 & k_3 & k_4 & k_5 & \cdots & k_{2f-2} & j \end{array}\right.$$

$$\left.\begin{array}{ccccccc} j_2 & j_4 & k_1' & j_6 & k_3' & \cdots & j_{2f} & k_{2f-3}' \\ j_3 & j_5 & k_2' & j_7 & k_4' & \cdots & j_1 & k_{2f-2}' \\ k_1' & k_2' & k_3' & k_4' & k_5' & \cdots & k_{2f-2}' & j \end{array}\right\},$$

$$(8.119)$$

$$\left\{\Delta_{T_{2f}}(\mathbf{j}_2\,\mathbf{k}_2)_j\,\big|\,\Delta_{T_{2f}}(\mathbf{j}_1\,\mathbf{k}_1)_j\right\} = \left\{\begin{array}{ccccccc} j_2 & j_4 & k_1' & j_6 & k_3' & \cdots & j_{2f} & k_{2f-3}' \\ j_3 & j_5 & k_2' & j_7 & k_4' & \cdots & j_1 & k_{2f-2}' \\ k_1' & k_2' & k_3' & k_4' & k_5' & \cdots & k_{2f-2}' & j \end{array}\right.$$

$$\left.\begin{array}{ccccccc} j_1 & j_3 & k_1 & j_5 & k_3 & \cdots & j_{2f-1} & k_{2f-3} \\ j_2 & j_4 & k_2 & j_6 & k_4 & \cdots & j_{2f} & k_{2f-2} \\ k_1 & k_2 & k_3 & k_4 & k_5 & \cdots & k_{2f-2} & j \end{array}\right\}.$$

We next evaluate the matrix elements of each of the four factors in the right-hand side of relation (8.118), take into account the common forks as determined by the shapes (8.115)-(8.116), and use the reduction rule (5.134) to obtain the following sequence of relations:

$$\left(R^{\text{Sh}_{2f}(2);\,\text{Sh}_{2f}(3)}\right)_{\mathbf{k}',j;\,\mathbf{k}'',j} = \Big(\prod_{i=1}^{2f-6}\delta_{k'_i,k''_i}\Big)\delta_{k'_{2f-5},k''_{2f-3}}\delta_{k'_{2f-4},k''_{2f-5}}$$

$$\times \delta_{k'_{2f-2},k''_{2f-4}}\left\{\begin{array}{ccc|cc} k'_{2f-5} & k'_{2f-3} & k'_{2f-4} & k'_{2f-5} \\ k'_{2f-4} & k'_{2f-2} & k''_{2f-2} & k''_{2f-2} \\ k'_{2f-3} & j & k''_{2f-2} & j \end{array}\right\};$$

$$\left(R^{\text{Sh}_{2f}(3);\,\text{Sh}_{2f}(4)}\right)_{\mathbf{k}'',j;\,\mathbf{k}''',j} = \Big(\prod_{i=1}^{2f-6}\delta_{k''_i,k'''_i}\Big)\delta_{k''_{2f-3},k'''_{2f-3}}\delta_{k''_{2f-2},k'''_{2f-2}}$$

$$\times \left\{\begin{array}{ccc|ccc} j_{2f-2} & j_{2f} & k''_{2f-5} & j_{2f-2} & j_{2f-1} & k'''_{2f-5} \\ j_{2f-1} & j_1 & k''_{2f-4} & j_1 & j_{2f} & k'''_{2f-4} \\ k''_{2f-5} & k''_{2f-4} & k''_{2f-2} & k'''_{2f-5} & k'''_{2f-4} & k'''_{2f-2} \end{array}\right\};\quad (8.120)$$

$$\left(R^{\text{Sh}_{2f}(4);\,\text{Sh}_{2f}(5)}\right)_{\mathbf{k}''',j;\,\mathbf{k}'''',j} = \Big(\prod_{i=1}^{2f-6}\delta_{k'''_i,k''''_i}\Big)\delta_{k'''_{2f-5},k''''_{2f-4}}\delta_{k'''_{2f-4},k''''_{2f-2}}$$

$$\times \delta_{k'''_{2f-3},k''''_{2f-5}}\left\{\begin{array}{ccc|cc} k'''_{2f-5} & k'''_{2f-3} & k''''_{2f-5} & k''''_{2f-3} \\ k'''_{2f-4} & k'''_{2f-2} & k''''_{2f-4} & k''''_{2f-2} \\ k'''_{2f-2} & j & k''''_{2f-3} & j \end{array}\right\};$$

$$\left(R^{\text{Sh}_{2f}(5);\,\text{Sh}_{2f}(1)}\right)_{\mathbf{k}'''',j;\,\mathbf{k},j} = \delta_{k''''_{2f-2},k_{2f-2}}\delta_{k''''_{2f-3},k_{2f-3}}$$

$$\times \left\{\Delta_{T_{2f-2}}(j_2,\ldots,j_{2f-2},j_1;k'_1,\ldots,k'_{2f-4})_{k_{2f-3}}\right.$$

$$\left.\Big|\Delta_{T_{2f-2}}(j_1,\ldots,j_{2f-2};k_1,\ldots,k_{2f-4})_{k_{2f-3}}\right\}.$$

We now take into account all of the Kronecker delta factors in the five relations (8.120). The idea here is to relate as many as possible to the $(k'_1, k'_2, \ldots, k'_{2f-2})$ and $(k_1, k_2, \ldots, k_{2f-2})$, since these are specified by the left-hand side of relation (8.118), and to assign the remaining k–type labels as the summation parameters. Thus, we arrive at

$$k''''_i = k'''_i = k''_i = k'_i,\ i = 1, 2, \ldots, 2f-6;\ k''_{2f-5} = k'_{2f-4},\ k''_{2f-4} = k'_{2f-2};$$

$$k''''_{2f-5} = k'''_{2f-3} = k''_{2f-3} = k'_{2f-5};\ k'''_{2f-2} = k''_{2f-2} = r;\qquad (8.121)$$

$$k''''_{2f-3} = k_{2f-3},\ k''''_{2f-2} = k_{2f-2};\ k'''_{2f-5} = s,\ k'''_{2f-4} = t.$$

8.4. HEISENBERG HAMILTONIAN: EVEN n

Combining relations (8.118)-(8.121), we obtain the following result:

The $3(2f-1)-j$ coefficient recursion relation:

$$\left\{\Delta_{T_{2f}}(j_2,\ldots,j_{2f},j_1;k'_1,\ldots,k'_{2f-2})_j \middle| \Delta_{T_{2f}}(j_1,\ldots,j_{2f};k_1,\ldots,k_{2f-2})_j\right\}$$

$$= \sum_{r,s,t} \left\{\begin{array}{ccc|cc} k'_{2f-5} & k'_{2f-3} & k'_{2f-4} & k'_{2f-3} \\ k'_{2f-4} & k'_{2f-2} & k'_{2f-2} & r \\ k'_{2f-3} & j & r & j \end{array}\right\}$$

$$\times \left\{\begin{array}{ccc|ccc} j_{2f-2} & j_{2f} & k'_{2f-4} & j_{2f-2} & j_{2f-1} & s \\ j_{2f-1} & j_1 & k'_{2f-2} & j_1 & j_{2f} & t \\ k'_{2f-4} & k'_{2f-2} & r & s & t & r \end{array}\right\}$$

$$\times \left\{\begin{array}{ccc|ccc} s & k'_{2f-5} & k'_{2f-5} & k_{2f-3} \\ t & r & t & k_{2f-2} \\ r & j & k_{2f-2} & j \end{array}\right\}$$

$$\times \left\{\Delta_{T_{2f-2}}\left(j_2,\ldots,j_{2f-2},j_1;k'_1,\ldots,k'_{2f-5},s\right)_{k_{2f-3}}\right.$$

$$\left. \middle| \Delta_{T_{2f-2}}\left(j_1,\ldots,j_{2f-2};k_1,\ldots,k_{2f-4}\right)_{k_{2f-3}}\right\}. \tag{8.122}$$

The detailed expression for the triangle coefficient of order $2f-2$ (the $3(2f-1)-j$ coefficient) on the left in this result is given by either of relations (8.119), while the similar detailed expression for the triangle coefficient of order $2f-4$ the $3(2f-3)-j$ coefficient under the summation is given by the relation:

$$\left\{\Delta_{T_{2f-2}}\left(j_2,\ldots,j_{2f-2},j_1;k'_1,\ldots,k'_{2f-5},s\right)_{k_{2f-3}}\right.$$

$$\left. \middle| \Delta_{T_{2f-2}}\left(j_1,\ldots,j_{2f-2};k_1,\ldots,k_{2f-4}\right)_{k_{2f-3}}\right\}$$

$$= \left\{\begin{array}{ccccccc} j_2 & j_4 & k'_1 & j_6 & k'_3 & \cdots & j_{2f-2} & k'_{2f-5} \\ j_3 & j_5 & k'_2 & j_7 & k'_4 & \cdots & j_1 & s \\ k'_1 & k'_2 & k'_3 & k'_4 & k'_5 & \cdots & s & k_{2f-3} \end{array}\right. \tag{8.123}$$

$$\left. \middle| \begin{array}{ccccccc} j_1 & j_3 & k_1 & j_5 & k_3 & \cdots & j_{2f-3} & k_{2f-5} \\ j_2 & j_4 & k_2 & j_6 & k_4 & \cdots & j_{2f-2} & k_{2f-4} \\ k_1 & k_2 & k_3 & k_4 & k_5 & \cdots & k_{2f-4} & k_{2f-3} \end{array}\right\}.$$

Observe that the summation index s occurs in this expression, and the coupling is to the "total" angular momentum $j' = k_{2f-3}$. We also note that the triangle coefficient of order six under the summation in (8.122) is given in terms of Wigner's $9-j$ coefficient by (8.113) above:

$$\left\{\begin{array}{ccc|ccc}j_{2f-2} & j_{2f} & k'_{2f-4} & j_{2f-2} & j_{2f-1} & s \\ j_{2f-1} & j_1 & k'_{2f-2} & j_1 & j_{2f} & t \\ k'_{2f-4} & k'_{2f-2} & r & s & t & r\end{array}\right\} = (-1)^{j_1+j_{2f}-k'_{2f-2}}$$

$$\times \sqrt{(2s+1)(2t+1)(2k'_{2f-4}+1)(2k'_{2f-2}+1)} \left\{\begin{array}{ccc}j_{2f-2} & j_1 & s \\ j_{2f-1} & j_{2f} & t \\ k'_{2f-4} & k'_{2f-2} & r\end{array}\right\}.$$

(8.124)

The starting point of the iteration of relation (8.122) is $f = 3$; that is, with the binary tree T_4 in (8.42), which gives the $9-j$ triangle coefficient of order six for the shape transformation w calculated in (8.109).

Independently of the general derivation of relation (8.122), we also derived the special case for $f = 4$, with the following result:

$$\left\{\begin{array}{ccccccc|ccccccc}j_2 & j_4 & k'_1 & j_6 & k'_3 & j_8 & k'_5 & j_1 & j_3 & k_1 & j_5 & k_3 & j_7 & k_5 \\ j_3 & j_5 & k'_2 & j_7 & k'_4 & j_1 & k'_6 & j_2 & j_4 & k_2 & j_6 & k_4 & j_8 & k_6 \\ k'_1 & k'_2 & k'_3 & k'_4 & k'_5 & k'_6 & j & k_1 & k_2 & k_3 & k_4 & k_5 & k_6 & j\end{array}\right\}$$

$$= \sum_{r,s,t} \left\{\begin{array}{cc|cc}k'_3 & k'_5 & k'_4 & k'_5 \\ k'_4 & k'_6 & k'_6 & r \\ k'_5 & j & r & j\end{array}\right\} \left\{\begin{array}{ccc|ccc}j_6 & j_8 & k'_4 & j_6 & j_7 & s \\ j_7 & j_1 & k'_6 & j_1 & j_8 & t \\ k'_4 & k'_6 & r & s & t & r\end{array}\right\}$$ (8.125)

$$\times \left\{\begin{array}{cc|cc}s & k'_3 & k'_3 & k_5 \\ t & r & t & k_6 \\ r & j & k_6 & j\end{array}\right\} \left\{\begin{array}{cccc|cccc}j_2 & j_4 & k'_1 & j_6 & k'_3 & j_1 & j_3 & k_1 & j_5 & k_3 \\ j_3 & j_5 & k'_2 & j_1 & s & j_2 & j_4 & k_2 & j_6 & k_4 \\ k'_1 & k'_2 & k'_3 & s & k_5 & k_1 & k_2 & k_3 & k_4 & k_5\end{array}\right\}.$$

The agreement of this result with the general result for $f = 4$ provides confirmation of the correctness of the general result.

We could now actually iterate relation (8.122), but it is, perhaps, better to leave it in recursive form.

An explicit method for calculating the $3(2f-1)-j$ coefficients has now been given for $n = 2f$; hence, the matrix elements of the recoupling matrix that occurs in the Heisenberg Hamiltonian $H_{\mathbf{B}_1}$ in relation (8.55) are fully known and calculable, in principal. The symmetric matrix $H_{\mathbf{B}_1}$ of order $N_j(\mathbf{j})$ can be diagonalized by a real orthogonal matrix, as indicated in relation (8.61). Cyclic symmetry is considered in Sect. 8.7.

8.5 The Heisenberg Ring Hamiltonian: Odd n

We outline in this section a method that writes the Heisenberg Hamiltonian as a sum of three Hamiltonians $H_1 + H_2 + H_3$, each part of which is diagonalizable on a separate binary coupling scheme. This method utilizes as much of the structure of the n even case as possible. The general Hamiltonian (8.1) for $n = 2f + 1, f \geq 2$, can be written as

$$H = H_1 + H_2 + H_3, \tag{8.126}$$

where the parts are defined by

$$H_1 = \sum_{\text{all odd } i \leq 2f-1} \mathbf{J}(i) \cdot \mathbf{J}(i+1),$$

$$H_2 = \sum_{\text{all even } i \leq 2f} \mathbf{J}(i) \cdot \mathbf{J}(i+1), \tag{8.127}$$

$$H_3 = \mathbf{J}(2f+1) \cdot \mathbf{J}(1).$$

The Hamiltonians H_1 and H_2 each contain f dot product terms, each of which is a Hermitian operator on the space $\mathcal{H}_\mathbf{j}$: H_1 is identical with that for the even case; and H_2 has $f - 1$ terms in common with the even case. The individual Hamiltonians are noncommuting, but each is diagonalizable on a separate binary coupling scheme. Most significantly: *The method of (8.50)-(8.52) extends to three such Hamiltonians.* The relation is given in (8.151) below, after developing the background.

We now require the binary trees of order $n = 2f + 1$ that are obtained from the family of binary trees T_{2f} in diagrams (8.42) by adjoining another $\circ$ point, as shown in the diagram:

$$T_{2f+1} = \quad \underset{T_{2f}}{\bigvee} \tag{8.128}$$

The assignment of the external quantum numbers to this shape entails angular momentum sequences of odd and even lengths: We have the sequences of odd length $2f + 1$ and even length $2f$ defined by

$$\mathbf{j}_1 = (j_1, j_2, \cdots, j_{2f+1}), \ \mathbf{j}_2 = (j_2, j_3, \cdots, j_{2f+1}, j_1),$$
$$\mathbf{j}_3 = (j_3, j_4, \cdots, j_{2f+1}, j_1, j_2);$$

$$\tag{8.129}$$

$$\mathbf{j}_1^e = (j_1, j_2, \cdots, j_{2f}), \ \mathbf{j}_2^e = (j_2, j_3, \cdots, j_{2f+1}),$$
$$\mathbf{j}_3^e = (j_3, j_4, \cdots, j_{2f+1}, j_1).$$

The even length sequences are the j_i labels assigned to the binary subtree T_{2f} in (8.128); the cyclical odd length sequences include the adjoined ○ point. Relations (8.129) give the labeling of the ○ points for three distinct binary coupling schemes, which we refer to as schemes I_o, II_o, III_o for which H_1, H_2, H_3 are, respectively, diagonal. We next describe the pairwise compounding of angular momenta associated with these three coupling schemes:

1. Binary coupling scheme I_o:

   (i) External angular momenta: First, there is the pairwise addition of angular momenta given by $\mathbf{J}(2i-1) + \mathbf{J}(2i), i = 1, 2, \ldots, f$, as assigned left-to-right respectively, to the (○ ○) endpoints of the f forks of type f° belonging to T_{2f}. Second, there is the addition of these $2f$ angular momenta to the intermediate angular momentum defined by

   $$\mathbf{K}(2f-1) = \mathbf{J}(1) + \mathbf{J}(2) + \cdots + \mathbf{J}(2f). \qquad (8.130)$$

   Third, there is the addition of the last angular momentum $\mathbf{J}(2f+1)$ to obtain the total angular momentum

   $$\mathbf{J} = \mathbf{K}(2f-1) + \mathbf{J}(2f+1). \qquad (8.131)$$

   (ii) Intermediate angular momenta:
   The $2f$ roots of the forks in (8.128) are labeled by the standard rule by $\mathbf{k} = (k_1, k_2, \ldots, k_{2f-1})$ and j. Since there are two • points at each level $\leq f$, except levels 1 and 0, the pair of labels (k_{2i-1}, k_{2i}) occurs at level $f-i+1$ for $i = 1, 2, \ldots, f-1$; the single label k_{2f-1} at level 1 for $i = f$; and the label j at level 0 for $i = f+1$. The addition of angular momenta associated with this standard labeling is

   $$\mathbf{K}(1) = \mathbf{J}(1) + \mathbf{J}(2); \; \mathbf{K}(2i) = \mathbf{J}(2i+1) + \mathbf{J}(2i+2),$$
   $$i = 1, 2, \ldots, f-1,$$
   $$\mathbf{K}(2i+1) = \mathbf{K}(2i-1) + \mathbf{K}(2i), i = 1, 2, , \ldots, f-1, \qquad (8.132)$$
   $$\mathbf{J} = \mathbf{K}(2f-1) + \mathbf{J}(2f+1).$$

2. Binary coupling scheme II_o:

   (i) External angular momenta: First, there is the pairwise addition of angular momenta given by $\mathbf{J}(2i) + \mathbf{J}(2i+1), i = 1, 2, \ldots, f$, as assigned left-to-right respectively, to the (○ ○) endpoints of the f forks of type f° belonging to T_{2f}. Second, there is the addition of these $2f$ angular momenta to the intermediate angular momentum defined by

   $$\mathbf{K}'(2f-1) = \mathbf{J}(2) + \mathbf{J}(3) + \cdots + \mathbf{J}(2f+1). \qquad (8.133)$$

Third, there is the addition of the last angular momentum $\mathbf{J}(1)$ to obtain the total angular momentum

$$\mathbf{J} = \mathbf{K}'(2f-1) + \mathbf{J}(1). \tag{8.134}$$

(ii) Intermediate angular momenta:

The $2f$ roots of the forks in (8.128) are labeled by the standard rule by $\mathbf{k}' = (k'_1, k'_2, \ldots, k'_{2f-1})$ and j. Since there are two • points at each level $\leq f$, except levels 1 and 0, the pair of labels (k'_{2i-1}, k'_{2i}) occurs at level $f-i+1$ for $i = 1, 2, \ldots, f-1$; the single label k'_{2f-1} at level 1 for $i = f$; and the label j at level 0 for $i = f+1$. The addition of angular momenta associated with this standard labeling is

$$\mathbf{K}'(1) = \mathbf{J}(2) + \mathbf{J}(3); \quad \mathbf{K}'(2i) = \mathbf{J}(2i+2) + \mathbf{J}(2i+3),$$
$$i = 1, 2, \ldots, f-1,$$
$$\mathbf{K}'(2i+1) = \mathbf{K}'(2i-1) + \mathbf{K}'(2i), i = 1, 2, , \ldots, f-1, \tag{8.135}$$
$$\mathbf{J} = \mathbf{K}'(2f-1) + \mathbf{J}(1).$$

3. Binary coupling scheme III_o:

(i) External angular momenta: First, there is the pairwise addition of angular momenta is given by $\mathbf{J}(2i+1) + \mathbf{J}(2i+2), i = 1, 2, \ldots, f-1$, and $\mathbf{J}(2f+1) + \mathbf{J}(1)$, as assigned left-to-right respectively, to the (∘ ∘) endpoints of the f forks of type f° belonging to T_{2f}. Second, there is the addition of these $2f$ angular momenta to the intermediate angular momentum defined by

$$\mathbf{K}''(2f-1) = \mathbf{J}(3) + \mathbf{J}(4) + \cdots + \mathbf{J}(2f+1) + \mathbf{J}(1). \tag{8.136}$$

Third, there is the addition of the last angular momentum $\mathbf{J}(2)$ to obtain the total angular momentum

$$\mathbf{J} = \mathbf{K}''(2f-1) + \mathbf{J}(2). \tag{8.137}$$

(ii) Intermediate angular momenta:

The $2f$ roots of the forks in (8.128) are labeled by the standard rule by $\mathbf{k}'' = (k''_1, k''_2, \ldots, k''_{2f-1})$ and j. Since there are two • points at each level $\leq f$, except levels 1 and 0, the pair of labels (k''_{2i-1}, k''_{2i}) occurs at level $f-i+1$ for $i = 1, 2, \ldots, f-1$; the single label k''_{2f-1} at level 1 for $i = f$; and the label j at level 0 for $i = f+1$. The addition of angular momenta associated with this standard labeling is

$$\mathbf{K}''(2i-1) = \mathbf{J}(2i+1) + \mathbf{J}(2i+2), \ i = 1, 2, \ldots, f-1,$$
$$\mathbf{K}''(2f-2) = \mathbf{J}(2f+1) + \mathbf{J}(1), \qquad (8.138)$$
$$\mathbf{J} = \mathbf{K}''(2f-1) + \mathbf{J}(2).$$

This coupling scheme overlaps maximally with coupling scheme I_o in its sharing of the compounding of the same $f-1$ pairs of external angular momenta. This has strong implications for the structure of the triangle coefficients relating the sets of two coupled state vectors.

The coupling schemes I_o, II_o, III_o are fully explicit. Throughout this chapter, we use the following notations for the sequences of external and internal (intermediate) angular momenta, where $n = 2f+1$:

$$\begin{aligned}
\mathbf{j}_1 &= (j_1, j_2, \ldots, j_n), \ \mathbf{m}_1 = (m_1, m_2, \ldots, m_n), \\
\mathbf{k}_1 &= \mathbf{k} = (k_1, k_2, \ldots, k_{n-2}); \\
\mathbf{j}_2 &= (j_2, j_3, \ldots, j_n, j_1), \ \mathbf{m}_2 = (m_2, m_3, \ldots, m_n, m_1), \qquad (8.139)\\
\mathbf{k}_2 &= \mathbf{k}' = (k'_1, k'_2, \ldots, k'_{n-2}). \\
\mathbf{j}_3 &= (j_3, j_4, \ldots, j_n, j_1, j_2), \ \mathbf{m}_3 = (m_3, m_4, \ldots, m_n, m_1, m_2), \\
\mathbf{k}_3 &= \mathbf{k}'' = (k''_1, k''_2, \ldots, k''_{n-2}).
\end{aligned}$$

We now have the following coupled state vectors corresponding to schemes I_o, II_o, III_o, where we now write $T = T_{2f+1}$, since it is always the points of the binary tree in diagram (8.128) that are being labeled:

scheme I_o:

$$|T(\mathbf{j}_1 \mathbf{k}_1)_{j\,m}\rangle = \sum_{\mathbf{m} \in \mathbb{C}_T(\mathbf{j})} C_{T(\mathbf{j}_1\,\mathbf{m}_1;\mathbf{k}_1)_{j\,m}} |\mathbf{j}\,\mathbf{m}\rangle. \qquad (8.140)$$

scheme II_o:

$$|T(\mathbf{j}_2 \mathbf{k}_2)_{j\,m}\rangle = \sum_{\mathbf{m} \in \mathbb{C}_T(\mathbf{j})} C_{T(\mathbf{j}_2\,\mathbf{m}_2;\mathbf{k}_2)_{j\,m}} |\mathbf{j}\,\mathbf{m}\rangle. \qquad (8.141)$$

scheme III_o:

$$|T(\mathbf{j}_3 \mathbf{k}_3)_{j\,m}\rangle = \sum_{\mathbf{m} \in \mathbb{C}_T(\mathbf{j})} C_{T(\mathbf{j}_3\,\mathbf{m}_3;\mathbf{k}_3)_{j\,m}} |\mathbf{j}\,\mathbf{m}\rangle. \qquad (8.142)$$

8.5. HEISENBERG HAMILTONIAN: ODD n

Just as for the n even case, the C-coefficients in the right-hand sides of (8.140)-(8.142) are the generalized WCG coefficients (products of ordinary WCG coefficients) that are read-off directly from the standard labeled binary tree T_{2f+1} defined by (8.128).

The complete set of mutually commuting Hermitian operator and eigenvector-eigenvalue relations for the coupling schemes I_o, II_o, III_o can be described as follows: The set of angular momentum operators

$$\mathbf{J}^2, J_3, \mathbf{J}^2(i), i = 1, 2, \cdots, 2f + 1, \qquad (8.143)$$

is diagonal on each of the binary coupled state vectors $|T(\mathbf{j}_h\,\mathbf{k}_h)_{jm}\rangle$, $h = 1, 2, 3$; that is, the following eigenvalue-eigenvector relations hold for each $h = 1, 2, 3$:

$$\mathbf{J}^2|T(\mathbf{j}_h\,\mathbf{k}_h)_{jm}\rangle = j(j+1)|T(\mathbf{j}_h\,\mathbf{k}_h)_{jm}\rangle,$$
$$J_3|T(\mathbf{j}_h\,\mathbf{k}_h)_{jm}\rangle = m|T(\mathbf{j}_h\,\mathbf{k}_h)_{jm}\rangle, \qquad (8.144)$$
$$\mathbf{J}^2(i)|T(\mathbf{j}_h\,\mathbf{k}_h)_{jm}\rangle = j_i(j_i+1)|T(\mathbf{j}_h\,\mathbf{k}_h)_{jm}\rangle, \ i = 1, 2, \ldots, 2f+1.$$

The full set of intermediate angular momenta for all $2f - 1$ forks for the coupling of the external angular momenta $\mathbf{J}(1), \mathbf{J}(2), \ldots, \mathbf{J}(2f+1)$ in accordance with their standard assignment to the shape of the binary tree T_{2f+1} is the following:

scheme I_o:

$$\mathbf{K}^2(i)|T(\mathbf{j}_1\,\mathbf{k}_1)_{jm}\rangle = k_i(k_i+1)|T(\mathbf{j}_1\,\mathbf{k}_1)_{jm}\rangle, \ i = 1, 2, \ldots, 2f-1; \qquad (8.145)$$

$$\mathbf{k}_1 = (k_1, k_2, \ldots, k_{2f-1}) \in \mathbb{K}_T^{(j)}(\mathbf{j}_1).$$

scheme II_o:

$$\mathbf{K}'^2(i)|T(\mathbf{j}_2\,\mathbf{k}_2)_{jm}\rangle = k'_i(k'_i+1)|T(\mathbf{j}_2\,\mathbf{k}_2)_{jm}\rangle, \ i = 1, 2, \ldots, 2f-1; \qquad (8.146)$$

$$\mathbf{k}_2 = (k'_1, k'_2, \ldots, k'_{2f-1}) \in \mathbb{K}_T^{(j)}(\mathbf{j}_2).$$

scheme III_o:

$$\mathbf{K}''^2(i)|T(\mathbf{j}_3\,\mathbf{k}_3)_{jm}\rangle = k''_i(k''_i+1)|T(\mathbf{j}_3\,\mathbf{k}_3)_{jm}\rangle, \ i = 1, 2\ldots, 2f-1; \qquad (8.147)$$

$$\mathbf{k}_3 = (k''_1, k''_2, \ldots, k''_{2f-1}) \in \mathbb{K}_T^{(j)}(\mathbf{j}_3).$$

8.5.1 Matrix Representations of H

We next give the diagonal matrices representing the three parts H_1, H_2, H_3 of the Hamiltonian on the respective binary coupling schemes I_o, II_o, III_o in preparation for using these results in the calculation of the diagonal matrix representing H on the orthonormal basis of scheme I_o. The results below are calculated directly from the expressions given above by (8.132), (8.135), (8.138) for the three coupling schemes that express the relevant $\mathbf{K}(i)$ in terms of the pairwise compounding of the external angular momenta $\mathbf{J}(1), \mathbf{J}(2), \ldots, \mathbf{J}(2f+1)$:

1. Eigenvalues of H_1 on coupling scheme I_o :

$$H_1 \, | \, T(\mathbf{j}_1 \, \mathbf{k}_1)_{jm}\rangle = \lambda(\mathbf{j}_1 \, \mathbf{k}_1) \, | \, T(\mathbf{j}_1 \, \mathbf{k}_1)_{jm}\rangle, \qquad (8.148)$$

$$2\lambda(\mathbf{j}_1 \, \mathbf{k}_1) = -\sum_{i=1}^{2f} j_i(j_i+1) + k_1(k_1+1) + \sum_{i=1}^{f-1} k_{2i}(k_{2i}+1).$$

These are the same eigenvalues as for even $n = 2f$ in (8.48).

2. Eigenvalues of H_2 on coupling scheme II_o :

$$H_2 \, | \, T(\mathbf{j}_2 \, \mathbf{k}_2)_{jm}\rangle = \lambda(\mathbf{j}_2 \, \mathbf{k}_2) \, | \, T(\mathbf{j}_2 \, \mathbf{k}_2)_{jm}\rangle, \qquad (8.149)$$

$$2\lambda(\mathbf{j}_2 \, \mathbf{k}_2) = -\sum_{i=2}^{2f+1} j_i(j_i+1) + k'_1(k'_1+1) + \sum_{i=1}^{f-1} k'_{2i}(k'_{2i}+1).$$

These are of a form similar to the eigenvalues for even $n = 2f$ in (8.49); they are obtained by accounting for the replacement of $\mathbf{J}(1)$ by $\mathbf{J}(2f+1)$ in going from H_2 for $n = 2f$ to the modified H_2 for $n = 2f + 1$.

3. Eigenvalues of H_3 on coupling scheme III_o :

$$H_3 \, | \, T(\mathbf{j}_3 \, \mathbf{k}_3)_{jm}\rangle = \lambda(\mathbf{j}_3 \, \mathbf{k}_3) \, | \, T(\mathbf{j}_3 \, \mathbf{k}_3)_{jm}\rangle,$$

$$\qquad (8.150)$$

$$2\lambda(\mathbf{j}_3 \, \mathbf{k}_3) = -j_1(j_1+1) - j_{2f+1}(j_{2f+1}+1) + k''_{2f-2}(k''_{2f-2}+1).$$

The derivation of relation (8.55) extends easily to $H = H_1 + H_2 + H_3$, thus giving the matrix $H_{\mathbf{B}_1(o)}$ of H on the orthonormal basis $\mathbf{B}_1(o)$ for coupling scheme I_o given in (8.140):

$$H_{\mathbf{B}_1(o)} = D^{\mathbf{j}_1} + R^{\mathbf{j}_2;\mathbf{j}_1 \; tr} D^{\mathbf{j}_2} R^{\mathbf{j}_2;\mathbf{j}_1} + R^{\mathbf{j}_3;\mathbf{j}_1 \; tr} D^{\mathbf{j}_3} R^{\mathbf{j}_3;\mathbf{j}_1}. \qquad (8.151)$$

8.5. HEISENBERG HAMILTONIAN: ODD n

We summarize the definitions of all quantities entering relation (8.151). To avoid proliferation of primes, we use $\bar{\mathbf{k}}, \mathbf{k}$ to denote the (row, column) indexing of quantum numbers with the same domain of definition:

1. The matrix representation of $H = H_1 + H_2 + H_3$ on the basis

$$\mathbf{B}_1(o) = \left\{ |T(\mathbf{j}_1\,\mathbf{k}_1)_{jm}\rangle \mid \mathbf{k}_1 \in \mathbb{K}_T^{(j)}(\mathbf{j}_1) \right\} \qquad (8.152)$$

is denoted by $H_{\mathbf{B}_1(o)}$ and defined by

$$\begin{aligned}
\left(H_{\mathbf{B}_1(o)}\right)_{\bar{\mathbf{k}}_1,j;\mathbf{k}_1,j} &= \langle T(\mathbf{j}_1\,\bar{\mathbf{k}}_1)_{jm} \mid H \mid T(\mathbf{j}_1\,\mathbf{k}_1)_{jm} \rangle, \\
\bar{\mathbf{k}}_1 &= (\bar{k}_1, \bar{k}_2, \ldots, \bar{k}_{2f-1}) \in \mathbb{K}_T^{(j)}(\mathbf{j}_1), \qquad (8.153) \\
\mathbf{k}_1 &= (k_1, k_2, \ldots, k_{2f-1}) \in \mathbb{K}_T^{(j)}(\mathbf{j}_1).
\end{aligned}$$

(a) The diagonal matrix $D^{\mathbf{j}_1}$ is the matrix representation of H_1 on the basis $\mathbf{B}_1(o)$; it has matrix elements given by (8.148):

$$\left(D^{\mathbf{j}_1}\right)_{\bar{\mathbf{k}}_1,j;\mathbf{k}_1,j} = \langle T(\mathbf{j}_1\,\bar{\mathbf{k}}_1)_{jm} \mid H_1 \mid T(\mathbf{j}_1\,\mathbf{k}_1)_{jm} \rangle = \delta_{\bar{\mathbf{k}}_1,\mathbf{k}_1}\,\lambda(\mathbf{j}_1\,\mathbf{k}_1). \qquad (8.154)$$

(b) The diagonal matrix $D^{\mathbf{j}_2}$ is the matrix representation of H_2 on the basis $\mathbf{B}_2(o)$ defined by

$$\mathbf{B}_2(o) = \left\{ |T(\mathbf{j}_2\,\mathbf{k}_2)_{jm}\rangle \mid \mathbf{k}_2 \in \mathbb{K}_T^{(j)}(\mathbf{j}_2) \right\}, \qquad (8.155)$$

with matrix elements given by

$$\begin{aligned}
\left(D^{\mathbf{j}_2}\right)_{\bar{\mathbf{k}}_2,j;\mathbf{k}_2,j} &= \delta_{\bar{\mathbf{k}}_2;\mathbf{k}_2}\,\langle T(\mathbf{j}_2\,\bar{\mathbf{k}}_2)_{jm} \mid H_2 \mid T(\mathbf{j}_2\,\mathbf{k}_2)_{jm} \rangle, \\
\bar{\mathbf{k}}_2 &= (\bar{k}'_1, \bar{k}'_2, \ldots, \bar{k}'_{2f-1}) \in \mathbb{K}_T^{(j)}(\mathbf{j}_2), \qquad (8.156) \\
\mathbf{k}_2 &= (k'_1, k'_2, \ldots, k'_{2f-1}) \in \mathbb{K}_T^{(j)}(\mathbf{j}_2).
\end{aligned}$$

(c) The diagonal matrix $D^{\mathbf{j}_3}$ is the matrix representation of H_3 on the basis $\mathbf{B}_3(o)$ defined by

$$\mathbf{B}_3(o) = \left\{ |T(\mathbf{j}_3\,\mathbf{k}_3)_{jm}\rangle \mid \mathbf{k}_3 \in \mathbb{K}_T^{(j)}(\mathbf{j}_3) \right\}, \qquad (8.157)$$

with matrix elements given by

$$\begin{aligned}
\left(D^{\mathbf{j}_3}\right)_{\bar{\mathbf{k}}_3,j;\mathbf{k}_3,j} &= \delta_{\bar{\mathbf{k}}_3;\mathbf{k}_3}\,\langle T(\mathbf{j}_3\,\bar{\mathbf{k}}_3)_{jm} \mid H_3 \mid T(\mathbf{j}_3\,\mathbf{k}_3)_{jm} \rangle, \\
\bar{\mathbf{k}}_3 &= (\bar{k}''_1, \bar{k}''_2, \ldots, \bar{k}''_{2f-1}) \in \mathbb{K}_T^{(j)}(\mathbf{j}_3), \qquad (8.158) \\
\mathbf{k}_3 &= (k''_1, k''_2, \ldots, k''_{2f-1}) \in \mathbb{K}_T^{(j)}(\mathbf{j}_3).
\end{aligned}$$

2. The eigenvalues in Items 1a-1c are summarized here in one place:

$$\lambda(\mathbf{j}_1\,\mathbf{k}_1) = \tfrac{1}{2}\Big(-\sum_{i=1}^{2f} j_i(j_i+1) + k_1(k_1+1) + \sum_{i=1}^{f-1} k_{2i}(k_{2i}+1)\Big),$$

$$\lambda(\mathbf{j}_2\,\mathbf{k}_2) = \tfrac{1}{2}\Big(-\sum_{i=2}^{2f+1} j_i(j_i+1) + k'_1(k'_1+1) + \sum_{i=1}^{f-1} k'_{2i}(k'_{2i}+1)\Big),$$

$$\lambda(\mathbf{j}_3\,\mathbf{k}_3) = \tfrac{1}{2}\Big(-j_1(j_1+1) - j_{2f+1}(j_{2f+1}+1) + k''_{2f-2}(k''_{2f-2}+1)\Big).$$
(8.159)

3. The matrices $R^{\mathbf{j}_2;\mathbf{j}_1}$ and $R^{\mathbf{j}_3;\mathbf{j}_1}$ are the recoupling matrices whose matrix elements give the triangle coefficients that relate the following respective orthonormal bases:

$$|T(\mathbf{j}_1\,\mathbf{k}_1)_{jm}\rangle = \sum_{\mathbf{k}_2 \in \mathbb{K}_T^{(j)}(\mathbf{j}_2)} \Big(R^{\mathbf{j}_2;\mathbf{j}_1}\Big)_{\mathbf{k}_2,j\,\mathbf{k}_1,j}|T(\mathbf{j}_2\,\mathbf{k}_2)_{jm}\rangle,$$

$$\text{each } \mathbf{k}_1 \in \mathbb{K}_T^{(j)}(\mathbf{j}_1), \qquad (8.160)$$

$$\Big(R^{\mathbf{j}_2;\mathbf{j}_1}\Big)_{\mathbf{k}_2,j;\mathbf{k}_1,j} = \big\{\Delta_T(\mathbf{j}_2\,\mathbf{k}_2)_j\big|\Delta_T(\mathbf{j}_1\,\mathbf{k}_1)_j\big\};$$

$$|T(\mathbf{j}_1\,\mathbf{k}_1)_{jm}\rangle = \sum_{\mathbf{k}_3 \in \mathbb{K}_T^{(j)}(\mathbf{j}_3)} \Big(R^{\mathbf{j}_3;\mathbf{j}_1}\Big)_{\mathbf{k}_3,j;\mathbf{k}_1,j}|T(\mathbf{j}_3\,\mathbf{k}_3)_{jm}\rangle,$$

$$\text{each } \mathbf{k}_1 \in \mathbb{K}_T^{(j)}(\mathbf{j}_1), \qquad (8.161)$$

$$\Big(R^{\mathbf{j}_3;\mathbf{j}_1}\Big)_{\mathbf{k}_3,j;\mathbf{k}_1,j} = \big\{\Delta_T(\mathbf{j}_3\,\mathbf{k}_3)_j\big|\Delta_T(\mathbf{j}_1\,\mathbf{k}_1)_j\big\}.$$

It is these transformations that carry the diagonal representation of H_2 (resp., H_3) on the basis $\mathbf{B}_2(o)$ (resp., $\mathbf{B}_3(o)$) back to the representation of H_2 (resp., H_3) on the basis $\mathbf{B}_1(o)$. These recoupling matrices are, of course, those associated with the standard labeled binary tree $T = T_{2f+1}$, as displayed in (8.42) and (8.128).

We need to calculate two sets of $3(n-1)-j$ coefficients for $n = 2f+1$, one set defined by the triangle coefficients for the matrix elements of the recoupling matrix $R^{\mathbf{j}_2;\mathbf{j}_1}$ defined by the transformation (8.160), and one set defined by the triangle coefficients for the matrix elements of the recoupling matrix $R^{\mathbf{j}_3;\mathbf{j}_1}$ defined by the transformation (8.161). We deal with these separately in Sects. 8.5.2 and 8.5.3, since they entail

8.5. HEISENBERG HAMILTONIAN: ODD n

triangle coefficients of different orders because coupling scheme III_o has at the outset many common forks with coupling scheme I_o; it reduces immediately to irreducible triangle coefficients of lower order.

8.5.2 Matrix Elements of R^{j_2,j_1}: The $6f - j$ Coefficients

The shape transformation we require is that associated with the two labelings of binary tree in diagram (8.128) corresponding to coupling schemes I_o and II_o. If we define an abbreviated shape by $X_{2f-2} = \text{Sh}_{2f-2}(2, 3, \ldots, 2f-1)$, then the shape transformation in question can be written in terms of binary bracketings as

$$\left(\left(X_{2f-2}(j_{2f}\, j_{2f+1})\right)j_1\right) \xrightarrow{w_{2f+1}^1} \text{Sh}_{2f+1}(1, 2, \ldots, 2f+1). \quad (8.162)$$

We now easily infer the following sequence of shape transformations:

$$\left(\left(X_{2f-2}(j_{2f}\, j_{2f+1})\right)j_1\right) \xrightarrow{A} \left(X_{2f-2}\left((j_{2f}\, j_{2f+1})j_1\right)\right)$$

$$\xrightarrow{C} \left(X_{2f-2}\left(j_1(j_{2f}\, j_{2f+1})\right)\right) \xrightarrow{A} \left(X_{2f-2}\left((j_1\, j_{2f})j_{2f+1}\right)\right)$$

$$\xrightarrow{C} \left(X_{2f-2}\left((j_{2f}\, j_1)j_{2f+1}\right)\right) \xrightarrow{A} \left(\left(X_{2f-2}(j_{2f}\, j_1)\right)j_{2f+1}\right)$$

$$\xrightarrow{w_{2f}} \text{Sh}_{2f+1}(1, 2, \ldots, 2f+1). \quad (8.163)$$

Thus, we obtain the recurrence relation:

$$w_{2f+1} = w_{2f}\, ACACA. \quad (8.164)$$

The shape transformation $ACACA$ again engenders the Wigner $9-j$ coefficient in yet another form. While the letter sequence is the same as that given by (5.245), the shapes on which these elementary shape transformations act are different. Thus, we have

$$\left\{ \begin{array}{c} b \quad c \\ a \quad k_1 \\ k_2 \quad d \\ j \end{array} \middle| \begin{array}{c} b \quad d \\ a \quad k_1' \\ k_2' \quad c \\ j \end{array} \right\} = \left(R^{((a(bc))d);((a(bd))c)}\right)_{k_1,k_2,j;\, k_1',k_2',j}$$

$$= \left\{ \begin{array}{ccc|ccc} b & a & k_2 & b & a & k_2' \\ c & k_1 & d & d & k_1' & c \\ k_1 & k_2 & j & k_1' & k_2' & j \end{array} \right\}, \quad (8.165)$$

270   CHAPTER 8. THE HEISENBERG MAGNETIC RING

where we have introduced the abbreviations $a = X_{2f-2}, b = j_{2f}, c = j_{2f+1}, d = j_1$ into (8.165). We are unable to obtain this result from the earlier ones for the $9-j$ coefficient given in Chapter 5. We therefore give the full derivation of this version of the $9-j$ coefficient by implementing our standard method base on common fork reduction and phase transformations as given by (5.134)-(5.135). In terms of the abbreviated notation, relations (8.163) become:

$$\Big((a(bc))d\Big) \xrightarrow{A} \Big(a((bc)d)\Big) \xrightarrow{C} \Big(a(d(bc))\Big)$$
$$\xrightarrow{A} \Big(a((db)c)\Big) \xrightarrow{C} \Big(a((bd)c)\Big) \xrightarrow{A} \Big((a(bd))c\Big);$$
(8.166)

$$R^{((a(bc))d);((a(bd))c)} = R^{((a(bc))d);(a(d(bc)))} R^{(a(d(bc)));(a((bd)c))} \times R^{(a((bd)c));((a(bd))c)}.$$

This product of recoupling matrices incorporates the transformations corresponding to phase-change operations, as remarked on earlier (see (5.281)).

The matrix elements of the recoupling matrix product in (8.166) now give the following sequence of relations:

$$\left(R^{((a(bc))d);((a(bd))c)}\right)_{k_1,k_2,j;k_1',k_2',j}$$
$$= \sum_{k_1'',k_2'',k_1''',k_2'''} \left(R^{((a(bc))d);(a(d(bc)))}\right)_{k_1,k_2,j;k_1'',k_2'',j}$$
$$\times \left(R^{(a(d(bc)));(a((bd)c))}\right)_{k_1'',k_2'',j;k_1''',k_2''',j}$$
$$\times \left(R^{(a((bd)c));((a(bd))c)}\right)_{k_1''',k_2''',j;k_1',k_2',j}; \qquad (8.167)$$

$$\left(R^{((a(bc))d);(a(d(bc)))}\right)_{k_1,k_2,j;k_1'',k_2'',j} = \begin{Bmatrix} b & a & k_2 \\ c & k_1 & d \\ k_1 & k_2 & j \end{Bmatrix} \begin{Bmatrix} b & d & a \\ c & k_1'' & k_2'' \\ k_1'' & k_2'' & j \end{Bmatrix}$$
$$= \delta_{k_1,k_1''} \begin{Bmatrix} a & k_2 & d & a \\ k_1 & d & k_1 & k_2'' \\ k_2 & j & k_2'' & j \end{Bmatrix};$$

$$\left(R^{(a(d(bc)));(a((bd)c))}\right)_{k_1'',k_2'',j;k_1''',k_2''',j} = \begin{Bmatrix} b & d & a \\ c & k_1'' & k_2'' \\ k_1'' & k_2'' & j \end{Bmatrix} \begin{Bmatrix} b & k_1''' & a \\ d & c & k_2''' \\ k_1''' & k_2''' & j \end{Bmatrix}$$

8.5. HEISENBERG HAMILTONIAN: ODD n

$$= \delta_{k_2'', k_2'''} \begin{Bmatrix} b & d & b & k_1''' \\ c & k_1'' & d & c \\ k_1'' & k_2'' & k_1''' & k_2'' \end{Bmatrix} ; \qquad (8.168)$$

$$\left(R^{(a((bd)c));((a(bd))c)} \right)_{k_1''',k_2''',j;k_1',k_2',j} = \begin{Bmatrix} b & k_1''' & a & b & a & k_2' \\ d & c & k_2''' & d & k_1' & c \\ k_1''' & k_2''' & j & k_1' & k_2' & j \end{Bmatrix}$$

$$= \delta_{k_1''',k_1'} \begin{Bmatrix} k_1' & a & a & k_2' \\ c & k_2''' & k_1' & c \\ k_2''' & j & k_2' & j \end{Bmatrix} .$$

We now assemble these results, account for the Kronecker delta factors, set $r = k_2''$ (summation parameter), and obtain the Wigner $9-j$ coefficient in its irreducible triangle coefficient form as follows:

$$\begin{Bmatrix} b & c & b & d \\ a & k_1 & a & k_1' \\ k_2 & d & k_2' & c \\ j & & j & \end{Bmatrix} = \left(R^{((a(bc))d);((a(bd))c)} \right)_{k_1,k_2,j;k_1',k_2',j}$$

$$= \begin{Bmatrix} b & a & k_2 & b & a & k_2' \\ c & k_1 & d & d & k_1' & c \\ k_1 & k_2 & j & k_1' & k_2' & j \end{Bmatrix} = \begin{Bmatrix} b & a & k_2' & b & a & k_2 \\ d & k_1' & c & c & k_1 & d \\ k_1' & k_2' & j & k_1 & k_2 & j \end{Bmatrix}$$

$$= \sqrt{(2k_1+1)(2k_2+1)(2k_1'+1)(2k_2'+1)} \begin{Bmatrix} b & d & k_1' \\ c & j & k_2' \\ k_1 & k_2 & a \end{Bmatrix}$$

$$= \sum_r \begin{Bmatrix} a & k_2 & d \\ k_1 & d & k_1 & r \\ k_2 & j & r & j \end{Bmatrix} \begin{Bmatrix} b & d & b & k_1' \\ c & k_1 & d & c \\ k_1 & r & k_1' & r \end{Bmatrix} \begin{Bmatrix} k_1' & a & a & k_2' \\ c & r & k_1' & c \\ r & j & k_2' & j \end{Bmatrix} . \quad (8.169)$$

The phase factor in the relation for the Wigner $9-j$ coefficient in relation (8.169) is unity. We present the details, since it provides independent verification of relation (8.169). First, from the symmetry of the $9-j$ coefficient and relation (5.268), we have the identity:

$$\begin{Bmatrix} b & d & k_1' \\ c & j & k_2' \\ k_1 & k_2 & a \end{Bmatrix} = \begin{Bmatrix} a & k_1' & k_2' \\ k_1 & b & c \\ k_2 & d & j \end{Bmatrix}$$

$$= \sum_r (-1)^{2r} (2r+1) \begin{Bmatrix} a & k_1 & k_2 \\ d & j & r \end{Bmatrix} \begin{Bmatrix} k_1' & b & d \\ k_1 & r & c \end{Bmatrix} \begin{Bmatrix} k_2' & c & j \\ r & a & k_1' \end{Bmatrix} .$$

$$(8.170)$$

Second, from relations (5.142) and (5.139), we have the following three relations of $6-j$ coefficients to triangle coefficients of order four:

$$\sqrt{2k_2+1)(2r+1)}\left\{\begin{array}{ccc} a & k_1 & k_2 \\ d & j & r \end{array}\right\}$$

$$= (-1)^{d+k_1-r}(-1)^{a+d+k_1+j}\left\{\begin{array}{cc|cc} a & k_2 & d & a \\ k_1 & d & k_1 & r \\ k_2 & j & r & j \end{array}\right\},$$

$$\sqrt{2k_1+1)(2k'_1+1)}\left\{\begin{array}{ccc} d & b & k'_1 \\ c & r & k_1 \end{array}\right\}$$

$$= (-1)^{b+d-k'_1}(-1)^{b+c+d+r}\left\{\begin{array}{cc|cc} b & d & b & k'_1 \\ c & k_1 & d & c \\ k_1 & r & k'_1 & r \end{array}\right\}, \qquad (8.171)$$

$$\sqrt{2k'_2+1)(2r+1)}\left\{\begin{array}{ccc} a & k'_1 & k'_2 \\ c & j & r \end{array}\right\} = (-1)^{a+c+j+k'_1}\left\{\begin{array}{cc|cc} k'_1 & a & a & k'_2 \\ c & r & k'_1 & c \\ r & j & k'_2 & j \end{array}\right\}.$$

By symmetry, each of the $6-j$ coefficients on the left-hand side of these three relations is exactly the one appearing, respectively, in relation (8.170). Use of these relations in (8.170) now gives exactly (8.169), since the phase factors combine to give exactly $(-1)^{2r}$ as shown by $(-1)^{-d-k_1+r}(-1)^{a+d+k_1+j}(-1)^{-b-d+k'_1}(-1)^{b+c+d+r}(-1)^{-a-c-j-k'_1} = (-1)^{2r}$. We thus have a second verification of relation (8.169).

The next step is to derive a recurrence relation for the relevant $6f-j$ coefficients by use of relation (8.164):

$$w_{2f+1} = w_{2f}\, ACACA. \qquad (8.172)$$

The transformation $ACACA$ is from the initial shape in (8.163) to the next to last shape; it is diagrammed by

$$\left\{\begin{array}{c} \begin{array}{c} j_{2f}\ \ j_{2f+1} \\ X_{2f-2}\ \circ\ \ \circ \\ k'_{2f-3}\bullet \diagdown\!\!\!\diagup\bullet k'_{2f-2} \\ k'_{2f-1}\bullet\diagup\ \ \ \circ j_1 \\ j\bullet \end{array} \Bigg| \begin{array}{c} j_{2f}\ \ j_1 \\ X_{2f-2}\ \circ\ \ \circ \\ k''_{2f-3}\bullet\diagdown\!\!\!\diagup\bullet k''_{2f-2} \\ k''_{2f-1}\bullet\diagup\ \ \ \circ j_{2f+1} \\ j\bullet \end{array} \end{array}\right\}. \qquad (8.173)$$

8.5. HEISENBERG HAMILTONIAN: ODD n

The full shape transformation from the initial shape to the final shape (8.163) is effected by w_{2f} :

$$S_{2f+1}(2) \xrightarrow{ACACA} S_{2f+1}(3) \xrightarrow{w_{2f}} S_{2f+1}(1), \qquad (8.174)$$

where these abbreviated shapes are defined by (see (8.163)):

$$\text{Sh}_{2f+1}(2) = \text{Sh}_{2f+1}(j_2, j_3, \ldots, j_{2f}, j_{2f+1}, j_1) = \Big(\big(X_{2f-2}(j_{2f}\, j_{2f+1})\big)j_1\Big),$$

$$\text{Sh}_{2f+1}(3) = (j_2, j_3, \ldots, j_{2f-1}, j_{2f}, j_1, j_{2f+1}) = \Big(\big(X_{2f-2}(j_{2f}\, j_1)\big)j_{2f+1}\Big),$$

$$\text{Sh}_{2f+1}(1) = \text{Sh}_{2f+1}(j_1, j_2, \ldots, j_{2f}, j_{2f+1}), \qquad (8.175)$$

$$X_{2f-2} = \underbrace{\Big(\cdots\Big(\big((23)(45)\big)(67)\Big)\cdots\Big)}_{f-2}(2f-2\ 2f-1)\Big).$$

Thus, corresponding to the shape transformation (8.174), we have the following product of recoupling matrices:

$$R^{\text{Sh}_{2f}(2);\,\text{Sh}_{2f}(1)} = R^{\text{Sh}_{2f}(2);\,\text{Sh}_{2f}(3)}\, R^{\text{Sh}_{2f}(3);\,\text{Sh}_{2f}(1)}. \qquad (8.176)$$

We next take matrix elements of relation (8.176) with the following result:

$$\left(R^{\text{Sh}_{2f+1}(2);\,\text{Sh}_{2f+1}(1)}\right)_{\mathbf{k}',j;\mathbf{k},j} \qquad (8.177)$$

$$= \left\{\Delta_{T_{2f+1}}(\mathbf{j}_2\,\mathbf{k}_2)_j\Big|\Delta_{T_{2f+1}}(\mathbf{j}_1\,\mathbf{k}_1)_j\right\} = \left\{\Delta_{T_{2f+1}}(\mathbf{j}_1\,\mathbf{k}_1)_j\Big|\Delta_{T_{2f+1}}(\mathbf{j}_2\,\mathbf{k}_2)_j\right\}$$

$$= \sum_{\mathbf{k}''} \left(R^{\text{Sh}_{2f+1}(2);\,\text{Sh}_{2f+1}(3)}\right)_{\mathbf{k}',j;\mathbf{k}'',j} \left(R^{\text{Sh}_{2f+1}(3);\,\text{Sh}_{2f+1}(1)}\right)_{\mathbf{k}'',j;\mathbf{k},j}.$$

The explicit forms of the irreducible triangle coefficients are:

$$\left\{\Delta_{T_{2f+1}}(\mathbf{j}_1\,\mathbf{k}_1)_j\Big|\Delta_{T_{2f+1}}(\mathbf{j}_2\,\mathbf{k}_2)_j\right\}$$

$$= \left\{\begin{matrix} j_1 & j_3 & k_1 & j_5 & k_3 & \cdots & j_{2f-1} & k_{2f-3} & k_{2f-1} \\ j_2 & j_4 & k_2 & j_6 & k_4 & \cdots & j_{2f} & k_{2f-2} & j_{2f+1} \\ k_1 & k_2 & k_3 & k_4 & k_5 & \cdots & k_{2f-2} & k_{2f-1} & j \end{matrix} \right.$$

$$\left.\begin{matrix} j_2 & j_4 & k'_1 & j_6 & k'_3 & \cdots & j_{2f} & k'_{2f-3} & k'_{2f-1} \\ j_3 & j_5 & k'_2 & j_7 & k'_4 & \cdots & j_{2f+1} & k'_{2f-2} & j_1 \\ k'_1 & k'_2 & k'_3 & k'_4 & k'_5 & \cdots & k'_{2f-2} & k'_{2f-1} & j \end{matrix}\right\}, \qquad (8.178)$$

274 CHAPTER 8. THE HEISENBERG MAGNETIC RING

$$\left\{\Delta_{T_{2f+1}}(\mathbf{j}_2\,\mathbf{k}_2)_j \middle| \Delta_{T_{2f+1}}(\mathbf{j}_1\,\mathbf{k}_1)_j\right\}$$

$$= \left\{ \begin{array}{cccccccc} j_2 & j_4 & k'_1 & j_6 & k'_3 & \cdots & j_{2f} & k'_{2f-3} & k'_{2f-1} \\ j_3 & j_5 & k'_2 & j_7 & k'_4 & \cdots & j_{2f+1} & k'_{2f-2} & j_1 \\ k'_1 & k'_2 & k'_3 & k'_4 & k'_5 & \cdots & k'_{2f-2} & k'_{2f-1} & j \end{array} \right.$$

$$\left. \begin{array}{cccccccc} j_1 & j_3 & k_1 & j_5 & k_3 & \cdots & j_{2f-1} & k_{2f-3} & k_{2f-1} \\ j_2 & j_4 & k_2 & j_6 & k_4 & \cdots & j_{2f} & k_{2f-2} & j_{2f+1} \\ k_1 & k_2 & k_3 & k_4 & k_5 & \cdots & k_{2f-2} & k_{2f-1} & j \end{array} \right\}. \quad (8.179)$$

We next evaluate the matrix elements of each of the two factors in the right-hand side of relation (8.177), take into account the common forks as determined by the shapes (8.175), and use the reduction rule (5.134) to obtain the following relations:

$$\left(R^{\mathrm{Sh}_{2f+1}(2);\,\mathrm{Sh}_{2f+1}(3)}\right)_{\mathbf{k}',j;\mathbf{k}'',j} = \prod_{i=1}^{2f-3} \delta_{k'_i,k''_i}$$

$$\times \left\{ \begin{array}{ccc|ccc} j_{2f} & k'_{2f-3} & k'_{2f-1} & j_{2f} & k''_{2f-3} & k''_{2f-1} \\ j_{2f+1} & k'_{2f-2} & j_1 & j_1 & k''_{2f-2} & j_{2f+1} \\ k'_{2f-2} & k'_{2f-1} & j & k''_{2f-2} & k''_{2f-1} & j \end{array} \right\}; \quad (8.180)$$

$$\left(R^{\mathrm{Sh}_{2f+1}(3);\,\mathrm{Sh}_{2f+1}(1)}\right)_{\mathbf{k}'',j;\mathbf{k},j} = \delta_{k''_{2f-1},k_{2f-1}}$$

$$\times \left\{ \begin{array}{ccccccc} j_2 & j_4 & k''_1 & j_6 & k''_3 & \cdots & j_{2f} & k''_{2f-3} \\ j_3 & j_5 & k''_2 & j_7 & k''_4 & \cdots & j_1 & k''_{2f-2} \\ k''_1 & k''_2 & k''_3 & k''_4 & k''_5 & \cdots & k''_{2f-2} & k_{2f-1} \end{array} \right.$$

$$\left. \begin{array}{ccccccc} j_1 & j_3 & k_1 & j_5 & k_3 & \cdots & j_{2f-1} & k_{2f-3} \\ j_2 & j_4 & k_2 & j_6 & k_4 & \cdots & j_{2f} & k_{2f-2} \\ k_1 & k_2 & k_3 & k_4 & k_5 & \cdots & k_{2f-2} & k_{2f-1} \end{array} \right\}.$$

The Kronecker delta factors require that $k''_i = k'_i, i = 1, 2, \ldots, 2f-3$, and $k''_{2f-1} = k_{2f-1}$ in these two relations, and leaves $r = k''_{2f-2}$ as the summation parameter. Thus, we obtain the following recurrence relation:

8.5. HEISENBERG HAMILTONIAN: ODD n

The $6f - j$ coefficient recursion relation:

$$\left\{ \begin{array}{l} \Delta_{T_{2f+1}}(j_2,\ldots,j_{2f+1},j_1;k'_1,\ldots,k'_{2f-1})_j \\ \left| \Delta_{T_{2f+1}}(j_1,\ldots,j_{2f+1};k_1,\ldots,k_{2f-1})_j \right. \end{array} \right\}$$

$$= \sum_r \left\{ \begin{array}{l} \Delta_{T_{2f}}(j_2,\ldots,j_{2f},j_1;k'_1,\ldots,k'_{2f-3},r)_{k_{2f-1}} \\ \left| \Delta_{T_{2f}}(j_1,\ldots,j_{2f};k_1,\ldots,k_{2f-2})_{k_{2f-1}} \right. \end{array} \right\}$$

$$\times \left\{ \begin{array}{ccc|ccc} j_{2f} & k'_{2f-3} & k'_{2f-1} & j_{2f} & k'_{2f-3} & k_{2f-1} \\ j_{2f+1} & k'_{2f-2} & j_1 & j_1 & r & j_{2f+1} \\ k'_{2f-2} & k'_{2f-1} & j & r & k_{2f-1} & j \end{array} \right\} . \quad (8.181)$$

The detailed expression for the triangle coefficient of order $2f$ under the summation, which is a $3(2f-1) - j$ coefficient, in this expression is given by the relation:

$$\left\{ \begin{array}{l} \Delta_{T_{2f}}\left(j_2,\ldots,j_{2f},j_1;k'_1,\ldots,k'_{2f-3},r\right)_{k_{2f-1}} \\ \left| \Delta_{T_{2f}}\left(j_1,\ldots,j_{2f};k_1,\ldots,k_{2f-2}\right)_{k_{2f-1}} \right. \end{array} \right\}$$

$$= \left\{ \begin{array}{cccccccc} j_2 & j_4 & k'_1 & j_6 & k'_3 & \cdots & j_{2f} & k'_{2f-3} \\ j_3 & j_5 & k'_2 & j_7 & k'_4 & \cdots & j_1 & r \\ k'_1 & k'_2 & k'_3 & k'_4 & k'_5 & \cdots & r & k_{2f-1} \end{array} \right.$$

$$\left. \begin{array}{cccccccc} j_1 & j_3 & k_1 & j_5 & k_3 & \cdots & j_{2f-1} & k_{2f-3} \\ j_2 & j_4 & k_2 & j_6 & k_4 & \cdots & j_{2f} & k_{2f-2} \\ k_1 & k_2 & k_3 & k_4 & k_5 & \cdots & k_{2f-2} & k_{2f-1} \end{array} \right\} . \quad (8.182)$$

We bring attention to the occurrence of the summation index r in this expression, as well as to the coupling to the "total" angular momentum $j' = k_{2f-1}$ in accord with diagram ((8.128). The triangle coefficient of order six under the summation in (8.181) is given in terms of Wigner's $9 - j$ coefficient by (8.165) by the following identification of parameters:

$$a = k'_{2f-3},\ b = j_{2f},\ c = j_{2f+1},\ d = j_1, j = j$$
$$k_1 = k'_{2f-2},\ k_2 = k'_{2f-1},\ k'_1 = r,\ k'_2 = k_{2f-1}. \quad (8.183)$$

We leave (8.181) in recursive form.

Independently of the general derivation of relation (8.181), we have also derived the special case for $f = 2$, which reads:

$$\left\{ \begin{array}{ccc|ccc} j_2 & j_4 & k'_1 & k'_3 & j_1 & j_3 & k_1 & k_3 \\ j_3 & j_5 & k'_2 & j_1 & j_2 & j_4 & k_2 & j_5 \\ k'_1 & k'_2 & k'_3 & j & k_1 & k_2 & k_3 & j \end{array} \right\}$$

$$= \sum_r \left\{ \begin{array}{ccc|ccc} j_2 & j_4 & k'_1 & j_1 & j_3 & k_1 \\ j_3 & j_1 & r & j_2 & j_4 & k_2 \\ k'_1 & r & k_3 & k_1 & k_2 & k_3 \end{array} \right\}$$

$$\times \left\{ \begin{array}{ccc|ccc} j_4 & k'_1 & k'_3 & j_4 & k'_1 & k_3 \\ j_5 & k'_2 & j_1 & j_1 & r & j_5 \\ k'_2 & k'_3 & j & r & k_3 & j \end{array} \right\}. \qquad (8.184)$$

The agreement of this result with the general result for $f = 2$ provides confirmation of the correctness of the general result.

8.5.3 Matrix Elements of $R^{j_3;j_1}$: The $3(f+1)-j$ Coefficients

We still need to determine the matrix elements of the recoupling matrix $R^{j_3;j_1}$; these are needed for the calculation of the Hamiltonian matrix $H_{\mathbf{B}_1(o)}$ in relation (8.150). This calculation has its own subtleties relating to the fact that the fully labeled shapes $\mathrm{Sh}_{2f+1}(j_3, j_4, \ldots, j_{2f+1}, j_1, j_2)$ and $\mathrm{Sh}_{2f+1}(j_1, j_2, \ldots, j_{2f+1})$ already contain $f - 1$ common forks. At the outset, the matrix elements of $R^{j_3;j_1}$ reduce in accordance with the reduction rule (5.134) to the following:

$$\left(R^{j_3;j_1} \right)_{\mathbf{k}',j;\mathbf{k}_1,j} = \delta_{k'_1,k_2} \left(\prod_{i=1}^{f-2} \delta_{k'_{2i},k_{2i+2}} \right)$$

$$\times \left\{ \begin{array}{cccccc} k'_1 & k'_3 & \cdots & k'_{2f-5} & j_{2f+1} & k'_{2f-3} & k'_{2f-1} \\ k'_2 & k'_4 & \cdots & k'_{2f-4} & j_1 & k'_{2f-2} & j_2 \\ k'_3 & k'_5 & \cdots & k'_{2f-3} & k'_{2f-2} & k'_{2f-1} & j \end{array} \right.$$

$$\left. \begin{array}{cccccc} j_1 & k_1 & k_3 & \cdots & k_{2f-3} & k_{2f-1} \\ j_2 & k_2 & k_4 & \cdots & k_{2f-2} & j_{2f+1} \\ k_1 & k_3 & k_5 & \cdots & k_{2f-1} & j \end{array} \right\}, \quad f \geq 2, \qquad (8.185)$$

where for $f = 2$, the left pattern consists of the last three columns; hence, the triangle coefficient is of order 6 and is a $9-j$ coefficient. The general triangle coefficient (8.185), after accounting for the Kronecker delta factors, is an irreducible triangle coefficient of order $2(f+1)$; it determines uniquely a $3(f+1) - j$ coefficient.

8.5. HEISENBERG HAMILTONIAN: ODD n

The triangle coefficient (8.185) carries a nonstandard labeling in consequence of its initial reduction, and it is convenient to introduce the standard labeling. The pair of binary trees with standard labels that gives the triangle coefficient (8.185) (with appropriate identification of labels) is the following:

$$\left\{ \begin{array}{c} \text{(pair of binary trees diagram)} \end{array} \right\}$$

$$= \left\{ \begin{array}{cccccccc} a_1 & k'_1 & k'_2 & \cdots & k'_{f-3} & j_{2f+1} & k'_{f-2} & k'_f \\ a_2 & a_3 & a_4 & \cdots & a_{f-1} & j_1 & k'_{f-1} & j_2 \\ k'_1 & k'_2 & k'_3 & \cdots & k'_{f-2} & k'_{f-1} & k'_f & j \end{array} \right. $$

$$\left| \begin{array}{cccccc} j_1 & k_1 & k_2 & \cdots & k_{f-1} & k_f \\ j_2 & a_1 & a_2 & \cdots & a_{f-1} & j_{2f+1} \\ k_1 & k_2 & k_3 & \cdots & k_f & j \end{array} \right\}. \qquad (8.186)$$

The reversal of (8.186) to obtain back the nonzero matrix elements in (8.185) (set $k'_1 = k_2, k'_{2i} = k_{2i+2}, i = 1, 2, \ldots, f-2$) is as follows:

$$a_i = k_{2i}, \; i = 1, 2, \ldots, f-1; \; k_i \mapsto k_{2i-1}, \; i = 1, 2, \ldots, f;$$

$$k'_i \mapsto k'_{2i+1}, \; i = 1, 2, \ldots, f-2; k'_{f-1} \mapsto k'_{2f-2}, \; k'_f \mapsto k'_{2f-1}. \qquad (8.187)$$

We make this reversal after computing the $3(f+1) - j$ coefficient associated with the pair (8.186) of labeled binary trees of order $f+2$ with irreducible triangle coefficient of order $2(f+1)$.

The simplest example of relation (8.186) occurs for $f = 2$, where it reduces to yet another presentation of the Wigner $9-j$ coefficient:

$$\left\{ \begin{array}{c} \text{(binary trees diagram)} \end{array} \right\} \qquad (8.188)$$

$$= \left\{ \begin{array}{ccc} j_5 & a_1 & k'_2 \\ j_1 & k'_1 & j_2 \\ k'_1 & k'_2 & j \end{array} \right| \left. \begin{array}{ccc} j_1 & k_1 & k_2 \\ j_2 & a_1 & j_5 \\ k_1 & k_2 & j \end{array} \right\} = \left\{ \begin{array}{ccc} j_1 & k_1 & k_2 \\ j_2 & a_1 & j_5 \\ k_1 & k_2 & j \end{array} \right| \left. \begin{array}{ccc} j_5 & a_1 & k'_2 \\ j_1 & k'_1 & j_2 \\ k'_1 & k'_2 & j \end{array} \right\}.$$

We calculate this form of the $9 - j$ coefficient by our standard procedure, using shape transformations, recoupling matrices, and associated reduction formulas.

We have the following relations between shape transformations and recoupling matrices:

$$\Big((a_1(j_5\,j_1))j_2\Big) \xrightarrow{A} \Big(a_1((j_5\,j_1)j_2)\Big) \xrightarrow{A} \Big(a_1(j_5(j_1\,j_2))\Big)$$
$$\xrightarrow{C} \Big(a_1((j_1\,j_2)j_5)\Big) \xrightarrow{A} \Big((a_1(j_1\,j_2))j_5\Big) \xrightarrow{C} \Big(((j_1\,j_2)a_1)j_5\Big);$$
$$\Big((a_1(j_5\,j_1))j_2\Big) \xrightarrow{CACAA} \Big(((j_1\,j_2)a_1)j_5\Big); \qquad (8.189)$$

$$R^{((a_1(j_5\,j_1))j_2);(((j_1\,j_2)a_1)j_5)} = R^{((a_1(j_5\,j_1))j_2);(a_1((j_5\,j_1)j_2))}$$
$$\times R^{(a_1((j_5\,j_1)j_2);(a_1((j_1\,j_2)j_5))} R^{(a_1((j_1\,j_2)j_5));(((j_1\,j_2)a_1)j_5)}.$$

The matrix elements of the recoupling matrix product in (8.189) now give the following sequence of relations:

$$R^{((a_1(j_5\,j_1))j_2);(((j_1\,j_2)a_1)j_5)}_{k'_1,k'_2,j;\,k_1,k_2,j}$$
$$= \sum_{k''_1,k''_2\,k'''_1,k'''_2} \Big(R^{((a_1(j_5\,j_1))j_2);(a_1((j_5\,j_1)j_2))}\Big)_{k'_1,k'_2,j;\,k''_1,k''_2,j}$$
$$\times R^{(a_1((j_5\,j_1)j_2);(a_1((j_1\,j_2)j_5))}_{k''_1,k''_2,j;\,k'''_1,k'''_2,j}$$
$$\times R^{(a_1((j_1\,j_2)j_5));(((j_1\,j_2)a_1)j_5)}_{k'''_1,k'''_2,j;\,k_1,k_2,j}; \qquad (8.190)$$

$$R^{((a_1(j_5\,j_1))j_2);(a_1((j_5\,j_1)j_2))}_{k'_1,k'_2,j;\,k_1,k_2,j}$$
$$= \begin{Bmatrix} j_5 & a_1 & k'_2 \\ j_1 & k'_1 & j_2 \\ k'_1 & k'_2 & j \end{Bmatrix} \begin{Bmatrix} j_5 & k''_1 & a_1 \\ j_1 & j_2 & k''_2 \\ k''_1 & k''_2 & j \end{Bmatrix} = \delta_{k'_1,k''_1} \begin{Bmatrix} a_1 & k'_2 & k'_1 & a_1 \\ k'_1 & j_2 & j_2 & k''_2 \\ k'_2 & j & k''_2 & j \end{Bmatrix},$$

$$R^{(a_1((j_5\,j_1)j_2);(a_1((j_1\,j_2)j_5))}_{k''_1,k''_2,j;\,k'''_1,k'''_2,j} \qquad (8.191)$$
$$= \begin{Bmatrix} j_5 & k''_1 & a_1 \\ j_1 & j_2 & k''_2 \\ k''_1 & k''_2 & j \end{Bmatrix} \begin{Bmatrix} j_1 & k'''_1 & a_1 \\ j_2 & j_5 & k'''_2 \\ k'''_1 & k'''_2 & j \end{Bmatrix} = \delta_{k''_2,k'''_2} \begin{Bmatrix} j_5 & k''_1 & j_1 & k'''_1 \\ j_1 & j_2 & j_2 & j_5 \\ k''_1 & k''_2 & k'''_1 & k''_2 \end{Bmatrix},$$

$$\Big(R^{(a_1((j_1\,j_2)j_5));(((j_1\,j_2)a_1)j_5)}\Big)_{k'''_1,k'''_2,j;\,k_1,k_2,j}$$
$$= \begin{Bmatrix} j_1 & k'''_1 & a_1 \\ j_2 & j_5 & k'''_2 \\ k'''_1 & k'''_2 & j \end{Bmatrix} \begin{Bmatrix} j_1 & k_1 & k_2 \\ j_2 & a_1 & j_5 \\ k_1 & k_2 & j \end{Bmatrix} = \delta_{k'''_1,k_1} \begin{Bmatrix} k_1 & a_1 & k_1 & k_2 \\ j_5 & k'''_2 & a_1 & j_5 \\ k'''_2 & j & k_2 & j \end{Bmatrix}.$$

8.5. HEISENBERG HAMILTONIAN: ODD n

We now assemble these results, account for the Kronecker delta factors, set $r = k_2''$ (summation parameter) and obtain the Wigner $9-j$ coefficient in its irreducible triangle coefficient form as follows:

$$\left\{ \begin{array}{c} j_5 \\ a_1 \\ k_2' \end{array} \!\!\!\! \begin{array}{c} j_1 \\ k_1' \\ j_2 \end{array} \middle| \begin{array}{c} j_1 \\ k_1 \\ k_2 \end{array} \!\!\!\! \begin{array}{c} j_2 \\ a_1 \\ j_5 \end{array} \right\}_j$$

$$= \left\{ \begin{array}{ccc} j_5 & a_1 & k_2' \\ j_1 & k_1' & j_2 \\ k_1' & k_2' & j \end{array} \middle| \begin{array}{ccc} j_1 & k_1 & k_2 \\ j_2 & a_1 & j_5 \\ k_1 & k_2 & j \end{array} \right\} = \left\{ \begin{array}{ccc} j_1 & k_1 & k_2 \\ j_2 & a_1 & j_5 \\ k_1 & k_2 & j \end{array} \middle| \begin{array}{ccc} j_5 & a_1 & k_2' \\ j_1 & k_1' & j_2 \\ k_1' & k_2' & j \end{array} \right\}$$

$$= (-1)^{j_1+j_5-k_1'}(-1)^{a_1+k_1-k_2}$$

$$\times \sqrt{(2k_1+1)(2k_2+1)(2k_1'+1)(2k_2'+1)} \left\{ \begin{array}{ccc} j_1 & j_5 & k_1' \\ j_2 & j & k_2' \\ k_1 & k_2 & a_1 \end{array} \right\}$$

$$= \sum_r \left\{ \begin{array}{cc|cc} a_1 & k_2' & k_1' & a_1 \\ k_1' & j_2 & j_2 & r \\ k_2' & j & r & j \end{array} \right\} \left\{ \begin{array}{cc|cc} j_5 & k_1' & j_1 & k_1 \\ j_1 & j_2 & j_2 & j_5 \\ k_1' & r & k_1 & r \end{array} \right\} \left\{ \begin{array}{cc|cc} k_1 & a_1 & k_1 & k_2 \\ j_5 & r & a_1 & j_5 \\ r & j & k_2 & j \end{array} \right\}. \quad (8.192)$$

The phase factor in front of the Wigner $9-j$ coefficient in this relation is obtained by transposing the Wigner $9-j$ coefficient in (8.169) and identifying $a = a_1, b = j_1, c = j_5, d = j_2$.

We proceed next to give the derivation of the general $3(f+1) - j$ coefficient associated with the pair of labeled binary trees (8.186). For this, we introduce the binary subtrees of the left and right binary trees in (8.186) having shapes defined by

$$X'_{f-1} = \underbrace{\Big(\cdots \Big(a_1\,a_2\Big)a_3\Big)\cdots \Big)a_{f-1}}_{f-2},$$

$$X_{f-1} = \underbrace{\Big(\cdots \Big(j_1\,j_2\Big)a_1\Big)a_2\Big)\cdots \Big)a_{f-3}}_{f-2}. \quad (8.193)$$

In terms of these notations, the shape transformation that maps the left pattern to the right pattern in (8.186) is expressed by

$$\Big((X'_{f-1}(j_{2f+1}\,j_1))j_2\Big) \xrightarrow{w'_{f+2}} \Big(\Big(\Big(X_{f-1}\Big)a_{f-2}\Big)a_{f-1}\Big)j_{2f+1}\Big), \quad (8.194)$$

where the shape transformation itself is given by

$$w'_{f+2} = A^{f-2}CACAA. \quad (8.195)$$

Proof. The shape transformation proceeds in two steps:

$$\left((X'_{f-1}(j_{2f+1}\,j_1))j_2\right) \xrightarrow{CACAA} \left(((j_1\,j_2)X'_{f-1})j_{2f+1}\right)$$

$$\xrightarrow{A^{f-2}} \left(\left(\left(\left(X_{f-1}\right)a_{f-2}\right)a_{f-1}\right)j_{2f+1}\right), \tag{8.196}$$

where the first shape transformation is a consequence of (8.189) upon replacing a_1 by X'_{f-1} and j_5 by j_{2f+1}. □

It is useful to illustrate the application of relations (8.194)-(8.195) for the case $f = 3$, since from this result we can immediately infer the general result. For this purpose, we introduce the following abbreviated shapes, their transformations, and the associated product of recoupling matrices:

$$\text{Sh}_1 = \left(((a_1\,a_2)(j_7\,j_1))j_2\right),\ \text{Sh}_2 = \left(((j_1\,j_2)(a_1\,a_2))j_7\right),$$

$$\text{Sh}_3 = \left(((j_1\,j_2)\,a_1)\,a_2)\,j_7\right);$$

(8.197)

$$\text{Sh}_1 \xrightarrow{CACAA} \text{Sh}_2 \xrightarrow{A} \text{Sh}_3;$$

$$R^{\text{Sh}_1;\text{Sh}_3} = R^{\text{Sh}_1;\text{Sh}_2} R^{\text{Sh}_2;\text{Sh}_3}.$$

The matrix elements of this product of recoupling matrices gives by the standard reduction process the elements of the $12-j$ coefficient:

$$\left(R^{\text{Sh}_1;\text{Sh}_3}\right)_{\mathbf{k}',j;\mathbf{k},j} = \sum_{\mathbf{k}''} \left(R^{\text{Sh}_1;\text{Sh}_2}\right)_{\mathbf{k}',j;\mathbf{k}'',j} \left(R^{\text{Sh}_2;\text{Sh}_3}\right)_{\mathbf{k}'',j;\mathbf{k},j}$$

(8.198)

$$\left(R^{\text{Sh}_1;\text{Sh}_2}\right)_{\mathbf{k}',j;\mathbf{k}'',j} = \left\{\begin{array}{cccc|cccc} a_1 & j_7 & k'_1 & k'_3 & j_1 & a_1 & k''_1 & k''_3 \\ a_2 & j_1 & k'_2 & j_2 & j_2 & a_2 & k''_2 & j_7 \\ k'_1 & k'_2 & k'_3 & j & k''_1 & k''_2 & k''_3 & j \end{array}\right\}$$

$$= \delta_{k'_1,k''_2} \left\{\begin{array}{ccc|ccc} j_7 & k'_1 & k'_3 & j_1 & k''_1 & k''_3 \\ j_1 & k'_2 & j_2 & j_2 & k''_1 & j_7 \\ k'_2 & k'_3 & j & k''_1 & k''_3 & j \end{array}\right\}, \tag{8.199}$$

$$\left(R^{\text{Sh}_2;\text{Sh}_3}\right)_{\mathbf{k}'',j;\mathbf{k},j} = \left\{\begin{array}{cccc|cccc} j_1 & a_1 & k''_1 & k''_3 & j_1 & k_1 & k_2 & k_3 \\ j_2 & a_2 & k''_2 & j_7 & j_2 & a_1 & a_2 & j_7 \\ k''_1 & k''_2 & k''_3 & j & k_1 & k_2 & k_3 & j \end{array}\right\}$$

$$= \delta_{k''_1,k_1}\delta_{k''_3,k_3} \left\{\begin{array}{cc|cc} a_1 & k''_1 & k_1 & k_2 \\ a_2 & k''_2 & a_1 & a_2 \\ k''_2 & k_3 & k_2 & k_3 \end{array}\right\}. \tag{8.200}$$

8.5. HEISENBERG HAMILTONIAN: ODD n

These results combine to give the factored form:

$$\left(R^{Sh_1;Sh_3}\right)_{k_1',k_2',k_3';j;\,k_1,k_2,k_3,j} = \left\{\begin{array}{ccc|ccc} a_1 & j_7 & k_1' & k_3' & j_1 & k_1 & k_2 & k_3 \\ a_2 & j_1 & k_2' & j_2 & j_2 & a_1 & a_2 & j_7 \\ k_1' & k_2' & k_3' & j & k_1 & k_2 & k_3 & j \end{array}\right\}$$

$$= \left\{\begin{array}{ccc|ccc} j_7 & k_1' & k_3' & j_1 & k_1 & k_3 \\ j_1 & k_2' & j_2 & j_2 & k_1' & j_7 \\ k_2' & k_3' & j & k_1 & k_3 & j \end{array}\right\}\left\{\begin{array}{cc|cc} a_1 & k_1 & k_1 & k_2 \\ a_2 & k_1' & a_1 & a_2 \\ k_1' & k_3 & k_2 & k_3 \end{array}\right\}. \quad (8.201)$$

This case for $f = 3$ factors because $k_1'' = k_1, k_2'' = k_1', k_3'' = k_3$; there is no summation parameter left over. It is, perhaps, an unexpected result.

We can confirm the factoring of the matrix elements of the recoupling matrix $R^{Sh_1;Sh_3}$ by examining the associated cubic graph and applying the Yutsis theorem (p. 159). The cubic graph of order eight associated with the pair of labeled binary trees (8.186) for $f = 3$ and the triangle coefficient of order eight in (8.201) is given by

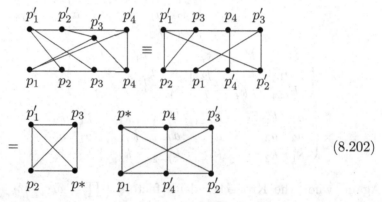

(8.202)

The cubic graph on the left is the one written out directly from the triangles (columns) of the triangle coefficient of order eight on the left in (8.201), and the second cubic graph is isomorphic to it, with points:

$p_1' = (a_1, a_2, k_1'),\ ,\ p_2' = (k_1', k_2', k_3'),\ p_3' = (j_5, j_1, k_2'),\ p_4' = (k_3', j_2, j),$
$p^* = (k_1, k_1', k_3),$ (8.203)
$p_1 = (j_1, j_2, k_1),\ p_2 = (k_1, a_1, k_2),\ p_3 = (k_2, a_2, k_3),\ p_4 = (k_3, j_5, j).$

The extra pair of points resulting from the cut and joining operation is labeled by $p^* = (k_1, k_1', k_3)$. These two cubic graphs are isomorphic because the two sets of labels of the points preserve adjacency of points. The isomorphic cubic graph in the center of the diagram has the property that three lines join two subgraphs, with the additional property that the cutting and joining of the three lines gives a product of two cubic graphs, each of angular momentum type — indeed, of a cubic graph corresponding to a $6-j$ coefficient and a $9-j$ coefficient, exactly as emerges from the factoring (8.201) by our application of the standard reduction method for recoupling matrices.

The general calculation for the shape transformations (8.193)-(8.196) of diagram (8.186) proceeds in exactly the same way as for $f = 3$ in (8.197) for all $f \geq 4$, where now we have

$$\text{Sh}_1 \xrightarrow{CACAA} \text{Sh}_2 \xrightarrow{A^{f-2}} \text{Sh}_3; \qquad (8.204)$$

$$R^{\text{Sh}_1;\text{Sh}_3} = R^{\text{Sh}_1;\text{Sh}_2} R^{\text{Sh}_2;\text{Sh}_3}.$$

Thus, we obtain:

$$\left(R^{\text{Sh}_1;\text{Sh}_3}\right)_{\mathbf{k}',j;\mathbf{k},j} = \left\{\begin{array}{cccccccc} a_1 & k'_1 & k'_2 & \cdots & k'_{f-3} & j_{2f+1} & k'_{f-2} & k'_f \\ a_2 & a_3 & a_4 & \cdots & a_{f-1} & j_1 & k'_{f-1} & j_2 \\ k'_1 & k'_2 & k'_3 & \cdots & k'_{f-2} & k'_{f-1} & k'_f & j \end{array}\right.$$

$$\left|\begin{array}{cccccc} j_1 & k_1 & k_2 & \cdots & k_{f-1} & k_f \\ j_2 & a_1 & a_2 & \cdots & a_{f-1} & j_{2f+1} \\ k_1 & k_2 & k_3 & \cdots & k_f & j \end{array}\right\}$$

$$= \left\{\begin{array}{ccc|ccc} j_{2f+1} & k'_{f-2} & k'_f & j_1 & k_1 & k_f \\ j_1 & k'_{f-1} & j_2 & j_2 & k'_{f-2} & j_{2f+1} \\ k'_{f-1} & k'_f & j & k_1 & k_f & j \end{array}\right\} \qquad (8.205)$$

$$\times \left\{\begin{array}{cccccc|cccc} a_1 & k'_1 & \cdots & k'_{f-4} & k'_{f-3} & k_1 & k_1 & k_2 & \cdots & k_{f-1} \\ a_2 & a_3 & \cdots & a_{f-2} & a_{f-1} & k'_{f-2} & a_1 & a_2 & \cdots & a_{f-1} \\ k'_1 & k'_2 & \cdots & k'_{f-3} & k'_{f-2} & k_f & k_2 & k_3 & \cdots & k_f \end{array}\right\}.$$

Again, when the Kronecker delta factors $\left(\prod_{i=1}^{f-3} \delta_{k'_i,k''_i}\right)\delta_{k'_{f-2},k''_{f-1}}$ and $\delta_{k''_{f-2},k_1}\delta_{k''_f,k_f}$ are taken into account in the detailed matrix element expansion of the product (8.204), we obtain $k''_i = k'_i$, $i = 1, 2, \ldots, f-3$; $k''_{f-2} = k_1, k''_{f-1} = k'_{f-2}, k''_f = k_f$; that is, *there are no summation parameters left over.* The matrix elements factorize into the product shown.

The general relation (8.205) gives for $f = 3$ and $f = 4$ the results:

$$\left\{\begin{array}{cccc|cccc} a_1 & j_7 & k'_1 & k'_3 & j_1 & k_1 & k_2 & k_3 \\ a_2 & j_1 & k'_2 & j_2 & j_2 & a_1 & a_2 & j_7 \\ k'2 & k'_3 & k'_3 & j & k_1 & k_2 & k_3 & j \end{array}\right\}$$

$$= \left\{\begin{array}{cccc|cccc} j_7 & k'_1 & k'_3 & j_1 & k_1 & k_3 \\ j_1 & k'_2 & j_2 & j_2 & k'_1 & j_7 \\ k'_2 & k'_3 & j & k_1 & k_3 & j \end{array}\right\}\left\{\begin{array}{cc|cc} a_1 & k_1 & k_1 & k_2 \\ a_2 & k'_1 & a_1 & a_2 \\ k'_1 & k_3 & k_2 & k_3 \end{array}\right\}; \qquad (8.206)$$

8.5. HEISENBERG HAMILTONIAN: ODD n

$$\left\{\begin{array}{ccccc|ccccc} a_1 & k'_1 & j_9 & k'_2 & k'_4 & j_1 & k_1 & k_2 & k_3 & k_4 \\ a_2 & a_3 & j_1 & k'_3 & j_2 & j_2 & a_1 & a_2 & a_3 & j_9 \\ k'_1 & k'_2 & k'_3 & k'_4 & j & k_1 & k_2 & k_3 & k_4 & j \end{array}\right\} \quad (8.207)$$

$$= \left\{\begin{array}{ccc|ccc} j_9 & k'_2 & k'_4 & j_1 & k_1 & k_4 \\ j_1 & k'_3 & j_2 & j_2 & k'_2 & j_9 \\ k'_3 & k'_4 & j & k_1 & k_4 & j \end{array}\right\} \left\{\begin{array}{ccc|ccc} a_1 & k'_1 & k_1 & k_1 & k_2 & k_3 \\ a_2 & a_3 & k'_2 & a_1 & a_2 & a_3 \\ k'_1 & k'_2 & k_4 & k_2 & k_3 & k_4 \end{array}\right\}.$$

The first result (8.206) for $f = 3$ confirms the Wigner $9-j$ coefficient (8.201), while the first triangle coefficient of order six in the right-hand side of (8.207) has exactly this same structure under appropriate change of parameters; the second triangle coefficient of order six on the right-hand side of (8.207) has the cubic graph as follows:

$$\text{[cubic graph diagram]} \quad (8.208)$$

$$p'_1 = (a_1, a_2, k'_1), \quad p'_2 = (k'_1, a_3, k'_2), \quad p'_3 = (k_1, k'_2, k_4),$$
$$p^* = (k'_1, k_1, k_3), \qquad (8.209)$$
$$p_1 = (k_1, a_1, k_2), \quad p_2 = (k_2, a_2, k_3), \quad p_3 = (k_3, a_3, k_4).$$

Thus, again, we have the factoring of the cubic graph (Yutsis rule):

$$\left\{\begin{array}{ccc|ccc} a_1 & k'_1 & k_1 & k_1 & k_2 & k_3 \\ a_2 & a_3 & k'_2 & a_1 & a_2 & a_3 \\ k'_1 & k'_2 & k_4 & k_2 & k_3 & k_4 \end{array}\right\}$$

$$= \left\{\begin{array}{cc|cc} a_1 & k'_1 & k_1 & k_2 \\ a_2 & k_1 & a_1 & a_2 \\ k'_1 & k_3 & k_2 & k_3 \end{array}\right\} \left\{\begin{array}{cc|cc} k'_1 & k_1 & k'_1 & k_3 \\ a_3 & k'_2 & k_1 & a_3 \\ k'_2 & k_4 & k_3 & k_4 \end{array}\right\}. \quad (8.210)$$

Relation (8.207) now transforms into the factored form:

$$\left\{\begin{array}{ccccc|ccccc} a_1 & k'_1 & j_9 & k'_2 & k'_4 & j_1 & k_1 & k_2 & k_3 & k_4 \\ a_2 & a_3 & j_1 & k'_3 & j_2 & j_2 & a_1 & a_2 & a_3 & j_9 \\ k'_1 & k'_2 & k'_3 & k'_4 & j & k_1 & k_2 & k_3 & k_4 & j \end{array}\right\} = \quad (8.211)$$

$$\left\{\begin{array}{ccc|ccc} j_9 & k'_2 & k'_4 & j_1 & k_1 & k_4 \\ j_1 & k'_3 & j_2 & j_2 & k'_2 & j_9 \\ k'_3 & k'_4 & j & k_1 & k_4 & j \end{array}\right\} \left\{\begin{array}{cc|cc} a_1 & k'_1 & k_1 & k_2 \\ a_2 & k_1 & a_1 & a_2 \\ k'_1 & k_3 & k_2 & k_3 \end{array}\right\} \left\{\begin{array}{cc|cc} k'_1 & k_1 & k'_1 & k_3 \\ a_3 & k'_2 & k_1 & a_3 \\ k'_2 & k_4 & k_3 & k_4 \end{array}\right\}.$$

It is evident from (8.206), (8.211), and the transformation of shapes (8.204) that the general expression for the triangle coefficient (8.205) is:

The $3(2f+1) - j$ coefficient in standard labels:

$$\left\{ \begin{array}{cccccccc} a_1 & k'_1 & k'_2 & \cdots & k'_{f-3} & j_{2f}|1 & k'_{f-2} & k'_f \\ a_2 & a_3 & a_4 & \cdots & a_{f-1} & j_1 & k'_{f-1} & j_2 \\ k'_1 & k'_2 & k'_3 & \cdots & k'_{f-2} & k'_{f-1} & k'_f & j \end{array} \right.$$
$$\left. \begin{array}{|cccccc} j_1 & k_1 & k_2 & \cdots & k_{f-1} & k_f \\ j_2 & a_1 & a_2 & \cdots & a_{f-1} & j_{2f+1} \\ k_1 & k_2 & k_3 & \cdots & k_f & j \end{array} \right\}$$

$$= \left\{ \begin{array}{ccc|ccc} j_{2f+1} & k'_{f-2} & k'_f & j_1 & k_1 & k_f \\ j_1 & k'_{f-1} & j_2 & j_2 & k'_{f-2} & j_{2f+1} \\ k'_{f-1} & k'_f & j & k_1 & k_f & j \end{array} \right\} \left\{ \begin{array}{cc|cc} a_1 & k'_1 & k_1 & k_2 \\ a_2 & k_1 & a_1 & a_2 \\ k'_1 & k_3 & k_2 & k_3 \end{array} \right\}$$

$$\times \prod_{s=1}^{f-3} \left\{ \begin{array}{cc|cc} k'_s & k_s & k'_s & k_{s+2} \\ a_{s+2} & k'_{s+1} & k_s & a_{s+2} \\ k'_{s+1} & k_{s+3} & k_{s+2} & k_{s+3} \end{array} \right\}. \qquad (8.212)$$

Proof. The cubic graph corresponding to the second factor in (8.205) is the following:

$$\begin{array}{cccccc} p'_1 & p'_2 & p'_3 & p'_{f-3} & p'_{f-2} & p'_{f-1} \\ \bullet & \bullet & \bullet & \cdots & \bullet & \bullet & \bullet \\ & & & \vdots & & & \\ \bullet & \bullet & \bullet & \cdots & \bullet & \bullet & \bullet \\ p_1 & p_2 & p_3 & p_{f-3} & p_{f-2} & p_{f-1} \end{array} \qquad (8.213)$$

The points of the cubic graph are read off the columns of the left and right patterns of the triangle coefficient of order $2(f-1)$ in (8.205):

$$p'_1 = (a_1, a_2, k'_1), \ p'_2 = (k'_1, a_3, k'_2), \ \ldots, \ p'_{f-2} = (k'_{f-3}, a_{f-1}, k'_{f-2}),$$
$$p'_{f-1} = (k_1, k'_{f-2}, k_f); \qquad (8.214)$$
$$p_1 = (k_1, a_1, k_2), \ p_2 = (k_2, a_2, k_3), \ \ldots, \ p_{f-2} = (k_{f-2}, a_{f-2}, k_{f-1}),$$
$$p_{f-1} = (k_{f-1}, a_{f-1}, k_f).$$

Thus, because of the occurrence of the joining of subgraphs by three lines at $f-3$ distinct locations, the triangle coefficient of order $2(f-1)$ on the right-hand side of (8.205) factors into a product of $f-2$ triangle coefficients of order four. The labels of these triangle coefficients are read successively off the columns of the triangle coefficient of order $2(f-1)$ on

8.5. HEISENBERG HAMILTONIAN: ODD n

the left-hand side of (8.205) as follows: (col 1, col 4) of the first triangle coefficient of order four are $(a_1\,a_2\,k_1'), (k_2\,a_2\,k_3)$; of the second triangle coefficient of order four $(k_1'\,a_3\,k_2'), (k_3\,a_3\,k_4)$; $\cdots$; of the $f-2$ triangle coefficient of order four $(k_{f-3}'\,a_{f-1}\,k_{f-2}'), (k_{f-1}\,a_{f-1}\,k_f)$. Then, the other columns (col 2, col 3) of each of these triangle coefficients of order four are filled-in by the standard rule for cutting and joining the lines that makes them into irreducible triangle coefficients. This gives the result (8.212). Equivalently, the cut and join process produces $f-3$ pairs of virtual points p^*, as in the diagrams (8.202) and (8.208), that give the factoring of the cubic graph (8.213) into the product of $f-2$ triangle coefficients of order four that appear in (8.212). □

We complete the calculation by restoring the original quantum labels in (8.185) by the making replacements (8.187) in (8.212), accounting for $k_1' = k_2, k_{2i}' = k_{2i+2}, i = 1, 2 \ldots, f-2$ from the Kronecker delta factors:

The $3(f+1)-j$ coefficient in final form:

$$\left(R^{\mathbf{j}_3;\mathbf{j}_1}\right)_{\mathbf{k}',j;\mathbf{k}_1,j} = \left(R^{\mathrm{Sh}_3;\mathrm{Sh}_1}\right)_{\mathbf{k}',j;\mathbf{k}_1,j} = \delta_{k_1',k_2}\left(\prod_{i=1}^{f-2}\delta_{k_{2i}',k_{2i+2}}\right)$$

$$\times\begin{Bmatrix} k_1' & k_3' & \cdots & k_{2f-5}' & j_{2f+1} & k_{2f-3}' & k_{2f-1}' \\ k_2' & k_4' & \cdots & k_{2f-4}' & j_1 & k_{2f-2}' & j_2 \\ k_3' & k_5' & \cdots & k_{2f-3}' & k_{2f-2}' & k_{2f-1}' & j \end{Bmatrix}$$

$$\begin{Bmatrix} j_1 & k_1 & k_3 & \cdots & k_{2f-3} & k_{2f-1} \\ j_2 & k_2 & k_4 & \cdots & k_{2f-2} & j_{2f+1} \\ k_1 & k_3 & k_5 & \cdots & k_{2f-1} & j \end{Bmatrix}$$

$$= \delta_{k_1',k_2}\left(\prod_{i=1}^{f-2}\delta_{k_{2i}',k_{2i+2}}\right) \qquad (8.215)$$

$$\times\begin{Bmatrix} j_{2f+1} & k_{2f-3}' & k_{2f-1}' & j_1 & k_1 & k_{2f-1} \\ j_1 & k_{2f-2}' & j_2 & j_2 & k_{2f-3}' & j_{2f+1} \\ k_{2f-2}' & k_{2f-1}' & j & k_1 & k_{2f-1} & j \end{Bmatrix}$$

$$\times\begin{Bmatrix} k_1' & k_3' & k_1 & k_3 \\ k_2' & k_1 & k_1' & k_2' \\ k_3' & k_5 & k_3 & k_5 \end{Bmatrix} \prod_{s=1}^{f-3}\begin{Bmatrix} k_{2s+1}' & k_{2s-1} & k_{2s+1}' & k_{2s+3} \\ k_{2s+2}' & k_{2s+3} & k_{2s-1} & k_{2s+2}' \\ k_{2s+3}' & k_{2s+5} & k_{2s+3} & k_{2s+5} \end{Bmatrix}.$$

The nonzero matrix elements of the recoupling coefficient $R^{\mathbf{j}_3;\mathbf{j}_1}$ is product of $f-1$ triangle coefficients: one of order six and $f-2$ of order four.

We have now completed the calculation of the matrix elements of all quantities needed to obtain the matrix elements of the Heisenberg

Hamiltonian $H_{\mathbf{B}_1(o)}$ in (8.151): These are the $3(2f-1)-j$ coefficients in (8.122), which are the matrix elements of the recoupling matrix $R^{j_2;j_1}$, and the matrix elements of $R^{j_3;j_1}$ in (8.215), as described above. As in the even case, it remains to determined the real orthogonal matrix R that effects the diagonalization $R^{tr} H_{\mathbf{B}_o} R = D$ of $H_{\mathbf{B}_o}$ given by (8.151).

We conclude this section on the Heisenberg Hamiltonian H for odd $n = 2f + 1$ by giving the maximum eigenvalues, since, just as with the case of even n $= 2f$ in Sect. 8.4.2, the space is one-dimensional.

Maximum eigenvalues for $n = 2f + 1$

We have that $N_{j_{\max}}(\mathbf{j}) = 1$ for $j_{\max} = j_1 + j_2 + \cdots + j_n$. Thus, all matrices in (8.151) are one-dimensional, and the recoupling matrices are unity. The eigenvalues of the Hamiltonian $H = H_1 + H_2 + H_3$ are:

$$H_{\mathbf{B}_1(o)}(\max) = \lambda_{\max} = D^{j_1}(\max) + D^{j_2}(\max) + D^{j_3}(\max), \quad (8.216)$$

where $D^{j_h}(\max), h = 1, 2, 3$, denote the eigenvalues in relations (8.159) evaluated on the maximum values of the intermediate quantum numbers, and for $j = j_{\max}$. The maximum values of the intermediate quantum numbers that enter into the eigenvalues of the respective terms in (8.216) are obtained from the coupling schemes I_o, II_o, III_o in relations (8.132), (8.135), (8.138) to be the following:

$$k_1(\max) = j_1 + j_2, \ k_{2i}(\max) = j_{2i+1} + j_{2i+2}, \ i = 1, 2, \ldots, f-1;$$
$$k'_1(\max) = j_2 + j_3, \ k'_{2i}(\max) = j_{2i+2} + j_{2i+3}, \ i = 1, 2, \ldots, f-1;$$
$$k''_{2f-2}(\max) = j_1 + j_{2f+1}. \quad (8.217)$$

Substitution of these maximal values into the three forms for the respective eigenvalues in (8.159) gives the following values for the three terms on the right in (8.216):

$$2D^{j_1}(\max) = -\sum_{i=1}^{2f} j_i(j_i + 1) + (j_1 + j_2)(j_1 + j_2 + 1)$$

$$+ \sum_{i=1}^{f-1} (j_{2i+1} + j_{2i+2})(j_{2i+1} + j_{2i+2} + 1); \quad (8.218)$$

$$2D^{j_2}(\max) = -\sum_{i=2}^{2f+1} j_i(j_i + 1) + (j_2 + j_3)(j_2 + j_3 + 1)$$

$$+ \sum_{i=1}^{f-1} (j_{2i+2} + j_{2i+3})(j_{2i+2} + j_{2i+3} + 1); \quad (8.219)$$

$$2D^{\mathbf{j_3}}(\max) = -j_1(j_1+1) - j_{2f+1}(j_{2f+1}+1)$$
$$+(j_1+j_{2f+1})(j_1+j_{2f+1}+1). \tag{8.220}$$

Thus, the addition of these terms gives the value of $2\lambda_{\max}$ in (8.216). We leave the result of $\lambda_{\max}$ in unsimplified form, since it shows vividly the underlying structure of the maximal eigenvalues:

$$H|\mathbf{j}\ \mathbf{j}\rangle = \lambda_{\max}|\mathbf{j}\ \mathbf{j}\rangle. \tag{8.221}$$

Of course, the eigenvector for this maximal eigenvalue is the tensor product vector:

$$|\mathbf{j}\ \mathbf{j}\rangle = |j_1\ j_1\rangle \otimes |j_2\ j_2\rangle \otimes \cdots \otimes |j_n\ j_n\rangle. \tag{8.222}$$

8.5.4 Properties of Normal Matrices

The symmetric matrix representing the Heisenberg magnetic ring Hamiltonian is now known, in principle, for all values of n. We have denoted this Hamiltonian symmetric matrix by $H_{\mathbf{B}_1}$ and $H_{\mathbf{B}_1(o)}$ for $n = 2f$ and $n = 2f+1$, respectively. To complete a method of determining all the eigenvalues and a set of orthonormal vectors that fully diagonalize this Hamiltonian without invoking any properties of the cyclic group (see the next section), it is useful to recall the method of principal idempotents that applies to any normal matrix A of arbitrary order. (The order of A in this discussion is chosen to be a generic n; it is not the even and odd integer of the previous sections.)

The matrix A is *diagonable* if it is similar to a diagonal matrix D, otherwise, it is *nondiagonable*. The class of diagonable matrices includes the class of all normal matrices, where a normal matrix N is any matrix that commutes with its Hermitian conjugate: $NN^\dagger = N^\dagger N$. The class of normal matrices includes the following types (Perlis [60, p. 195]): Hermitian, skew-Hermitian, real symmetric, real skew-symmetric, unitary, orthogonal, diagonal, and all matrices unitarily similar to normal matrices. Indeed, a complex matrix is *unitarily similar* to a diagonal matrix if and only if it is normal. By definition, a diagonable matrix A can be brought to diagonal form, but it need not be normal. It is normal if and only if its eigenvectors are linearly independent, which is always the case if its eigenvectors are distinct.

Let A be a normal matrix of order n with distinct eigenvalues $\lambda_1, \lambda_2, \ldots, \lambda_k$ with eigenvalue λ_i repeated m_i times, $i = 1, 2, \ldots, k$, with $m_1 + m_2 + \cdots + m_k = n$. The principal idempotents E_i of A are defined by

$$E_i = \prod_{j \neq i}^{k} \frac{A - \lambda_j I_n}{\lambda_i - \lambda_j}, \quad i = 1, 2, \ldots, k. \tag{8.223}$$

These principal idempotents of A have the following properties:

1. Hermitian: $E_i = E_i^\dagger$, $i = 1, 2, \ldots, k$.

2. Orthogonal projections: $E_i E_j = \delta_{i,j} E_i$.

3. Trace: $\mathrm{tr} E_i = m_i =$ rank of $E_i =$ the number of linearly independent columns (rows) of E_i, $i = 1, 2, \ldots, k$.

4. Resolution of the identity: $I_n = E_1 + E_2 + \cdots + E_k$.

5. Complete: $A = \lambda_1 E_1 + \lambda_2 E_2 + \cdots + \lambda_k E_k$.

The following relations are consequences of these basic properties:

(i). Eigenvectors of A:

$$AE_i = E_i A = \lambda_i E_i, \quad i = 1, 2, \ldots, k,$$
$$A u_j(E_i) = \lambda_j u_j(E_i), \text{ each column } u_j(E_i) \text{ of } E_i, \quad (8.224)$$
$$j = 1, 2, \ldots, k, \text{ where the columns are orthogonal:}$$
$$u_j(E_i) \perp u_{j'}(E_{i'}), j \neq j', i \neq i'.$$

(ii). A unitary matrix that diagonalizes a normal matrix is constructed as follows: First apply the Gram-Schmidt process, using the rejection algorithm, to the columns of E_1, thereby constructing m_1 orthonormal columns $u_1, \ldots, u_{m_1}$; repeat the procedure for E_2, thereby constructing m_2 orthonormal columns $u_{m_1+1}, \ldots, u_{m_1+m_2}$, each of which is necessarily $\perp$ to the first m_1 columns; $\ldots$; repeat the procedure for E_k, thereby constructing m_k orthonormal columns $u_{m_1+\cdots+m_{k-1}+1}, \ldots, u_{m_1+\cdots+m_k}$, each of which is necessarily $\perp$ to all previously constructed columns. The unitary matrix U defined by $U = (u_1 \ u_2 \ \cdots \ u_n)$ then diagonalizes A.

(iii). Functions of A: If $F(x)$ is any well-defined function of a single variable x, then

$$F(A) = \sum_{i=1}^{k} F(\lambda_i) E_i. \quad (8.225)$$

Special applications of this result are: A matrix U is unitary if and only if there exists a Hermitian matrix H such that $U = \exp(iH)$; a matrix R is real orthogonal if and only if there exists a real skew-symmetric matrix A such that $R = \exp A$.

For a discussion of orthonormalization methods and, in particular, the rejection algorithm referred to above, we refer to [L, Sect. 10.3].

The idempotent matrices of a normal matrix are very powerful for dealing with many matrix problems that arise in physical systems. The book by Perlis [60] offers a concise and very readable account of their properties.

The idempotent method can be applied to each of the symmetric matrices $H_{\mathbf{B}_1}$ or $H_{\mathbf{B}_1(o)}$ to obtain all of their eigenvalues, which are necessarily real, and a real orthogonal matrix that fully diagonalizes each. In this method, no complex numbers whatsoever appear. The real orthogonal matrix, is of course, not unique. If the real orthogonal matrix R transforms H to real diagonal form D; that is, $R^{tr}HR = D$, then the columns of R are the eigenvectors of H. If an eigenvalue has multiplicity 1, then the corresponding column of R is unique up to $\pm$ sign; if an eigenvalue λ is of multiplicity m_λ, then the m_λ columns of R corresponding to eigenvalue λ can be transformed among themselves by arbitrary real transformations that preserve their perpendicularity. Effecting such operations for all roots gives a new orthogonal matrix S that also diagonalizes H; that is, $S^{tr}HS = D$. Equivalently, if T is an arbitrary real orthogonal matrix such that $DT = TD$, then also $S = RT^{tr}$ diagonalizes H. This construction gives the class of all real orthogonal matrices that diagonalize H. *In many respects, this method is superior to the classification of states under the action of the cyclic group discussed in Sect. 8.7 below.*

8.6 Recount, Synthesis, and Critique

The Heisenberg magnetic ring Hamiltonian H is an invariant under the $SU(2)$ quantum-mechanical group of frame rotations. This means that the viewpoint can be adopted that the system is a composite whole whose quantum states can be characterized by the total angular momentum of the constituent parts and the squares of the angular momenta of each part. Moreover, the corresponding set of mutually commuting Hermitian operators $\mathbf{J}^2, J_3, \mathbf{J}^2(1), \mathbf{J}^2(2), \ldots, \mathbf{J}^2(n)$ becomes a complete set when augmented with the squares of any set of intermediate angular momenta $\mathbf{K}^2(1), \mathbf{K}^2(2), \ldots, \mathbf{K}^2(n-2)$ corresponding to any binary coupling scheme of the angular momenta $\mathbf{J}(1), \mathbf{J}(2), \ldots, \mathbf{J}(n)$ of the constituent parts. It is this observation that sets the stage for the determination of eigenvalues of H. The motivation for this approach is that it is known from the outset that this will split the Hilbert space of states into subspaces of dimension equal to the Clebsch-Gordan number $N_j(\mathbf{j})$, hence, the determination of the eigenvectors and eigenvalues of H is framed as one of diagonalizing the matrix of H on each such subspace. This, in turn, is a problem of diagonalizing symmetric matrices of order $N_j(\mathbf{j})$ by real orthogonal matrices.

We have set forth thus far in this lengthy chapter the special structures required from the binary coupling theory of angular momenta, as developed in [L], that enter into the implementation of the viewpoint above. We synthesize the principal concepts that emerge and the resulting recast of problems to be solved:

1. The first results are the exact eigenvectors and eigenvalues of H for $n = 2, 3, 4$, together with their respective relationships to ordinary WCG coefficients, Racah (or $6 - j$) coefficients and Wigner $9 - j$ coefficients.

2. The exact diagonalization of H does not extend to $n \geq 5$. Instead, a new structure emerges: First, for all even $n \geq 4$, the Hamiltonian H can be split into two noncommuting parts, $H = H_1 + H_2$, where H_1 and H_2 are separately diagonalizable on two distinct binary coupling schemes, called schemes I and II. Using this fact, the matrix representing H can be given on, say, the set of orthonormal basis vectors $\mathbf{B}_1$ that define binary coupling scheme I. It is this symmetric matrix $H_{\mathbf{B}_1}$ of order $N_j(\mathbf{j})$ that needs to be diagonalized by a real orthogonal matrix to determine the eigenvectors and eigenvalues of H itself. Second, for all odd $n \geq 5$, the Hamiltonian H can be split into three noncommuting parts, $H = H_1 + H_2 + H_3$, and $H_1, H_2,$ and H_3 are separately diagonalizable on three distinct binary coupling schemes, called schemes $I_o, II_o,$ and III_o. Again, the symmetric matrix $H_{\mathbf{B}_1(o)}$ of order $N_j(\mathbf{j})$ of H on the basis $\mathbf{B}_1(o)$ can be given. It is this symmetric matrix $H_{\mathbf{B}_1(o)}$ of order $N_j(\mathbf{j})$ that needs to be diagonalized by a real orthogonal matrix to determine the eigenvectors and eigenvalues of H itself.

3. The simplicity of description in Items 1 and 2 belies the underlying binary coupling theory that must be put in place to arrive at the symmetric matrix of order $N_j(\mathbf{j})$ representing the Heisenberg Hamiltonian H, respectively, on the binary coupled bases $\mathbf{B}_1$ and $H_{\mathbf{B}_1(o)}$. This is what most of this chapter has been about. Briefly, the pathway has proceeded along the directions as follows:

(a) Enumeration of the labeled binary tree that encode a given coupling scheme, together with the rule for reading off the elementary triangles — triplets of angular momenta — that define the orthonormal bases set of states associated with the labeled binary tree in terms of generalized WCG coefficients, which are well-defined products of ordinary WCG coefficients. The extension of this to pairs of labeled binary trees that gives the full set of triangle coefficients that determines the real orthogonal transformation between the pair of orthonormal basis sets corresponding to each labeled binary tree. Once this pair relation is established, all the remaining developments have to do with the remarkable properties that the triangle coefficients possess, much of which is made transparent under the concept of a recoupling matrix, which is

8.6. RECOUNT, SYNTHESIS, AND CRITIQUE

nothing more than a matrix of order $N_j(\mathbf{j})$ with matrix elements defined in terms of the left and right patterns of a triangle coefficient, or equivalently, directly in terms of the inner product of the orthonormal vectors associated with the pair of labeled binary trees.

(b) It is the shape of a binary tree that controls its structure, not only in the determination of the configuration of the binary tree itself as a diagram, but also in the manner in which labeled shapes relate to one another through the elementary transformations by commutation and association, as implemented in detail through shape transformations. This manifests itself through the reduction of triangle coefficients to ones of lower order, or to irreducible triangle coefficients — ones that cannot be reduced — that define the so-called $3(n-1)-j$ coefficients of Wigner, Racah, and many others, as well as the special $3(n-1)-j$ coefficients of Yutsis type that factor. Indeed, it through the concept of a path from one binary coupled set of basis vectors to a second such set, such a path being effected by commutation and association of symbols in a labeled binary tree, that gives a product of recoupling matrices, which leads irrevocably to the unique determination of all irreducible triangle coefficients in terms of irreducible triangle coefficients of order four.

(c) The preceding sections identify in considerable detail (with many examples) those special labeled binary trees from Items (i) and (ii) that enter into obtaining the symmetric matrices $H_{\mathbf{B}_1}$ for even $n = 2f$ and $H_{\mathbf{B}_1(o)}$ for odd $n = 2f+1$ needed, and which still need to diagonalized. The cubic graphs associated with each coupling scheme can also be identified. Cubic graphs serve to identify the various types of $3(n-1)-j$ coefficients (including those that factor) that can arise in the general theory of binary coupling of angular momentum. This is particularly important in the Heisenberg problem.

(d) The full structure underlying the Heisenberg Hamiltonian has thus been set forth from the viewpoint that it is a composite system as described in the introductory paragraph above. While the combinatorial aspects of this structure are quite straightforward, the computational aspects are, by any measure, very complicated. While the $3(2f-1)-j, 6f-j, 3(f+1)-j$ coefficients that arise can, in principle, be calculated, this itself is a major effort, even if efficient computational algorithms can be put in place. And then there remains the problem of diagonalizing the symmetric matrices $H_{\mathbf{B}_1}$ and $H_{\mathbf{B}_1(o)}$, which is a formidable undertaking, even when given the complicated calculations that precede it. Perhaps, since one deals with matrices of order $N_j(\mathbf{j})$, some progress can be made for small values of the $(j_1, j_2, \ldots, j_n)$ and of n. But considerable scepticism remains for implementing these many-faceted computational features.

(e) The determination of the class of all real orthogonal matrix that diagonalizes the real symmetric matrices representing the Heisenberg

Hamiltonian is in itself a fully implementable procedure, once the Hamiltonian matrix is known. The general method is the method of principal idempotents, as described above. The Heisenberg Hamiltonian is, of course, invariant under the action of the cyclic group C_n. As such, its set of eigenvectors can be classified as transforming irreducibly under the action of C_n. But this classification cannot change the eigenvalues; it can at most effect a unitary transformation V that commutes with the diagonal matrix D in $R^{tr}HR = D$; that is, $VD = DV$. We conclude:

The most general class of unitary matrices that diagonalizes H is given by $U = RV^\dagger$, where R is any real orthogonal matrix that diagonalizes H, and V is any unitary matrix that commutes with H.

Complex numbers can now enter the classification because the irreducible representations of the cyclic group are given in terms of the complex roots of unity. We consider next the action of this invariance group in the space of coupled state vectors.

8.7 Action of the Cyclic Group

Given the invariance group of a Hamiltonian, it is always possible to classify the eigenvectors of the Hamiltonian in the state vector space $\mathcal{H}$ in which H acts by their transformation properties under the action of the invariance group. For the case at hand, the Hamiltonian is the Heisenberg Hamiltonian given by

$$H = \mathbf{J}(1)\cdot\mathbf{J}(2) + \mathbf{J}(2)\cdot\mathbf{J}(3) + \cdots + \mathbf{J}(n)\cdot\mathbf{J}(1). \tag{8.226}$$

This Hamiltonian is clearly invariant under the group of n substitutions:

$$i \mapsto i+1,\ i+1 \mapsto i+2,\ \ldots,\ n \mapsto 1,\ 1 \mapsto 2,\ \ldots,\ i-1 \mapsto i, \tag{8.227}$$

where $i = 1, 2, \ldots, n$. This group of substitutions is called the *cyclic group* and is denoted C_n.

The cyclic group $C_n \subset S_n$ may be described as an abelian subgroup of the symmetric group from which it inherits its multiplication rule:

$$\begin{aligned}
C_n &= \{g_1, g_2, \ldots, g_k, \ldots, g_n\}, \\
g_1 &= (1,2,\ldots,n), \ldots, g_k = (k, k+1, \ldots, n, 1, 2, \ldots, k-1), \\
&\ldots, g_n = (n, 1, 2, \ldots, n-1); \tag{8.228} \\
g_i g_j &= g_{i+j-1},\ i+j = 2, 3, \ldots, n+1, \\
g_i g_j &= g_{i+j-n-1},\ i+j = n+2, n+3, \ldots, 2n.
\end{aligned}$$

The group element g_2 is of order n, and it is also a generator of the group; that is, its powers give all the elements of the group: $g_i = (g_2)^{i-1}$, $i = 1, 2, \ldots, n$, with $(g_2)^0 = g_1$.

8.7. ACTION OF THE CYCLIC GROUP

For applications to physical problems, the irreducible representations of groups are particularly important: The irreducible representations of the abelian group C_n are constructed from the complex roots of unity:

$$\omega_j = e^{2\pi i(j-1)/n}, j = 1, 2, \ldots, n, \text{ with } \omega_1 = 1. \tag{8.229}$$

We write these distinct n-th roots of unity as a sequence with n parts:

$$\omega = (\omega_1, \omega_2, \omega_3, \ldots, \omega_n). \tag{8.230}$$

The $(r-1)$-th power of ω is given by

$$\omega^{r-1} = (\omega_1^{r-1}, \omega_2^{r-1}, \omega_3^{r-1}, \ldots, \omega_n^{r-1}), r = 1, 2, \ldots, n, \tag{8.231}$$

where we write each $\omega_j^{r-1} = e^{2\pi i(j-1)(r-1)/n} = e^{2\pi i q_j/n}$, where q_j is defined by $(j-1)(r-1) = q_j \pmod{n}$; that is, we reduce, when possible, the number $e^{2\pi i(j-1)(r-1)/n}$ by factoring out all powers $\omega^n = 1$. For uniform expression of results for all n, we make no further identification of the powers of ω as simplified expressions (by Euler's formula) of the complex number in question. For this same reason, we also write the character table as the $n \times n$ array of roots of unity given by the table with the following entries:

$$\chi(C_n) = \begin{array}{c|ccccc} & g_1 & g_2 & g_3 & \cdots & g_n \\ \hline g_1 & 1 & 1 & 1 & \cdots & 1 \\ g_2 & 1 & \omega_2 & \omega_3 & \cdots & \omega_n \\ g_3 & 1 & \omega_2^2 & \omega_3^2 & \cdots & \omega_n^2 \\ \vdots & \vdots & \vdots & \vdots & \cdots & \vdots \\ g_n & 1 & \omega_2^{n-1} & \omega_3^{n-1} & \cdots & \omega_n^{n-1} \end{array} \tag{8.232}$$

The entry in row r and column j of this table is the character $\chi_r(g_j)$ of group element g_j in the irreducible representation r, where this nomenclature is appropriate for an abelian group, since each element is in a class by itself, and the number of irreducible representations equals the number of classes. The entire row r, denoted χ_r, of this table constitutes the r-th irreducible representation with $r = 1$ being the identity representation. Thus, we have for each $r = 1, 2, \ldots, n$:

$$\chi_r = (1, ,\omega_2^{r-1}, \omega_3^{r-1}, \ldots, \omega_n^{r-1}) = \left(1, \chi_r(g_2), \chi_r(g_3), \ldots, \chi_r(g_n)\right), \tag{8.233}$$

$$\chi_r(g_j) = \omega_j^{r-1} = \omega_j^{q_{r,j}}, (r-1)(j-1) = q_{r,j} \bmod n, j = 1, 2, \ldots, n.$$

These characters then multiply according to the group multiplication
$$\chi_r(g_i)\chi_r(g_j) = \chi_r(g_i g_j). \tag{8.234}$$
since for an abelian group each element is in a class by itself. The presentation of the character table in the form (8.232) is unconventional, but suits well our purpose. (Tables for $n = 2, 3, \ldots, 6$ for the characters of C_n are given in Wilson et al. [86, p. 322], where the rows of (8.230) are arranged in a more conventional form, in which the complex conjugate pairs of rows are identified. See also Littlewood [42, p. 273] for a brief statement on the power of a character.)

In quantum theory, it is the action of a symmetry group G in the Hilbert space $\mathcal{H}$ of state vectors of a given physical system that manifests the role of abstract groups in classifying state vectors by their transformation properties under such actions. Corresponding to the definition of the action of each element of the group, there is also defined an action with respect to other operators that act in the Hilbert space, in particular, the Hamiltonian. We use the notation $A_g, g \in G$, to denote such a group action: Each A_g is then an operator acting in the relevant Hilbert space. Such operators are to satisfy the same multiplication rules as the abstract group itself: $A_e A_g = A_g A_e = A_e = \mathbb{I} =$ identity operator; $A_{g'} A_g = A_{g'g}$; $A_{g'''}(A_{g'} A_g) = (A_{g'''} A_{g'})A_g$; $A_{g^{-1}} = A_g^{-1}$, for all $g, g', g'' \in G$. The relations between the action of the group G on the Hilbert space $\mathcal{H}$ and its actions on operators that act in the Hilbert space are the following:

arbitrary operator action in $\mathcal{H}$: $|\psi\rangle \mapsto A|\psi\rangle$,

group action in $\mathcal{H}$: $|\psi\rangle \mapsto A_g|\psi\rangle$ (8.235)

combined operator action in $\mathcal{H}$; first A, then A_g :
$$A_g(A|\psi\rangle) = (A_g A)|\psi\rangle = (A_g A A_{g^{-1}})A_g|\psi\rangle.$$

Thus, under the state vector transformation $|\psi\rangle \mapsto A_g|\psi\rangle$ corresponding to the group action, an arbitrary operator A on $\mathcal{H}$ undergoes the operator similarity transformation $A \mapsto A_g A A_{g^{-1}}$. Since we will only be interested in transformations of Hermitian operators, we realize the group action by unitary operators on $\mathcal{H}$; that is, $A_g^\dagger = A_{g^{-1}}$.

We next adapt the above situation for a general finite symmetry group to a Hermitian Hamiltonian H acting in a finite-dimensional Hilbert space $\mathcal{H}_d$. Thus, U_g is a unitary operator that acts in $\mathcal{H}_d$ and commutes with H; that is, $U_g H = H U_g$, each $g \in G$. The degeneracy of the eigenvalues of H and the most general linear action of the group G can then be described in terms of a Hilbert space $\mathcal{H}_d$ that has an orthonormal basis given by

$$\mathbf{B}_d = \{|\Psi_{i1}\rangle, |\Psi_{i2}\rangle, \ldots, |\Psi_{im_i}\rangle \,|\, i = 1, 2, \ldots, k\},$$
$$d = m_1 + m_2 + \cdots + m_k. \tag{8.236}$$

8.7. ACTION OF THE CYCLIC GROUP

The action of H and U_g on this basis is given by

$$H|\Psi_{ij}\rangle = \lambda_i|\Psi_{ij}\rangle, \, i = 1, 2, \ldots, k;$$
$$j = 1, 2, \ldots, m_i,$$
$$HU_g|\Psi_{ij}\rangle = \lambda_i U_g|\Psi_{ij}\rangle, \, i = 1, 2, \ldots, k; \quad (8.237)$$
$$j = 1, 2, \ldots, m_i,$$
$$U_g|\Psi_{ij}\rangle = \sum_{j'=1}^{m_i} \left(U_g^{(k)}\right)_{j'j} |\Psi_{ij'}\rangle, \, i = 1, 2, \ldots, k.$$

The eigenvalues λ_i, $i = 1, 2, \ldots, k$, in the first relation are the real distinct eigenvalues of the Hermitian operator H, each occurring with multiplicity m_i. An important point is: Because the eigenvalues λ_i are distinct, the group action can only effect transformations in each of the degenerate subspaces, giving rise to a unitary matrix representation $U_g^{(k)}$ of order k of the unitary operator U_g in each such subspace.

The additional structure provided by the symmetry group G is:

Each unitary matrix representation $U_g^{(k)}, g \in G$, can be reduced by a unitary similarity transformation into a direct sum of irreducible representations of G, thus providing a more detailed classification of the eigenvectors of H. The classification can still leave behind degeneracies because the same irreducible representation of G can be repeated.

The structure exhibited by relations (8.236)-(8.237) is, by design, that encountered in the Heisenberg ring problem.

8.7.1 Representations of the Cyclic Group

We next consider the application of relations (8.235)-(8.237) to the Heisenberg ring Hamiltonian H with the invariance group $G = C_n$, the cyclic group of permutations described above. For definiteness, we assume that H has been diagonalizes along the lines outlined on the idempotent method (or other methods); that is, that we have found a real orthogonal matrix R such that
$$R^{tr} H_\mathbf{B} R = D. \quad (8.238)$$

We now employ the notation $H_\mathbf{B}$ to denote either of the symmetric matrices $H_{\mathbf{B}_1}$ or $H_{\mathbf{B}_1(o)}$, for which we have provided the algorithms for their calculation in the previous sections. The symmetric matrix $H_\mathbf{B}$ is always of order equal to the CG number $d = N_\mathbf{j}(\mathbf{j})$, where $\mathbf{j} = (j_1, j_2, \cdots, j_n)$ and $j \in \{j_{\min}, j_{\min} + 1, \ldots, j_{\max}\}$ are specified.

We next present in a uniform notation the background of vector space structures, the binary coupled orthonormal basis sets, on which H is represented by the symmetric matrix $H_\mathbf{B}$:

$$\mathbf{B} = \{|T(\mathbf{j\,k})_{j\,m}\rangle \mid \mathbf{k} \in \mathbb{K}\}. \qquad (8.239)$$

The vectors in $\mathbf{B}$ are the orthonormal coupled state vectors for a standard labeled binary tree $T \in \mathbb{T}_n$, in which $\mathbf{j} = (j_1, j_2, \ldots, j_n)$ are assigned to the shape of the binary tree. The total angular momentum labels $j\,m$ are regarded as specified from their respective domains of definition, $j \in \{j_{\min}, j_{\min}+1, \ldots, j_{max}\}$ and $m \in \{j, j-1, \ldots, -j\}$. In every usage of the symbol $\mathbf{B}$, the vectors in $\mathbf{B}$ are enumerated by $\mathbf{k} \in \mathbb{K}$. This defines a vector space $\mathcal{H}_{\mathbf{j},j}$ with orthonormal basis $\mathbf{B}$ of dimension $N_j(\mathbf{j})$. The intermediate angular momenta have the standard labels of the binary tree: $\mathbf{k} = (k_1, k_2, \ldots, k_{n-2}) \in \mathbb{K}$, $|\mathbb{K}| = N_j(\mathbf{j})$. This general situation then captures both the n even and n odd cases for the Heisenberg Hamiltonian.

We next cast the problem into a format adapted to relations (8.237), beginning with the statement of the form of the eigenvectors themselves as presented by relation (8.238):

$$\begin{aligned}
|\mathbf{j}; p, q; j\,m\rangle &= \sum_{\mathbf{k}\in\mathbb{K}} R_{\mathbf{k};p,q}|T(\mathbf{j\,k})_{j\,m}\rangle, \\
H|\mathbf{j}; p, q; j\,m\rangle &= \lambda_{j,p}(\mathbf{j})|\mathbf{j}; p, q; j\,m\rangle, \qquad (8.240) \\
H A_{g_k}|\mathbf{j}; p, q; j\,m\rangle &= \lambda_{j,p}(\mathbf{j})\, A_{g_k}|\mathbf{j}; p, q; j\,m\rangle, \\
A_{g_k}|\mathbf{j}; p, q; j\,m\rangle &= \sum_{q'} \left(A_{g_k}\right)_{q',q} |\mathbf{j}; p, q'; j\,m\rangle.
\end{aligned}$$

The (row, column) elements of R are enumerated by

$$R = (R)_{\mathbf{k};p,q}, \ \mathbf{k} \in \mathbb{K} \text{ and } p, q \in \mathbb{D}_{p,q},\ |\mathbb{D}_{p,q}| = N_j(\mathbf{j}). \qquad (8.241)$$

The indices (p, q) result from the diagonalization by R in relation (8.238): p enumerates the distinct eigenvalues, and q the degeneracy of each eigenvalue $\lambda_{j,p}(\mathbf{j})$ for given p. The indices (p, q) belong to some unspecified domain of definition $\mathbb{D}_{p,q}$ with cardinality equal to the CG number $N_j(\mathbf{j})$; they are determined by the explicit diagonalization of $H_\mathbf{B}$ in (8.238), and are not among the class of angular momentum quantum number otherwise present. We also define the domain of definition of q by $\mathbb{Q}_q(p)$, noting that it depends on p (and the other quantum labels that we suppress). Thus, the dimension of the degeneracy space for given p is $|\mathbb{Q}_q(p)|$. The form of relations (8.240) presents the most general situation that can occur for the diagonalization of $H_\mathbf{B}$. It is then the case that the action of the unitary operator A_{g_k}, each $g_k \in C_n$, can only effect a unitary matrix transformation with coefficients $\left(A_{g_k}\right)_{q',q}$ among these degenerate states, just as in (8.237).

8.7. ACTION OF THE CYCLIC GROUP

The eigenvectors determined by the orthogonal transformation in (8.240) satisfy the following orthogonality relations:

$$\langle j; p', q'; j\,m | j; p, q; j\,m \rangle = \delta_{p',p}\delta_{q',q}. \tag{8.242}$$

It follows from this relation and the last relation in (8.240) that

$$\langle j; p, q'; j\,m | A_{g_k} | j; p, q; j\,m \rangle = \left(A_{g_k}\right)_{q',q} = \left(U^{(k)}\right)_{q',q}, \tag{8.243}$$

where $\{U^{(k)} \,|\, g_k \in C_n\}$ denotes a unitary matrix representation of order $|Q_q(p)|$ of the cyclic group C_n. It is this unitary matrix representation of C_n on each degeneracy space of order $|Q_q(p)|$ that must be reduced into the irreducible representations in order to classify the eigenvectors of H as eigenvectors that are irreducible under the action of the group C_n.

Then, since every unitary matrix representation of C_n is completely reducible by a unitary transformation, there must exist a new orthonormal basis of the degeneracy space of the following form:

$$|j; p; j\,m\rangle_r = \sum_q V_{q,r} |j; p, q; j\,m\rangle, \tag{8.244}$$

where the complex coefficients $V_{q,r}$ are the elements of a unitary matrix V, and the action of A_{g_k} is irreducible on the new basis:

$$A_{g_k}|j; p; j\,m\rangle_r = \chi_r(g_k)|j; p; j\,m\rangle_r = \omega_k^{q_r,k}|j; p; j\,m\rangle_r. \tag{8.245}$$

Thus, the problem is to determined the unitary matrix V in so far as possible.

The substitution of relation (8.244) into both sides of relation (8.245), followed by multiplication from the left by $V^*_{q',r}$, summation over r, and use of the unitary conditions $\left(VV^\dagger\right)_{q,q'} = \sum_r V^*_{q',r} V_{q,r} = \delta_{q,q'}$ gives the relation:

$$A_{g_k}|j; p, q'; j\,m\rangle = \sum_q \left(U^{(k)}\right)_{q,q'} |j; p, q; j\,m\rangle, \tag{8.246}$$

$$\left(U^{(k)}\right)_{q,q'} = \sum_r V_{q,r} \chi_r(g_k) V^\dagger_{r,q'},$$

where $U^{(k)}$ is the unitary matrix defined in (8.243). Thus, if we define the diagonal matrix $\Gamma^{(k)}$ to be the matrix with elements given by

$$\left(\Gamma^{(k)}\right)_{r,r'} = \delta_{r,r'} \chi_r(g_k), \tag{8.247}$$

then relation (8.246) is just the expression of the unitary similarity transformation

$$V\Gamma^{(k)}V^\dagger = U^{(k)}; \text{ equivalently, } V^\dagger U^{(k)} V = \Gamma^{(k)}. \qquad (8.248)$$

This relation makes explicit that the significance of the unitary similarity transformation by V is to bring the reducible representation $U(k)$ of the cyclic group C_n to its fully reduced form.

We next summarize the results obtained so far, at the risk of being repetitious, so that attention can shift to what remains to be done:

Summary. *The simultaneous eigenvector-eigenvalue relations for the Heisenberg ring Hamiltonian H and the irreducible cyclic group action A_{g_k}, $g_k \in C_n$, are of the following form for all $n \geq 5$:*

$$\begin{aligned}
H|\mathbf{j};p;jm\rangle_r &= \lambda_{j,p}(\mathbf{j})|\mathbf{j};p;jm\rangle_r, \\
A_{g_k}|\mathbf{j};p;jm\rangle_r &= \chi_r(g_k)|\mathbf{j};p;jm\rangle_r = \omega_k^{q_{r,k}}|\mathbf{j};p;jm\rangle_r, \\
|\mathbf{j};p;jm\rangle_r &= \sum_q V_{q,r}|\mathbf{j};p,q;jm\rangle, \qquad (8.249) \\
V^\dagger U^{(k)} V &= \Gamma^{(k)}, \\
U^{(k)} &= R^{tr} R^{\mathbf{j};\mathbf{j}_k} U_{g_k}^{\mathbf{j}_k;\mathbf{j}_1} R = R^{tr}\left(R^{\mathbf{j};\mathbf{j}_2} U_{g_2}^{\mathbf{j}_2;\mathbf{j}_1}\right)^{k-1} R.
\end{aligned}$$

We have taken the liberty of placing the last relation in (8.249) with the other relations, although it is derived in the next section. The matrix $U_{g_k}^{\mathbf{j}_k;\mathbf{j}_1}$ of order $N_j(\mathbf{j})$ is the unitary matrix representing the action of the unitary operator A_{g_k} on the basis $\mathbf{B}_1$, as given by relations (8.256)-(8.257) below. The unitary matrix $U_{g_k}^{\mathbf{j}_k;\mathbf{j}_1}$ is fully arbitrary, subject only to the rules that it satisfy all the group multiplication rules of the group of unitary operators $\mathbb{A}_{C_n} = \{A_{g_k}\,|\,g_k \in C_n\}$, since it is the unitary matrix representation of this group of operators on the basis $\mathbf{B}_1$.

The results summarized by relations (8.249) show that there are many sets of simultaneous eigenvectors $|\mathbf{j};p;jm\rangle_r$ of the Heisenberg Hamiltonian and the group of unitary operators $\mathbb{A}_{C_n}$: There is no internal structure that determines the unitary matrix representation $\{U_{g_k}^{\mathbf{j}_k;\mathbf{j}_1}\,|\,g_k \in C_n\}$ of the group C_n — these matrices are not even known, hence, there is no possibility of bringing them to irreducible diagonal form. *The cyclic group does not remove the degeneracy of the states of the Heisenberg Hamiltonian.*

8.7.2 The Action of the Cyclic Group on Coupled State Vectors

The completion of the description of the simultaneous eigenvector-eigenvalue relations given by relations (8.249) for the Heisenberg ring Hamiltonian H and the irreducible cyclic group action $A_{g_k}, g_k \in C_n$ requires still the derivation of the last relation $U^{(k)} = R^{tr} R^{\mathbf{j};\mathbf{j}_k} U_{g_k}^{\mathbf{j}_k;\mathbf{j}_1} R$. Direct substitution of the first relation (8.240) into relation (8.243) gives

$$U^{(k)} = R^{tr} W^{(k)} R, \tag{8.250}$$

$$\left(W^{(k)}\right)_{\mathbf{k}',j;\mathbf{k},j} = \langle T(\mathbf{j}\,\mathbf{k}')_{jm} | A_{g_k} | T(\mathbf{j}\,\mathbf{k})_{jm} \rangle.$$

Thus, we need to evaluate the matrix elements of the group action A_{g_k} on the coupled state vector $|T(\mathbf{j}\,\mathbf{k})_{jm}\rangle$. For this, we use the cyclic permutation realization of C_n given by relations (8.6).

The labeled binary trees of interest are all generated from the one that gives the coupled state vectors $|T(\mathbf{j}\,\mathbf{k})_{jm}\rangle$. It is convenient now to call this scheme 1, and correspondingly define schemes $2, 3, \ldots, n$ as follows:

$$\text{scheme } i : \mathbf{j}_i = (j_i, j_{i+1}, \ldots, j_n, j_1, \ldots, j_{i-1}),$$

$$\mathbf{k}_i = (k_1^{(i)}, k_2^{(i)}, \ldots, k_{n-2}^{(i)}), \ i = 1, 2, \ldots, n, \tag{8.251}$$

where for $i = 1$ the sequence $\mathbf{j} = \mathbf{j}_1$ ends with j_n. The domain of definition of the intermediate quantum labels is

$$\mathbf{k}_i \in \mathbb{K}^{(i)} = \mathbb{K}_T^{(j)}(\mathbf{j}_i). \tag{8.252}$$

Each such domain $\mathbb{K}^{(i)}$ is, of course, a uniquely defined set of quantum numbers for prescribed $\mathbf{j}$ and $j \in \{j_{\min}, j_{\min}+1, \ldots, j_{\max}\}$, it being described as the set of values such that the columns of the corresponding triangle pattern are all angular momentum triangles satisfying the simple CG rule for the addition of two angular momenta. We denote the corresponding set of orthonormal basis vectors by

$$\mathbf{B}_i = \{|T(\mathbf{j}_i\,\mathbf{k}_i)_{jm}\,|\,\mathbf{k}_i \in \mathbb{K}^{(i)}\}. \tag{8.253}$$

Each basis set $\mathbf{B}_i$ of vectors then spans one and the same vector space $\mathcal{H}_{\mathbf{j},j}$ and contains $|\mathbb{K}^{(i)}| = N_j(\mathbf{j})$ orthonormal vectors.

The n basis sets of orthonormal vectors $\mathbf{B}_i, i = 1, 2, \ldots, n$, are important for describing the action of A_{g_i} on the coupled state vectors $|T(\mathbf{j}_1\,\mathbf{k}_1)_{jm}\rangle$ defining the basis $\mathbf{B}_1$, which is the basis that occurs in the matrix elements for $U^{(i)}$ in (8.250) (for clarity of notation below, we replace k by i in that relation). Thus, we have that

$$g_i : \mathbf{j}_1 \mapsto \mathbf{j}_i = (j_i, j_{i+1}, \ldots, j_n, j_1, \ldots, j_{i-1}), \ i = 1, 2, \ldots, n, \tag{8.254}$$

$$g_1 = (1, 2, \ldots, n), \ g_2 = (2, 3, \ldots, n, 1), \ \ldots, \ g_n = (n, 1, \ldots, n-1).$$

The action of each $A_{g_i}, g_i \in C_n$, on each of the angular momentum operator constituents themselves is given by the unitary operator similarity transformation in relation (8.235):

$$A_{g_i} \mathbf{J}(k) A_{g_i}^{-1} = \mathbf{J}(k+i), \ k = 1, 2, \ldots, n, \tag{8.255}$$

where all integers are to be taken modulo n; that is, $n+1 \mapsto 1, n+2 \mapsto 2, \ldots, n+n \mapsto n$. These permutations of the n constituent angular momenta gives the corresponding permutation of the angular momentum quantum numbers given by (8.254). Thus, the most general action of A_{g_i} on the state vector $|T(\mathbf{j}_1\,\mathbf{k}_1)_{jm}\rangle$ is given by a unitary transformation of the basis vectors $|T(\mathbf{j}_i\,\mathbf{k}_i)_{jm}\rangle$:

$$A_{g_i}|T(\mathbf{j}_1\,\mathbf{k}_1)_{jm}\rangle = \sum_{\mathbf{k}_i \in \mathbb{K}^{(i)}} \left(U_{g_i}^{\mathbf{j}_i;\mathbf{j}_1}\right)_{\mathbf{k}_i,j;\mathbf{k}_1,j} |T(\mathbf{j}_i\,\mathbf{k}_i)_{jm}\rangle$$

$$= \sum_{\mathbf{k}'_1 \in \mathbb{K}^{(1)}} \left(R^{\mathbf{j}_1;\mathbf{j}_i} U_{g_i}^{\mathbf{j}_i;\mathbf{j}_1}\right)_{\mathbf{k}'_i,j;\mathbf{k}_1,j} |T(\mathbf{j}_1\,\mathbf{k}'_1)_{jm}\rangle, \tag{8.256}$$

where $R^{\mathbf{j}_1;\mathbf{j}_i}$ is one of our standard recoupling matrices. Thus, the matrix elements of the group action A_{g_i} on the basis $\mathbf{B}_1$ are given by

$$\langle T(\mathbf{j}_1\,\mathbf{k}'_1)_{jm} | A_{g_i} | T(\mathbf{j}_1\,\mathbf{k}_1)_{jm}\rangle = \left(R^{\mathbf{j}_1;\mathbf{j}_i} U_{g_i}^{\mathbf{j}_i;\mathbf{j}_1}\right)_{\mathbf{k}'_1,j;\mathbf{k}_1,j}. \tag{8.257}$$

This relation gives the most general action of A_{g_i}, which then has the matrix representation on the basis $\mathbf{B}_1$ given by (see (8.250)): $W^{(i)} = R^{\mathbf{j}_1;\mathbf{j}_i} U_{g_i}^{\mathbf{j}_i;\mathbf{j}_1}$. Because $A_{g_i} = (A_{g_2})^{i-1}$, $i = 1, 2, \ldots, n$, repeated application of A_{g_2} to relation (8.256) gives $W^{(i)} = \left(R^{(\mathbf{j}_1;\mathbf{j}_2)} U_{g_2}^{\mathbf{j}_2;\mathbf{j}_1}\right)^{i-1}$. This result, in turn, gives $U^{(i)}$ in relation (8.250), and then $U^{(k)}$ in relation (8.249), upon restoring index k in place of i.

The matrix elements of the recoupling matrix $R^{\mathbf{j}_1;\mathbf{j}_2}$ are the $3(2f-1) - j$ coefficients (irreducible triangle coefficients of order $2(2f-1)$) given for even $n = 2f$ by the recursion formula (8.122), and the $6j - j$ coefficients (irreducible triangle coefficients of order $2f$) given for odd

8.7. ACTION OF THE CYCLIC GROUP

$n = 2f + 1$ by the recursion formula (8.181). Thus, these quantities, which appear in relations (8.249) are, in principle, known.

But the unitary matrix $U_{g_2}^{j_2;j_1}$ representation of A_{g_2} is an arbitrary unitary matrix subject only to the same multiplication properties as A_{g_2}; that is, as the generator g_2 of C_n. Relation (8.256) gives the most general action of A_{g_i}.

In particular, there is no control over the structure of $U_{g_2}^{j_2;j_1}$, beyond its group multiplication properties — it is not determined. Unless there are special reasons for classifying the energy states by their cyclic group properties, it may even be advisable to ignore the classification (8.249).

The exact cases $n = 2, 3, 4$

The simultaneous eigenvectors and eigenvalues of H are given exactly for $n = 2, 3, 4$, respectively, by relations (8.28)-(8.29), (8.30)-(8.31), (8.34)-8.38). Only the action of the group C_n on the eigenvectors need be provided.

For $n = 2$, the action of the group C_2 on the simultaneous eigenvectors of H is given by

$$A_{g_1} |(j_1 \, j_2)j\, m\rangle = |(j_1 \, j_2)j\, m\rangle,$$
$$A_{g_2} |(j_1 \, j_2)j\, m\rangle = (-1)^{j_1+j_2-j} |(j_1 \, j_2)j\, m\rangle. \tag{8.258}$$

The eigenvectors $|(j_1 \, j_2)j\, m\rangle$ of H are already eigenvectors of the group action, and this action is irreducible, since the character table for C_2 is

$$\chi(C_2) = \begin{array}{c|cc} & g_1 & g_2 \\ \hline g_1 & 1 & 1 \\ g_2 & 1 & -1 \end{array} \tag{8.259}$$

For $n = 3$ (see (8.30)-(8.31)), the action of the $A_{g_i}, i = 1, 2, 3$, is still correctly given by relations (8.256)-(8.257), Because the eigenvalues $\lambda_j(\mathbf{j}) = \frac{1}{2}\Big(j(j+1) - j_1(j_1+1) - j_2(j_2+1) - j_3(j_3+1)\Big)$ of H are symmetric in j_1, j_2, j_3, and do not depend on the intermediate quantum number k_1, the simultaneous eigenvectors of $H, A_{g_i}, i = 1, 2, 3$, now take the form:

$$|j_1, j_2, j_3; j\, m\rangle_r = |\mathbf{j}, j\, m\rangle_r = \sum_{k_i} V_{k_i, r} |(j_i, k_i)j\, m\rangle, \tag{8.260}$$

where the matrix V is unitary of order $N_j(\mathbf{j})$. With this form for the simultaneous eigenvectors, relations (8.249) still hold for $n = 3$, upon setting $R = I_{N_j(\mathbf{j})}$:

$$H|\mathbf{j}; j\, m\rangle_r = \lambda_j(\mathbf{j})|\mathbf{j}; j\, m\rangle_r,$$
$$A_{g_k}|\mathbf{j}; j\, m\rangle_r = \chi_r(g_k)|\mathbf{j}, j\, m\rangle_r = \omega_k^{q_{r,k}}|\mathbf{j}; j\, m\rangle_r,$$
$$|\mathbf{j}; j\, m\rangle_r = \sum_{\mathbf{k}\in\mathbb{K}} V_{\mathbf{k},r}|T(\mathbf{j}\,\mathbf{k})_{jm}\rangle, \qquad (8.261)$$
$$V^\dagger U^{(k)} V = \Gamma^{(k)},$$
$$U^{(k)} = R^{\mathbf{j};\mathbf{j}_k}\, U_{g_k}^{\mathbf{j}_k;\mathbf{j}_1} = \left(R^{\mathbf{j};\mathbf{j}_2}\, U_{g_2}^{\mathbf{j}_2;\mathbf{j}_1} \right)^{k-1}.$$

Thus, already for $n = 3$, the cyclic group classification does not remove the degeneracy of eigenvectors because of the occurrence of the unitary matrix $U_{g_2}^{\mathbf{j}_2;\mathbf{j}_1}$. The character table accompanying relations (8.261) is:

$$\chi(C_3) = \begin{array}{c|ccc} & g_1 & g_2 & g_3 \\ \hline g_1 & 1 & 1 & 1 \\ g_2 & 1 & \omega_2 & \omega_3 \\ g_3 & 1 & \omega_3 & \omega_2 \end{array} \qquad \begin{pmatrix} \omega_2 = e^{2\pi i/3} \\ \omega_3 = e^{4\pi i/3} \end{pmatrix} \qquad (8.262)$$

The group elements of C_3 are here realized by the cyclic permutions:

$$g_1 = (1,2,3),\ g_2 = (2,3,1),\ g_3 = (3,1,2). \qquad (8.263)$$

For $n = 4$ (see (8.34)-8.38)), the action of the $A_{g_i}, i = 1,2,3,4$ is still correctly given by relations (8.256)-(8.257). Because the eigenvalues $\lambda_j(\mathbf{k}') = \frac{1}{2}\left(j(j+1) - k'_1(k'_1+1) - k'_2(k'_2+1)\right)$ of H are independent of j_1, j_2, j_3, j_4, the simultaneous eigenvectors of H and $A_{g_i}, i = 1,2,3,4$, must now be of the form:

$$|\mathbf{j}; \mathbf{k}'; j\, m\rangle_r = \sum_{q=1}^{4} V_{\mathbf{j}'_q, r}|(\mathbf{j}'_q\, \mathbf{k}')_{jm}\rangle, \qquad (8.264)$$

where the same intermediate angular momenta $\mathbf{k}' = (k'_1, k'_2)$ occur in both sides of this relation. For brevity of expression, we have also have introduced the following notations, where, for clarity, the $\mathbf{j}'_q$ are written

8.7. ACTION OF THE CYCLIC GROUP

in the shape notation of the binary tree (8.34) in question:

$$\mathbf{j} = (j_1, j_2, j_3, j_4), \quad \mathbf{j}'_1 = \Big((j_1\, j_3)(j_2\, j_4)\Big), \quad \mathbf{j}'_2 = \Big((j_2\, j_4)(j_3\, j_1)\Big),$$

$$\mathbf{j}'_3 = \Big((j_3\, j_1)(j_4\, j_2)\Big), \quad \mathbf{j}'_4 = \Big((j_4\, j_2)(j_1\, j_3)\Big). \tag{8.265}$$

These $\mathbf{j}'_q$ sequences are just the cyclic permutations of the group C_4 given by $g_1 = (1,2,3,4)$, $g_2 = (2,3,4,1)$, $g_3 = (3,4,1,2)$, $g_4 = (4,1,2,3)$, applied, in turn, to $\mathbf{j}'_1 = \Big((1\,3)(2\,4)\Big)$. Thus, the summation on the right-hand side of relation (8.264) is over all these cyclic permutations. Because the eigenvalue of H is independent of the j_i, and the same $\mathbf{k}' = (k'_1, k'_2)$ appear on both sides, the Hamiltonian H has the same eigenvalue on each of the coupled state vectors in (8.264); that is, the vector $|\mathbf{j}; \mathbf{k}'; j\,m\rangle_r$ is an eigenvector of H with eigenvalue $\lambda_j(\mathbf{k}')$ defined above, for arbitrary linear combinations. Attention must be paid to the triangle conditions associated with each fork in the four coupled state vectors — a state vector is taken to be the zero vector (with 0 coefficient) if these conditions are violated. Thus, we find again the simultaneous eigenvectors of H and the cyclic group operators A_{g_k}, $k = 1, 2, 3, 4$ are of similar form (8.249), where now the details are as follows:

$$H|\mathbf{j}; \mathbf{k}'; j\,m\rangle_r = \lambda_j(\mathbf{k}')|\mathbf{j}; \mathbf{k}'; j\,m\rangle_r,$$

$$A_{g_k}|\mathbf{j}; \mathbf{k}'; j\,m\rangle_r = \chi_r(g_k)|\mathbf{j}; \mathbf{k}'; j\,m\rangle_r = \omega_k^{q_{r,k}}|\mathbf{j}; \mathbf{k}'; j\,m\rangle_r,$$

$$|\mathbf{j}; \mathbf{k}'; j\,m\rangle_r = \sum_{q=1}^{4} V_{\mathbf{j}'_q, r}|(\mathbf{j}'_q\, \mathbf{k}')_{j\,m}\rangle, \tag{8.266}$$

$$V^\dagger U^{(k)} V = \Gamma^{(k)}, \quad U^{(k)} = R^{\mathbf{j}; \mathbf{j}'_k}.$$

Proof. The last relation giving $U^{(k)}$ requires proof. The action of each A_{g_i} on each coupled state vector $|(\mathbf{j}'_q\, \mathbf{k}')_{j\,m}\rangle$ cannot now effect a general unitary transformation, because it must preserve the intermediate angular momentum labels $\mathbf{k}'$; the action is simply to permute the four state vectors among themselves. Indeed, the action of A_{g_i} on the eigenvector (8.34) of H is given by $A_{g_i}|T(\mathbf{j}'_1\,(k'_1, k'_2))_{j\,m}\rangle = |T(\mathbf{j}'_i\,(k'_1, k'_2))_{j\,m}\rangle$. This result and relation (8.256) show that the matrix $U_{g_i}^{\mathbf{j}'_i; \mathbf{j}'_1}$ whose elements effect the transformation from $A_{g_i}|T(\mathbf{j}'_1\,(k'_1, k'_2))_{j\,m}\rangle$ to the coupled state vector $|T(\mathbf{j}'_1\,(k''_1, k''_2))_{j\,m}\rangle$ is the unit matrix of order $I_{N_j(\mathbf{j})}$. Thus, $U^{(k)}$ does not contain the factor $U_{g_k}^{\mathbf{j}'_k; \mathbf{j}'_1}$ that would otherwise be present. □

The matrix elements of the recoupling matrix $R^{\mathbf{j}; \mathbf{j}'_k}$ also reduce to a product of Kronecker delta and phase factors because of common forks.

The properties of the simultaneous eigenvectors of H and the C_4 group action operators $A_{g_k}, k = 1,2,3,4$, is completed by the group character table:

$$\chi(C_4) = \begin{array}{c|cccc} & g_1 & g_2 & g_3 & g_4 \\ \hline g_1 & 1 & 1 & 1 & 1 \\ g_2 & 1 & i & -1 & -i \\ g_3 & 1 & -1 & 1 & -1 \\ g_4 & 1 & -i & -1 & i \end{array} \qquad (8.267)$$

8.8 Concluding Remarks

I. **Probabilistic interpretation.** The probabilistic interpretation of the triangle coefficients, which are $3(2f-1)-j, 6f-j$, and $3(f+1)-j$ coefficients of various types, that give the coupled angular momentum states for the various complete sets of commuting Hermitian operators can be read off from the general rules stated in Chapters 1 and 5.

II. **Solved problem.** A full collection of structural relationships underlying the Heisenberg ring Hamiltonian H from the viewpoint of a composite system in which the total angular momentum is conserved has been presented. Formulas have been developed for the calculation of all binary coupled states that enter into the determination of the simultaneous eigenvectors and eigenvalues of the complete sets of commuting Hermitian operators that characterize the states of H, and a method (principal idempotents) given for diagonalizing H, including also the classification of states into irreducible states under the action of the cyclic invariance group C_n. In this sense, the problem is solved.

III. **Computational challenges.** The implementation of Item II requires computation of complicated $3(2f-1)-j, 6f-j$, and $3(f+1)-j$ coefficients of various types, all of which have been identified, and formulas developed for their calculation. By any measure, the actual computations will be difficult. Methods developed by Wei [79] may be useful.

IV. **Bethe Ansatz.** We have made no attempt to formulate the relationship between the Bethe Ansatz approach to the Heisenberg ring problem and that presented in this Chapter.

Appendix A

Counting Formulas for Compositions and Partitions

We summarize in this Appendix the notations used for various sets of compositions and partitions, where most of this information is synthesized from Andrews [1], Stanley [75], and [L].

Compositions are sequences of integers that sum to a prescribed integer, often coming with other conditions as well, such as all positive integers, all nonnegative integers, etc. The entries in the sequence are called *parts*. *Partitions* can be viewed as compositions with highly restrictive conditions on the parts. It is often the custom in the mathematics literature to consider only partitions having positive parts, while in the physics literature zero is admitted as a part. Considerable care must be taken in making the transition from one such class to the other. The symbols C and Par (with extra embellishments) are often used to denote classes of compositions and partitions with only positive parts admitted; we will modify these symbols to the form $\mathbb{C}$ and $\mathbb{P}ar$ when 0's are admitted as parts.

A.1 Compositions

We list the following classical results for compositions from Andrews [1, p.54] and Stanley [75, p. 15]):

1. The number of solutions $x = (x_1, x_2, \ldots, x_k)$ to the linear relation $x_1 + x_2 + \cdots + x_k = n \in \mathbb{N}$ in nonnegative integers is the binomial

coefficient given by

$$N(n,k) = \binom{n+k-1}{k-1}. \qquad (A.1)$$

We denote the set of all solutions by $\mathbb{C}(n,k)$ with cardinality $N(n,k) = |\mathbb{C}(n,k)|$.

2. The number of solutions $x = (x_1, x_2, \ldots, x_k)$ to $x_1 + x_2 + \cdots + x_k = n \in \mathbb{P}$ in positive integers is the binomial coefficient given by

$$p(n,k) = \binom{n-1}{k-1} = N(n-k, k). \qquad (A.2)$$

We denote the set of all solutions by $P(n,k)$ with cardinality $p(n,k) = |P(n,k)|$.

3. The number of solutions $x = (x_1, x_2, \ldots, x_k)$ to $x_1 + x_2 + \cdots + x_k \leq n \in \mathbb{N}$ in nonnegative integers is the binomial coefficient given by

$$\widehat{N}(n,k) = \binom{n+k}{k} = N(n, k+1). \qquad (A.3)$$

We denote the set of all solutions by $\widehat{\mathbb{C}}(n,k)$ with cardinality $\widehat{N}(n,k) = |\widehat{\mathbb{C}}(n,k)|$.

4. The number of solutions $x = (x_1, x_2, \ldots, x_k)$ to $x_1 + x_2 + \cdots + x_k = n \in \mathbb{P}$ in positive integers such that each part satisfies $x_i \leq m$ is the coefficient $c(m, k, n)$ of t^n given by

$$(1 + t + \cdots + t^{m-1})^k = \sum_{n=k}^{(m-1)k} c(m, k, n) t^{n-k}. \qquad (A.4)$$

We denote the set of all solutions by $C(m, k, n)$ with cardinality $c(m, k, n) = |C(m, k, n)|$.

5. The number of solutions $x = (x_1, x_2, \ldots, x_k)$ to $x_1 + x_2 + \cdots + x_k = n \in \mathbb{N}$ in nonnegative integers such that each part satisfies $x_i \leq m$ is the coefficient $N(m, k, n)$ of t^n given by

$$(1 + t + \cdots + t^m)^k = \sum_{n=0}^{mk} N(m, k, n) t^n. \qquad (A.5)$$

We denote the set of all solutions by $\mathbb{C}(m, k, n)$ with cardinality $N(m, k, n) = |\mathbb{C}(m, k, n)|$.

The cardinalities defined by these relations are related by

$$c(m, k, n) = p(n, k), \text{ for all } m \geq n,$$
$$N(n, k) = \widehat{N}(n, k-1) = p(n+k, k), \quad (A.6)$$
$$N(m, k, n) = c(m+1, k, n+k).$$

The relation $N(m, k, n) = c(m+1, k, n+k)$ is obtained from (A.4) by comparing coefficients in the expansion of the left-hand side of each of (A.4) and (A.5); it must have the stated interpretation because the shifts $m \to m+1$ and $n \to n+k$ are exactly what is needed to make the transition from the nonnegative condition to the positive condition, as is also evident in the relation $N(n, k) = p(n+k, k)$. Moreover, it follows from the multinomial expansion of relation (A.5) that the explicit form of $N(m, k, n)$ is given by the following expression in terms of multinomial coefficients:

$$N(m, k, n) = \sum_{\substack{(k_0, k_1, \ldots, k_m) \text{ such that} \\ k_0 + k_1 + \cdots + k_m = k;\ k_1 + 2k_2 + \cdots + mk_m = n}} \binom{k}{k_0, k_1, \ldots, k_m}. \quad (A.7)$$

A.2 Partitions

In the mathematical literature (see, for example, Andrews [1], Macdonald [50], Stanley [75]), partitions are defined to be sequences of positive integers $(\lambda_1, \lambda_2, \ldots, \lambda_n)$ satisfying the conditions

$$\lambda_1 \geq \lambda_2 \geq \cdots \geq \lambda_n > 0, \ n \in \mathbb{P}, \quad (A.8)$$

where the length $l(\lambda) = n$ of the sequence is any positive integer. The positive integer λ_i is called the i-th *part* of the partition. If $\sum_{i=1}^{n} \lambda_i = k$ and $l(\lambda) = n$, then λ is said to be a partition of k into n parts. (A partition is therefore also a composition in the set $P(n, k)$ in which the conditions $x_i \geq x_{i+1}, i = 1, 2 \ldots, n-1$, are enforced.) This set of partitions is denoted by

$$\text{Par}_n(k) = \{\lambda = (\lambda_1, \lambda_2, \ldots, \lambda_n) \mid \lambda \vdash k\}, \ 1 \leq n \leq k, \quad (A.9)$$

where $\lambda \vdash k$ denotes that the parts of λ sum to k. The cardinality of the set $\text{Par}_n(k)$ is denoted by

$$p_n(k) = |\text{Par}_n(k)|. \quad (A.10)$$

By convention, we take $p_n(k) = 0$, for all $k < n$. The number $p_n(k)$ has the special values $p_1(k) = p_k(k) = 1, k \geq 1$, and $p_{k-1}(k) = 1, k \geq 2$, since the corresponding partitions are $(k), (1^k)$, and $(2, 1^{[k-2]}), k \geq 2$.

The notation $\alpha^{[h]}$ for h a nonnegative integer means that the symbol α is to be written sequentially h times with $h = 0$ denoting no occurrence, but we often omit the [ ]. For all other values of the pair (k, n), the number $p_n(k)$ can be calculated from the recurrence relation (Stanley [75, p. 28]):

$$p_n(k) = p_{n-1}(k-1) + p_n(k-n), 2 \leq n \leq k, \qquad (A.11)$$

where, by convention, $p_n(k-n) = 0$, for $k < 2n$. The set Par(k) of all partitions of k is then given by

$$\text{Par}(k) = \bigcup_{n=1}^{k} \text{Par}_n(k), \ k \geq 1. \qquad (A.12)$$

For example, the set Par(4) is given by

$$\text{Par}(4) = \{(4), (3,1), (2,2), (2,1,1), (1,1,1,1)\}. \qquad (A.13)$$

The cardinality of the set Par(k) is given by

$$|\text{Par}(k)| = \sum_{n=1}^{k} p_n(k). \qquad (A.14)$$

The set of all nonempty partitions of arbitrary length is given by

$$\text{Par} = \bigcup_{k \geq 1} \text{Par}(k). \qquad (A.15)$$

By definition, Par(0) is the empty partition, which we denote by $\lambda_\emptyset$, and could be included in (A.15), although it the usual custom not to place the symbol $\lambda_\emptyset$ explictly in the set. (By convention, the empty set is considered to be a member of every set, and it is assigned cardinality 0.)

Our applications of partitions to unitary symmetry are most conveniently carried out by allowing partitions to have zeros as parts, for reasons related to Gelfand-Tsetlin patterns, as discussed in detail in [L].

A *partition with zero as part* is a partition λ to which a sequence of zeros $(0, 0, \ldots)$ has been adjoined at the right-hand end:

$$(\lambda, 0^{n-l}) = (\lambda_1, \lambda_2, \cdots, \lambda_l, \underbrace{0, 0, \ldots, 0}_{n-l}), \qquad (A.16)$$

where $l = l(\lambda)$ denotes the number of nonzero parts. This situation corresponds to modifying definition (A.8) of a partition to the following:

A.2. PARTITIONS

A partition with zero as a part is a sequence of nonnegative integers $(\lambda_1, \lambda_2, \ldots, \lambda_n)$ satisfying the conditions

$$\lambda_1 \geq \lambda_2 \geq \cdots \geq \lambda_n \geq 0, \ n \in \mathbb{P}. \tag{A.17}$$

We use notations analogous to those above for the following sets of partitions with zero as a part:

$$\begin{aligned}
\mathbb{P}\mathrm{ar}_n(k) &= \{\lambda = (\lambda_1, \ldots, \lambda_n) | \lambda_1 \geq \cdots \geq \lambda_n \geq 0, \lambda \vdash k\}, \\
\mathbb{P}\mathrm{ar}_n &= \bigcup_{k \geq 0} \mathbb{P}\mathrm{ar}_n(k), \\
\mathbb{P}\mathrm{ar} &= \bigcup_{n \geq 1} \mathbb{P}\mathrm{ar}_n.
\end{aligned} \tag{A.18}$$

The sets $\mathbb{P}\mathrm{ar}_n(k)$ and $\mathbb{P}\mathrm{ar}_n$ of partitions always has n parts, where we now count each repeated zero as a part; $\mathbb{P}\mathrm{ar}_n(k)$ contains a finite number of elements, but $\mathbb{P}\mathrm{ar}_n$ is countably infinite, as is $\mathbb{P}\mathrm{ar}$. The set $\mathbb{P}\mathrm{ar}_n$ includes $\mathbb{P}\mathrm{ar}_n(0) = (0^n)$. In general, the number of zero parts is unspecified. For example, the set $\mathbb{P}\mathrm{ar}_3(5)$ is given by

$$\mathbb{P}\mathrm{ar}_3(5) = (5,0,0), (4,1,0), (3,2,0), (3,1,1), (2,2,1). \tag{A.19}$$

We will refer to the sequences defined by (A.16) simply as "partitions," occasionally amplifying the description of the parts for clarity, but generally letting the notation $\mathbb{P}$ be the reminder that such sets of partitions contain all those for which 0 qualifies as a part, as in (A.19).

The development of the properties of partitions is a fairly sophisticated subject with a long and interesting history (see Andrews [1]). The cardinality of various classes of partitions is often available from generating functions or recursion relations, cases at hand being relations (A.4), (A.5) and (A.11). Here we note several further classical results, taken from Andrews.

The simplest case is the generating function of the cardinality $|\mathrm{Par}(k)|$ of the set of partitions $\mathrm{Par}(k)$ defined by (A.12) (see the example (A.13)):

$$\prod_{j \geq 1} (1 - t^j)^{-1} = \sum_{k \geq 0} |\mathrm{Par}(k)| t^k. \tag{A.20}$$

A very general result that finds many applications in unexpected ways was already discovered by Gauss; these polynomial generating functions, now called Gaussian binomial polynomials, generalize the binomial coefficients in a nontrivial way, as we next describe.

The polynomial $(t)_k, k \in \mathbb{N}$, of degree k is defined by

$$(t)_k = (1-t)(1-t^2) \cdots (1-t^k), k = 1, 2, \ldots; (t)_0 = 1, \tag{A.21}$$

where t is an arbitrary indeterminant (variable). The Gaussian binomial polynomials are denoted by $\begin{bmatrix} m \\ k \end{bmatrix}_t$, and are defined in terms of the polynomials $(t)_k$ for all $m, k \in \mathbb{N}$ by the following ratios of numerator and denominator polynomials, all giving the same Gaussian polynomial because of cancelation of numerator and denominator factors:

$$\begin{bmatrix} m \\ k \end{bmatrix}_t = \begin{bmatrix} k \\ m \end{bmatrix}_t = \frac{(t)_{m+k}}{(t)_m (t)_k}$$

$$= \frac{(1-t^{m+k})(1-t^{m+k-1})\cdots(1-t^{m+1})}{(1-t^m)(1-t^{m-1})\cdots(1-t)} \quad (m \geq 1)$$

$$= \frac{(1-t^{k+m})(1-t^{k+m-1})\cdots(1-t^{k+1})}{(1-t^k)(1-t^{k-1})\cdots(1-k)} \quad (k \geq 1), \quad (A.22)$$

$$\begin{bmatrix} 0 \\ k \end{bmatrix}_t = 1, k \geq 0; \quad \begin{bmatrix} m \\ 0 \end{bmatrix}_t = 1, m \geq 0.$$

These two forms result from the cancelation of common factors from the numerator and denominator polynomials in $(t)_{m+k}/(t)_m (t)_k$. After this cancelation of identical factors, the remaining denominator factors still divide the numerator factors to give a polynomial of degree mk; hence, we have the identity:

$$\begin{bmatrix} m \\ k \end{bmatrix}_t = \begin{bmatrix} k \\ m \end{bmatrix}_t = \sum_{n=0}^{mk} p(m, k, n) t^n = \sum_{n=0}^{mk} p(k, m, n) t^n. \quad (A.23)$$

It is these *Gaussian coefficients* $p(m, k, n) = p(k, m, n)$, which are positive integers defined for all $m, k \in \mathbb{N}$ and all $n = 1, \ldots, km$ that are of interest because of their interpretation in the theory of partitions:

The coefficient $p(m, k, n)$ is the number of (positive) partitions $\lambda = (\lambda_1, \lambda_2, \ldots, \lambda_j)$, for $j = 1, 2, \ldots, k \leq n$, such that each $\lambda_i \leq m$; that is, all partitions such that

$$\lambda_1 + \lambda_2 + \cdots + \lambda_j = n, j = 1, 2, \ldots, k; 1 \leq \lambda_i \leq m. \quad (A.24)$$

Boundary coefficients have the values $p(m, 0, n) = p(0, k, n) = 1$, if $m = k = n = 0$; $p(m, 0, n) = p(0, k, n) = 0$, if $n \geq 1$; $p(m, k, n) = 0$, if $n > mk$.

We denote by $\mathrm{Par}(m, k, n)$ the set of partitions defined by (A.24). The partitions in this set are called *restricted partitions*. The set $\mathrm{Par}(m, k, n)$ of restricted partitions is a subset of $\mathrm{Par}(m)$ — the subset such that each $\lambda \vdash n$ has at most k parts, each part $\leq m$. Relations (A.22)-(A.23) show

A.2. PARTITIONS

that we have the identity of cardinalities of the two sets $\operatorname{Par}(m,k,n)$ and $\operatorname{Par}(k,m,n)$ given by

$$p(m,k,n) = |\operatorname{Par}(m,k,n)| = |\operatorname{Par}(k,m,n)| = p(k,m,n). \qquad (A.25)$$

We refer to Andrews [1, Chapter 3] for many more properties of Gaussian binomial polynomials and of the Guassian coefficients $p(m,k,n) = p(k,m,n)$. (Partitions can, of course, be restricted in many other ways, and the theory of such partitions is a rich, deep, and difficult subject.) We note here the following analogies of Gaussian binomial polynomials with ordinary binomial coefficients. For these analogies, we use only the binomial polynomials $\begin{bmatrix} m \\ k \end{bmatrix}_t$ with $k = 0, 1, \ldots, m$:

$$\begin{bmatrix} m \\ 0 \end{bmatrix}_t = \begin{bmatrix} m \\ m \end{bmatrix}_t = 1,$$

$$\begin{bmatrix} m \\ k \end{bmatrix}_t = \begin{bmatrix} m \\ m-k \end{bmatrix}_t,$$

$$\begin{bmatrix} m \\ k \end{bmatrix}_t = \begin{bmatrix} m-1 \\ k \end{bmatrix}_t + t^{m-k} \begin{bmatrix} m-1 \\ k-1 \end{bmatrix}_t, \qquad (A.26)$$

$$\begin{bmatrix} m \\ k \end{bmatrix}_t = \begin{bmatrix} m-1 \\ k-1 \end{bmatrix}_t + t^k \begin{bmatrix} m-1 \\ k \end{bmatrix}_t,$$

$$\lim_{t \to 1} \begin{bmatrix} m \\ k \end{bmatrix}_t = \binom{m}{k} \quad (t = 1 \text{ after division}).$$

It is interesting that the generating functions for the various cardinalities given above also give the number of partitions in the set $\mathbb{P}\mathrm{ar}_n(k)$ that includes zero parts. Thus, for $k \leq n$, the set of partitions $\mathbb{P}\mathrm{ar}_n(k)$ with zero parts is obtained from the set $\operatorname{Par}(k)$ with positive parts simply by adjoining zeros as needed to obtain the n parts, including the zeros. Since this operation does not change the counting, we have the relation:

$$|\mathbb{P}\mathrm{ar}_n(k)| = |\operatorname{Par}(k)|, \ k \leq n. \qquad (A.27)$$

Thus, the generating function (A.19) gives also the positive integers $|\mathbb{P}\mathrm{ar}_n(k)|$. For $k \geq n$, it is the Gaussian binomial polynomials that enter. Thus, we proceed as follows: We add a 1 to each part of the n parts of a partition $\lambda \in \mathbb{P}\mathrm{ar}_n(k)$, thus obtaining a partition $\lambda' \vdash k+n$ having n parts with all positive parts and the restriction that each $\lambda'_i \leq k$. This is a partition in the set $\operatorname{Par}(k+1, n, n+k)$, since the condition "at most n parts" reduces now to exactly n parts because the condition $\lambda'_i \leq k+1$ does not allow partitions $\lambda' \vdash n+k, k > n$, of length

$l(\lambda') = 1, 2, \ldots, n-1$ to occur. Thus, the addition of 1 to each part of a partition in the set $\mathrm{Par}_n(k), k \geq n$, is a bijection (one-to-one onto mapping) to the set $\mathrm{Par}(k+1, n, n+k)$. Thus, we have the identity between cardinalities:

$$|\mathrm{Par}_n(k)| = |\mathrm{Par}(k+1, n, n+k)| = p(k+1, n, n+k), \; k \geq n > 1. \quad \text{(A.28)}$$

Relation (A.28) is important for physics, where partitions $\lambda \in \mathrm{Par}_n(k)$ find frequent applications (the relation seems not to appear in the literature). We therefore note in full how it appears in the expansion of the Gaussian binomial function:

$$\begin{bmatrix} k+1 \\ n \end{bmatrix}_t = \frac{(1-t^{k+n+1})(1-t^{k+n})\cdots(1-t^{k+2})}{(1-t^{k+1})(1-t^k)\cdots(1-t)} = \sum_{s=0}^{(k+1)n} p(k+1, n, s) t^s, \quad \text{(A.29)}$$

where $n \geq 1$. Thus, $|\mathrm{Par}_n(k)|, k \geq n \geq 1$, is the coefficient of t^{k+n} in this relation. Together, this result and (A.23) give the number of partitions in the set $\mathrm{Par}_n(k)$. Thus, *partitions with zero parts having $k \geq n \geq 1$ may be viewed as a special class of restricted partitions.*

Appendix B

No Single Coupling Scheme for $n \geq 5$

B.1 No Single Coupling Scheme Diagonalizing H for $n \geq 5$

The most general Hamiltonian H that can be diagonalized on a binary tree $T \in \mathbb{T}_n$ is a linear combination with arbitrary real coefficients of the set of mutually commuting Hermitian operators (5.202):

$$H = a_T \mathbf{J}^2 + \sum_{i=1}^{n} b_T(i) \mathbf{J}^2(i) + \sum_{i=1}^{n-2} c_T(i) \mathbf{K}_T^2(i). \qquad (\text{B.1})$$

Each intermediate angular momentum $\mathbf{K}_T(i)$ is itself a sum of angular momenta from the set $\{\mathbf{J}(1), \mathbf{J}(2), \ldots, \mathbf{J}(n)\}$; say,

$$\mathbf{K}_T(i) = \mathbf{J}(k_{i,1}) + \mathbf{J}(k_{i,2}) + \cdots + \mathbf{J}(k_{i,t_i}),$$
$$\text{each } \mathbf{J}(k_{i,t}) \in \{\mathbf{J}(1), \mathbf{J}(2), \ldots, \mathbf{J}(n)\}, \qquad (\text{B.2})$$

where the indices $k_{i,t}, t = 1, 2, \ldots, t_i$, depend on the binary tree $T \in \mathbb{T}_n$. Relation (B.2) can be used to eliminate the squared intermediate angular momenta from the Hamiltonian H defined by (B.1), since the following identity holds:

$$\mathbf{K}_T^2(i) = \sum_{t=1}^{t_i} \mathbf{J}^2(k_{i,t}) + 2 \sum_{1 \leq t < t' \leq n}^{t_i} \mathbf{J}(k_{i,t}) \cdot \mathbf{J}(k_{i,t'}). \qquad (\text{B.3})$$

Also, since $\mathbf{J} = \mathbf{J}(1) + \cdots + \mathbf{J}(n)$, the Hamiltonian H defined by (B.1) can always be re-expressed in the form:

$$H = \sum_{k=1}^{n} \alpha_T(k)\,\mathbf{J}^2(k) + \sum_{k=1}^{n} \beta_T(k)\,\mathbf{J}(k) \cdot \mathbf{J}(k+1). \qquad (B.4)$$

for some real coefficients $\alpha_T(k)$ and $\beta_T(k)$. This Hamiltonian is clearly a Hermitian $SU(2)$ rotational invariant. Similarly, given the form (B.4), we can reverse it back to the form (B.1).

We next show that the Heisenberg ring Hamiltonian (8.1) cannot be written in the form (B.1) for $n \geq 5$ in any binary coupling scheme. This result is not difficult to prove, once a key property of binary trees of order n is recognized, as exhibited by the following two diagrams:

$$T = \begin{array}{c} \mathbf{J}(i_1) \;\; \mathbf{J}(i_2) \\ \mathbf{K}_T(1) \qquad \mathbf{J}(i_3), \\ \mathbf{K}_T(2) \end{array} \quad T' = \begin{array}{c} \mathbf{J}(i_1) \;\; \mathbf{J}(i_2) \;\; \mathbf{J}(i_3) \;\; \mathbf{J}(i_4) \\ \mathbf{K}_{T'}(1) \qquad \mathbf{K}_{T'}(2) \\ \mathbf{K}_{T'}(3) \end{array} \qquad (B.5)$$

The indices i_1, i_2, i_3, i_4 are any distinct set selected from $\{1, 2, \ldots, n\}$; that is, each $\mathbf{J}(i_h) \in \{\mathbf{J}(1), \ldots, \mathbf{J}(n)\}$. The vertical dots indicate the continuation of the diagrams to the full binary tree $T \in \mathbb{T}_n$. The key property is: *For $n \geq 5$, every binary tree $T \in \mathbb{T}_n$ has its top three levels of either the form in the left diagram or the form in the right diagram.* (Actually, we must consider also the reflection of the left diagram through the vertical line containing the point $\mathbf{K}_T(2)$, but the proof given below applies also to such reflection-equivalent binary trees.)

The addition of angular momenta encoded in the two respective diagrams (B.5) is:

$$\mathbf{K}_T(1) = \mathbf{J}(i_1) + \mathbf{J}(i_2), \;\; \mathbf{K}_T(2) = \mathbf{K}_T(1) + \mathbf{J}(i_3),$$
$$\mathbf{K}_{T'}(1) = \mathbf{J}(i_1) + \mathbf{J}(i_2), \;\; \mathbf{K}_{T'}(2) = \mathbf{J}(i_3) + \mathbf{J}(i_4), \qquad (B.6)$$
$$\mathbf{K}_{T'}(3) = \mathbf{K}_{T'}(1) + \mathbf{K}_{T'}(2).$$

The left diagram in (B.5) continues with $n-3$ ○ points labeled by distinct angular momenta in $\Big\{\{\mathbf{J}(1), \mathbf{J}(2), \ldots, \mathbf{J}(n)\} - \{\mathbf{J}(i_1), \mathbf{J}(i_2), \mathbf{J}(i_3)\}\Big\}$, and with $n - 4$ ● points labeled by the standard rule with the intermediate angular momenta in $\{\mathbf{K}_T(3), \mathbf{K}_T(4), \ldots, \mathbf{K}_T(n-2)\}$. The right diagram continues with $n - 4$ ○ points labeled by distinct angular momenta in $\Big\{\{\mathbf{J}(1), \mathbf{J}(2), \ldots, \mathbf{J}(n)\} - \{\mathbf{J}(i_1), \mathbf{J}(i_2), \mathbf{J}(i_3), \mathbf{J}(i_4)\}\Big\}$, and with $n - 5$ ●

B.1. DIAGONALIZING H

points labeled by the standard rule with the intermediate angular momenta in $\{\mathbf{K}_{T'}(4), \mathbf{K}_{T'}(5), \ldots, \mathbf{K}_{T'}(n-2)\}$.

We use next an important property of any binary coupling scheme:

The dot products of pairs of angular momenta from the set $\mathbf{J}(1), \ldots \mathbf{J}(n)$ that occur can only arise from the squares of the coupled intermediate angular momenta in the set $\{\mathbf{K}_T^2(1), \mathbf{K}_T^2(2), \ldots, \mathbf{K}_T^2(n-2)\}$.

We now derive the following result: The Heisenberg Hamiltonian H defined by (8.1) is expressible in terms of the coupling scheme for the binary tree $T \in \mathbb{T}_n$, if and only if it is expressible in the following form:

$$\begin{aligned} H &= \sum_{i=1}^{n-2} a_i \, \mathbf{K}_T^2(i) \\ &= a_1 \Big(\mathbf{J}^2(i_1) + \mathbf{J}^2(i_2) + 2\mathbf{J}(i_1) \cdot \mathbf{J}(i_2)\Big) \\ &+ a_2 \Big(\mathbf{J}^2(i_1) + \mathbf{J}^2(i_2) + \mathbf{J}^2(i_3) \\ &+ 2\mathbf{J}(i_1) \cdot \mathbf{J}(i_2) + 2\mathbf{J}(i_1) \cdot \mathbf{J}(i_3) + 2\mathbf{J}(i_2) \cdot \mathbf{J}(i_3)\Big) + H'. \end{aligned} \quad (B.7)$$

Here H' denotes all terms in H not included in those displayed to the left of H'. The important property is: H' can contain no dot product terms between the angular momenta $\mathbf{J}(i_1), \mathbf{J}(i_2), \mathbf{J}(i_3)$ since these angular momenta do not occur in the continuation of the binary tree $T \in \mathbb{T}_n$ in (B.7). The three dot products multiplying a_2 occur as a unit; if one occurs, all three occur. Thus, if one term is contained in H as presented by (8.1), then all three occur in H; moreover, the coefficients of each of the separate terms $\mathbf{J}(i_1) \cdot \mathbf{J}(i_2), \mathbf{J}(i_1) \cdot \mathbf{J}(i_3), \mathbf{J}(i_2) \cdot \mathbf{J}(i_3)$ must be equal to 1. This requires that $a_1 = 0, a_2 = 1/2$. But it must also be the case that each of the integer pairs $(i_1, i_2), (i_1, i_3), (i_2, i_3)$ be an adjacent pair; that is, the integers in each of the three pairs differ by unity, which is impossible. Thus, H cannot be written in the form (B.7), unless $a_1 = a_2 = 0$. We conclude that relation (B.7) can only be satisfied if $H = H'$. But then H' must, for example, include the term $\mathbf{J}(i_1) \cdot \mathbf{J}(i_1 + 1)$, which was precluded at the outset ii the expansion (B.7). We conclude: *The Heisenberg Hamiltonian (8.1) cannot be written in the form (B.1) for any $T \in \mathbb{T}_n$ of the first form in (B.5). (This same argument applies to the reflection of T.)*

A similar argument applies to the binary tree $T' \in \mathbb{T}_n$ of the second form in diagram in (B.5). Because $n \geq 5$ the • point labeled by $\mathbf{K}_{T'}(3)$ cannot be the root of the full tree $T' \in \mathbb{T}_n$, which is always labeled by the total angular momentum $\mathbf{J}$. The Heisenberg Hamiltonian H defined by (8.1) is expressible in terms of the coupling scheme for the binary tree

$T' \in \mathbb{T}_n$, if and only if it is expressible in the following form:

$$\begin{aligned} H &= \sum_{i=1}^{n-2} a_i \mathbf{K}_{T'}^2(i) \\ &= a_1\Big(\mathbf{J}^2(i_1) + \mathbf{J}^2(i_2) + 2\mathbf{J}(i_1)\cdot\mathbf{J}(i_2)\Big) \\ &+ a_2\Big(\mathbf{J}^2(i_3) + \mathbf{J}^2(i_4) + 2\mathbf{J}(i_3)\cdot\mathbf{J}(i_4)\Big) \\ &+ a_3\Big(\mathbf{J}^2(i_1) + \mathbf{J}^2(i_2) + \mathbf{J}^2(i_3) + \mathbf{J}^2(i_4) \\ &+ 2\mathbf{J}(i_1)\cdot\mathbf{J}(i_2) + 2\mathbf{J}(i_1)\cdot\mathbf{J}(i_3) + 2\mathbf{J}(i_1)\cdot\mathbf{J}(i_4) \\ &+ 2\mathbf{J}(i_2)\cdot\mathbf{J}(i_3) + 2\mathbf{J}(i_2)\cdot\mathbf{J}(i_4) + 2\mathbf{J}(i_3)\cdot\mathbf{J}(i_4)\Big) + H'. \end{aligned}$$
(B.8)

As above, dot products of pairs of the angular momenta $\mathbf{J}(i_1), \mathbf{J}(i_2), \mathbf{J}(i_3)$, $\mathbf{J}(i_4)$ cannot occur in any of the continuing terms in H'. The six dot products multiplying a_3 occur as a unit; if one occurs, all six occur. Thus, if one term is contained in H as presented by (8.1), then all six must occur. Moreover, the coefficient of each of the six terms $\mathbf{J}(i_1)\cdot\mathbf{J}(i_2)$, $\mathbf{J}(i_1)\cdot\mathbf{J}(i_3)$, $\mathbf{J}(i_1)\cdot\mathbf{J}(i_4)$, $\mathbf{J}(i_2)\cdot\mathbf{J}(i_3)$, $\mathbf{J}(i_2)\cdot\mathbf{J}(i_4)$, $\mathbf{J}(i_3)\cdot\mathbf{J}(i_4)$ must be equal to 1. This requires that $a_1 = 0, a_2 = 0, a_3 = 1/2$. But it must also be the case that each of the integer pairs $(i_1, i_2), (i_1, i_3), (i_1, i_4), (i_2, i_3), (i_2, i_4)(i_3, i_4)$ be an adjacent pair, which is impossible. Thus, H cannot be written in the form (B.8), unless $a_1 = a_2 = a_3 = 0$. We conclude that relation (B.8) can only be satisfied only if $H = H'$. But then H' must, for example, include the term $\mathbf{J}(i_1)\cdot\mathbf{J}(i_1 + 1)$, which was precluded at the outset in the expansion (B.8). We conclude: *The Heisenberg Hamiltonian (8.1) cannot be written in the form (B.1) for any $T' \in \mathbb{T}_n$ of the second form in (B.5)*

We observe that for $n = 4$ the last three lines in (B-8) are not present because $\mathbf{K}_{T'}(3) = \mathbf{J}$. In this case, the values $i_1 = 1, i_2 = 3, i_3 = 2, i_3 = 4$ give the binary coupling scheme used earlier for which the Heisenberg Hamiltonian is exactly diagonalizable.

Appendix C

Generalization of Binary Coupling Schemes

C.1 Generalized Systems

The abstract structure of what may be called binary coupling schemes can be delineated. It is useful to outline this generalization, since such coupling schemes could well have application to complex systems going beyond many-body angular momentum theory.

We begin with a single system — system i — of a collection of n systems whose "states" can be described by a finite-dimension Hilbert space $\mathcal{A}_i$ of dimension $\mathrm{Dim}\mathcal{A}_i$ with an orthonormal basis given by

$$\mathbf{A}_i = \left\{ |a_i, \alpha_i\rangle \,\middle|\, \alpha_i \in \mathbb{D}_i \right\}, i = 1, 2, \ldots, n, \tag{C.1}$$

where each a_i is a sequence of parameters that gives the generic characterization of an object of interest whose detailed structure is described by an additional family of sequences of parameters α_i that belong to a fully defined domain of definition $\mathbb{D}_i$; that is, $\alpha_i \in \mathbb{D}_i, i = 1, 2, \ldots, n$. We refer to generic parameters a_i as "*state parameters.*" Then, the collection of n systems is to be describable by vectors in the tensor product space $\mathcal{A}$ with basis $\mathbf{B}$ defined by

$$\mathcal{A} = \mathcal{A}_1 \otimes \mathcal{A}_2 \otimes \cdots \otimes \mathcal{A}_n,$$

$$\mathbf{B} = \left\{ |a_1, \alpha_1\rangle \otimes |a_2, \alpha_2\rangle \otimes \cdots \otimes |a_n, \alpha_n\rangle \,\middle|\, \text{each } \alpha_i \in \mathbb{D}_i \right\}, \tag{C.2}$$

$$\mathrm{Dim}\mathbf{A} = \prod_{i=1}^{n} \mathrm{Dim}\mathcal{A}_i.$$

The notion of a composite system or collective system in which we form vectors in the tensor product space with basis (C.2) by unitary transformations of the basis vectors requires two additional concepts:

1. Multiplication rule for state parameters (closure rule): There must be a composition rule for states, denoted $\otimes$ that specifies how the states are to be multiplied pairwise to give back the states in the system; that is,

$$a_i \otimes a_j = \sum_{k=1}^{n} \oplus K(i,j,k) a_k, \text{ all } i, j = 1, 2 \ldots, n, \qquad (C.3)$$

where the right-hand side is the direct sum associated with the tensor product space. It is anticipated here that each state in the tensor product will occur with multiplicity $K(i,j,k)$ (a nonnegative integer). It is furthermore specified that the multiplicity integer $K(i,j,k)$ is **symmetric** in all permutations i,j,k; that is, in all permutations of the state vector parameters, although this might not always be required:

$$((a_i \otimes a_j) \otimes a_k) = (a_i \otimes (a_j \otimes a_k)) = a_i \otimes a_j \otimes a_k, \qquad (C.4)$$

in which no association parenthesis pairs are needed, and the product is invariant under all six permutations of i,j,k. This multiplication of states is called the *tensor product of states;* it is to satisfy the association rule (C.4).

2. Multiplication rule for multiplicities of states (completion rule): Multiplicity free systems are the simplest (all multiplicity integers 0 or 1), but, in general, the multiple occurrence of states must be admitted in the $\otimes$ rule. Thus, we require not only a multiplicity parameter set $\mathbb{K}_{i,j}$, but also a rule for the pairwise multiplication of these multiplicity parameter sets in order to obtain a complete description of all tensor product states in which no degeneracy remains. We denote the multiplicity set rule of multiplication by $\boxtimes$; it is expected that it should satisfy the association rule in analogy to (C.4), but we do not enforce this. Thus, we require:

$$\mathbb{K}_{i,j} \boxtimes \mathbb{K}_{k,l} = \sum_{p,q} \oplus \mathbb{K}_{p,q}, \qquad (C.5)$$

where $i \neq j \neq k \neq l \in \{1,2,\ldots,n\}$, and the summation is over a subset of $p \neq q \in \{1,2,\ldots,n\}$, *with no multiplicity in the right-hand direct sum.* The product rule $\boxtimes$ is called the *multiplicity product of states.*

We next give the labeling of a binary tree $T \in \mathbb{T}_n$ corresponding to the state coupling scheme (C.3). We illustrate this labeling by two examples which can be extended to the full tree, and also use the standard labeling of a binary tree in place of the double index notation in (C.5):

C.1. GENERALIZED SYSTEMS

$$T = \begin{array}{c} a_1 \quad a_2 \\ \diagdown \diagup \\ k_1, \kappa_1 \quad \begin{array}{c} a_3 \\ \diagdown \diagup \\ k_2, \kappa_2 \\ \vdots \end{array} \end{array} \qquad T' = \begin{array}{c} a_1 \quad a_2 \quad a_3 \quad a_4 \\ \diagdown \diagup \quad \diagdown \diagup \\ k'_1, \kappa'_1 \quad k'_2, \kappa'_2 \\ \diagdown \diagup \\ k'_3, \kappa'_3 \\ \vdots \end{array} \qquad (C.6)$$

In the binary tree T, we have the standard labeling:

$$k_1 \in a_1 \otimes a_2 \text{ with multiplicity } \kappa_1 \in \mathbb{K}_1;$$
$$k_2 \in k_1 \otimes a_3 \text{ with multiplicity } \kappa_2 \in \mathbb{K}_2; \ldots, \qquad (C.7)$$

where $\mathbb{K}_1 \boxtimes \mathbb{I}_2 = \mathbb{K}_1, \ldots$. In the binary tree T', we have the standard labeling:

$$k'_1 \in a_1 \otimes a_2 \text{ with multiplicity } \kappa'_1 \in \mathbb{K}'_1;$$
$$k'_2 \in a_3 \otimes a_4 \text{ with multiplicity } \kappa'_2 \in \mathbb{K}'_2; \qquad (C.8)$$
$$k'_3 \in k'_1 \otimes k'_2 \text{ with multiplicity } \kappa'_3 \in \mathbb{K}'_3, \ldots,$$

where $\mathbb{K}'_1 \boxtimes \mathbb{K}'_2 = \mathbb{K}'_3, \ldots$.

In our use of binary trees for coupling angular momenta, it is the existence of a complete set of mutually commuting Hermitian operators that implies completeness (no degeneracy). In general systems, as described above, completeness means that the $\boxtimes$ rule, whatever its form, but one structured to govern the composition of multiplicities, must leave no degeneracy behind. (In some instances, it may be necessary to deal with incomplete systems.)

The construction of coupled states associated with an arbitrary binary tree $T \in \mathbb{T}_n$ now proceeds as follows:

$$|T(\mathbf{a}, \mathbf{k}, \boldsymbol{\kappa}))\rangle = \sum_{\boldsymbol{\alpha} \in \mathbb{D}(\mathbf{a})} U_T(\mathbf{a}, \boldsymbol{\alpha}; \mathbf{a}, \mathbf{k}, \boldsymbol{\kappa})|\mathbf{a}, \boldsymbol{\alpha}\rangle, \qquad (C.9)$$

where the boldface notation denotes the sequences of sequences (in this general setting, each a_i, etc., is itself a sequence) given by the standard assignment to the full binary tree:

$$\mathbf{a} = (a_1; a_2; \ldots; a_n), \ \mathbf{k} = (k_1; k_2; \ldots; k_{n-1}), \ \boldsymbol{\kappa} = (\kappa_1; \kappa_2; \ldots; \kappa_{n-1}), \qquad (C.10)$$

where the pair (k_{n-1}, κ_{n-1}) labels the root of the full binary tree T. The construction of coupled states associated with a second binary tree $T' \in \mathbb{T}_n$ proceeds just as in (C.9):

$$|T'(\mathbf{a}', \mathbf{k}', \boldsymbol{\kappa}'))\rangle = \sum_{\boldsymbol{\alpha} \in \mathbb{D}(\mathbf{a})} U_{T'}(\mathbf{a}, \boldsymbol{\alpha}; \mathbf{a}', \mathbf{k}', \boldsymbol{\kappa}')|\mathbf{a}, \boldsymbol{\alpha}\rangle, \qquad (C.11)$$

where the boldface notation denotes the sequences of sequences given by the standard assignment to the full binary tree:

$$\mathbf{a}' = (a_{\pi_1}; a_{\pi_2}; \ldots; a_{\pi_n}), \; \mathbf{k} = (k'_1; k'_2; \ldots; k'_{n-1}, \; \boldsymbol{\kappa}' = (\kappa'_1; \kappa'_2; \ldots; \kappa'_{n-1}), \tag{C.12}$$

in which $\pi \in S_n$ denotes an arbitrary permutation. We usually choose the same labels $(k'_{n-1}, \kappa'_{n-1}) = (k_{n-1}, \kappa_{n-1})$ for the root of each binary tree T' and T. While arbitrary permutations of the $a_1, a_2, \ldots, a_n$ are allowed in (C.11), these parameters are not to be permuted in the basis state vectors $|\mathbf{a}, \boldsymbol{\alpha}\rangle$ of the tensor product space $\mathcal{A}$, as the notation in (C.11) exhibits.

The coefficients in relations (C.9) and (C.11) are given, respectively, by the inner products

$$U_T(\mathbf{a}, \boldsymbol{\alpha}; \mathbf{a}, \mathbf{k}, \boldsymbol{\kappa}) = \langle \mathbf{a}, \boldsymbol{\alpha} | T(\mathbf{a}, \mathbf{k}, \boldsymbol{\kappa}) \rangle,$$
$$U_{T'}(\mathbf{a}, \boldsymbol{\alpha}; \mathbf{a}', \mathbf{k}', \boldsymbol{\kappa}') = \langle \mathbf{a}, \boldsymbol{\alpha} | T'(\mathbf{a}', \mathbf{k}', \boldsymbol{\kappa}') \rangle. \tag{C.13}$$

Each of relations (C.9) and (C.11) is invertible. Substitution of the inversion of the (C.9) for $|\mathbf{a}, \boldsymbol{\alpha}\rangle$ into relation (C.11) gives the relation between coupled state vectors that span the same vector space (same tree root labels):

$$|T'(\mathbf{a}', \mathbf{k}', \boldsymbol{\kappa}')\rangle = \sum_{\mathbf{k}, \boldsymbol{\kappa}} W_{\mathbf{k}, \boldsymbol{\kappa}; \mathbf{k}', \boldsymbol{\kappa}'} |T(\mathbf{a}, \mathbf{k}, \boldsymbol{\kappa})\rangle, \tag{C.14}$$

where the unitary matrix W has elements given by

$$W_{\mathbf{k}, \boldsymbol{\kappa}; \mathbf{k}', \boldsymbol{\kappa}'} = \left(U_T^{(\mathbf{a}) \dagger} U_{T'}^{(\mathbf{a}')} \right)_{\mathbf{k}, \boldsymbol{\kappa}; \mathbf{k}', \boldsymbol{\kappa}'}. \tag{C.15}$$

The matrix of the transformation coefficients between the two coupled states in (C.15) is given by

$$W_{T, T'}^{(\mathbf{a}; \mathbf{a}')} = U_T^{(\mathbf{a}) \dagger} U_{T'}^{(\mathbf{a}')}. \tag{C.16}$$

This matrix is the unitary *recoupling matrix* between these coupled states. Its matrix elements are given in terms of the inner product of coupled state vectors by

$$W_{\mathbf{k}, \boldsymbol{\kappa}; \mathbf{k}', \boldsymbol{\kappa}'} = \left(U_T^{(\mathbf{a}) \dagger} U_{T'}^{(\mathbf{a}')} \right)_{\mathbf{k}, \boldsymbol{\kappa}; \mathbf{k}', \boldsymbol{\kappa}'} \tag{C.17}$$
$$= \langle T(\mathbf{a}, \mathbf{k}, \boldsymbol{\kappa}) | T'(\mathbf{a}', \mathbf{k}', \boldsymbol{\kappa}') \rangle = \langle T'(\mathbf{a}',', \boldsymbol{\kappa}') | T(\mathbf{a}, \mathbf{k}, \boldsymbol{\kappa}) \rangle^*.$$

The preceding results are valid for each pair of binary trees $T_1, T_2 \in \mathbb{T}_n$, and for each standard labeling of the ∘ points given, respectively, by

$\mathbf{a}_1$ and $\mathbf{a}_2$. In this case, the recoupling matrix between the corresponding state vectors $|T_1(\mathbf{a}_1,\mathbf{k}_1,\boldsymbol{\kappa}_1)\rangle$ and $|T_2(\mathbf{a}_2,\mathbf{k}_2,\boldsymbol{\kappa}_2)\rangle$ is given by

$$W_{T_1,T_2}^{(\mathbf{a}_1;\mathbf{a}_2)} = U_{T_1}^{(\mathbf{a}_1)\dagger} U_{T_2}^{(\mathbf{a}_2)}, \tag{C.18}$$

$$\left(W_{T_1,T_2}^{(\mathbf{a}_1;\mathbf{a}_2)}\right)_{\mathbf{k}_1,\boldsymbol{\kappa}_1;\mathbf{k}_2,\boldsymbol{\kappa}_2} = \langle T_1(\mathbf{a}_1,\mathbf{k}_1,\boldsymbol{\kappa}_1)\,|\,T_2(\mathbf{a}_2,\mathbf{k}_2,\boldsymbol{\kappa}_2)\rangle.$$

Recoupling matrices have the two important properties:

$$W_{T_1,T_2}^{(\mathbf{a}_1;\mathbf{a}_2)\dagger} = W_{T_2,T_1}^{(\mathbf{a}_2;\mathbf{a}_1)}, \tag{C.19}$$

$$W_{T_1,T_3}^{(\mathbf{a}_1;\mathbf{a}_3)} W_{T_3,T_2}^{(\mathbf{a}_3;\mathbf{a}_2)} = W_{T_1 T_2}^{(\mathbf{a}_1;\mathbf{a}_2)}. \tag{C.20}$$

Relation (C.18) is, of course, the Landé form for doubly stochastic matrices. Because of the multiplication property (C.20) for arbitrary pairs of recoupling matrices, it is expected that any composite system that satisfies the above rules will exhibit very nice properties with respect to constructions involving several binary tree coupled state vectors.

The question arises: Are there systems that possess the abstract properties set forth above? This structure is, in fact, just what is needed for extending to the general unitary group $U(n)$ the generalizations of $3(n-1)-j$ coefficients for $SU(2)$-multiplets that we have set forth in [L] and in the present volume. We outline below how this method can be put into effect, but we cannot list here the many relevant references in Ref. [6] and [L]. But it is our burden to point out the potential of this method for carrying forward a problem that has for the most part remained beyond the realm of conventional methods.

C.2 The Composite $U(n)$ System Problem

We enumerate several basic features of our approach to this problem, showing how it fits with the general structure set forth above:

1. The a_i quantities in the relations (C.1) are partitions $\lambda \in \mathbb{P}\mathrm{ar}_n$; that is, partitions of length n with zeros admitted as a part, which are counted in the length (see Appendix A). But now the binary tree is take to be the set $\mathbb{T}_t$; that is, it consists of $t \circ$ endpoints, $t-1$ interior $\bullet$ points of degree 3, and 1 tree root $\bullet$ point of degree 2, as assembled from $t-1$ fork constituents. (We have chosen to retain n in defining the partitions $\lambda \in \mathbb{P}\mathrm{ar}_n$ and modified the previous notation to $T \in \mathbb{T}_t$ for the order of the binary trees.) Thus, we now select t partitions $\lambda^{(1)}, \lambda^{(2)}, \ldots, \lambda^{(t)}$

from the set Par_n and assign them by the standard rule to the shape of a binary tree $T \in \mathbb{T}_t$. We often use $\lambda, \mu, \nu \in \text{Par}_n$ for partitions when more detail is not required.

2. The state vectors in (C.1) now become the famous Gelfand-Tsetlin orthonormal basis vectors denoted by

$$\mathbf{B}_\lambda = \left\{ \left| \begin{matrix} \lambda \\ m \end{matrix} \right\rangle \,\Big|\, \lambda \in \text{Par}_n;\ \text{all lexical patterns } m \right\}. \tag{C.21}$$

Thus, the state vectors in (C.1) are replaced by the orthonormal basis vector set given by

$$\mathbf{B}_{\boldsymbol{\lambda}} = \left\{ \left| \begin{matrix} \lambda^{(1)} \\ m^{(1)} \end{matrix} \right\rangle, \left| \begin{matrix} \lambda^{(2)} \\ m^{(2)} \end{matrix} \right\rangle, \ldots, \left| \begin{matrix} \lambda^{(t)} \\ m^{(t)} \end{matrix} \right\rangle \,\Big|\, \text{each} \left| \begin{matrix} \lambda^{(i)} \\ m^{(i)} \end{matrix} \right\rangle \in \mathbf{B}_{\lambda^{(i)}} \right\},$$

$$\boldsymbol{\lambda} = (\lambda^{(1)}, \lambda^{(2)}, \ldots, \lambda^{(t)}). \tag{C.22}$$

The GT state vectors are an orthonormal basis for the irreducible action of the Lie algebra of $U(n)$; correspondingly, the action of the unitary group on this basis gives the irreducible representations of the unitary group $U(n)$. Indeed, the complete set of operators and their irreducible action on the GT basis is fully known. This is in full analogy to the results given earlier for the $SU(2)$-multiplets.

3. The tensor product of states rule is now replaced by the *tensor product of partitions rule* as given by

$$\lambda \otimes \mu = \sum_\nu \oplus K(\lambda, \nu - \mu)\, \nu, \tag{C.23}$$

where the nonnegative integer $K(\lambda, \nu - \mu)$ is the Kostka number. The Kostka number is usually written in the form $K(\lambda, \alpha)$, where α is the multiplicity of weight α in the set of all GT patterns $\mathbb{G}_\lambda, \lambda \in \text{Par}_n$ (see Sect. 4.3). That $\nu - \mu$, for $\nu \in \lambda \otimes \mu$ is always a weight $\alpha \in w\binom{\lambda}{m}$ is discussed at length in [L]. Thus, the multiplication rule for partitions is well-defined, although there is no known closed formula for the Kostka numbers, except for $n = 3$. Relation (C.23) has the additional property as follow: Consider any two GT patterns $\binom{\lambda}{m}$ and $\binom{\mu}{m'}$ that have the partitions occurring in the left-hand side of relation (C.23), and any pattern $\binom{\nu}{m''}$ that corresponds to a ν in the summation on the right-hand side. Then, relation (C.23) conserves weights in the sense that

$$w\binom{\lambda}{m} + w\binom{\mu}{m'} = w\binom{\nu}{m''}, \tag{C.24}$$

for every term on the right-hand side, and for arbitrary GT patterns on the left-hand side. It is useful to illustrate this by an example:

C.2. THE COMPOSITE $U(n)$ SYSTEM PROBLEM

Example. Let $\lambda = (2\,1\,0)$, $\mu = (3\,1\,0)$; $|\mathbb{G}_{(2\,1\,0)}| = 8$, $|\mathbb{G}_{(3\,1\,0)}| = 15$. The tensor product of partitions is given by

$$(2\,1\,0) \otimes (3\,1\,0) = (5\,2\,0) \oplus (5\,1\,1) \oplus (4\,3\,0) \oplus 2(4\,2\,1) \oplus (3\,3\,1) \oplus (3\,2\,2). \tag{C.25}$$

See the rule (11.174) in [L] for the best method of calculating this result. The number $|\mathbb{G}_\nu|$ (Weyl dimension formula) of GT patterns in the right-hand side of (C.25) is, respectively, $42, 15, 24, 2(15), 6, 3$; hence, the dimensionality of the two sides of relation (C.25) is equal, as required. We select the following patterns in the left-hand side:

$$\binom{\lambda}{m} = \begin{pmatrix} 2 & & 1 & & 0 \\ & 1 & & 0 & \\ & & 1 & & \end{pmatrix}, \; w\binom{\lambda}{m} = (1,0,2);$$

$$\binom{\mu}{m'} = \begin{pmatrix} 3 & & 1 & & 0 \\ & 2 & & 0 & \\ & & 1 & & \end{pmatrix}, \; \binom{\mu}{m'} = (1,1,2). \tag{C.26}$$

For this selection, weight conservation with $w = (2\,1\,4)$ then admits the following patterns in the right-hand side of (C.25):

$$\begin{pmatrix} 5 & & 2 & & 0 \\ & 3 & & 0 & \\ & & 2 & & \end{pmatrix}, \begin{pmatrix} 5 & & 2 & & 0 \\ & 2 & & 1 & \\ & & 2 & & \end{pmatrix}, \begin{pmatrix} 5 & & 1 & & 1 \\ & 2 & & 1 & \\ & & 2 & & \end{pmatrix}, \tag{C.27}$$

$$2\begin{pmatrix} 4 & & 3 & & 0 \\ & 3 & & 0 & \\ & & 2 & & \end{pmatrix}, \begin{pmatrix} 4 & & 2 & & 1 \\ & 2 & & 1 & \\ & & 2 & & \end{pmatrix}. \quad \square$$

4. The multiplicities of states product rule, the completeness rule in (C.5), is placed in the context of the composite $U(n)$ system problem as follows. We require some preliminary notions and nomenclature. It is here that a remarkable result discovered by Biedenharn comes into play — the concept of an *operator pattern*. An operator pattern is associated with each partition $\lambda \in \mathbb{P}\mathrm{ar}_n$. Indeed, in all appearances of its entries, it is exactly a GT pattern

$$\mathbb{B}_\lambda = \left\{ \left| \binom{\lambda}{\gamma} \right\rangle \,\Big|\, \lambda \in \mathbb{P}\mathrm{ar}_n; \text{ all lexical patterns } \gamma \right\}. \tag{C.28}$$

But this set of patterns does not label basis sets of vectors in any sense; it is an enumerative object. Its principal property is its weight, which is defined in terms of the entries in contiguous rows of the pattern exactly as for an ordinary GT pattern (see relation (4.12)) in which the $m_{i,j}$ are replaced by $\gamma_{i,j}$. We now denote the set of all weights corresponding to partition λ by $\mathbb{W}_\lambda$ and the weight of a particular pattern by $\Delta =$

$\Delta\binom{\lambda}{\gamma}$, where the abbreviated form Δ is used whenever the full pattern is unambiguous. We call Δ the *shift of the pattern* $\binom{\lambda}{\gamma}$, since its role is that of shifting the representation labels of a state vector μ to $\mu + \Delta$. Given a partition $\mu \in \text{Par}_n$, only certain shift patterns Δ qualify for supplying a partition $(\mu + \Delta) \in \text{Par}_n$, since the inequalities in the definition of a partition might be violated — this nonlexical problem is easily handled. But the profound role that operator patterns play in the multiplicity problem requires careful analysis.

The $\boxtimes$ multiplication rule is now stated for arbitrary partitions $\lambda, \mu \in \text{Par}_n$ by

$$\mathbb{B}_\lambda \boxtimes \mathbb{B}_\mu = \sum_{\nu \in \lambda \otimes \mu} \oplus \mathbb{B}_\nu, \tag{C.29}$$

where each $\nu \in \lambda \otimes \mu$ occurs exactly once in the summation. It is still required, as in (C.23), that all operator patterns associated with the $\boxtimes$ multiplication rule conserve the shift of the operator patterns; that is,

$$\Delta\binom{\lambda}{\gamma} + \Delta\binom{\mu}{\kappa} = \Delta\binom{\nu}{\tau}, \text{ each } \nu \in \lambda \otimes \mu. \tag{C.30}$$

But this rule, as it now stands, is ambiguous: It requires further restrictions that assign exactly what single operator pattern $\Delta\binom{\nu}{\tau}$ is to be assigned to the partition ν, which stands alone in the right-hand side of (C.29). Thus, in the example given above in (C.25), a supplementary rule must be given for eliminating all but one of the patterns having the same partition ν. It is this part of the rule associated with the Biedenharn operator patterns that has lead to many alternative methods in the literature for resolving the multiplicity problem. We address the Biedenharn-Louck (BL) solution, or, at worst, partial solution, of the multiplicity problem below. We insist that relation (C.29)-(C.30) with its single occurrence of ν and weight conservation must prevail. (The concept of intermediate angular momenta fails for $U(n), n \geq 3$; hence, a new concept must replace completeness of state vectors.)

It is the above inter-related structures that lead to the concept of a unit tensor operator as set forth in the BL approach to the construction of the WCG coefficients of the general unitary group $U(n)$. This approach is synthesized in Chapter 9 of [L], as drawn from the many references to the published literature given there. The notion of a unit tensor operator itself brings together the two concepts, the basic irreducible transformation of state vectors in the basis set $\mathbf{B}_\lambda$ defined by (C.21), and the notion of a shift transformation in the $\boxtimes$ multiplication rule (C.29), by combining the two patterns into a single form:

$$\text{irreducible } U(n) \text{ tensor operator: } \left\langle \begin{matrix} \gamma \\ \lambda \\ m \end{matrix} \right\rangle. \tag{C.31}$$

C.2. THE COMPOSITE $U(n)$ SYSTEM PROBLEM

Here the common shared partition $\lambda \in \mathbb{P}\mathrm{ar}_n$ is written only once, and the Biedenharn shift pattern $\binom{\lambda}{\gamma}$ is inverted over the lower GT patterns for convenience of display, but with no change in its shift action on an arbitrary state vector $\left| \begin{smallmatrix} \nu \\ m' \end{smallmatrix} \right\rangle$. This action is the following:

$$\left\langle \begin{smallmatrix} \gamma \\ \lambda \\ m \end{smallmatrix} \right| \left| \begin{smallmatrix} \mu \\ m' \end{smallmatrix} \right\rangle = \sum_{m''} \left\langle \begin{smallmatrix} \mu+\Delta \\ m'' \end{smallmatrix} \right| \left\langle \begin{smallmatrix} \gamma \\ \lambda \\ m \end{smallmatrix} \right| \left| \begin{smallmatrix} \mu \\ m' \end{smallmatrix} \right\rangle \left| \begin{smallmatrix} \mu+\Delta \\ m'' \end{smallmatrix} \right\rangle. \tag{C.32}$$

It is the transformation coefficients, the matrix elements of an irreducible $U(n)$ tensor operator, as given by

$$\left\langle \begin{smallmatrix} \mu+\Delta \\ m'' \end{smallmatrix} \right| \left\langle \begin{smallmatrix} \gamma \\ \lambda \\ m \end{smallmatrix} \right| \left| \begin{smallmatrix} \mu \\ m' \end{smallmatrix} \right\rangle, \tag{C.33}$$

that are the subject of extensive review in [L]; it is the determination of these $U(n)$ WCG coefficients that is the first basic step in the subject of the binary coupling theory outlined above for a composite $U(n)$ system. By definition, all such coefficients are defined to be 0 unless all patterns that appear in it are lexical. *But this is not enough.*

The introduction of tensor operators into the analysis allows now for a new structure to have a role in resolving the multiplicity problem: the notion of *null space* of an operator. It is useful to show how this works in the example given by (C.25). The reason that there are two copies of the partition (4 3 1) in the tensor product relation (C.25) is the occurrence of the two GT patterns $\begin{pmatrix} 2 & 1 & 0 \\ & 2 & 0 \\ & & 1 \end{pmatrix}$ and $\begin{pmatrix} 2 & 1 & 0 \\ & 1 & 1 \\ & & 1 \end{pmatrix}$ that have the same shift $\Delta = (1\,1\,1)$. This number can be reduced to one, giving uniqueness in (C.29), by requiring the state vector with partition $\mu = (3\,1\,0)$ to belong to the null space of one of the tensor operators with shift pattern $\Delta = (1\,1\,1)$:

$$\left\langle \begin{smallmatrix} & 1 & & \\ 1 & & 1 & \\ 2 & & 1 & 0 \\ & m & & \end{smallmatrix} \right| \left| \begin{smallmatrix} 3 & 1 & 0 \\ & m' & \end{smallmatrix} \right\rangle \tag{C.34}$$

$$= \sum_{m''} \left\langle \begin{smallmatrix} 4 & 2 & 1 \\ & m'' & \end{smallmatrix} \right| \left\langle \begin{smallmatrix} & 1 & & \\ 1 & & 1 & \\ 2 & & 1 & 0 \\ & m & & \end{smallmatrix} \right| \left| \begin{smallmatrix} 3 & 1 & 0 \\ & m' & \end{smallmatrix} \right\rangle \left| \begin{smallmatrix} 4 & 2 & 1 \\ & m'' & \end{smallmatrix} \right\rangle.$$

$$\left\langle \begin{smallmatrix} & 1 & & \\ 2 & & 0 & \\ 2 & & 1 & 0 \\ & m & & \end{smallmatrix} \right| \left| \begin{smallmatrix} 3 & 1 & 0 \\ & m' & \end{smallmatrix} \right\rangle = \mathbf{0} \text{ (zero vector)}. \tag{C.35}$$

Of course, there must be given a specific rule, or further structural elements, for making this pair of assignments, since they could equally well be reversed. In any case, by introducing the null space concept, we have achieved the goal of removing any repetition of the same partition in the $\boxtimes$ multiplication rule (C.29).

It is a general result for $n = 3$ that an appropriate ordering of the operator patterns having the same Δ shift renders the null space concept fully operative in resolving the multiplicity problem. For $n \geq 4$, further supplementary rules must be introduced to obtain a full resolution of the multiplicity problem for all possible cases. Nonetheless, the null space concept always gives a partial resolution, and the corresponding properties of irreducible tensor operators are an important aspect of this resolution. As noted earlier, it is not our purpose here to continue beyond what is presented in [L] on the null space aspects of the $U(n)$ tensor operators (C.32), where various aspects of the null space properties of unit tensor operators are given and extended, and methods given for their full calculation. Our purpose here is to fold these coefficients, taken as fully defined and given, into a composite viewpoint of the binary coupling of such states. These coefficients then give the full result for the coupling of two systems (binary tree of order 2) from which the fork structure of an arbitrary binary tree allows the full theory of the composite $U(n)$ system to be constructed, as we discuss briefly in conclusion.

The composite $U(n)$ system theory continues with the assignment of the basic $U(n)$–multiplets to the endpoints of the binary trees $T \in \mathbb{T}_t$; their coupling is fully controlled by the structure of the binary tree. But this is the same as the situation for angular momentum theory, where the basic $SU(2)$–multiplets are assigned to the endpoints of a binary tree $T \in \mathbb{T}_n$. (Recall that we replaced n by t in the analysis above so as to retain n in $U(n)$.) It is the fork structure that dictates the formation of the intermediate states in each instance. *The labels of every fork can be put into one-to-one correspondence with the pairwise coupling of the basic $U(n)$–multiplets.* This means that state vectors, inner products of state vectors, recoupling matrices, triangle coefficients and corresponding cubic graphs are all one-to-one. It is only the distinct meaning of the labels and the systems to which they refer that places the theory in the given context. The fact that completeness is resolved by unrelated methods does not affect the manner in which the composite systems are built-up. This structure of $U(n)$ triangle coefficients and their associated recoupling matrices would seem to be a structure of extraordinary complexity and beauty.

But this is not all: There is an almost structurally identical theory for the coupling of matrix Schur functions, which includes in its specialization the above binary coupling for the irreducible $U(n)$ representations functions themselves; indeed, the binary coupling for the integer representations of the general linear group $GL(n, \mathbb{C})$.

Bibliography

[1] G. E. Andrews, The Theory of Partitions, Encyclopedia of Mathematics and Its Applications, Vol. 2, edited by G.-C. Rota, Cambridge University Press, Cambridge, 1976. [1] [303], [305]–[307], [309]

[2] G. E. Andrews, Plane partitions V: The TSSCPP conjecture, J. Combin. Theory Ser. A **66** (1994) 28–39. [196], [222]

[3] J. S. Bell, Speakable and unspeakable in quantum mechanics, Cambridge University Press, Cambridge, New York, 1987. [103]–[104]

[4] I. Bengtsson and K. Życkowski, Geometry of Quantum States, Cambridge, 2006. [57], [174]

[5] H. Bethe, Sur Theorie der Metalle. I. Eigenwerte und Eigenfuncktionen der linearen Atomkette, Z. Physik **17** (1931) 205–226. (Translation in: D. C. Mattis, The Many-Body Problem (World Scientific, Singapore, 1991, pp. 6889–716). [xxiv], [223]

[6] L. C. Biedenharn and J. D. Louck, Angular Momentum in Quantum Physics, Vol.8: The Racah-Wigner Algebra in Quantum Theory, Vol. 9: Encyclopedia of Mathematics and Its Applications, edited by G.-C. Rota, Cambridge University Press, Cambridge, 1981. [vii], [xxii], [xxvi], [1], [3], [95]–[96], [100], [117], [124], [153], [166], [172], [180], [196], [321], [343]

[7] L. C. Biedenharn and H. van Dam, Quantum Theory of Angular Momentum, Academic Press, New York, 1965. [1]

[8] G. Birkhoff, Tres ovservaciones sobre el algebra lineal, Univ. Nac. Tucuman Rev. Ser. A (1946) 147-151. [xxiv], [19], [30]–[31]

[9] D. J. Bohm, Quantum Theory, Prentice-Hall, New York, 1951. [104]

[10] D. Bohm and B. Hiley, The Undivided Universe, Routledge, London, 1993. [104]

---

[1]The page numbers where the reference is made are enclosed in brackets.

[11] A. Bohr and O. Ulfbeck, Primary manifestation of symmetry. Origin of quantal indeterminacy, Rev. Mod. Phys. (1995) 1–35. [92]

[12] D. Bressoud and J. Propp, How the alternating sign matrix conjecture was solved, Notices of the AMS **46** (1999) 637-646. [196]

[13] Bressoud D M 1999 *Proofs and Confirmations: The Story of the Alternating Sign Matrix Conjecture* (Cambridge: Cambridge University Press) [196]

[14] R. A. Brualdi and H. J. Ryser, Combinatorial Matrix Theory, Encyclopedia of Mathematics and Its Applications, edited by G.-C. Rota, Cambridge University Press, Cambridge, 1991. [19], [30], [36], [77], [219]

[15] E. Cartan, Thesis, Sur la Structure des Groupes Finis and Continus, Paris, Nony, 1894, Ouevres Complète, Part 1, Gauthier-Villars, Paris (1952) [vii], 137–287.

[16] J. H. Carter and J. D. Louck, Permutation matrices and the representation of matrices with fixed line-sum (LAUR 04-5399, Los Alamos National Laboratory, Los Alamos, NM 87545, unpublished). [57]

[17] J. H. Carter and J. D. Louck, Magic squares: symmetry and combinatorics, Mol. Phys. **102** (2004) 1243-1267. [57], [180]

[18] W. J. Caspers, Spin Systems, World Scientific, Singapore, 1989. [226]

[19] A. Cayley, On the analytical forms called trees, with application to the theory of chemical combinations, Report of the British Associaton for the Adv. of Science (1875) 257–305. [xxii], [245]

[20] W. Y. C. Chen, and J. D. Louck, Enumeration of cubic graphs by inclusion-exclusion, J. Comb. Theory (1999) 151-164. [172]

[21] D. Chruściński, Positive maps, doubly stochastic matrices, and new family of spectral conditions, *Proc. Conf. Symmetry and Structural Properties of Condensed Matter*, editors, T. Lulek, B. Lulek, and A. Wal, J. Physics, Conference Series **213** (2010) 012003. [57], [174]

[22] L. Comtet, Advanced Combinatorics, D. Reidel Publishing Company,, Dordrecht, 1974. [169]

[23] P. Désarménien, J. P. S. Kung, and G.-C. Rota, Invariant theory, Young bitableaux, and combinatorics, Adv. Math. **27** (1978) 63–92. [xxvi]

[24] P. A. M. Dirac, The Principles of Quantum Mechanics, Oxford University Press, 4th ed., London, 1958. [xiv], [2], [82]

[25] A. Einstein, P. Podolsky, and N. Rosen, Can the quantum-mechanical description of physical reality be considered complete? Phys. Rev. **37** (1935) 777–780 (reprinted in [81]). [104]

[26] I. M. Gelfand, Lectures on Linear Algebra, Interscience, New York, 1961. [32]–[33]

[27] M. Gell-Mann, The Quark and the Jaguar, W. H. Freeman, New York, 1994. [104]

[28] R. L. Graham, M. Grötschel, L. Lovász, Editors, Handbook of Combinatorics, The MIT Press, Cambridge, 1995, Elsevier Science B. V., Amsterdam, The Netherlands. [xxvi]

[29] R. B. Griffiths, Consistent Quantum Theory, Cambridge University Press, Cambridge, 2002. [104]

[30] M. Hamermesh, Group Theory and Its Applications to Physical Problems, Addison-Wesley, 1962. [59], [61], [63], [68]–[69]

[31] L. Hulthén, Arkiv.Nat. Astron. Fys, **26A** (1938) 1-x. [226]

[32] P. Jakubczyk, J. Topolewicz, A. Wal, and T. Lulek, Schwinger geometry, Bethe Ansatz, and a magnonic qudit, Open Systems & Information Dynamics **16** (2009) 221-233. [57], [69]

[33] A. N. Kirillov and N. Yu. Reshetikhin, The Bethe ansatz and the combinatorics of Young tableaux, Plenum Publishing Corporation (1988) 925-955 (Translation from: Zap. Nauch. Semin. LOMI **155** (1986) 65–115. [226]

[34] D. E. Knuth, Permutations, matrices, and generalized Young tableaux, Pac. J. Math. **34** (1970) 709–727. [190]

[35] T. S. Kuhn, The Structure of Scientific Revolutions, The University of Chicago Press, third edition, Chicago, 1996. [xxii]

[36] J. P. S. Kung and G. C. Rota, The invariant theory of binary forms, Bull. Amer. Math. Soc. **10** (1984) 27–85. [xxvi]

[37] J. P. S. Kung, Editor, Gian-Carlo Rota on Combinatorics, Birhaäuser, Bostin, 1995. [xxvi]

[38] G. Kuperberg, Another proof of the alternating sign matrix conjecture, Internat. Math. Res. Notices (3) (1996) 139–150. [222]

[39] A. Landé, New Foundations of Quantum Mechanics, Cambridge, 1965. [xxiv], [22], [82], [90]

[40] A. J. Leggett, The Quantum Measurement Problem, Science **307** (2005) 871–872. [104]

[41] C. Lévi-Strauss, Structural Anthropology, Basic Books, Perseus, 1963. (see Ideas and Trends, The New York Times, Sunday, November 8, 2009). [103]

[42] D. E. Littlewood, The Theory of Group Characters, Oxford University Press, London, 1950, reprinted 1958. [69], [294]

[43] J. D. Louck, MacMahon's master theorem, double tableau polynomials, and representations of groups, Adv. Appl. Math. **17** (1996) 143–168.

[44] J. D. Louck, Doubly stochastic matrices in quantum mechanics, Found. of Phys. **27** (1997) 1085–1104. [22], [94], [98]–[99], [102]

[45] J. D. Louck, Survey of zeros of $3j$ and $6j$ coefficients by Diophantine equation methods, Group Theory and Special Symmetries in Nuclear Physics, Proceedings of the International Symposium in Honor of K. T. Hecht, editors, J. P. Draayer and J. Jänecke, World Scientific, Singapore, 1992, pp. 38-48. [163]

[46] J. D. Louck, Unitary Symmety and Combinatorics, World Scientific, Singapore, 2008. [vii]

[47] J. D. Louck, Matrix Schur functions, permutation matrices, and Young operators as inner product spaces, *Proc. Conf. Symmetry and Structural Properties of Condensed Matter*, editors, T. Lulek, B. Lulek, and A. Wal, J. Physics, Conference Series **213** (2010) 012010. [60]

[48] T. Lulek, Bethe ansatz, Young tableaux, and the spectrum of Jucys-Murphy operators, Symmetry and Structural Properties of Condensed Matter, editors, T. Lulek, W. Florek, B. Lulek, World Scientific, Singapore, 2003, pp. 279–291. [69], [71], [223], [226]

[49] T. Lulek, B. Lulek, D. Jakubczyk, and J. Jakubczyk, Rigged strings, Bethe Ansatz, and the geometry of the classical configuration space of the Heisenberg magnetic ring, Physica B: Phys. Condensed Matter **382** (2006) 162-180. [69]

[50] I. Macdonald, Symmetric Functions and Hall Polynomials, Oxford University Press, Oxford, 1979, second edition, 1995. [189], [307]

[51] S. Mac Lane. Mathematics: Form and Function, Springer-Verlag, New York, 1986. [xxv]

[52] P. A. MacMahon, Combinatory Analysis, Chelsea, New York, 1915. [192]

[53] M. A. Méndez, Directed graphs and the combinatorics of the polynomial representations of $GL(n, \mathbb{C})$, Annals of Comb. **5** (2001) 459–478. [67]

[54] M. A. Méndez, Towards a combinatorial description of the matrices corresponding to irreducible representations of the unitary and general linear groups, Symmetry and Structural Properties of Condensed Matter, editors, T. Lulek, A. Wal, and B. Lulek, World Scientific, Singapore, 2003, pp. 265–273. [67]

[55] L. Michel and B. I. Zhilinskii, Symmetry, Invariants, Topology, Physics Reports, North-Holland, Amsterdam **341** (2001) 11-84. [67]

[56] W. H. Mills, D. P. Robbins, and H. Rumsey, Self-complementary, totally symmetric plane partitions, J. Combin Theory Ser. A **42**(1986) 277–292. [202], [222]

[57] M. A. Nielsen and I. L. Chuang, Quantum Computation and Quantum Information, Cambridge, 2000. [57]

[58] R. Omnès, The Interpretation of Quantum Mechanics, Princeton University Press, Princeton, 1994. [104]

[59] R. Penrose, The Road to Reality, Knopf, New York, 2005. [104]

[60] S. Perlis, Theory of Matrices, Addison-Wesley, Cambridge, Massachusetts, 1952. [84], [86], [262], [269], [287], [289]

[61] J. Propp, *Discrete mathematics and Theoretical Computer Proceedings* **AA** **(DM-CCG)** 43 (2001). [196]

[62] G. Racah, Theory of complex spectra. I, Phys. Rev. **61** (1942) 186-197; II, *ibid.* **62** (1942) 438-462; III, *ibid.* **63** (1943) 367-382. [123]

[63] G. Racah, Group Theory and Spectroscopy, Ergebnisse der exakten Naturwissenschaften **37** (1965) 28-84, Springer-Verlag, Berlin (based on Princeton lectures, 1951). [123]

[64] R. C. Read, The enumeration of locally restricted graphs. I, J. London Math. Soc. **34** (1959) 417-436; II, *ibid.* **35** (1960) 344-351. [172]

[65] T. Regge, Symmetry properties of Clebsch-Gordan coefficients, Nuovo Cimento **10** (1958) 544-545; Symmetry properties of Racah's coefficients, *ibid.* **11** (1959) 116-117. [180]

[66] D. P. Robbins, W. H. Mills, and H. Rumsey, Jr, Alternating sign matrices and descending plane partitions, J. Comb. Theory Ser. A **34** (1983) 340-359. [202], [222]

[67] G. de B. Robinson, Representation Theory of the Symmetric Group, University of Toronto Press, 1961. [1], [60]

[68] S. M. Roman and G.-C Rota, The umbral calculus, Advances in Math. **27** (1978) 95-188. [xxvi]

[69] M. E. Rose and C. N. Yang, Eigenvalues and eigenvectors of a symmetric matrix of $6j$ coefficients, J. Math. Phys. **31** (1962) 106. [253]

[70] G.-C. Rota, Finite Operator Calculus, Academic Press, 1975. [xxvi]

[71] G.-C. Rota, Hopf algebra methods in combinatorics, Colloques Internationaux, C.N.R.S., N° 260—Pròblemes Combinatores et Théorie des Graphes, 363-365 (Reprinted in Ref. [98]). G.-C. Rota, Invariant theory, old and new, The American Math. Soc. Colloquium Series, Lecture 2, Baltimore, 1998. [xxvi]

[72] G-C. Rota, D. Kahaner, and A. Odlyzko, Finite operator calculus, J. Math. Anal. Appl. **42** (1973) 685-760 (Reprinted in [37]). [xxvi]

[73] E. Schrödinger, Probability relations between separated systems, Cambridge University Press, Cambridge **35** (1935) 555-563. [3], [104]

[74] J. Schwinger, On Angular Momentum, U.S. Atomic Energy Commission Report NYO-3071, 1952 (Reprinted in Ref. [7]). [67]

[75] R. P. Stanley, Enumerative Combinatorics, Vols. 1,2, Cambridge University Press, Cambridge, 1997. [177], [179]–[181], [186], [189], [191], [303], [305], [307]–[308]

[76] D. Stanton, Editor, Invariant Theory and Tableau, The IMA Volumes in Mathematics and Its Applications, Vol. 19, Springer-Verlag, New York, 1988.

[77] Symmetry and Structural Properties of Condensed Matter. Proceedings of the International School of Theoretical Physics. Proceedings of the first-seventh published by World Scientific, Singapore in: 1991, 1993, 1995, 1997, 1999, 2001, 2003; Proceedings of the eighth-tenth published by Journal of Physics Conference Series, IOP Publishing, Philadelphia, **30** (2006), **104** (2008), **213** (2010), online at jpconf.iop.org (editors, W. Florek, T. Lulek, B. Lulek, M. Mucha, D. Lipiński, S. Wa*l*cerz, A. Wa1). [266]

[78] J. Von Neumann, Mathematical Foundations of Quantum Mechanics, Princeton University Press, Princeton University, 1955. [22]–[23]

[79] L. Wei, Unified approach for calculation of angular momentum coupling and recoupling coefficients, Comput. Phys. Commun. **120** (1999) 222-230. [304], [344]

[80] H. Weyl, The Theory of Groups and Quantum Mechanics, Methuen, London, 1931; reissued by Dover, New York, 1949. [1], [60]

[81] J. A. Wheeler and W. H. Zurek, Editors, Quantum Theory and Measurement, Princeton University Press, Princeton, New Jersey, 1983. [3], [104]

[82] E. P. Wigner, Group Theory and Its Application to the Quantum Mechanics of Atomic Spectra. Academic Press, New York, 1959 (Original German edition: Gruppentheorie und ihre Anwendung auf die Quanten-mechanik der Atomspektren, Braunschweig, 1931). [xxiv], [1]

[83] E. P. Wigner, On the matrices which reduce the Kronecker products of representations of S. R. groups, 1940. (Reprinted in [7]: Quantum Theory of Angular Momentum, editors, L. C. Biedenharn and H. van Dam, Academic Press, New York, 1965, pp. 87-133). [125]

[84] E. P. Wigner, Interpretation of Quantum Mechanics, Princeton University Press, Princeton, New Jersey (1983) 260-314. [103]

[85] E. P. Wigner, On hidden variables in quantum mechanical probabilities, Quantum Mechanics, Determinism, Causality and Particles, editors, M. Flato, Z. Maric, A. Milojevic, D. Sterheimer, and J. P. Vigier, Reidel, Dordrecht, 1976, pp. 31–41. [103]

[86] E. B. Wilson, Jr., J. C. Decius and P. C. Cross, Molecular Vibrations, MacGraw-Hill, New York, 1955. [294]

[87] B. G. Wybourne, Symmetric functions and their application to physics, Symmetry and Structural Properties of Condensed Matter, editors, W. Florek, D. Lipskiński, T. Lulek, World Scientific, Singapore, 1993, pp. 79–100. [1], [59]

[88] The Collected Papers of Alfred Young, Mathematical Expositions No. 21, Editorial Board, H. S. M. Coexeter, G. F. D. Duff, D. A. S. Fraser, G. de B. Robinson, P. G. Rooney, Univeristy of Toronto Press, Toronto, 1977. [60]

[89] A. P. Yutsis (Jucys), I. B. Levinson, and V. V. Vanagas, The Theory of Angular Momentum, Vilnius, 1960. [159], [172]

[90] A. P. Yutsis (Jucys) and A. A. Bandzaitis, Angular Momentum Theory in Quantum Physics, Moksias, Vilnius 1977. [159], [172]

[91] D. Zeilberger, Proof of the alternating sign matrix conjecture, Electronic J. of Combinatorics **3** (2) (1996).[196]

[92] D. Zeilberger, Proof of the refined alternating sign matrix conjecture, J. Math. **2**, New York (1996) 59-68. [196]

Index

A
adjacent points of a graph 172
A-expansion rule 45
algebra of permutation
    matrices 23-40
alternating sign matrices xxiv,
    195-222
    and incidence matrices 219
    -222
    and permutation matrices 31
    Andrews-Zeilberger numbers
        196-197
    and strict Gelfand-Tsetlin patterns 202
    and Young tableaux 96, 211
A-matrix array 199-200
Andrews-Zeilberger number 196
angular momentum theory 1
    bra-ket vectors 2
    compounding (addition) 107
    coupled basis 8-11, 141-142
    irreducible unitary representation 11
    orthonormality of basis vectors 7
    pairwise addition xiii
    pairwise addition rule ix, 107
    standard action of components 5
    standard unitary action 11
    state vectors of a composite system 4-11
association A xii, 155-157,
    167-171
    and commutation C xii, 155-157, 161-171
    uncoupled basis 6-7, 140-142

B
basis sets 2
    basis $\Sigma_n(e)$ 39
    basis $\Sigma_n(e,p)$ 40, 42
    cardinality b_n 27-28
    coupled binary 145-151
    of permutations 31-42
    of permutation matrices 31-42
    of Π-matrix 36-38
    uncoupled binary 142
basis theorem 207
Biedenharn-Elliott identity xiii,
    154-155, 158
    cubic graph 158-159
    path transformations 155
    recoupling matrices 154
    transition probability amplitude 154
    triangle coefficients 157-159
bijection of patterns 200, 202, 216
    of GT patterns 200
    of skew GT patterns 216-218
    of strict GT patterns 202
binary bracketings 109-114
    and binary trees 109-114
    definition 110
    number of 113
binary coupling theory 104-172
    associated generalized WCG coefficients 119-120, 126, 129, 143-144
    build-up rule ix, 2-3
    coupling schemes viii
    coupled state vectors 14-19, 87, 145-148
    for two, three, and four an-

gular momenta 115–132
recoupling matrices xvii–xx,
14–19, 147–151
reduction of Kronecker products 11–14
uncoupled state vectors 141–142
binary tree(s) viii, 111, 235, 261
and binary bracketings viii,
110
Catalan number a_n 113
common fork xx, 115
forks viii, 112
fork array (matrix) 112–113
fork pasting viii, 112
fork root viii
four types of forks viii, 112
irreducible labeled pairs 124
labeled and unlabeled x–xii
shape xi, 109–114
shape transformation xii,
167–171
standard labeling xiii, 114
Birkhoff theorem on doubly
stochastic matrices 21
and permutation matrices 21
build-up rule 3–4
for binary trees vii–ix
for composite systems 3–4

C

Cardinality relations 10, 201–202,
204–205
invariance properties 199, 213
recurrence for GT patterns
[199]
Catalan numbers 113
C–coefficient mapping rule 116,
143
Clebsch-Gordon numbers 105–109
multiset generation 107
Clebsch-Gordon series 107
common fork 121, 143
complete set of commuting Hermitian operators 7–10, 82–83, 88, 104–109
and Dirac 82, 88
coupled basis 8–11, 105–106

doubly stochastic matrices
81–82, 85–87
uncoupled basis 7–8
unitary frame rotations vii
composite quantal systems 3–4
compounding of angular momenta examples 99, 104,
108, 236, 263–264
coordinates of fixed-line sum matrices 32, 48, 51
coupled bases 9–11, 87, 115–117,
119–121, 123, 126–129,
145–149, 237–238, 264–265
cubic graphs xx, 171–172
adjacent points 172
angular momentum type 171
cutting and rejoining of lines
158–159
factoring properties 159–160,
281–284
isomorphisms 172
labeled cubic graphs 172
of Heisenberg ring 244–245,
281–284
of $6-j$ and $9-j$ coefficients
158–159
relation to triangle coefficients 158
symbolic relations 159, 281,
283–284
tetrahedron and Racah coefficients 159–160
Yutsis factor theorem 159
cyclic group C_4 78–80, 302
cyclic group C_n 78–80
actions 294–295
action on coupled states 298–302
as permutations 292
character table 293, 301–302
invariance group of H 292
irreducible action 297–298

D

density matrices 22
measured states 22–24

INDEX

prepared states 22–24
doubly stochastic matrices xxiv, 19, 22–24, 31, 81–104, 132–140
  and binary couplings 140–174
  and composite systems 97–104
  and density matrices 22–24
  and permutation matrices 91
  and transition probability amplitudes 94, 99, 101–104
  and transition probabilities 94, 99, 101–104
  as sums of permutation matrices 31
  four angular momenta 125–128
  three angular momenta 123–125
  two angular momenta 92–97
D^λ–polynomials 63
  as matrix Schur functions 63
  basic properties 63–67
  double patterns 64
  Gelfand-Tsetlin polynomials 63–64, 197–199
  group property 66
  invertibility 66
  lexical patterns 64
  multiplication property 66
  orthogonality 65
  relation to Robinson-Schensted Knuth identity 66
  relation to Schur functions 67
  shape 65
  weight 65
domains of definition 9,
  of angular momentum quantum numbers j_i 7–8,
  of intermediate quantum numbers k_i 107–108, 127, 130, 144–145, 148
  of magnetic quantum numbers m_i 5, 7–8, 142
  of total angular momentum quantum number j 14, 106, 116
  of magnetic quantum number m 8, 105–106, 137, 145
dual space of permutation matrices 47–55
  basis sets 47–49, 51–55
  of Young operators 62–63

E

elementary operations A and C xii, xviii, 156
  and B-E identity 157–158
  and Racah sum-rule 155, 159–160
  and Wigner $9-j$ coefficients 155, 160, 246, 253–254, 269–270, 277
  of shape transformation 156–157, 169–171
entanglement 3, 91

F

factorization assumption 4
finite invariance groups 294–295
forks viii, 112, 143, 148
  and triangle coefficients xix, 121–122, 148, 151–152
  common fork 121
  endpoints viii, 112
  fork array (matrix) 112, 148
  fork pasting ix
  fork root viii, 112
  four types of forks viii, 112, 143
  standard labeling 143
  two types for magnetic ring 235
  unlabeled and labeled 111, 143, 148

G

Gelfand-Tsetlin (GT) polynomials 63–67
  see D^λ–polynomials and matrix Schur functions 63
Generalized WCG coefficients 119–120, 126, 136–137, 143–144

Gram matrix 33, 42, 48–53
  symmetries 48, 51–52
GT skew patterns 211–218
  and alternating sign matrices 214–216
  sign-reversal-shift symmetry 216–219
GT strict patterns 202–204
  and alternating sign matrices 202–204
  and incidence matrices 219–222
  sign-reversal-shift symmetry 204–205

H

Heisenberg Hamiltonian 223, 235, 261
  course of action 225–226
  exact solutions $n = 2, 3, 4$ 230–234
  invariance under cyclic group 224, 298
  recount and synthesis 289–292
  uncoupled and coupled states 226–230
  irreducible C_n states 298
Heisenberg Hamiltonian for n even 235–260
  and $3(2f - 1) - j$ coefficients 253–260
  binary coupling schemes 236–238
  binary trees 235–236
  commuting Hermitian operators for schemes I and II 237–238
  cubic graph 245
  diagonalization of H_1 and H_2 237–238
  diagonalization of H 240, 289, 295
  eigenvalue-eigenvectors 237–238
  eigenvalues and eigenvectors of intermediate states 237–238
  eigenvalues and eigenvectors of parts H_1 and H_2. 237–238
  irreducible C_n state vectors 298
  matrix representation of H on coupled states 239
  maximum eigenvalues 243–244
  parts H_1 and H_2 235
  recoupling matrices 240–242
  shape transformations 245–249
  unlabeled binary trees 235–236
Heisenberg Hamiltonian for n odd 261–292
  and $6f - j$ coefficients 269–276
  and $3(f + 1) - j$ coefficients 276–286
  binary coupling schemes 261–264
  binary trees 261
  commuting Hermitian operators for schemes I_o, II_o, III_o 265
  cubic graphs of $3(f + 1)$-coefficients 281, 283, 284
  diagonalization of H_1, H_2, H_3 266
  diagonalization of H 240, 286, 289, 295
  eigenvalue-eigenvectors 265–266
  eigenvalues and eigenvectors of intermediate states 265
  eigenvalues and eigenvectors of parts H_1, H_2, H_3 266
  factorization of $3(f + 1) - j$ coefficients 284
  irreducible C_n state vectors 298
  matrix representation of H on coupled states 266
  maximum eigenvalues 286–287

INDEX

parts H_1, H_2, H_3 261
recoupling matrices 268
shape transformations 269, 272
unlabeled binary trees 235, 261
Hermitian operators 5, 7, 82, 105
   complete sets of mutually commuting Hermitian operators xiv–xv, 82–88, 98–99, 105, 226
   Heisenberg ring Hamiltonian 223
Hilbert space for angular momentum systems xv–xvii, 10
   composite system basis 2–4, 105
   constituent system basis 2-4, 7
   direct sums of total angular momentum spaces xvi
   incomplete total angular momentum set 8
   models of physical state spaces 2–4
   orthonormal basis 2
   $SU(2)$ irreducible multiplets 6, 11
   tensor product $\mathcal{H}_\mathbf{j}$ xv, 2
   transformation under group action 10–11

I

inner product of matrices 25
inner product of permutations 35
inner product of permutation matrices 33–34
incidence matrices 35, 219–222
   and Andrews-Zielberger number 222
   bijection with alternating sign matrices 119–222
   bijection with strict GT patterns 119–222
interpolation rule for sign-reversal-shift polynomials 208

irreducible representations of $GL(n, \mathbb{C})$ 67
irreducible representations of S_n 67, 59, 67–71
irreducible $SU(2)$–multiplets 6, 11, 94
irreducible triangle coefficients xx, 121–122, 148, 151
   cubic graphs xx–xxi, 171–172

J

Jucy Murphy commuting operators 59, 69–70, 80
   and irreducible representations of S_n 68–69
   eigenvalues 79–71

K

Klein's four-group 78, 80
Kostka numbers 66, 190
   recurrence relation 190
Kronecker product 11–14, 97
   of doubly-stochastic matrices 97
   multiplicity structure 13
   reduction of 12–14
   standard direct sum 11–12

L

labeled shapes 113–114, 118–119
   standard rule 114
Landé form 89
lexical GT patterns 198
lexicographic ordering 26
   of permutations and permutation matrices 26, 48, 50
   of sequences 26
linear matrix forms 25–30
   independent forms 27–30
linear operator action in Hilbert space 294
line-sums of matrices 29–30, 46–50
linear vector space of Young operators 60–62

M

MacMahon master theorem 67
magic squares xxiv, 31, 177–194

adjunct 184
and addition of angular momentum 180
and permutation matrices 31
counting formulas 181, 188–193
Diophantine equations 177–179
dominant 183
generalized Regge form 182–186
number of order three 181
Regge arrays 180–181
magic square polynomials 188–192
algorithm for computing 191–193
generating function for coefficients 186
Kostka numbers 190
RSK formula 190
explicit forms 192
zeros 188
matrix types
of fixed-line sum xxiv, 29–30
characteristic equation 70
diagonal and nondiagonable 287
exponentiated 288
Kronecker product 12, 97
normal 86, 287
permutation matrices 21
magnetic ring of Heisenberg xx, 223
matrix Schur functions 59–60, 63–67
measured state 88–89, 136, 140
M-matrix 182–184
Regge form 180–184
mixed state 99, 103
multiplication rule for recoupling matrices xxiii, 170

N

Newtonian physics 9
normal matrices 86, 287
idempotent method 287

O

orthonormalization methods 288–287
Gram matrix 48–52

P

parenthesis pair viii, 110
and shapes 110–112, 118
as abstract symbol viii, 111
basic role viii, 111
partitions 303–310
pair for skew GT patterns 212
stacked 64
path of a shape transformation xix, 157, 167–171
associations and commutations xii, xviii, 157
length of a path xx, 170
permutation matrices, algebra of xxiv, 20–21, 24, 25–43, 59
alternating sign matrices 31
Birkhoff's theorem 21
doubly stochastic matrices 31, 91
inner product 33
left and right 72–78
magic squares 31
of an arbitrary finite group 72–78
phase rule for triangle coefficients 122
phase transformations 122
Π–matrix 36
prepared state 22–24, 88, 135, 139
primitive phase factor 122
pure state 99, 103

Q

Q-matrix 182–184
magic squares 177–194
Regge form 180–184

R

Racah coefficients 118–121, 123–125, 134–136, 153, 159
addition of three angular momenta 118–121

INDEX

classification 123–125
  in terms of paths 153
  in terms of recoupling matrices 153, 253
  in terms of 6−j coefficients 125
  in terms of triangle coefficients 123
  new identity 253
sum rule 155
  cubic graph 159–160
  path transformations 155
  probabilistic interpretation 135–136
  transition probability amplitude 153
recoupling matrices xvii–xix, 14–19, 87–88
  and triangle coefficients xix–xxi, 168
  association and commutation operations xviii–xix
recursion relations:
  $3(2f − 1) − j$ coefficient 259
  $6f − j$ coefficient 275
  $3(f + 1) − j$ coefficient 284–285
reduction property for triangle coefficients 122
Regge matrix 180–186
  and two angular momenta 180
  M-matrix 182–184
  Q-matrix 182–184
regular representation 72–78
  left and right 72
  rotation by $SU(2)$ vii
Robinson-Knuth-Schensted identity 66, 68, 190

S

Santa Fe Institute xxii
Schur functions 67
Schwinger's master theorem 67
shape of a binary tree xi, 110–113
  labeled xii–xiii, 113
  unlabeled 110–112, 115, 118

shape transformation rule between binary trees xii–xiii, 167–172
  and recoupling matrix multiplication xviii, 169–171
  path viv, 157, 167–171
  sequence of A and C transformations xvii–xix
shift operator action on permutations 38–41
sign-reversal-shift conjugation 199, 213
  basis theorem 207
  interpolation rule 208
  polynomials 206
skew Gelfand-Tsetlin patterns 212
  and d_n 216
  cardinality invariance 213
  conjugates 213
  embedded GT pattern 214–216
  semistrict type 216
stacked partitions 64, 212–216
$SU(2)$ frame rotation vii
$SU(n)$ and $U(n)$ representations 66
standard C−coefficient mapping rule 116
standard labeling rule for binary trees xiii, 114
standard reduction of Kronecker product 11–12 itemstate vectors: α-coupled, β-coupled 9–18
  of commuting sets of observables 8
  transition probabilities 18–19
  transition probability amplitudes 18–19
strict GT patterns 202
  and d_n 202
superposition of state vectors 2
symmetries of GT patterns 199
  of strict patterns 204–205
  of skew patterns 213–214
symmetric group 1, 20–21

synthesis of magnetic ring, 289–292

T

tensor product space(s) 2–4
  and angular momentum state vectors 4–11, 97–98
  of separable Hilbert spaces spatial and spin $\mathcal{H}_\mathbf{j}$ of dimension $N(\mathbf{j})$ x–xi
time evolution of states 88–89
time reversal 90
triangle coefficients of order $2(n-1)$ xix, 121–122, 151, 167–172
  and doubly stochastic matrices 19–22, 151–153
  and left and right triangle patterns xix, 151
  and $3(n-1)-j$ coefficients 171
  and pairs of labeled binary trees 151, 157
  and recoupling matrices 19, 167–172
  irreducible xx, 122, 151
  phase-equivalent 122
transition probability amplitude 18, 91, 151–153
  and Racah sum-rule 153
  and B-E identity 157–158
transition probability 18, 91
triangle pattern 108–109, 151
triangle rule 107

U

unlabeled shapes of binary trees x–xii, 110–111
unitary action of $SU(2)$ 10–11
unitary group $U(n)$ representations 66
unitary similarity transformations 11, 294
unsolved problems 174

V

vector spaces: model Hilbert spaces 2–3

of GT polynomials (matrix Schur functions) 65–66
tensor product spaces 2, 143, 145

W

weight of GT pattern 198
Weyl's dimension formula 66
Wigner D^j–functions 95
Wigner $9-j$ coefficient 125–132
  different forms, 125–132, 160–165, 253–256, 270, 278–279
  new identity, 252
  in terms of paths and recoupling matrices 155–156, 164, 253–254, 270, 278
  general form 166
  orthogonality relations 138–139
  symmetries 166
Wigner-Clebsch-Gordan (WCG) coefficients ix, 115–117
  complete sets of operators 115
  orthogonality relations 116
  symmetries 117
words in A and C of shape transformations xiii, 156–157

X

$\otimes$ as tensor product
$\times$ as split multiplication

Y

Young operators (duals) 59–63
Young tableaux [198], [211]
Yutsis factor theorem 159, 171

Z

Zeilberger polynomials 196–197
  complete set of zeros 197, 205
  proof of zeros 219
  requirement for zeros 211

Errata and Related Notes

We list here corrections to formulas in [BL] and [L].

In [BL]:

Because of the importance of the Wigner $9-j$ coefficient, we reproduce an explicit formula for its value:

$$\begin{Bmatrix} a & b & c \\ d & e & f \\ h & i & j \end{Bmatrix} = (-1)^{c+f-j} \frac{(dah)(bei)(jhi)}{(def)(bac)(jcf)}$$

$$\times \sum_{xyz} \frac{(-1)^{x+y+z}}{x!y!z!} \times \frac{(2f-x)!(2a-z)!}{(2i+1+y)!(a+d+h+1-z)!}$$

$$\times \frac{(d+e-f+x)!(c+j-f+x)!}{(e+f-d-x)!(c+f-j-x)!}$$

$$\times \frac{(e+i-b+y)!(h+i-j+y)!}{(b+e-i-y)!(h+j-i-y)!}$$

$$\times \frac{(b+c-a+z)!}{(a+d-h-z)!(a+c-b-z)!}$$

$$\times \frac{(a+d+j-i-y-z)!}{(d+i-b-f+x+y)!(b+j-a-f+x+z)!},$$

$$(abc) = \sqrt{\frac{(a-b+c)!(a+b-c)!(a+b+c+1)!}{(b+c-a)!}}.$$

Note that this corrects relation (3.326) in [6]. That this relation was in error was pointed out by private correspondence by numerous colleagues beginning in 1983. S. J. Alisauskas provided the correct result, September 25, 1984, which was subsequently confirmed by independent numerical calculations by R. N. Zare and D. Zhao in early 1988. In many ways, such formulas are obsolete, since far more efficient means can be devised for their calculation based on generating functions and the fact that $3n - j$ coefficients are square-root factors times a summation that always gives an integer. (See Wei [79] and [L].)

Vol.8, p 100, Eq. (3.240): change j''' to j'.

Vol.9, p. 473, item 4: remove top-front horizontal line from cubic graph.

Vol.9, p. 471, item 2: replace phase factor by $(-1)^x$.

Vol.9, p. 476, second graph: left-hand diagram is wrong.

In [L]:

p. 91: j is not bold on the left-hand side of relation (2.34).

p. 218: Eq. (3.125), sixth graph from top; the line joining point 4 in bottom line to point 5 in top line should join to point 4 in top line.

There are, no doubt, other errors, some too trivial to correct, since they are obvious; others, unobserved.